Nature and Dynamics of Forest–Savanna Boundaries

Edited by

P.A. FURLEY
Department of Geography
University of Edinburgh
UK

J. PROCTOR
School of Molecular
and Biological Sciences
University of Stirling
UK

J.A. RATTER
Royal Botanic Gardens
Edinburgh
UK

CHAPMAN & HALL
London · Glasgow · New York · Tokyo · Melbourne · Madras

Published by Chapman & Hall, 2–6 Boundary Row, London SE1 8HN

Chapman & Hall, 2–6 Boundary Row, London SE1 8HN, UK

Blackie Academic & Professional, Wester Cleddens Road, Bishopbriggs, Glasgow G64 2NZ, UK

Chapman & Hall, 29 West 35th Street, New York NY10001, USA

Chapman & Hall Japan, Thomson Publishing Japan, Hirakawacho Nemoto Building, 6F, 1-7-11 Hirakawa-cho, Chiyoda-ku, Tokyo 102, Japan

Chapman & Hall Australia, Thomas Nelson Australia, 102 Dodds Street, South Melbourne, Victoria 3205, Australia

Chapman & Hall India, R. Seshadri, 32 Second Main Road, CIT East, Madras 600 035, India

First edition 1992

Typeset in 10/12 pt Palatino by Best-Set Typesetter Ltd., Hong Kong
Printed in Great Britain by St Edmundsbury Press,
Bury St Edmunds, Suffolk

ISBN 0 412 44370 8

A catalogue record for this book is available from the British Library

Library of Congress Cataloging-in-Publication data
Nature and dynamics of forest–savanna boundaries / edited by
P.A. Furley, J. Proctor, J.A. Ratter.
p. cm.
Papers originally presented at a symposium held in Glasgow in January 1990.
Includes bibliographical references and index.
ISBN 0-412-44370-8
1. Vegetation boundaries – Tropics – Congresses. 2. Forest ecology – Tropics – Congresses. 3. Savanna ecology – Congresses. 4. Plant communities – Tropics–Congresses. I. Furley, Peter A. II. Ratter, J.A. III. Proctor, J.
QK936.N38 1992
574.5′2642′0913–dc20 92-23210
CIP

Contents

Preface

The contributions in this volume represent a part of the four-day symposium organized by the British Biogeography Research Group and the British Ecological Society, held in Glasgow in January 1990.

It was evident that, despite considerable advances in technical methods, in knowledge of humid tropical ecology and especially in process research, relatively little had been published on one of the world's critical tension zones – the forest–savanna boundary. Boundaries are always of interest because of their role in helping to understand the development and migration of plant communities. They represent the margins of vegetation systems and may change abruptly. Since they are rarely in stable equilibrium, monitoring the boundaries helps to track the advance or retreat of one ecosystem against another. This may reflect environmental change and/or human intervention. Boundaries can be regarded, therefore, as diagnostic criteria of the state of the environment and their dynamic quality alone justifies the present interest.

The volume is divided into six parts representing different approaches to the theme. Following an introduction which explores the nature of boundaries and the types of process at work in boundary zones (Part One), the book considers the physical environment (Part Two) of the forest–savanna interface with contributions examining satellite mapping and physical factors such as soil, water and nutrient dynamics. The second theme (Part Three) examines environmental change through time, and questions the age and mobility of the constituent ecosystems, including altitudinal margins. The major emphasis of the book looks at vegetational structure and floristics across the boundary (Part Four), and considers a number of case studies particularly in the neotropics. In Part Five, some of the unique features of animal life at the boundary are analysed. The volume is concluded in Part Six, with contributions concerned with modelling the forest–savanna interactions and the generation of predictive models of boundary change.

It is clear from the state of knowledge outlined in this volume that a number of important areas need further attention. Amongst the most important is a need for more experimental research, both in terms of acquiring field data and in the modelling of processes. Permanent transects across forest–savanna boundaries would provide long-term quantitative monitoring and provide the data, at present lacking, for accurate analysis of the nature and rates of change.

The editors would like to thank Malcolm Hadley and Unesco for a generous donation towards the Symposium which allowed distinguished visitors from overseas to participate, and also valuable financial and logistic help from the British Ecological Society and the Biogeography Research Group through the Institute of British Geographers.

P.A. Furley
J. Proctor
J.A. Ratter

Contributors

Dr LUC ABBADIE
Ecole Normale Supérieure, Laboratoire d'Ecologie, CNRS, URA 258, 46 Rue d'Ulm, 75230 Paris cedex 05, France.

Prof. J.O. ADEJUWON
Department of Geography, Obafemi Awolowo University, Ile-Ife, Nigeria.

F.A. ADESINA
Department of Geography, Obafemi Awolowo University, Ile-Ife, Nigeria.

ROBERT G. AMUNDSON
Boyce Thompson Institute at Cornell University, Ithaca, NY 14853, USA.

Dr A. JOY BELSKY
Cornell Plantations, Cornell University, Ithaca, NY 14853, USA.

Dr ROBERTO B. CALVACANTI
Departamento de Ecologia, Universidade de Brasília, Brasília 70910, DF, Brazil.

Prof. MONICA M. COLE
Department of Geography, Royal Holloway and Bedford New College, University of London, Egham Hill, Egham, Surrey TW20 0EX, UK.

R.W. CRITCHLEY
Department of Environment, Newcastle Polytechnic, Lipman Building, Newcastle-upon-Tyne NE1 8ST, UK.

Dr BALTHASAR DUBS
Institut für systematische Botanik, Zollikerstrasse 107, CH-8008, Zurich, Switzerland.

MICHAEL J. EDEN
Department of Geography, Royal Holloway and Bedford New College, University of London, Egham Hill, Egham, Surrey TW20 0EX, UK.

Dr W.N. ELLERY
Resource Ecology Group, Department of Botany, University of Witwatersrand, PO Wits 2050, South Africa.

JEANINE M. FELFILI
Oxford Forestry Institute, South Parks Road, Oxford, UK, and Departamento de Engenharia Florestal, Universidade de Brasília, CP 152807, Brasília DF, Brazil.

Prof. JOHN R. FLENLEY
Department of Geography, Massey University, Palmerston North, North Island, New Zealand.

Dr PETER A. FURLEY
Department of Geography, University of Edinburgh, Drummond Street, Edinburgh EH8 9XP, UK.

Dr PETER E. GIBBS
Department of Biology and Preclinical Medicine, Harald Mitchell Building, The University, St Andrews KY16 9TH, UK.

Dr JOHN GRACE
School of Ecological Sciences, Institute of Ecology and Resource Management, University of Edinburgh, The King's Buildings, Edinburgh EH9 3JU, UK.

Dr MUNDAYATAN HARIDASAN
Departamento de Ecologia, Universidade de Brasília, CP 153081, 70910 Brasília DF, Brazil.

Dr BRIAN HOPKINS
New England College, Arundel, West Sussex BN18 0DA, UK.

Dr PHILIP E. HOWSE
Chemical Entomology Unit, Department of Biology, Building 19, University of Southampton, Southampton SO9 5NH, UK.

Dr JAN JENÍK
Institute of Botany, Charles University, CS-12801 Prague, Czechoslovakia.

Dr JULIA ALLEN JONES
Department of Geography and Environmental Studies Program, University of California, Santa Barbara, California 93106, USA.

Contributors

Dr A. PETER KERSHAW
Department of Geography and Environmental Studies, Monash University, Clayton, Victoria 3168, Australia.

MICHAEL LEPAGE
Ecole Normal Supérieure, Laboratoire d'Ecologie, CNRS, URA 258, 46 Rue d'Ulm, 75230 Paris Cedex 05, France.

XAVIER LE ROUX
Ecole Normale Supérieure, Laboratoire d'Ecologie, CNRS, URA 258, 46 Rue d'Ulm, 75230 Paris Cedex 05, France.

Dr K. ALAN LONGMAN
Institute of Terrestrial Ecology, Edinburgh Research Station, Bush Estate, Penicuik, Midlothian EH26 0QB, UK.

FLÁVIO J. LUIZÃO
Department de Ecologia, INPA, CP 478, Manaus, Am., Brazil.

REGINA C.C. LUIZÃO
Department de Ecologia, INPA, CP 478, Manaus, Am., Brazil.

Dr DUNCAN F.M. MCGREGOR
Department of Geography, Royal Holloway and Bedford New College, University of London, Egham Hill, Egham, Surrey TW20 0EX, UK.

RODRIGO A. MEDELLIN
Department of Wildlife and Range Sciences, School of Forest Resources and Conservation, University of Florida, Gainesville, Florida 32611, USA.

Prof. M.T. MENTIS
Resource Ecology Group, Department of Botany, University of Witwatersrand, PO Wits 2050, South Africa.

Dr ANDREW C. MILLINGTON
Department of Geography, University of Reading, Whiteknights, Reading RG6 2AB, UK.

Dr ARNILDO POTT
Centro de Pesquisa Agropecuária do Pantanal (CPAP/EMBRAPA), Corumbá, MS, Brazil.

VAL J. POTT
Centro de Pesquisa Agropecuária do Pantanal (CPAP/EMBRAPA), Corumbá, MS, Brazil.

Dr DARIEN E. PRADO
Department of Biology and Preclinical Medicine, Harald Mitchell Building, The University, St Andrews KY16 9TH, UK.

Prof. GHILLEAN T. PRANCE
Royal Botanic Gardens, Kew, Richmond, Surrey TW9 3AB, UK.

Dr JOHN PROCTOR
Department of Biological and Molecular Sciences, University of Stirling, Stirling FK9 4LA, UK.

Dr JAMES A. RATTER
Royal Botanic Garden, Inverleith Row, Edinburgh EH3 5LR, UK.

Dr ANTHONY RAW
Departamento de Ecologia, Universidade de Brasília, Brasília 70910 DF, Brazil.

Dr KENT H. REDFORD
Department of Wildlife and Range Sciences, School of Forest Resources and Conservation, and Center for Latin American Studies, University of Florida, Gainesville, Florida 32611, USA.

Dr SHEILA M. ROSS
Department of Geography, University of Bristol, University Road, Bristol BS8 1SS, UK.

Dr GUILLERMO SARMIENTO
CIELAT, Facultad de Ciencias, Universidad de Los Andes, Nucleo La Hechicera, 5101 Mérida, Venezuela.

Dr R.J. SCHOLES
Resource Ecology Group, Botany Department, University of Witwatersrand, PO Wits 2050, South Africa.

DUNCAN A. SCOTT
Department of Biological and Molecular Sciences, University of Stirling, Stirling FK9 4LA, UK.

MANOEL CLÁUDIO DA SILVA Jr
Department of Geography, University of Edinburgh, Edinburgh EH8 9XP, UK, and Departamento de Engenharia Florestal, Universidade de Brasília, CP 152807, Brasília DF, Brazil.

P.J. STYLES
Earth Observation Sciences Ltd, Branksome Chambers, Branksomewood Road, Fleet GU13 8JS, UK.

Prof. MICHAEL F. THOMAS
Department of Environmental Science, University of Stirling, Stirling FK9 4LA, UK.

Dr JILL THOMPSON
Department of Biological and Molecular Sciences, University of Stirling, Stirling FK9 4LA, UK.

Dr MARTIN B. THORP
Department of Geography, University College, Dublin 4, Republic of Ireland.

VIRGÍLIO VIANA
University of São Paulo, São Paulo, Brazil.

PART ONE

Introduction: Boundaries and Processes

1

Forest–savanna boundaries: general considerations

K.A. LONGMAN and J. JENÍK

SUMMARY

1. Boundaries can be viewed at many different levels and scales, ranging from the sub-atomic to the whole biosphere. Their study needs the cooperation of different disciplines, experimental as well as descriptive.
2. From a general viewpoint, a boundary can mean 'an edge or limit of', or 'a barrier to'; in detailed studies it is usually 'a transition between'.
3. The ecological concept of transition or ecotone is both realistic and practical. Its analysis includes not only physical, abiotic aspects, but also biotic factors intrinsic to the organisms involved.
4. Among ecotones between world biomes, interest has long centred on those between forest and non-forest. Terminology and methods have been developed in temperate regions that may be of general relevance.
5. Special features of forest–savanna boundaries include a wider range of biotic factors, structural diversity, life-forms and human pressure.
6. Experimental approaches are both desirable and possible, despite technical problems. Growth physiology research on the shoot growth and seed dormancy of tropical forest trees indicates that temperature and light (including photoperiod) may have unexpectedly large effects and interactions, and may influence the establishment and persistence of dominant populations.
7. Increased human impact on forests and savannas and their ecotones raises severe problems of degradation, which scientists can help to alleviate, especially by stressing the vital roles played by trees – the more vulnerable and vanishing element of the vegetation.

INTRODUCTION

Boundaries of many different kinds allow the existence of life on planet Earth. Seen at the microcosmic level, the unique surface of the carbon atom permits the formation of an almost endless series of organic molecules – the primary source of biological diversity. Viewed as a macrocosm, on the other hand, the atmospheric layers of the globe, themselves influenced by the activity of past organisms, filter the flow of energy into and out of the biosphere, moderating extremes of ultraviolet radiation and surface temperature. Both regional climate and local weather result from interactions between incident solar radiation and the Earth's surface, again with physical factors modified by the activities of living organisms.

The evolution of life-forms has itself often been associated with interfaces between contrasting environments. For example, the coast-line has been the site not only of the emergence of land plants, but also of the great diversification of plant and animal species in coral reefs and mangrove/estuarine ecosystems. Hominids appear to have originated at the forest–savanna boundary, probably close to the edges of rivers or lakes. Before considering the specific theme of this volume, however, it will be appropriate to consider some more general questions concerning boundaries (Table 1.1), including the historical development of ecological ideas of forest/non-forest ecotones.

'Boundary' has several ecological meanings. Firstly, it can indicate the external limits (or edges) of a particular stand, vegetation type or area cleared by disturbance. Here, the view is essentially one-sided: where is the margin of closed-canopy forest? What is the coast-line like? How far does the distribution of an indigenous species or a crop plant extend? Secondly, it may denote the transition zone (or ecotone) between two kinds of formation, such two-sided interfaces occurring at various scales in a variety of forest/non-forest types throughout the world. These two

Table 1.1 Some general questions concerning boundaries

1. What different meanings does the word 'boundary' have?
2. How precise (or vague) should the definitions be?
3. Is there a need for a multidisciplinary approach?
4. Are the frequently studied forest/non-forest boundaries in the temperate zone relevant to the tropical forest/savanna ecotone?
5. Can they be defined just in terms of climate, topography or soil, or is there an important biotic component?
6. Is there scope for increasing the experimental component of ecophysiological research, despite the technical problems?
7. What general principles might prove useful, and what practical guidelines should there be for the local inhabitants of forest and savanna, and for tropical land-use planners and managers?
8. Should present boundaries be shifted in favour of the trees by human actions?

Table 1.2 Some tropical forest–savanna boundaries

Type of boundary	*Limits (one-sided, external edges)*	*Ecotones (two-sided, internal interfaces)*
Sharper (convergent)	Eroded edges of large rivers, lake margins, coast-lines; extensive swamps, flood plains; areas of large-scale natural disturbance (e.g. hurricanes, volcanic eruptions); large-scale clearance by man	Damage by fire; riverine vegetation, thickets, small swamps and rocky outcrops; treeless parts of forest, dense thickets in savanna, thicket on root-suckering species in forest; natural forest gaps (*chablis*); past or present termite mounds; small-scale shifting cultivation
More gradual (divergent)	Transitional zones with altitude, rainfall, latitude; major change in orography, ground-water table or soil characteristics	Soil and vegetation catenas, with topographical changes in soil drainage, number of species, height of dominants; relics of past agriculture

types of boundary may be more gradual (divergent) or abrupt (convergent) (Shimwell, 1971); and be natural or man-made (Table 1.2).

A third meaning of the term 'boundary' is a barrier to or restraint on movement of some kind. For example, oceans, large lakes or mountain ranges interpose a restriction on gene flow, as do extensive clearances or industrialized land. The concepts of limit and barrier (first and third meanings) require clear-cut delimitation of the individual formations on both sides, with reasonably precise criteria for working at different scales. In most ecological situations, however, including tropical forest–savanna boundaries, the most useful description is generally the ecotone (second definition).

VARIETY OF VIEWS

What scientists understand by the forest–savanna boundary will depend partly upon the concepts and scales of their discipline. A geographer may simply be speaking about the extent of the natural distribution of tropical forests and savannas. Someone preparing an environmental impact assessment may be more concerned with where the transitions between forest and savanna occur at that moment, and in predicting change. In descriptive ecological terms, one could be mapping an abrupt or gradual discontinuity in the dominant or characteristic plant species, and recording any differences in soil and topography that seemed relevant. A theoretical modeller might seek to compute a relationship between the main interacting factors that could help to explain the observed distribution.

On the other hand, a plant ecophysiologist would perhaps be measuring differences in microclimate, and comparing them with known responses of the same plant species when grown under controlled environments. An animal ecologist might well be dealing with competition between herbivores, predation and population dynamics. A stress physiologist could be more interested in the influence of relatively rare, extreme conditions on survival. A geneticist might be studying the frequency of adaptive traits across the transition, or the incidence of temporal barriers to gene flow, while a biophysicist could be examining the roughly horizontal boundaries above, within and below a stand.

Thus the task of appraising forest–savanna boundaries not only has to be multi-disciplinary but also demands a recognition that our own standpoint is one of many. Indeed, each boundary presumably reflects a series of discontinuities occurring at many different levels. These will range from processes occurring at molecular and membrane surfaces, through those at the boundary layers of the phyllosphere and rhizosphere (including widespread and even obligate associations with micro-organisms), to the spatial and temporal arrangement of the organisms in the two neighbouring ecosystems, and their competition and collaboration. All these types of boundary, and others, contribute to determining where 'the lines are drawn' between success and failure to survive and grow, and to reproduce enough to avoid extinction.

As will be seen, the importance of biotic factors in the development of tropical boundaries means that a key role is played by particular plant species, competing and achieving a temporary position which is recurrently shifted by growth, reproduction and death. Thus the forest–savanna boundary needs to be viewed as a dynamic phenomenon, with the life strategies of its dominant organisms to be clarified through the experimental science of whole-plant physiology. Precisely because ecological science deals with complex interactions between a very large number of factors, its approach can come to be seen as consisting solely of observation, description, measurement and theoretical calculation. Such a view would not have been likely during the great flowering of botany in central Europe 80–100 years ago, when the close bond between structure and function was sealed through physiological testing of hypotheses.

For instance, writing on evolution from the University of Glasgow, Professor F.O. Bower (1908) observes: '. . . the experimental morphology of to-day will serve as a most valuable guide, and even a check to any more speculative method, by limiting its exuberances within the lines of physiological probability'. He goes on to recognize that the experiment also needs the theory, but catches the special quality of the former in quoting Goebel's (1905) summing up of the relationship between these two branches of biological science: 'Morphology includes such phenomena as are not yet physiologically understood'.

All attempts to define vegetation types involve paradox. For instance, biological and geographical research could hardly proceed without general acceptance of the meaning of 'a species', but at close range its definition becomes blurred with provisos and exceptions. Similarly, drawing the outline of a leaf may be quite sufficient to identify a particular species, but is inadequate to give a picture of the range of variability in leaf shape. Moreover, the continuous pencil line depicting its margin becomes unhelpful when considering evaporation, since most of this occurs from more extensive internal surfaces. It is often necessary to fall back on quantitative categories and practical approximations to describe the boundary in terms of gas exchange or radiation balance. The same problems exist in defining grassy or wooded stands, and the plants that characterize them. For example, looking at a mountain flank from a distance may suggest clear-cut, convergent lines between the stands, while on the ground one finds a broad transition zone or ecotone. Thus defining and mapping forest–savanna boundaries is part of a general problem of specifying the limits of taxa, organs or vegetation dominated by different life-forms.

VARIETY OF ECOTONES

The conditions which favour a biome in its core area change towards its margins until it declines and ultimately ceases to exist. This applies equally to other ecological units used, such as formation, ecosystem type and association (in the sense of the Zürich/Montpellier School). Whatever the scale and precise definition, there is usually a transitional zone or ecotone. A striking example is that between aquatic and terrestrial ecosystems, developed at the shore-lines of seas and lakes, and along river banks. Investigation of the structure and dynamics of these boundaries is the objective of recent international collaboration (DiCastri *et al.*, 1988); studies of the ecotone between savanna and semi-desert biomes have been stimulated by concern at widespread desertification.

There has been long-lasting interest in forest/non-forest ecotones, arising from the striking differences in the respective dominant life-forms (trees and grasses). At the scale of biological zones, this boundary zone occurs between:

1. coniferous forest and tundra (for example 'forest-tundra' in Siberia (Polunin, 1960);
2. broadleaved deciduous forest and steppe (for example 'forest-steppe' in Eastern Europe);
3. tropical forest and savanna.

With regard to altitudinal zonation on mountains ('orobiomes' *sensu* Walter and Breckle, 1985–86), marked ecotones have been described for (a) forest and alpine (or sub-alpine) grasslands (occasionally via *krummholz*); and (b) forest and *páramo* (or *sub-páramo*) grasslands (sometimes via scrubby thickets). With respect to local environmental constraints, various local forest limits and forest/non-forest ecotones have been studied (Table 1.2), such as 'orographic forest boundaries' on exposed precipitous slopes or screes, 'xerothermic forest boundaries' on dry and sunny slopes, or 'hydric forest boundaries' around waterlogged areas.

In Europe, early attention was focused on identifying one-sided limits (Table 1.2) between the 'useful forest' and the less important non-forest formations, especially in mountainous regions (Schröter, 1926; Jeník and Lokvenc, 1962; Plesník, 1971). Here a very few dominant tree species create the boundary between the forests and the non-forest *krummholz* or alpine tundra formations: primarily the coniferous Norway spruce (*Picea abies*) and the broadleaved, deciduous European beech (*Fagus silvatica*). Despite this relative simplicity when compared with North America (Morriset and Payette, 1983) and the tropics, there is no uniformity in the basic views and concepts that have since been developed. Some authors have preferred to speak of a 'tree-line' (*Baumgrenze*), based on the distribution of individual mature trees; while others have seen the boundary as a 'forest limit' (*Waldgrenze*), an expression of the potential of the whole population or closed-canopy community. Some see such limits as an abrupt, thread-like line, while other workers have recognized a broader transitional zone (*zone contestée, Parklandschaft, Kampfzone*), corresponding with the ecotone concept.

European authors have also pointed out that there are certain prerequisites to successful field recognition and mapping of forest margins. To define a forest boundary, one has to ask what is a forest? To define a forest, one needs to say what constitutes a tree, its dominant life-form. These statements, far from being simplistic, have formed the basis for practical working criteria of 'a forest':

1. dominant trees attaining a minimum height of 5 m (8 m, according to the former Austro-Hungarian forestry department);

2. minimum crown cover (canopy density) of 50%; and
3. occupying a minimum area of 0.01 ha (0.1 ha, according to other authors).

Using these definitions together with aerial photographs, forest boundaries were mapped in many mountain ranges in the Alps, Sudeten region and Carpathians. North American workers have mostly used the broad term 'timberline' (Daubenmire, 1954), suggesting a certain size, height and economic value as their criterion; recently 'forest limit' and 'tree-line' have been distinguished (Morriset and Payette, 1983).

For nearly a century, both botanists and foresters have been fascinated by the limits to tree growth and to closed forest, seeking the explanations particularly in abiotic, physical factors. At both the boreal and alpine ecotones, the 'climatic tree limit' was considered to be an expression of combined environmental control by heat, moisture and duration of growing season. For example, along the arctic ecotone, a minimum average July temperature of 10 °C was often reported as decisive for adequate tree growth. Later, many microclimatic and ecophysiological measurements have been made, including growth physiology studies on *Picea abies*, *Pinus cembra* and *Pinus mugo* in the Alps (Tranquillini, 1979), as well as in other parts of the world (Scott *et al.*, 1987). Besides temperature, the effects of wind, snow drifts, aspect and avalanches have frequently been quoted as constraints on tree growth.

It was less common to take account of 'edaphic tree limits', the concept being important but more diffuse. Factors that have been considered include shallow and leached soil profiles; presence of toxic mineral elements, such as magnesium and nickel in ultramafic (serpentine) soils; and waterlogged soil, with oxygen deficiency (Kozlowski, 1971).

Although human interference, such as burning and grazing by domestic animals, has been widely studied at the arctic and alpine forest/non-forest ecotones, other biotic factors have seldom been allowed for. The most obvious evidence for this is the fact that plantations of introduced or indigenous species can grow outside the current forest margin. In the Scottish Highlands, for example, extensive planting of the western American *Picea sitchensis* has markedly raised the forest limit, locally even higher than the natural tree-line. This of course had been much lowered by clear-cutting, burning, grazing and accompanying leaching, as well as dominance by ericaceous and other species (Pearsall, 1968). Thus many authors distinguish between natural and artificial forest boundaries, and between actual and potential vegetation (Schröter, 1926; Jeník and Lokvenc, 1962).

FEATURES OF FOREST–SAVANNA BOUNDARIES

A number of the features and events at the forest–savanna boundary are comparable to ecotones in temperate regions. However there are generally:

1. a much wider range of abiotic factors involved;
2. a greater diversity of participant life-forms;
3. a more complex ecosystem organization; and
4. a greater human pressure from both sides of the boundary.

At low latitudes and altitudes in tropical and near-tropical regions there is relatively less fluctuation in photoperiod and temperature, and little, if any, influence of frost. Under these conditions, rainfall totals and rainfall seasonality play a dominant role in plant life, particularly through alternating periods of rain and drought – and subsequently through soil moisture deficit and waterlogging. Forest ecosystems can only tolerate moderate climatic fluctuations (Fig. 1.1); beyond this point various types of savanna develop, depending upon the length and extremes of the dry period(s). Existing climatic measurements, maps and diagrams (Walter and Lieth, 1960–67) suggest that the position of the forest/savanna ecotone usually occurs where mean monthly rainfall totals are still well over 100 mm, falling below 50 mm only once or twice in a year. Considered on a large scale, a gradient between various types of forests and savannas tends to develop, and the ecotone may cover a very wide transitional zone, as for example in West Africa (White, 1983). Seen in a detailed view, edaphic and biotic factors and fire make the same ecotone appear narrower and discontinuous.

Soil factors at forest–savanna boundaries are influenced by the usual factors of topography and parent material, but deeply penetrating chemical weathering and soil-forming processes create soil catenas that are never encountered in temperate regions (Markham and Babbedge, 1979; Longman and Jeník, 1987). The occurrence of a lateritic crust can favour savanna biomes both at the top and the foot of slopes and escarpments. Soils rich in heavy metals greatly favour savanna over forest, as in Katanga, Zaïre.

Soils are also affected by the levels of groundwater, direction of rivers and extent of flooding that are controlled by heavy rainfall in adjacent or distant mountains. Gallery forest along the flood plains can extend the forest–savanna boundary deep into arid regions. Climatic and edaphic factors combine in an array of habitat types, yet the general, large-scale pattern seems most frequently to be explained by the availability of water and the duration of physiological stress caused by soil dryness, or waterlogging, of root environments.

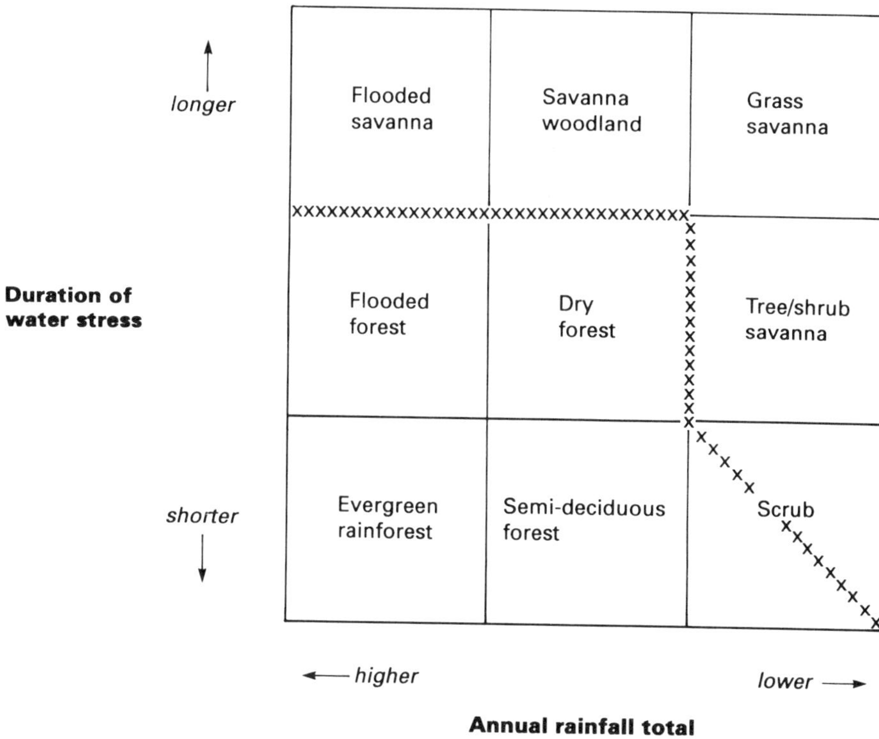

Fig. 1.1 Ecodiagram showing a broad classification of tropical and near-tropical formations, according to variation in annual rainfall totals and duration of stress periods, caused by waterlogged soil or drought or both. Notes: (a) Relatively less seasonality in temperature and photoperiod than at higher latitudes, (b) Soil variation and biotic factors play a larger part in determining local conditions, (c) X = the forest/savanna ecotone.

Species richness and the diversity of life-forms in tropical and near-tropical regions create numerous of problems in analysing forest/savanna ecotones. Increasing numbers of species of flora and fauna with decreasing latitude is a well recognized phenomenon. Among dominant plants, a variety of life-forms is encountered that have no counterparts in other forest/non-forest ecotones, for example palms, cycads, succulents, epiphytes, large woody climbers, stranglers, giant grasses and sedges. Trees and shrubs display a great range of above-ground architecture (Hallé *et al.*, 1978) and root systems (Jeník, 1978). Woody plants and graminoids can interlace, and their association is less stabilized than in equivalent temperate situations. It is thus particularly difficult to compare forest–savanna boundaries in different biogeographical regions and floristic realms.

Tropical animals also play a key role at these ecotones. Amongst the most obvious examples are sizeable African herbivores such as the large herds of ungulates that graze and browse in the marginal zones between savanna and forest, but there are also efficient consumers of American, Asian and Australasian vegetation. Insects are another important factor in delineating the pattern of tropical biomes – in some regions the activity of white ants appears to be responsible for a mosaic of grassy savanna patches and closed-canopy thickets, the so-called termite savanna. Early human populations in Africa, and subsequent inhabitants of other parts of the tropics, utilized fire as an omnipresent feature of their cultural systems. In fact, nowhere else on Earth is the forest/non-forest ecotone so profoundly disturbed by fire.

The great biodiversity causes fundamental problems in defining even the biomes on either side, let alone the ecotone itself. What is a tropical forest; what is a savanna? Unlike the situation in temperate regions, it is actually difficult to make a complete distinction into woody plants, and their counterparts – herbs and graminoids. Palms, for example, although they contain lignified tissues, are not strictly speaking a woody life-form. There are also numerous discrepancies when defining a tree in the tropics, which make imprecise the use of tree form and height as characteristic features of a forest or savanna formation. The ecotone itself may often be occupied by a typical tropical scrub (Fig. 1.1), composed of (a) suckering trees, large shrubs, polycormic palms and shrubs with large underground root systems; and on the other hand (b) tall grasses, large sedges, succulent plants and numerous specialized life-forms, such as members of the Velloziaceae.

For practical purposes, such as remote sensing, mapping and land-use classification, it is possible to base a rough definition of each formation on the mutual exclusion and overlapping of woody and graminoid plants, expressed as relative ground cover. A 'forest' could then be best described as a layered formation of prevailingly woody plants, with a canopy closed enough to prevent the growth of heliophilous graminoids and high enough to enable free movement of browsing animals and human hunters in the undergrowth. A 'savanna' would be any lowland tropical formation with dominant perennial grasses, typically photosynthesizing by the C_4 pathway, and interspersed with small trees or shrubs.

In many parts of the tropics, biomes near the forest–savanna boundary have been variously named. In Africa, the Yangambi conference of 1956, organized by the Scientific Council of Africa South of the Sahara, reached a consensus between francophone and anglophone workers on terms that have since been widely used. There appears to be lesser unity among authors in South America, while in Australia local names are frequently utilized. In Fig. 1.1, nine terms have been selected from the anglophone literature in order to emphasize the dominant categories of vegetation, the main physical factors involved and the approximate position of different forest–savanna boundaries.

THE GROWTH PHYSIOLOGY APPROACH

A study of forest–savanna boundaries needs to include appropriate experiments on the growth and development of formations and of key plant species. However, two kinds of mental barrier often hamper experimental research: the idea that it is unnecessary; and that it is impossible. The former is clearly misapprehension, for the experimental approach lies at the core of the scientific method, giving the extra precision and fresh viewpoints which come from controlled testing of null hypotheses (Heath, 1970, 1976). In an age when spectacular technological advances have allowed detailed satellite photography, automated sensing and recording of data, and easy computing of complex relationships, experimental research has perhaps been somewhat overlooked. However, its ability to correct too great a reliance upon the existence of conjunction and correlation between factors is perhaps even more valuable during a period of rapid advance by the descriptive and theoretical branches of the subject.

The second barrier is more substantial, but despite daunting technical and sometimes conceptual difficulties, there are at least four general classes of experimental ecophysiology possible at the forest-savanna boundary. Firstly, field experiments may be done 'indirectly' by selecting matched areas, such as watersheds, with and without the natural forest cover. Alternatively, selected parts of a study area might be fenced to exclude grazing animals or protected from fire, and then compared with unfenced or burnt areas. Secondly, 'direct' treatments, such as the application of fertilizers, or the planting, defoliation or pruning of plants, can be applied to randomly chosen plots in a more structured layout, with appropriate controls. Thirdly, small potted plants of relevant species, or Petri dishes seeds contained in, can be grown in controlled environments, to test the effects of one or more factors, other conditions being maintained as similar as possible in all treatments. Fourthly, laboratory experiments may be done, using detached organs, tissue cultures or enzyme preparations, to compare reactions under different environments.

Heath (1976) has argued that, because each of these four approaches has its inbuilt limitations, what is needed is 'to have experiments all up and down the range between field and test tube, and to argue back and forth from one to the other'. An analogous type of interplay is also required between the ecologist and growth physiologist, so that the experimental information can contribute to defining and understanding the ecological situation. Choosing a plant species found at the forest–savanna boundary could show, for instance, how sensitive it is to critical environmental factors, and whether their effects are likely to be additive or synergistic.

It can be misleading if factors are excluded from experimental testing on the basis of presuppositions. For example, it had often been predicted from the relatively small seasonal changes in tropical photoperiod and air temperature that these factors would be unimportant in controlling growth. However, a series of controlled environment experiments investigating their effects on shoot growth in potted seedlings, rooted cuttings and grafts of more than 20 tropical trees and shrubs showed that this was not so (Table 1.3). Most species were in fact rather sensitive to alteration in day length and temperature, and many different processes were shown to be affected. In these experiments, the plants under different day lengths all received approximately the same total daily light energy. However, it should be noted that because air humidity control was usually unavailable, higher temperatures were usually confounded with greater saturation water vapour deficits.

Reproductive activity is always more difficult to investigate than vegetative growth, but experiments on flower initiation have been achieved

Table 1.3 Effects of photoperiod and temperature on growth and development of tropical woody plants (from Longman, 1969, 1978; Longman and Jeník, 1987)

1. Photoperiod and air temperature differences can affect rates of stem elongation, node production, leaf enlargement and dry weight accumulation; and also influence final internode length and leaf size
2. Qualitative changes can also be stimulated, including bud-break, cessation of shoot elongation and leaf abscission
3. Unless the process is insensitive to photoperiod (for example node production in *Gmelina arborea*), shoot growth is greater in long days
4. Photoperiodic effects can be of a threshold or a quantitative type, and responses are often large: a 4 h increase in day length may double the rate of stem extension
5. Temperature optima are often around 30 °C, but some genera show different responses. Stem elongation of *Cassia mimosoides*, a small African/Asian shrub, typical of open grassy forest margins, showed a clear optimum at 29 °C day/22 °C night; whereas most growth was made at constant 36 °C in *Ceiba pentandra*, the pantropical forest tree
6. Day and night temperatures can have similar, or strongly contrasting effects. A sub-optimal day temperature can often be partially offset by a warmer night, even when under optimal conditions the night would be cooler. An increase of 5 °C in night temperature alone may increase shoot growth by 50% or more
7. Significant interactions between day length and night temperatures were often found, not all organs responded in the same way in each species, and adult tissue could be more responsive than juvenile tissue

by careful choice of amenable species and cultivars (Medina, 1982; Longman, 1985; Longman *et al.*, 1990). On the other hand, seeds are the simplest of study materials, and many tests of viability and storage have been carried out by applied scientists in the tropics. Surprisingly, however, there have been relatively few fundamental explorations of seed dormancy, despite the frequent observation that germination peaks tend to be more strongly seasonal than those of fruit and seed dispersal, in both the forest (Garwood, 1982) and savanna (Menaut, 1983).

One exception is a valuable series of laboratory and field experiments carried out by Vázquez-Yanes and Orozco Segovia (1984) with several genera of colonizing trees of Central American forests. They found that the great majority of seeds did not germinate unless certain conditions were met, and that these were often characteristic of forest gaps above a certain size, and of larger clearings. Several different treatments could break seed dormancy, including specific diurnal fluctuations in temperature, and a period of light with a higher red/far-red ratio than is characteristic of the forest interior, or the edges of gaps. Seed dormancy also tended to disappear under certain storage regimes, but to be prolonged in others. Each species showed a different combination of physiological responses, which appear to be involved in natural as well as in controlled conditions.

There are, of course, many other types of seed dormancy, while numerous tropical trees show prompt germination, without delay (Longman and Jeník, 1987). There appear to be few reports of seed research on savanna species, but two cosmopolitan herbaceous species, colonizers of man-made lawns, were found to have multiple and contrasting seed dormancies (Longman, 1969). Achenes of *Dactyloctenium aegyptium*, an annual grass occurring in African ecotones, did not germinate if kept imbibed in Petri dishes and only very sparsely when plots were watered during the middle of the dry season. Dormancy could, however, be broken experimentally by diurnal fluctuations in temperature of 8°C or especially 16°C, or by after-ripening in dry storage (at 20°C or particularly at 40°C). Dormancy involved the rate as well as the final percentage of germination, and there was also evidence for a light requirement and an inhibition at higher temperatures. On the other hand, dormancy in *Tridax procumbens* involves light and a sophisticated timing mechanism. Hardly any seeds germinated in the dark, or when given light continually for 24 hours, and nearly half still remained dormant after four days of continuous light. However, a high percentage germinated when four hours of light were given on four successive days, indicating that dark as well as light environments played a part in breaking dormancy. Further studies showed that far-red as well as red radiant energy could promote germination. These and other experiments indicate that plants may possess a formidable array of mechanisms whose ecological significance appears to be the limiting of germination peaks to a specific, more favourable season, while still permitting some seeds to start opportunistic growth at other times from reserves in the seed bank.

There is clearly much scope for comparative experimental studies on ecologically as well as economically important tropical and sub-tropical plants, involving collaboration between ecologists, plant physiologists and biochemists. One aim might be to carry out a large number of standard tests, in order to build up a basic picture of the major differences between groups of species. This kind of approach would be particularly relevant to the study of likely adaptive traits at forest–savanna boundaries, and could, for instance, usefully include experiments with genera having both forest and savanna species, such as *Khaya* and *Pterocarpus*.

HUMAN IMPACT AND THE FUTURE

The late Jacob Bronowski identified the ability to make plans as Man's most important distinguishing feature. Fortunately an extraordinary increase in public awareness of the fragility of global ecosystems has occurred during the last decade. It is hoped that this volume will help to stimulate active protection of habitats such as the *locus classicus* of *Homo sapiens* against Man himself. Indeed, because of our very rapid 'social and cultural evolution', we could clearly choose to replace short-sighted and short-term aims with more effective, sustainable goals that are based on our accumulated experience and science.

In seeking general principles that might provide practical guidelines for management of forest–savanna boundaries, it should nowadays be axiomatic that the more scientific information there is about an ecosystem, plant formation, stand or species, the more likely it is that a successful job can be made of managing it on a sustained yield basis. However, besides more research, there is a critical need for synthesis, undistorted explanation and clear communication. With emotive issues such as the disappearance of the rainforests, and the threat of 'savannization' and desertification, it is especially important for scientists to steer an independent course between the extremes of scare-mongering and complacency. Global responses to other critical issues have been concentrated upon warning of problems ahead, indicating that solutions are possible and noting the priorities, bearing in mind that ecology is complex and the evidence still incomplete (for example, Harwell and Hutchinson, 1985; WMO, 1989).

One useful guideline is that ecologically viable systems of tropical agriculture can become destructive simply by an increase in human population and available technology. Small-scale, partial and temporary openings in tropical forest, on a roughly 20 year cycle, may regenerate naturally by successional processes fairly similar to those in natural gaps, even when farms are burnt. Complete ground cover is often rapidly restored by the active regrowth of cut or burnt shoots, and through the breaking of seed dormancy in pioneer tree species. By contrast, more frequent, longer-lasting, larger-scale or more complete clearance alters the environment considerably such that the balance swings strongly away from woody plants and towards graminoid species.

However, the result is not necessarily a true savanna, although the term 'derived savanna' is often used. An unstable, positive feedback loop tends to be set up between rising human population, shortening periods under forest fallow, rapidly declining crop yields, loss of 'wild' food sources and the creation of useless scrub vegetation. Increased radiation reaching the ground, greater daytime temperature and desiccation at the soil surface, greatly accelerated soil erosion and leaching, and other changes in soil physico-chemical constituents and microflora may mean that the area no longer provides conditions in which non-pioneer tree species regenerate. Contributory factors include a lack of seed trees and invasion by rampant weeds, whose capacity for sexual or asexual reproduction or both and perennation may allow them to outgrow even the fast-growing colonizing trees. The availability of chainsaws, bulldozers and other types of heavy logging and road-making equipment can all contribute further to widespread degrading of the land available for farming. Large-scale ecosystem decline can also be caused in savanna areas (Huntley and Walker, 1982), for example by intense fires, overgrazing, frequently repeated cultivation and the use of heavy mechanized implements.

Perhaps the most important guideline for the management of tropical forest and savanna and their ecotones comes as a sharp contrast to common practice in the temperate zones. The presence of trees is nearly always vital for successful production of food and materials in the tropics, except for lowland rice cultivation on freshwater swamps. Both traditional and novel agroforestry systems offer a variety of alternatives for sustained annual production, while degraded forest and moister savanna sites can often be restored relatively rapidly by planting trees. Moreover, there is now a welcome trend away from complete clearance and the planting of large blocks of exotic monocultures. In the driest savannas and in semi-desert conditions, tree-planting schemes are also being extended, although these have to be planned carefully if they are not to lower groundwater levels.

Much depends upon wise decisions being taken on land management:

locally, regionally and globally (Dickinson, 1987; Ortiz-Crespo, 1989). Sooner or later, tropical countries will have to come to terms with reality – eating today has often been destroying tomorrow's larder. If there were neither forest nor savanna left, the boundaries for continued human survival might have been crossed in the very territories in which we evolved. Environmental scientists have a great responsibility, not only for making the 1990s the Decade of the Biosphere, but also for ensuring that our combined knowledge is used to help find positive solutions to these urgent problems.

REFERENCES

Bower, F.O. (1908) *The Origin of a Land Flora,* Macmillan, London.

Daubenmire, R. (1954) Alpine timberlines in the Americas and their interpretation. *Butler University Botanical Studies,* **11**, 119–36.

DiCastri, F., Hasen, A.J. and Holland, M.M.(eds) (1988) A new look at ecotones: emerging international projects on landscape boundaries. *Biology International,* special issue 17, IUBS, Paris.

Dickinson, R.E. (ed.) (1987) *The Geophysiology of Amazonia: Vegetation and Climate Interactions,* John Wiley, New York.

Garwood, N.G. (1982) Seasonal rhythm of seed germination in a semi-deciduous tropical forest, in *The Ecology of a Tropical Forest – Seasonal Rhythms and Long-term Changes,* (eds E.G. Leigh, Jr, A.S. Rand and D.M. Windsor), Smithsonian Inst., Washington, DC, pp. 173–85.

Goebel, K. (1905) Die Grundprobleme der heutigen Pflanzenmorpholgie. *Biol. Zentralblatt,* **25**(3), 65–83.

Hallé, F., Oldeman, R.A.A. and Tomlinson, P.B. (1978) *Tropical Trees and Forests: An Architectural Analysis,* Springer-Verlag, Berlin.

Harwell, M.A. and Hutchinson, T.C. (eds) (1985) *Environmental Consequences of Nuclear War. Vol. II. Ecological and Agricultural Effects,* Scientific Committee on Problems of the Environment (SCOPE 28), John Wiley, Chichester.

Heath, O.V.S. (1970) *Investigation by Experiment,* Studies in Biology 32, Edward Arnold, London.

Heath, O.V.S. (1976) In praise of experiments, in *Perspectives in Experimental Biology, Volume 2 – Botany,* (ed. N. Sunderland), Pergamon, Oxford, pp. 1–8.

Huntley, B.J. and Walker, B.H. (eds) (1982) *Ecology of Tropical Savannas, Ecological Studies 42,* Springer-Verlag, Berlin.

Jeník, J. (1978) Roots and root systems in tropical trees: morphologic and ecologic aspects, in *Tropical Trees as Living Systems* (eds P.B. Tomlinson and M.H. Zimmermann), Cambridge University Press, Cambridge, pp. 332–49.

Jeník, J. and Lokvenc, T. (1962) Die alpine Waldgrenze im Krkonose Gebirge. *Rozpravy Československé Akademie Věd, Prague*, **72**(1), 1–65.

Kozlowski, T.T. (1971) *Growth and Development of Trees*, Vol. II, Academic Press, New York.

Longman, K.A. (1969) The dormancy and survival of plants in the humid tropics. *Symposium of the Society for Experimental Biology*, **23**, 471–88.

Longman, K.A. (1978) Control of shoot extension and dormancy: external and internal factors, in *Tropical Trees as Living Systems*, (eds P.B. Tomlinson and M.H. Zimmermann), Cambridge University Press, Cambridge, pp. 465–95.

Longman, K.A. (1985) Tropical forest trees, in *Handbook of Flowering*, Vol. 1, (ed. A.H. Halevy), CRC Press, Boca Raton, Florida, pp. 23–39.

Longman, K.A. and Jeník, J. (1987) *Tropical Forest and its Environment*, 2nd edn, Longman, Harlow.

Longman, K.A., Manurung, R.M. and Leakey, R.R.B. (1990) Use of small, clonal plants for experiments on factors affecting flowering in tropical trees, in *Reproductive Ecology of Tropical Forest Plants* (eds K.S. Bawa and M. Hadley), Unesco, Paris, pp. 395–405.

Markham, R.H. and Babbedge, A.J. (1979) Soil and vegetation catenas on the forest–savanna boundary in Ghana. *Biotropica*, **11**, 224–34.

Medina, E. (1982) Physiological ecology of neotropical savanna plants, in *Ecology of Tropical Savannas, Ecological Studies 42* (eds B.J. Huntley and B.H. Walker), Springer-Verlag, Berlin, pp. 308–35.

Menaut, J.C. (1983) The vegetation of African savannas, in *Ecosystems of the World, 13: Tropical savannas*, (ed. F. Bourlière), Elsevier, Amsterdam, pp. 109–49.

Morriset, P. and Payette, S. (eds) (1983) *Tree-line Ecology. Proceedings of the Northern Tree-line Conference. Collection Nordicana* 47, Université Laval, Quebec.

Ortiz-Crespo, F. (1989) The future of tropical forests, in *Tropical Forests: Botanical Dynamics, Speciation and Diversity*, (eds L.B. Holm-Nielsen, I.C. Nielsen and H. Balslev), Academic Press, London, pp. 335–8.

Pearsall, W.H. (1968) *Mountains and Moorlands*, 2nd edn, Collins, London.

Plesník, P. (1971) *Horná kranica lesa*, Vydav. Slov. Akad. Vied, Bratislava.

Polunin, N. (1960) *Introduction to Plant Geography*, Longman, London.

Schröter, (1926) *Das Pflanzenleben der Alpen*, Verlag A. Raustein, Zürich.

Scott, P.A., Bentley, C.V., Fayle, D.C.F. and Hansell, R.I.C. (1987) Crown forms and shoot elongation of white spruce at the tree-line, Churchill, Manitoba, Canada. *Arctic and Alpine Research*, **19**, 175–86.

Shimwell, D.W. (1971) *The Description and Classification of Vegetation*, Sidgwick & Jackson, London.

Tranquillini, W. (1979) *Physiological Ecology of the Alpine Timberline*, Springer-Verlag, Berlin.

Vázquez-Yanes, C. and Orozco Segovia, A. (1984) Ecophysiology of seed germination in the tropical humid forests of the world: a review, in *Physiological Ecology of Plants of the Wet Tropics*, (eds E. Medina, H. Mooney and C. Vázquez-Yanes), Junk, The Hague, pp. 37–50.

Walter, H. and Breckle, S.-W. (1985–86) *Ecological Systems of the Geobiosphere*, Vol. 2, Springer-Verlag, Berlin.

Walter, H. and Lieth, H. (1960–67) *Klimmadiagramm-Weltatlas*, Fischer-Verlag, Jena.

White, F. (1983) *The Vegetation of Africa*, Unesco, Paris.

WMO (1989) The changing atmosphere: implications for global security. *Proceedings of the Toronto Conference June 1988* No. 710, World Meteorological Organization, Geneva.

2

Ecological processes at the forest–savanna boundary

BRIAN HOPKINS

INTRODUCTION

Dynamics of Forest–Savanna Boundaries, the title of this book, is an immense subject. In this chapter my aim is to present a broad view of the main processes which occur in the forest–savanna boundary and to relate these to the dynamics of the relationships between forest and savanna. I shall confine my remarks to Africa and, although I shall deal particularly with western Africa, I believe that most of the chapter is relevant to the whole continent. There is an enormous literature on this subject; see general and African region reviews of Schnell (1976), Unesco (1978, 1979), Huntley and Walker (1982), Bourlière (1983), White (1983), Tothill and Mott (1985) and Walker (1985).

The main factor controlling tropical terrestrial vegetation is dry-season severity. The climatic change from the very humid forest areas to the very arid desert is a gradual one. Yet the vegetation boundary between forest and savanna is very abrupt. Why is this?

Early in the Pleistocene there was almost certainly a natural vegetation continuum for forest, through woodland and savanna, to desert. The sequence of the zones in western Africa is simpler than elsewhere on the continent (because of less complex topography and, therefore, climate). From north to south it probably would have been:

Zone	Group
desert	desert
sub-desert	desert
grassland with trees	savanna
wooded grassland	savanna
woodland	forest
dry deciduous forest	forest
dry semi-deciduous forest	forest
dry evergreen forest	forest
moist semi-deciduous forest	forest
moist evergreen forest	forest

This sequence would, of course, be modified by soil type and other factors as discussed below.

During the Pleistocene these zones were displaced latitudinally and altitudinally, but the zonation probably remained unchanged. Owing to the action of fire (which was probably frequent after 50 000 BP) and (during the last 3000 years) to the effects of grazing by domesticated animals, and, especially, of cultivation, tree densities have been greatly reduced. Thus, vast areas have been converted into grassland with trees or into wooded grassland, i.e. into savanna which, consequently, abuts directly onto the forest (Table 2.1).

The zone surrounding the forest is the area with which this book is mainly concerned. It has various names; forest–savanna mosaic zone is as suitable as any, for it consists of patches of four types of vegetation: forest, savanna, farmland and fallow. The nature of each of these needs to be considered before the forest–savanna boundary itself can be discussed.

Forest consists almost entirely of woody plants; even most of the herbs are woody, and there are virtually no grasses. For most of the humid tropics, forest is the 'natural' ecosystem – the 'climatic climax'. By contrast, in savanna, grasses predominate in the herb layer, and trees and shrubs may or may not be present.

Some, if not many, of these savanna areas are because of cultivation followed by fire. No individual sere has been recorded, but what has happened is clear. A farm in the forest, but near the savanna, is invaded by the seeds of grasses. If these become established then, when cultivation ceases, they will probably be burnt. The fires will kill the invading forest species but not the grasses and so convert the area to grassland. Possibly, later, savanna trees and shrubs will invade and become established. This process is a secondary succession and it occurs during the

Table 2.1 Changes in vegetation from early Pleistocene to the present day

Original vegetation	*Present-day vegetation*
Desert	Desert
Sub-desert	Sub-desert
Grassland with trees	Savanna
Wooded grassland	Savanna
Woodland	Savanna
Dry deciduous forest	Savanna
Dry semi-deciduous forest	Savanna
Dry evergreen forest	Savanna
Moist semi-deciduous forest	Forest or savanna
Moist evergreen forest	Forest

fallow period (Aubréville, 1949a,b; Schnell, 1950; West, 1965; Hopkins, 1965, 1983; Bourlière and Hadley, 1970).

These savannas are maintained by annual fires and, as has been shown by experiments in western Africa, if protected from fire, they rapidly revert to forest by a secondary succession (MacGregor, 1937; Charter and Keay, 1960; Ramsay and Rose-Innes, 1963; Adjanohoun, 1964; Monnier, 1968; Vuattoux, 1970).

These, then, are the four phases of the mosaic: forest, farmland, fallow and savanna.

DISTRIBUTION OF THE VEGETATION TYPES

Overall distribution

The distribution, within the forest–savanna mosaic zone, of patches of these phases reflects, in part, the land-use history. Forest may occur on areas which have never been cultivated or where there has been no cultivation for a long time, whereas savanna may occur on land which has been farmed relatively recently.

Towards the forest zone, that is at lower latitudes, the patches are mainly of forest, or farm, or secondary succession to forest. Towards the savanna zone, that is at higher latitudes, the patches are mainly of savanna, or farm, or of secondary succession to savanna. This gradation is climatically determined although, between the two extremes, the overall distribution is very irregular. However, in general, when a forest farm is abandoned it reverts to forest, and when a savanna farm is abandoned it reverts to savanna.

The distribution of patches of forest and savanna is also related to habitat factors. Often, these are edaphic and are due to the texture and depth of the soil – factors which largely control the water relations of the soil which are so important for savanna vegetation (Latham and Dugerdi, 1970; Adejuwon, 1971a; Swaine *et al.*, 1976; Moss, 1977a,b; Moss and Morgan, 1977).

This relationship is causal in some places, although the soil also acts indirectly through the farming process (Moss, 1977b). In selecting an area to cultivate, the farmer bases his choice on the soil's fertility, which is greater in the valley bottoms where inwash occurs. However, fertility is often less important to the farmer than other factors. These include the ease of clearing the land, which results in areas with less dense woody vegetation being selected, or the ease of cultivation, which may result in sandy rather than clayey soils being chosen. These aspects were particularly important before metal tools became available. So, forest often remains on the clayey soils of valley bottoms whilst both forest and savanna may occur on the upper slopes. This pattern is refined by fire which travels uphill easily, so that the ridges and crests are burnt more severely than the valleys and lower slopes.

As soil–vegetation toposequences are common in tropical Africa, the soil type is related to the rock type and to the geomorphology. So, topography and geomorphology are related to the distribution of these vegetation patches (Cole, 1982, 1986). Here again, the causal factors act indirectly.

Sometimes, the valleys are not forest, but treeless grassland. The grassland is because the forest has been destroyed – probably by farming. Their higher soil moisture results in taller and denser grasses. This larger amount of fuel causes more frequent and more intense fires which kill the trees. So, from the same topographic and vegetation starting point, the end results may be quite different.

Savanna inliers

Savanna patches in the forest are common near the main forest–savanna boundary. Mostly, these have arisen from forest–savanna farms being converted into savanna as described previously. Where there is a general advance of savanna, they tend to coalesce into larger patches. Occasionally, savanna patches are due to the rapid advance of forest into the savanna. These usually disappear quickly as they are converted into forest.

Savanna inliers also occur throughout the forest zone, even where the rainfall exceeds $2000\,mm\,yr^{-1}$ (Le Testa, 1938). Most are treeless grasslands with an impoverished herbaceous flora because they are too far from other savanna areas with the necessary seed parents. The origin of these savanna patches has been disputed. There is no doubt that the frequency and severity of fire is related to the general increase in savanna (Thomas, 1941; Aubréville, 1966; Letouzey, 1968), but is it the only cause? Five explanations are available.

First, that fire alone causes the change. This is possible, especially for the conversion of woodland and dry forest. But, is it possible from evergreen forest, as has also been claimed (Aubréville, 1966)? An experiment at Adiopodoume (Ivory Coast), under an annual rainfall of 1950 mm, did produce a savanna-like vegetation after eight years, but only with enormous inputs of grass and other seed, and with cultivation and annual burning (Adjanohoun, 1964; Adjanohoun and Assi, 1968). Clearly, in addition to cultivation and fire, other factors are necessary for savanna to develop from evergreen forest.

Secondly, it has been claimed that cultivation alone can change forest into savanna (Sillans, 1958). It seems doubtful if the grasses invading a forest-zone farm could compete successfully against the forest species unless they were aided by fire. Even if cultivation could *cause* the change, fire would still be needed to *maintain* the savanna. However, it is possible that dry forest might be changed to savanna by cultivation alone.

As previously described, the change of forest into savanna is usually due to cultivation followed by burning – a combination of the first two of these explanations; see reviews of Aubréville (1949a,b), Schnell (1950) and West (1965).

Thirdly, there are edaphic causes (Ahn, 1959; Cole and Jarrett, 1969; Adejuwon, 1971b). The soil may be too shallow for tree roots. Alternatively, there may be a hard horizon near the surface, so that the soil becomes waterlogged in the wet season and desiccated in the dry season. Again, the soil may be waterlogged for at least part of the year, or it may be of a different texture. Not surprisingly, the vegetation of such soils differs from their surroundings. These edaphic savannas are usually burnt regularly, which helps to maintain them as savanna. Indeed, near the main forest–savanna boundary, it is only the annual burning which prevents the surrounding forest from over-running them; in these areas and under the conditions currently prevailing, it is the soil type which governs the type of vegetation.

Fourthly, there are sites of ancient cultivation, or other forest destruction, which resulted in soil changes which rendered them incapable of supporting forest (Aubréville, 1949a; Schnell, 1952).

Fifthly and finally, there are savanna patches which are relicts from the dry climate which existed before 12 000 BP (Portères, 1950; Schnell, 1952; Leneuf and Aubert, 1956; Adjanohoun, 1964; Aubréville, 1966; Adjanohoun and Assi, 1968; Bellier *et al.*, 1969). These too occur throughout the forest zones, and are particularly common in coastal areas. Although there is little doubt of this explanation of the *origin* of some of them, it is difficult to understand how they would be *maintained* were it not for the annual burning, unless there were also associated soil differences.

With these five opinions concerning the origin of savanna inliers deep within the forest zone, it is not surprising that disputes abound. Each explanation is probably correct for certain sites; other sites may well need different explanations, and generalizations are dangerous.

Forest outliers

Forest outliers are of two types. First, there are patches of forest which have escaped change during the general advance of savanna, and are relicts of the former forest cover. These are especially numerous and large near the forest edge of the forest–savanna mosaic zone, and the climate at the dry limit of such patches gives an indication of the natural climatic limit of forest. As has been discussed, their distribution is related to topography, geomorphology, soil and land-use history.

Secondly there are fringing (or gallery) forests along the watercourses (Nash and Page, 1953; Keay, 1960; Jones, 1963; Bonvallot *et al.*, 1970; Spichiger and Pamard, 1973; Devineau, 1976; Spichiger, 1977). These occur throughout the humid savanna zones; they are replaced by fringing woodland in the arid savannas. Fringing forests result from the valley soils having a greater water availability (and often a much higher proportion of clay) than the soils of the surrounding savanna. Fringing forests are rarely farmed. One important property is their very high length:area ratio, a feature which results in their enormous edge effect.

One consequence of this and of the microclimate of fringing forest being more humid than that of the surrounding savanna is of applied importance. Tstese (*Glossina* spp.) are mainly forest or woodland insects. However, the *G. palpalis* group of species uses fringing forest to 'rest' and 'breed'. From the fringing forest they make forays out into savanna to feed on the blood of vertebrates and, by doing so, are vectors for the transmission of trypanosomes.

THE BOUNDARY

Now that the natures of forest and of savanna have been discussed, attention can be focused on the forest–savanna boundary itself. Many profiles of the forest–savanna boundary have been published and the most detailed ones are from Olokemeji (Nigeria) (Clayton, 1958) and Lamto (Ivory Coast) (Devineau, 1975, 1976, 1984; Spichiger, 1977). The 'typical' boundary is very abrupt (Fig. 2.1); normally, the whole change from forest to savanna occurs in less than 50 m. Usually, separating the forest from the savanna is a central zone of 'transition' woodland in which both fire-tolerant savanna and fire-tender forest tree species occur. Within the transition, there is a dramatic change between the savanna, with its dense grasses, and the woodland, in which the grasses are sparse.

As with any boundary, there are three dynamic possibilities: the forest will advance into the savanna; or the savanna will advance into the forest; or that the boundary will remain stable – at least over a century or so.

Forest advance

If the area is protected from cultivation, as occurs in reserves of various types, the forest will advance into the savanna; the transition becomes a

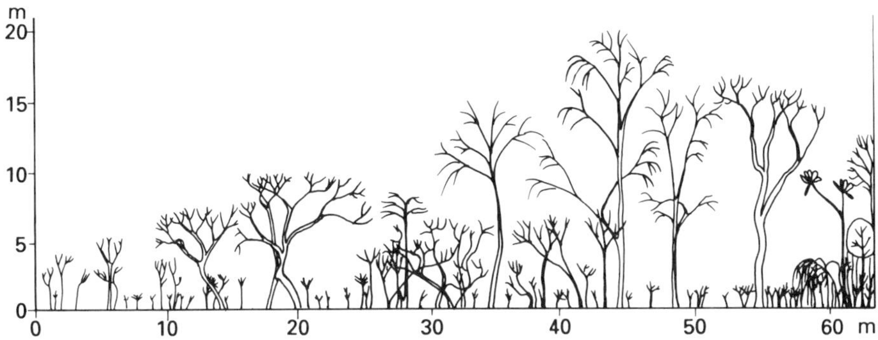

Fig. 2.1 Profile diagram of the vegetation 1.5 m or more tall on a 9 m wide strip across the forest–savanna boundary in the Olokemeji Forest Reserve, Nigeria. A well-developed central zone of transition woodland is shown and the forest (on the right) is advancing into the savanna (on the left). Modified from Clayton (1958).

secondary succession. As the forest trees grow, they form larger canopies which shade out the savanna grasses both in the transition woodland and on the edge of the savanna. Normally, the savanna fires die out where the grass density decreases – at the boundary between savanna and woodland. In some years, the woodland is burnt, but only rarely does the fire pass into the litter on the forest floor. With the gradual shading-out of the grasses on the boundary between the savanna and the transition woodland, this area becomes available for the colonization by, and the establishment of, forest trees and, so, the forest advances into savanna (Lebrun, 1936; Eggeling, 1947; Sillans, 1958; Hopkins, 1962; Miège, 1966; Letouzey, 1968; Spichiger, 1977).

The boundary is tortuous and the advance of forest into the savanna is erratic. The rate depends on the severity of the annual fires. Averages (over a decade or longer) of 2.5–3.5 m yr^{-1} have been reported from forest reserves in western Africa (Hopkins, 1962; Miège, 1966), but the rates are extremely variable. If the forest advance is pincer-like, then areas of savanna become surrounded and cut off from the general savanna. These may then escape burning and, so, rapidly change to forest. Occasionally, woodland 'islands' form in the savanna. An area of a few square metres escapes burning and more trees become established on it. These shade out the savanna grasses and enable forest trees to become established. Once such woodland clumps are formed, and provided conditions do not change, the whole of the affected area of savanna will rapidly change to forest.

Forest advance is often aided by wild or domesticated grazing animals. The amount of organic matter that the animals consume can so reduce

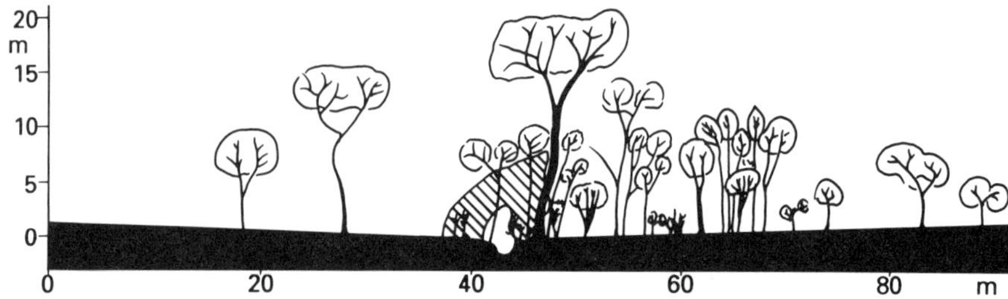

Fig. 2.2 Profile diagram of the fringing forest–savanna boundary in the Anara Forest Reserve, Nigeria. A zone of transition woodland is shown on the right-hand side of the stream where the forest is probably advancing into the savanna. On the left-hand side of the stream there is no transition woodland; just an abrupt change, with the savanna abutting onto a forest-climber tangle (shown cross-hatched). Here the boundary is probably stable due to reinforcing edaphic factors (see text). Modified from Keay (1960).

the amount of fuel that fires are very limited and forest trees become established. Termites also aid this process either as grass-consuming herbivores or by providing termitarium mounds which are favoured sites for tree colonization.

Whilst forest advance is particularly characteristic of the main forest–savanna boundary it is not confined to it. For example, fringing forest may advance (as 'ordinary' forest) into the savanna beyond the valley (Fig. 2.2).

Savanna advance

In most places the opposite has happened. As has been described previously, where there is no protection from cultivation and fire, savanna has advanced at the expense of forest. In these situations, the boundary is much more abrupt, there is no transition woodland zone, and the grassland abuts directly on to the forest. Also, the boundaries are often straight for short distances, being along the edges of former farms. Profile diagrams of such boundaries have been published (Fig. 2.2), but no site has been studied in the same detail as those where the forest is advancing.

The rate of savanna advance is largely governed by the frequency and intensity of cultivation and fire; so it occurs where the human population density is or has been high. In some places there is another important natural factor which can assist the conversion of forest into savanna – elephant. After man, elephant is the most important animal species for modifying tropical African vegetation. Alone, elephants cannot change forest into savanna; the process also needs fire. Nevertheless, substantial areas of forest were changed into savanna in national parks in eastern Africa when the density of elephant was high (Robyns, 1936; Buechner and Dawkins, 1961; Laws, 1970; Laws *et al.*, 1970, 1975).

Stable boundary

Finally, there are places where neither the forest nor the savanna is advancing and the boundary is stable. Stable boundaries are fairly abrupt and may be much more prevalent than was once thought.

They are common where a topographic soil-moisture sequence reinforces the vegetation pattern. Such cases occur where fringing forest occupies the valley bottom and sides (Fig. 2.2), which are on the moister soils with a higher clay content, and savanna occurs on the upper slopes where the soils are sandier and drier (Nash and Page, 1953; Keay, 1960; Jones, 1963; Bonvallot *et al.*, 1970; Spichiger and Pamard, 1973; Devineau, 1976, 1984; Spichiger, 1977). Stable boundaries also occur around most savanna inliers which are caused by edaphic factors.

There is one other situation where there is little, if any, boundary movement. Woodland and dry forest are relatively easy to convert to savanna and, in western Africa, most were converted some centuries ago. However, where the climatic humidity is higher, there are progressively moister and more evergreen types of forest which are much more difficult to convert to savanna. So, many of the boundaries between the moist evergreen forest and savanna are stable.

CONCLUSIONS

It is well known that the presence of savanna in the forest–savanna mosaic zone is, generally, due to fire following cultivation. If fire is prevented, the vegetation will change back to forest by a secondary succession. After climatic aridity, fire is the predominant factor governing the dynamics and, so, the physiognomy of savanna vegetation. To a large extent, it controls both the direction of movement and the rate of movement of the forest–savanna boundary.

Nevertheless, there is a considerable danger of over-generalizing about the importance of the role of fire. There are several other factors which, with or without fire, are often vital to an understanding of the relationships of forest and savanna. These factors include topography, geology and geomorphology, soil, past climate, cultivation history, elephant, herbivory and termites. Each can be crucially important in certain situations.

The general explanation of 'fire' will not fit all cases. Consequently, the factors governing the dynamics of the forest–savanna boundary need to be elucidated at each locality – and not all these factors are strictly ecological. As A.S. Thomas wrote in 1965: 'More history and less philosophy are needed for progress in ecological studies in Africa' (Thomas, 1965).

REFERENCES

Adejuwon, J.O. (1971a) Savanna patches within forest areas in western Nigeria: a study of the dynamics of forest savanna boundary. *Bulletin de l'Institut Francaise d'Afrique Noire*, **33(A)**, 327–44.

Adejuwon, J.O. (1971b) The ecological status of savannas associated with inselbergs in the forest areas of Nigeria. *Tropical Ecology*, **12**, 51–65.

Adjanohoun, E. (1964) Végétation des savanes et des rochers découverts en Côte d'Ivoire Centrale. *Mémoires Office de la Recherche Scientifique et Technique Outre-Mer*, **7**, 1–178.

Adjanohoun, E. and Assi, A.L. (1968) Essais de création de savanes incluses en Côte-d'Ivoire forestière. *Université d'Abidjan, Annales de la Faculté des Sciences*, **4**, 237–56.

Ahn, P. (1959) The savanna patches of Nzima, south-western Ghana. *Journal of the West African Science Association*, **5**, 10–25.

Aubréville, A. (1949a) *Climats, Forêts et Desertification de l'Afrique Tropicale*, Société d'éditions géographiques, maritimes et coloniales, Paris.

Aubréville, A. (1949b) *Contribution à la Paléohistoire des Forêts de l'Afrique Tropicale*, Société d'éditions géographiques, maritimes et coloniales, Paris.

Aubréville, A. (1966) Les lisières forêt-savane des régions tropicales. *Adansonia, Paris*, **6**, 175–87.

Bellier, L., Gillon, D., Gillon, Y. *et al.* (1969) Recherches sur l'origine d'une savane incluse dans le bloc forestier du Bas-Cavally (Côte d'Ivoire) par l'étude des sols et de la biocenose. *Cahiers Mémoires Office de la Recherche Scientifique et Technique Outre-Mer, série Biologique*, **10**, 65–94.

Bonvallot, J., Dugeril, M. and Duviard, D. (1970) Recherches écologiques dans la savane de Lamto (Côte d'Ivoire): repartition de la végétation dans la savane préforestière. *Terre et Vie*, **24**, 3–21.

Bourlière, F. (ed.) (1983) *Ecosystems of the World, 13: Tropical Savannas*, Elsevier, Amsterdam.

Bourlière, F. and Hadley, M. (1970) The ecology of tropical savannas. *Annual Review of Ecology and Systematics*, **1**, 125–52.

Buechner, H.K. and Dawkins, H.C. (1961) Vegetation changes induced by elephant and fire in Murchison Falls National Park, Uganda. *Ecology*, **42**, 752–66.

Charter, J.R. and Keay, R.W.J. (1960) Assessment of the Olokemeji fire-control experiment (investigation 254) 28 years after institution. *Nigerian Forestry Information Bulletin*, **3**, 1–32.

Clayton, W.D. (1958) Secondary vegetation and the transition to savanna near Ibadan, Nigeria. *Journal of Ecology*, **46**, 217–38.

Cole, M.M. (1982) The influence of soils, geomorphology and geology on the distribution of plant communities in savanna ecosystems, in *Ecology of Tropical Savanna*, (eds B.J. Huntley and B.H. Walker), Springer-Verlag, Berlin, pp. 145–74.

Cole, M.M. (1986) *The Savannas: Biogeography and Geobotany*, Academic Press, London.

Cole, N.H.A. and Jarrett, H.D. (1969) Tropical plant communities of limited occurrence. *Journal of the West African Science Association*, **14**, 95–102.

Devineau, J.L. (1975) *Etude quantitative des forêts-galeries de Lamto (Moyenne Côte d'Ivoire)*, These de III cycle, Université, Paris.

Devineau, J.L. (1976) Principales caractéristiques physiognomiques et floristiques des formations forestières de Lamto (moyenne Côte d'Ivoire). *Annales de la Université d'Abidjan, série E*, **9**, 273–303.

Devineau, J.L. (1984) Structure et dynamique de quelques forêts tropophiles de l'Ouest africain (Côte d'Ivoire). *Travaux des Chercheurs de la Station de Lamto (Côte d'Ivoire)*, **5**, 1–295.

Eggeling, W.J. (1947) Observations on the ecology of the Bundongo Rain Forest, Uganda. *Journal of Ecology*, **34**, 20–87.

Hopkins, B. (1962) Vegetation of the Olokemeji Forest Reserve, Nigeria I. General features and the research sites. *Journal of Ecology*, **50**, 559–98.

Hopkins, B. (1965) *Forest and Savanna*, Heinemann, London.

Hopkins, B. (1983) Successional processes, in *Ecosystems of the World 13: Tropical Savannas*, (ed. F. Bourlière), Elsevier, Amsterdam, pp. 605–16.

Huntley, B.J. and Walker, B.H. (eds) (1982) *Ecology of Tropical Savannas*, Springer-Verlag, Berlin.

Jones, E.W. (1963) The forest outliers in the guinea zone of Northern Nigeria. *Journal of Ecology*, **51**, 415–34.

Keay, R.W.J. (1960) An example of northern guinea zone vegetation in Nigeria. *Nigerian Forestry Information Bulletin*, **1**, 1–46.

Latham, M. and Dugerdi, M. (1970) Contribution à l'étude de l'influence du sol sur la végétation au contact forêt-savane dans l'ouest et le centre de la Côte-d'Ivoire. *Adansonia*, **10**, 553–76.

Laws, R.M. (1970) Elephants as agents of habitat and landscape change in East Africa. *Oikos*, **21**, 1–15.

Laws, R.M., Parker, I.S.C. and Johnstone, R.C.B. (1970) Elephants and habitats in North Bunyoro, Uganda. *East African Wildlife Journal*, **8**, 163–80.

Laws, R.M., Parker, I.S.C. and Johnstone, R.C.B. (1975) *Elephants and their Habitats: the Ecology of Elephants in North Bunyoro, Uganda*, Oxford University Press, Oxford.

Lebrun, J. (1936) La forêt équatoriale congolaise. *Bulletin Agricole du Congo Belge, Brussels*, **27**, 163–92.

Leneuf, N. and Aubert, G. (1956) Sur l'origine des savanes de la basse Côte d'Ivoire. *Compte Rendu d'Academie des Sciences, Paris*, **243**, 859–60.

Le Testa, G. (1938) Note sur la végétation dans de bassin de la Nyanga et de la Ngouné au Gabon. *Mémoires de la Société Linnéenne de Normandie (Calvados), Caen*, **1**, 83–108.

Letouzey, R. (1968) *Etude Phytogéographique de Cameroun*, Lechevalier, Paris.

MacGregor, W.D. (1937) Forest types and succession in Nigeria. *Empire Forestry Journal*, **16**, 234–42.

Miège, J. (1966) Observations sur les fluctuations des limites savannes-forêts en basse Côte d'Ivoire. *Annales de la Faculté des Sciences, Dakar*, **19**, 149–66.

Monnier, Y. (1968) Les effets des feux de brousse sur une savane préforiestière de Côte-d'Ivoire. *Etudes Eburnéennes*, **9**, 1–260.

Moss, R.P. (1977a) Soils, plants and farmers in West Africa. Part 2: Further examination of the role of edaphic factors and cropping patterns in south-west Nigeria, in *Human Ecology in the Tropics* (eds J.P. Garlick and R.W.J. Keay), Symposium, Society for the Study of Human Biology, 16, Taylor & Francis, London, pp. 59–77.

Moss, R.P. (1977b) Relations between tropical moist forest, edaphic factors and husbandry systems under limiting conditions in tropical West Africa, in *Land Use and Development*, (eds P.O. O'Keefe and B. Wisner), African Environment Special Reports, 5, International African Institute, London, pp. 6–17.

Moss, R.P. and Morgan, W.B. (1977) Soils, plants and farmers in West Africa. 1: A consideration of some aspects of their relationships, with special reference to continuous areas of forest and savanna in south-west Nigeria, in *Human Ecology in the Tropics*, (eds J.P. Garlick and R.W.J. Keay), Symposium, Society for the Study of Human Biology, 16, Taylor & Francis, London, pp. 27–57.

Nash, T.A.M. and Page, W.A. (1953) The ecology of *Glossina palpalis* in northern Nigeria. *Transactions of the Royal Entomological Society, London*, **104**, 71–169.

Portères, R. (1950) Problémes sur la végétation de Basse Côte d'Ivoire. *Bulletin de la Société Botanique de France*, **97**, 153–6.

Ramsay, J.M. and Rose-Innes, R. (1963) Some quantitative observations on the effects of fire on the guinea savanna vegetation of northern Ghana over a period of eleven years. *African Soils*, **8**, 41–86.

Robyns, W. (1936) Contribution à l'étude des formations herbeuses du district forestier central de Congo belge. *Mémoires de l'Institut Royale Colonial Belge, Section des Sciences Naturelles et Médicales*, **5**, 1–151.

Schnell, R. (1950) *La Foret Dense*, Lechevalier, Paris.

Schnell, R. (1952) Végétation et flore de la region montagneuse du Nimba (A.O.F.). *Mémoires de l'Institut Française d'Afrique Noire*, **22**, 1–604.

Schnell, R. (1976) *Introduction à la Phytogéographie des Pays Tropicaux, 3, La Flora et la Végétation de l'Afrique Tropicale* Gauthier-Villars, Paris.

Sillans, R. (1958) *Les Savanes de l'Afrique Centrale*, Encyclopédie Biologique, **55**, Lech evalier, Paris.

Spichiger, R. (1977) Contribution à l'étude de contact entre flores sèche et humide sur les lisières des formations forestières humides semi-décidues de V baoulé et de son extension nord-ouest. *Bulletin de Liaison des Chercheurs de Lamto, (Cote d'Ivoire)*, Numero Special 1977 fasc. 1.

Spichiger, R. and Pamard, C. (1973) Recherches sur le contact forêt–savane en Côte-d'Ivoire: étude du recrû forestier sur des parcelles cultivées en lisière d'un îlot forestier dans le sud du pays bauolé. *Candollea*, **28**, 21–37.

Swaine, M.D., Hall, J.B. and Lock, J.M. (1976) The forest-savanna boundary in west-central Ghana. *Ghana Journal of Science*, **16**, 35–52.

Thomas, A.S. (1941) The vegetation of the Sese Islands, Uganda. *Journal of Ecology*, **29**, 330–53.

Thomas, A.S. (1965) There is all Africa. *Nature, London*, **208**, 16–19.

Tothill, J.C. and Mott, J.J. (eds) (1985) *Ecology and Management of the World's Savannas*, Australian Academy of Sciences, Canberra.

Unesco (1978) *Tropical Forest Ecosystems*, Unesco, Paris.

Unesco (1979) *Tropical Grazing Land Ecosystems*, Unesco, Paris.

Vuattoux, R. (1970) Observations sur l'évolution des strates arborées et arbustives dans la savane de Lamto. *Annales de la Université d'Abidjan, serié E*, **3**(1), 285–315.

Walker, B.H. (ed.) (1985) *Determinants of Tropical Savannas*, International Union of Biological Sciences, Paris.

West, O. (1965) Fire in vegetation and its use in pasture management with special reference to tropical and sub-tropical Africa. *Commonwealth Bureau of Pastures and Field Crops Bulletin*, **1**, 1–53.

White, F. (1983) *The Vegetation of Africa*, Unesco, Paris.

PART TWO

Physical Environment at the Boundary

LIST OF ACRONYMS

AVHRR	Advanced Very High Resolution Radiometer
GT	Gigatonnes
UNDP	United Nations Development Programme
ESMAP	Energy Sector Management Assistance Programme
ETC	Education and Training Consultants (Dutch environmental consulting company)
NOAA	National Oceanic and Atmospheric Administration
NDVI	Normalized Difference Vegetation Index
GVI	Global Vegetation Index
NASA/GSFC	National Aeronautics and Space Administration/ Goddard Space Flight Center

3

Mapping forests and savannas in sub-Saharan Africa from advanced very high resolution radiometer (AVHRR) imagery

A.C. MILLINGTON, P.J. STYLES and R.W. CRITCHLEY

SUMMARY

1. Advanced very high resolution radiometer (AVHRR) data acquired during 1986 were used to map land cover classes in sub-Saharan Africa. Digital image processing algorithms were applied to multi-temporal AVHRR data to provide provisional land cover class maps which were field-verified. Final land cover class maps were produced by supervised classification. The importance of multi-temporal imagery in mapping forest and savanna woodland distributions is exemplified by the sub-division of Sudanian and Miombo woodlands on the basis of their 'AVHRR phenology'.
2. The land cover classes were defined in terms of their woody biomass characteristics and a data base of woody biomass stock and sustainable yield was derived from secondary data sources. This data base was applied to the land cover class maps in a simple information system enabling maps of growing stock and sustainable yield to be produced. The total growing stock of woody biomass is provisionally estimated at 32.81 Gt for savannas; 23.22 Gt for forests; 3.76 Gt for the transition zones between forests and savannas (Gigatonnes: t $\times$ 10^9). The annual woody biomass yields for these classes are provisionally estimated at 0.53 Gt, 2.78 Gt and 0.05 Gt respectively.

Key words Tropical forests, Savanna, Land cover, Remote sensing, AVHRR, Supervised classification, Growing stock, Sustainable yield, Africa.

INTRODUCTION

One method of examining forest–savanna boundaries is to take a broad-scale approach and to map the distribution of forests and savannas for entire continental areas. This approach is used extensively in the production of vegetation maps (e.g. White 1983). However, there are other reasons why a continental-scale analysis of forest and savanna distributions is important. In this study, commissioned by the World Bank/UNDP Energy Sector Management Assistance Programme (ESMAP), the rationale was to provide a first, and up-to-date map of land cover classes for energy policy assessments in sub-Saharan Africa with an accompanying database to provide information on growing stocks and sustainable yields.

The approach taken in the project was to rely heavily on remotely-sensed imagery as the primary data source with relatively little effort being expended on field data collection. Secondary data provided the bulk of the information on class identification and description, and for the establishment of the data base. The approach was therefore essentially synthetic, with the majority of the original research being expended on the analysis and interpretation of remotely-sensed data.

Consequently, in this chapter, emphasis is placed on the role that remotely-sensed data can play in mapping forest and savanna woodland at a continental scale, particularly in the context of wood energy planning. The chapter describes the analysis and interpretation of the AVHRR data used in this project; along with the creation of a woody biomass data base. The image processing and mapping is illustrated by reference to savanna woodland classes. The growing stock and sustainable yield from both forest and savanna woodland classes are analysed in the context of fuelwood energy planning. Obviously a number of the vegetation and land cover types mapped in this project do not concern savannas or forests and these are not considered in the present contribution. Further information on the project and the other classes can be found in ETC (1989) and Millington *et al.* (in press).

Advanced very high resolution radiometer imagery

The image data used in this research were acquired by the advanced very high resolution radiometer (AVHRR), a sensor carried on board the American NOAA series of meteorological satellites. The salient characteristics of this satellite-sensor combination are provided in Table 3.1. AVHRR data are acquired for any part of the earth's surface every 12 hours. One of the two data sets taken in each 24 hour period is at night. The daylight equatorial crossing time for the two satellites that provided

Table 3.1 Characteristics and status of NOAA weather satellites (a) and characteristics of the AVHRR sensor (b)

(a) Satellite

Satellite	*Launch date*	*Current status*
TIROS-N	Oct. 1978	Non operational
NOAA 6	Jun. 1979	Non operational
NOAA 7	Jun. 1981	Non operational
NOAA 8	Mar. 1983	Non operational
NOAA 9	Dec. 1984	Non operational
NOAA 10	Oct. 1986	Operational
NOAA 11	Sept. 1988	Operational

(b) AVHRR sensor

Coverage cycle	12 hours
Ground coverage	3000 km
Orbital height	870 km
Orbital period	102 minutes
Ground resolution	1.1 km^2 to 2.4 × 6.9 km^2
Spectral channels	1- 0.58–0.68 2- 0.725–1.10 3- 3.55–3.93 4- 10.30–12.5 5- 11.50–12.5 wavelength (μm)
Equatorial crossing time (local sun time)	(i) 0730 and 1930 (ii) 1400 and 0200

imagery for this study were 1400 hrs (local time) for NOAA-9 and 0730 hrs (local time) for NOAA-10. These data have a very coarse spatial resolution, a large areal coverage and a high temporal frequency of image acquisition (Table 3.1) to meet the weather forecasting requirements for which the satellite and sensor were designed. However, because data are collected daily, there is a high probability that in a given time period, a ground resolution element (the smallest area on the ground that can be resolved on the imagery) will be cloud-free at the time of data acquisition allowing land surface characteristics to be sensed.

Normalized difference vegetation index (NDVI)

The interaction of electromagnetic radiation with the structures in leaves means that remotely-sensed data acquired in the visible-red (600–700 nm) and near infra-red (800–1100 nm) parts of the electromagnetic spectrum are applicable to vegetation and land-use mapping. Detailed studies relating the physiology and biochemistry of plants to their spectral characteristics have been undertaken (e.g. Sellers, 1985, 1986, and Tucker and Sellers, 1986) concluding that spectral reflectance of vegetation is related to photosynthesis and respiration.

The configuration of the AVHRR sensor, particularly the wavelengths of the sensors in Channels 1 (visible red) and 2 (near infra-red) (Table 3.1), means that when there are no clouds, ecologically relevant information on vegetation canopy reflectance can be obtained for the ground resolution elements. These reflectance data are used to calculate a vegetation index (an indicator of vegetation status) for each ground resolution element. Vegetation indices are derived to reduce measurements of the vegetation canopy, ground cover and biomass to a single number (Perry and Lautenschlager, 1984). There are at least 48 different vegetation indices used by remote sensors, but the normalized difference vegetation index (NDVI) has been successfully and extensively used with AVHRR data (Justice *et al.*, 1986; Townshend *et al.*, 1987), principally because the normalizing nature of the equation reduces atmospheric effects. It is calculated by the following equation:

$$\text{NDVI} = \frac{\text{DN(Channel 2)} - \text{DN(Channel 1)}}{\text{DN(Channel 2)} + \text{DN(Channel 1)}}$$

where NDVI is the normalized difference vegetation index, DN is the digital number, Channel 1 is the reflectance from the visible red part of electromagnetic spectrum and Channel 2 is the reflectance from the near infra-red part of electromagnetic spectrum.

Advantage has been taken of NDVI data from meteorological satellites by a number of researchers since 1981 for vegetation and land-use mapping over large areas (Gray and McCrary, 1981; Schneider *et al.*, 1981; Gatlin *et al.*, 1983; Norwine and Greegor, 1983; Tarpley *et al.*, 1984; Townshend and Tucker, 1984; Tucker *et al.*, 1984a,b; Goward *et al.*, 1985). This work has led to a number of natural resource applications in the tropics, for instance, monitoring Lake Chad (Schneider *et al.*, 1985); mapping biomass production for browsing and grazing potential in Senegal (Tucker *et al.*, 1985) and Tunisia (Kennedy, 1989a,b); monitoring semi-arid rangelands in Botswana (Prince, 1986); monitoring cropping patterns in south east Asia (Malingreau, 1986); monitoring tropical deforestation (Malingreau and Tucker, 1988; Malingreau *et al.*, 1989), mapping forest fires (Malingreau *et al.*, 1985), and assessing woody biomass resources for woody energy planning in southern Africa (Millington *et al.*, 1989).

Raw AVHRR data are available from NOAA at different spatial resolutions:

1. the global vegetation index (GVI) is available at 8 × 8 km resolution, but can be resampled to the Polar Stereographic Projection at 15 × 15 km resolution;
2. global area coverage (GAC) data, which have a spatial resolution of 4 × 4 km;
3. local area coverage (LAC) data, which have a spatial resolution of 1.1 × 1.1 km.

Vegetation index data from both local and global area coverage data are available as composites over ten-day intervals. The NDVI data obtained from Goddard Space Flight Centre (NASA/GSFC) for this project were 'cloud screened'. This process is carried out by identifying pixels with clouds in the AVHRR images using thermal infra-red data from Channels 4 and 5 (Table 3.1) and excluding them during the production of temporal composites of NDVI. This procedure generally works well, but in areas where thin cirrus cloud is prevalent there are problems (Henderson-Sellers *et al.*, 1987). After cloud-screening the daily data are composited over ten-day intervals. For any ten-day period each pixel is examined and the day with the highest NDVI for that pixel is used to represent the ten-day period (Holben, 1986). This procedure reduces the atmospheric, cloud and sun-angle effects of vegetation index calculation.

METHODOLOGY

Image processing and land cover mapping

Land cover class mapping was based on the interpretation of digital AVHRR-NDVI data combined with selective field verification, and by reference to secondary data. All data were digitally processed, initially evaluated on-screen, and hard-copy imagery for further interpretation, field verification and publication was then obtained.

Data specifications and preliminary data inspection

NDVI data at 8 km spatial resolution were provided by NASA/GSFC for 1986 in ten-day and monthly time intervals. All imagery was registered to the Hammer-Aitoff Conic Equal Area Projection. The only problems encountered with the ten-day data were the occasional pixels without NDVI values.

The monthly data held on tape are configured into pixels in the image processing system; a pixel on-screen represents the ground resolution element of the sensor. An absence of data in a pixel can occur due to sensor, transmission or processing errors. In many cases this only involved single pixels and a digital filter was developed to eliminate empty single pixels (ETC, 1989; Millington *et al.*, in press). Larger, contiguous areas of empty pixels occurred in areas with a high probability of persistent cloud cover. This meant that during some months no cloud-free pixels had been found after temporal compositing and the NDVI for those pixels had not been calculated. These areas are mainly found in the coastal humid tropics and mountainous areas. They were not found in the same location in each month of the year but, because a complete sequence of NVDI values for each month of 1986 was used to classify land cover classes, the occurrence of pixels with cloud cover affects the accuracy of the maps. The smaller areas of cloud were eliminated after classification using an iterative median filter (ETC, 1989; Millington *et al.*, in press). However, there were very large areas of cloud in Cameroon, Ethiopia, Nigeria, South Africa and Zaire which could not be eliminated entirely. These areas were not included in the final classification because they would produce classes that were impossible to interpret and therefore a cloud mask was produced.

Imagery derived from AVHRR data

Various image products were derived from the monthly NDVI data and used to map provisional land cover classes; these were

1. individual monthly NDVI images;
2. a mean annual NDVI image;
3. difference images between pairs of monthly NDVI values;
4. unsupervised classification images; and
5. temporal plots of NDVI data.

A mean annual GVI image was created from all the monthly images (Fig. 3.1) and it provides a good indication of the amount of, and variation in, annual vegetation productivity. It is analogous to the integrated NDVI images that have been produced by other researchers but differs from them in that they are calculated on the basis of the area under the NDVI curve, whereas this image is a simple arithmetic mean.

Difference images were created by subtracting one set of monthly NDVI data from another and adding a constant, on a pixel-by-pixel basis, for corresponding pixels. The months used to create the difference images were chosen on the basis of known phenological characteristics of African vegetation and land use. Two types of difference images are particularly useful in the interpretation of biomass classes. First, those in which monthly images from the period of vegetation senescence are subtracted from the period of maximum greenness. Such images provide a good

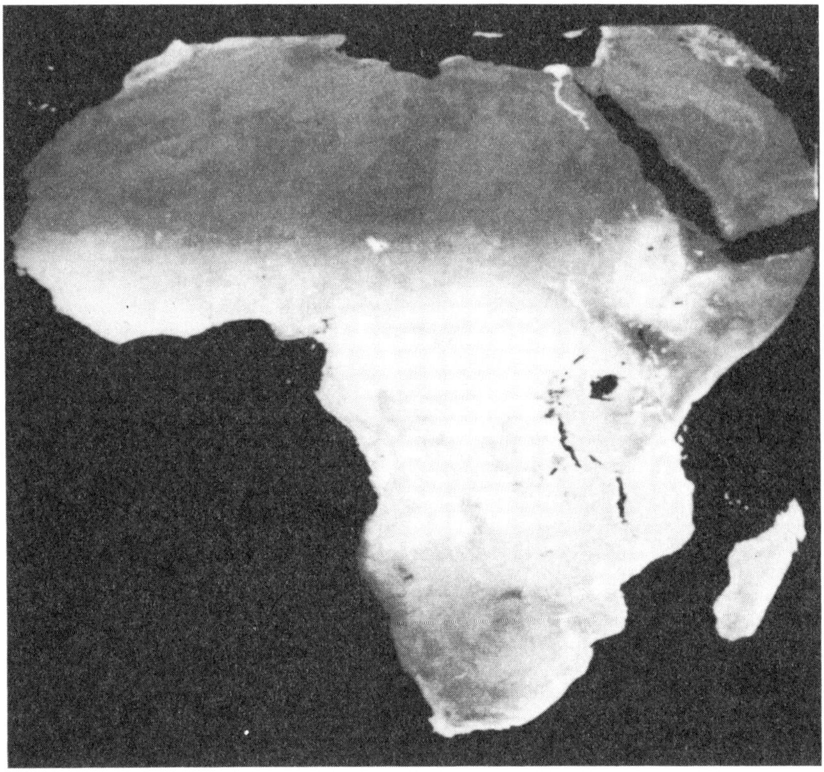

Fig. 3.1 Sub-Saharan Africa: mean annual GVI image for 1986. This image was created from 12 monthly images by the pixel-by-pixel addition of monthly GVI values for corresponding pixels; the total GVI values for each pixel were then divided by 12. The image shows low integrated NDVI values in dark grey tones and the highest values in light grey. The sea has been masked and only information on land is shown.

indication of vegetation die-back in response to declining seasonal rainfall, varying levels of residual soil moisture or the onset of winter conditions, such as cooler temperatures or increased frequency of frosts. Secondly, images where the high NDVI image is subtracted from the low NDVI period image to provide information on greening-up of vegetation.

In addition to these images, temporal profiles of NDVI were plotted. These profiles indicate the seasonal fluctuations of NDVI (the so-called NDVI phenology) and have been used indirectly to analyze seasonal variations in photosynthesis, leaf area index, and biomass production in a number of studies. Each point represents an average NDVI over the same

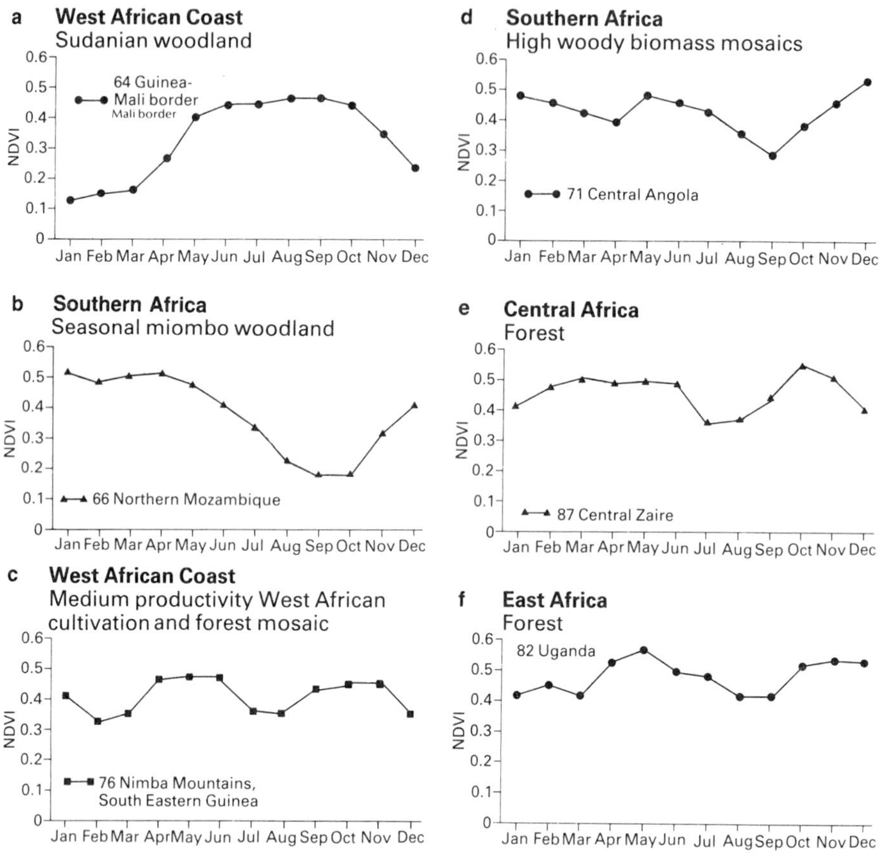

Fig. 3.2 Temporal profiles of monthly NDVI values from 1986 for forest, savanna and high woody biomass transitions. The locations of the profiles are as follows: Savanna Woodlands (a) Guinea-Mali border, (b) Northern Mozambique; High Woody Biomass Transitions (c) South Eastern Guinea, (d) Central Angola; Forests (e) Central Zaire, (f) Uganda.

3×3 pixel kernel in each of the monthly images. Examples of temporal profiles for selected African forests and savannas are shown in Fig. 3.2.

Mapping land cover (woody biomass) classes

The imagery was used to identify and map the vegetation types on 1:5 000 000 scale base maps for five regions in which field verification was undertaken. (In the project seven countries were visited – Congo, Ethiopia, Ghana, Ivory Coast, Mali, Senegal, South Africa, and Zaire). In addition to these areas, field verification data from other land cover mapping projects was available for The Gambia, Kenya, Malawi, Sierra Leone, and Swaziland. Identification of land cover types in these five areas was based on the following criteria:

1. vegetation phenology (obtained from temporal profiles of NDVI values and the seasonality/difference images);
2. monthly and annual vegetation productivity (based on monthly and mean NDVI images);
3. secondary data sources (particularly vegetation, land use and forestry maps) but also other environmental information, particularly geological, soils and topographic maps and climatic statistics.

The initial mapping of provisional land cover classes was carried out at a scale of 1:5 000 000 for two reasons. Firstly, this scale was successfully used for woody biomass class mapping from 1984 AVHRR imagery in southern Africa (Millington *et al.*, 1989). Secondly, the UNESCO vegetation map of Africa (White, 1983) was produced at this scale. The UNESCO map is one of the most recent vegetation maps covering all of Africa, although as its derivation and accuracy are questionable in places, it only provided a partial reference for checking image interpretations. Moreover, the classification scheme for land cover mapping in this project was a modified version of the scheme proposed by White (1983). However, it should be noted that the classification and distribution of land cover classes does not correspond exactly with any previous botanical or ecological study of the region, although they are broadly comparable to most regional vegetation maps.

The final land cover (woody biomass) class maps were produced by supervised classification. This is a powerful image-processing procedure in which the values of each pixel are compared to various sets of pixel values (training statistics) from preselected, known areas (training sites). The individual pixels are assigned to the most suitable class based on the training site statistics. Different algorithms can be used in supervised classification, but in this project a maximum likelihood classifier was used. Training sites were identified from the provisional land cover class maps and field observations. They were located in areas where the classes probably had their maximum areal extent, or were known to be distinctive. Supervised classification produced approximately 60 biomass classes for sub-Saharan Africa.

The supervised classification was based on six months of data (January, April, June, July, October and December). The decision to carry out the classification on six months of data was based on the following criteria. Firstly, there is a significant amount of temporal autocorrelation in the NDVI for adjacent months (Table 3.2), which would tend to lead to data redundancy if all 12 months were used in the classification. Secondly, two

Table 3.2 Correlation matrix of monthly NDVI values for 1986 for sub-Saharan Africa north of the Equator

	Jan	*Feb*	*Mar*	*Apr*	*May*	*Jun*	*Jul*	*Aug*	*Sep*	*Oct*	*Nov*	*Dec*
Jan	1	0.917	0.872	0.849	0.822	0.787	0.706	0.733	0.739	0.8	0.862	0.891
Feb		1	0.919	0.865	0.822	0.788	0.674	0.694	0.702	0.758	0.825	0.828
Mar			1	0.9	0.864	0.813	0.682	0.685	0.678	0.75	0.817	0.817
Apr				1	0.952	0.888	0.72	0.715	0.704	0.777	0.857	0.839
May					1	0.918	0.737	0.723	0.795	0.777	0.856	0.834
Jun						1	0.816	0.794	0.78	0.825	0.872	0.806
Jul							1	0.891	0.867	0.864	0.84	0.745
Aug								1	0.95	0.93	0.881	0.771
Sep									1	0.941	0.886	0.767
Oct										1	0.941	0.835
Nov											1	0.893
Dec												1

sequences of alternate months were classified – January to November and February to December – as well as classification using all 12 months. The differences between the classified images based on 12 month's data, and the January to November sequence were insignificant. Previous work on land cover mapping from AVHRR-NDVI data has shown that the more months used in the analysis, the greater the accuracy of classification (Townshend *et al.*, 1987). Whilst the present work does not question this finding, the decision to use six months of data was based on operational, rather than scientific criteria. It should not necessarily be assumed that accuracy will increase as the number of months used in the classification increases; but it would increase significantly if data from other years apart from 1986 were included. This is because NDVI is very sensitive to rainfall and soil moisture variations which vary considerably from year-to-year over much of Africa.

The land cover maps were reviewed by experts in regional workshops and recommendations made to merge some classes. Most of these mergers are of classes with statistically similar NDVI curves which, on the basis of floristic and ecological criteria, were 'incorrectly' split during the supervised classification. However, the special case of cultivation mosaics in west Africa and Ethiopia presented problems. In the west African coastal zone, the areas originally classified as different types of humid tropical forest and forest regrowth were merged into two classes:

1. West African mixed agriculture and forest fallow mosaic (high productivity phase);
2. West African mixed agriculture and forest fallow mosaic (moderate productivity phase).

This merger was based on the realization that in this region there is very little humid tropical forest remaining. That which is, lies in small blocks and reserves interspersed with rubber, cacao, and coffee plantations, farms, and a seral succession forest regrowth. This pattern of land use occurs at scales that cannot be mapped from 8 km AVHRR data, although a certain amount of success at mapping the remaining forest in this area has been made using 1.1 km resolution AVHRR data (Malingreau *et al.*, 1989).

In the Ethiopian highlands, a similar problem of small land parcels with different biomass levels and NDVI profiles exists, but here there is an additional problem of altitude. In this case it was decided to introduce altitude into the mapping procedure. The 2000 and 3500 m contours were digitized and overlain on the classified image. With the exception of the montane forests, all of the classes occurring between 2000 and 3499 m in the area were merged to form the Highland cultivation mosaic and those above 3500 m were merged to form the Ethiopian montane steppe.

The land cover classes are defined as areas with a vegetation or land-use type, or types, which are comparable in terms of the overall woody biomass stock, productivity and seasonality. They may, however, combine more than one type of ecological structure and include a number of floristic units. In this respect they differ slightly in places from the UNESCO vegetation map of Africa, but for the purposes of wood energy planning, a classification based on parameters which have been correlated with growing stock and productivity elsewhere, are of more use than a floristic map.

The map of land cover classes in sub-Saharan Africa is one of the final products of this project. In addition, a summary map was created by categorizing the land cover classes into nine groups (summary land cover classes) on the basis of the dominant ecological structures (Fig. 3.3).

Compilation of woody biomass database

Statistical information on woody biomass stocks and yields was compiled from a variety of secondary sources. These ranged from forest inventories to ecological studies. Consequently the secondary data available to create the database were variable in both emphasis and quality. The database was developed using dBase 3+ (disk or hard-copy format from ETC(UK), 37–39 Norfolk St, North Shields, Tyne and Wear, UK.) and comprised nine fields (Table 3.3), in addition to details of the study.

Interfacing the remotely sensed data with the database

The area of each land cover class in each country was calculated as follows. The coastline, national boundaries, and shorelines of major lakes

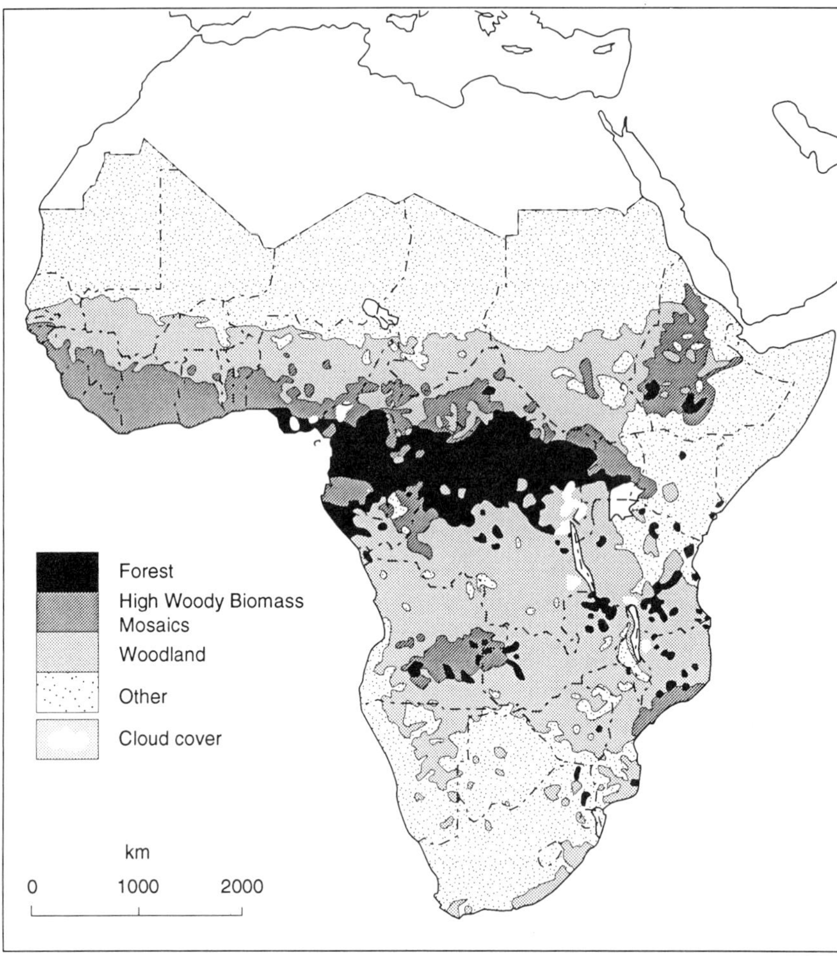

Fig. 3.3 Sub-Saharan Africa: Summary Woody Biomass Classes.

were digitized from a 1:10 000 000 scale map of Africa. These data were registered to the AVHRR imagery, and masks for the sea and lakes created. Each country was considered as a separate polygon and the pixels in each class, as well as the lake and cloud masks, counted. The file of pixel counts for each polygon was transferred to a spreadsheet and the pixel counts converted to areas. Accuracy checking on the area statistics for country areas and lakes was carried out by reference to gazetteers.

A sub-file was made from the database which consisted of growing stock and mean annual increment measurements (in dry weight units) and each measurement was geo-located with reference to the land cover

Table 3.3 Data fields in woody biomass database

Geographical location
Area of sample site
Vegetation type
Wood production, consumption or yield (depending on the type of study)
Standing stock
Tree density
Dominant tree or shrub species
Biomass Growing Sustainable yield

classes. This sub-file was interfaced with the area statistics for each country in a spreadsheet. The geo-referencing of land cover measurements was not done by latitude and longitude coordinates because a land cover type may have changed between the time of biomass measurement and image acquisition, but to land cover class by analysing vegetation and land use descriptions. Tables of provisional growing stock and sustainable yields were created for each country in sub-Saharan Africa on the basis of land cover and summary land cover classes. Provisional maps of growing stock (Fig. 3.4) and sustainable yield (Fig. 3.5) were made by using the data from the sub-file and the land cover class images.

RESULTS

Spatial distribution of forests and savannas

Only three of the nine summary land cover classes – Forests, Woodlands and High Woody Biomass Mosaics – are concerned with forest, savannas and the transitional zones between them respectively. The High Woody Biomass Mosaics in almost all instances represent the transitional zones between forest and savanna woodland. The other summary classes represent the drier savanna vegetation types, grasslands and desert and are not considered in this chapter. Twenty-one land cover classes were recognized in these three summary classes (Table 3.4), with seven in each of the three summary classes. If the areas covered by the summary classes are considered (Table 3.5) it can be seen that land cover classes dominated by savanna woodlands are very important in central Africa (40.9% of the region), the coastal region of west Africa (39.7%), and southern Africa (37.8%). In particular, large tracts of savanna woodlands are found in Zaire (1 084 415 km^2), Angola (728 248 km^2), Sudan (619 064 km^2) and Zambia (520 419 km^2). High woody biomass mosaics are important in the coastal region of west Africa (44.4% of the region) and central Africa (16.3%); and very large areas occur in Ethiopia (306 476 km^2), Ivory Coast (257 205 km^2) and Zaire (208 831 km^2). Woody biomass classes

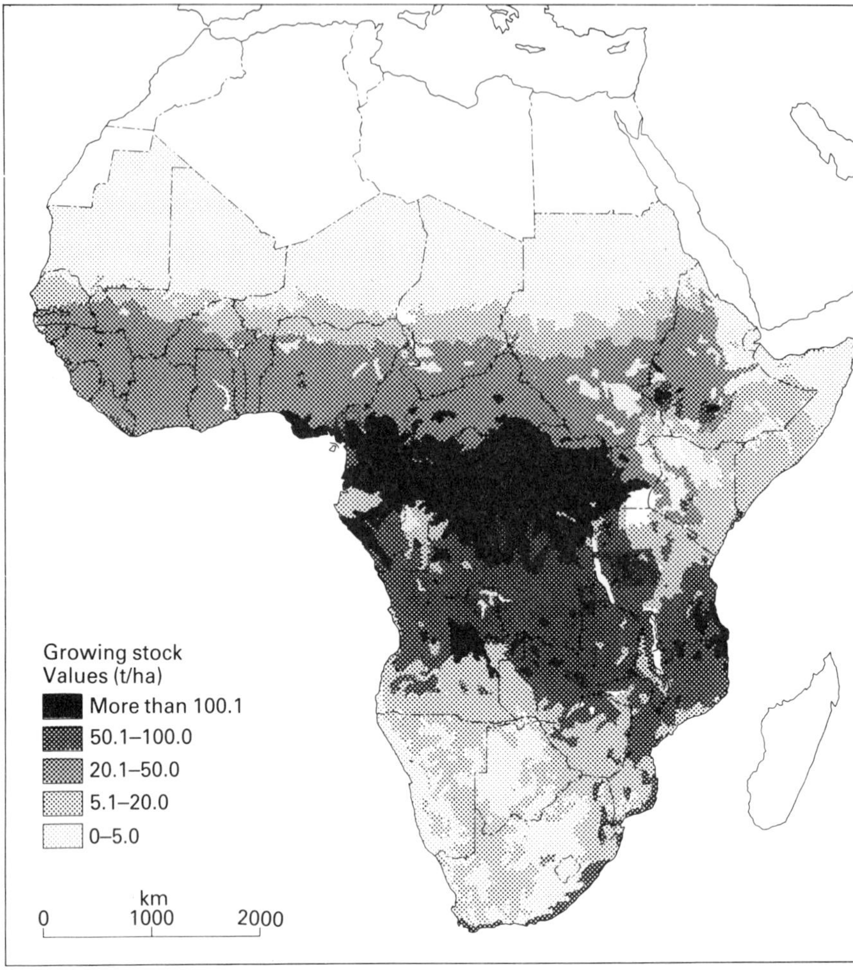

Fig. 3.4 Sub-Saharan Africa: stock (tonnes/ha).

dominated by forest formations are only important in central Africa where they account for 39.2% of the land area. Three countries exhibit large forest areas – Zaire (897 002 km^2), Cameroon (205 871 km^2) and Congo (181 798 km^2).

The importance of high woody biomass mosaics in west and central Africa, as well as Ethiopia, are indicative of extensive transitional zones between forested savanna vegetation in these areas. In west Africa and Ethiopia they mainly represent anthropogenically derived vegetation due to forest and, to a much lesser extent, woodland clearance. Their origins in central Africa are mostly conjectural, but anthropogenic factors are again of importance.

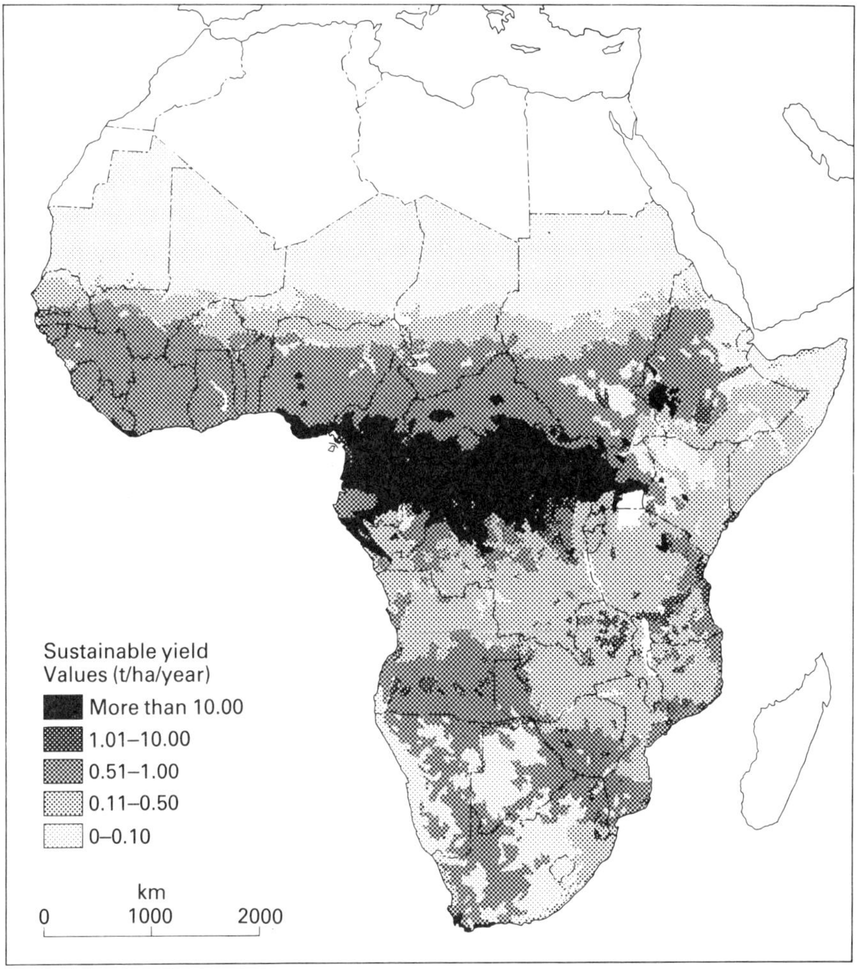

Fig. 3.5 Sub-Saharan Africa: sustainable yield (tonnes ha^{-1} $year^{-1}$).

Growing stocks and yields

The proportion of the growing stock and sustainable yield in the woodland land cover classes is greater than that in either the High Woody Biomass Mosaic classes or Forest classes (Tables 3.6 and 3.7). This indicates the importance of savanna woodlands over forests in sub-Saharan Africa in terms of area, growing stocks and sustainable yields. Furthermore, if the high woody transitional land cover classes are examined, it is apparent that some of these classes have previously been considered as moist savanna woodlands (e.g. Guinea Woodland in west Africa, the Evergreen Woodland Mosaic in the Miombo zone of southern Africa)

Table 3.4 Savanna, forest, and high woody biomass transitional land cover classes

Savanna
Open woodland
Dry Sudanian woodland
Sudanian woodland
Moist Sudanian woodland
Sudan/Ethiopian woodland and thicket
Seasonal Miombo
Wet Miombo
Forest
Montane forests
Coastal and gallery forest
Evergreen forests
Mesophilous humid tropical forest
Ombrophilous humid tropical forest
Humid tropical swamp forest
Mangrove
High Woody Biomass Transitions (Forest–Savanna Transition Zones)
Cultivation and forest/woodland mosaic
Cultivation and forest regrowth mosaic
High productivity West African cultivation and forests mosaic
Medium productivity West African cultivation and forest mosaic
High cultivation mosaic (Ethiopian Highlands)
Guinea woodland
Evergreen woodland mosaic

or are derived savannas (e.g. the west African cultivation mosaics) reinforcing the importance of savannas.

The total woody biomass growing stock in sub-Saharan Africa is estimated to be 59.80 Gt of which 32.81 Gt (54.9%) is in the savanna woodlands, 23.22 Gt (38.8%) in forests and 3.76 Gt (6.3%) in the High Woody Biomass Mosaics (Table 3.6). There is a similar, uneven distribution of woody biomass sustainable yields. The estimated total annual yield is 3.35 Gt; the largest proportion is found in the forests, 2.78 Gt/year (83.0%), followed by savanna woodlands, 0.53 Gt/year (15.8%) and high woody biomass transitions, 0.05 Gt/year (1.5%) (Table 3.7).

About 70% of the growing stock in savannas is found in the southern and central Africa regions, 42.6% and 28.4% respectively. The growing stock of the High Woody Biomass Mosaics is concentrated in the west African Coastal region (40.5%), central Africa (28.1%) the Horn (16.3%) and Southern Africa (9.6%). Almost all of the woody biomass stocks of forests (75.5%) are found in central Africa. The sustainable yield estimates are even more skewed than those of growing stock. The southern Africa region has 37.2% of the annual woody biomass sustainable yield from

Table 3.5 Areas of summary land cover classes by region (km^2). Values on upper line are areas in km^2. Values on lower line are proportions of summary biomass classes by region expressed as a percentage (%)

Region	*Savanna*	*High woody biomass mosaics*	*Forests*
Central	1 620 272	645 098	1 550 884
	(40.9)	(16.3)	(39.2)
Eastern	526 003	140 220	202 455
	(29.2)	(7.8)	(11.3)
Horn	755 016	363 598	52 695
	(17.4)	(8.4)	(1.2)
Southern	2 243 923	329 029	317 803
	(37.8)	(5.5)	(5.4)
W. Coast	809 981	906 252	64 611
	(39.7)	(44.4)	(3.2)
W. Sahel	730 777	8326	2793
	(13.9)	(0.2)	(0.1)

The low data values for high woody biomass mosaics (forest–savanna transition zones) in the West African Sahel Region (cf. Tables 5, 6 and 7) are due to the fact that the definition of this 'region' in the project comprises only countries in the north of West Africa (Senegal, The Gambia, Mauritania, Mali, Burkina Faso, Niger and Chad).

Table 3.6 Growing stocks of savannas and forests. Values on upper line are growing stocks in million tonnes. Values on lower line are growing stock in each region as a proportion of regional total, expressed as a percentage (%)

Region	*Savanna*	*High woody biomass mosaics*	*Forests*
Central	9 312	1058	17 526
	(33.3)	(3.8)	(62.6)
Eastern	3 581	194	1 557
	(55.3)	(3.0)	(24.0)
Horn	2 016	612	555
	(43.2)	(13.1)	(11.9)
Southern	13 790	360	1 928
	(73.8)	(1.9)	(10.3)
W. Coast	2 163	1526	1 584
	(40.3)	(28.5)	(29.5)
W. Sahel	1 951	14	69
	(78.1)	(0.6)	(2.8)

Table 3.7 Sustainable yield of savannas and forests. Values on upper line are sustainable yields in thousand tonnes per year. Values on lower line are sustainable yields in each region as a proportion of regional total, expressed as a percentage (%)

Region	*Savanna*	*High woody biomass mosaics*	*Forests*
Central	137 551	12 257	2 226 337
	(5.8)	(0.5)	(93.6)
Eastern	50 305	2 664	153 265
	(22.0)	(1.2)	(66.9)
Horn	47 645	6 908	38 798
	(34.8)	(5.0)	(28.3)
Southern	195 603	6 252	161 784
	(43.6)	(1.4)	(36.1)
W. Coast	43 683	17 219	188 654
	(17.0)	(6.7)	(73.4)
W. Sahel	50 397	158	8 237
	(50.5)	(0.2)	(8.3)

savannas, the west Africa Coastal and central Africa regions have almost two thirds of the High Woody Biomass Mosaics (37.9% and 27.0% respectively), while 80% of the woody biomass yield from forests is found in central Africa.

DISCUSSION

Savanna woodlands

Woody land cover classes dominated by Sudanian woodland formations are found in a broad belt stretching from the Atlantic coast to east-central Sudan, south of the Sahelian vegetation which fringes the southern Sahara. This belt is broadest in western Sudan and the eastern Central African Republic, where it is almost continuous with Miombo woodland formations which dominate central Africa and the northernmost parts of southern Africa.

The ecology of these woodlands have been the subject of a number of important studies and it is known that significant variations in the woody biomass distribution in these woodlands exist because of climatic variations and human disturbance. The temporal behaviour of NDVI (NDVI phenology) has been used to subdivide the woodlands into a series of classes with different levels of woody biomass stocks and sustainable yields, and to map these subdivisions.

Table 3.8 NDVI statistics for selected savanna woodland classes

Mean NDVI statistics*	*Dry Sudanian*	*Sudanian*	*Moist Sudanian*	*Seasonal Miombo*	*Wet Miombo*
			Woodland Classes		
Dry Season (Lowest NDVI)					
$\bar{x}$	0.12	0.15	0.21	0.22	0.27
σ	0.02	0.06	0.08	0.07	0.07
Month(s)	May	Mar	Mar	Aug	Sep
Wet Season (Highest NDVI)					
$\bar{x}$	0.43	0.49	0.51	0.51	0.51
σ	0.06	0.05	0.03	0.05	0.06
Month(s)	Sep	Sep	Sep	Dec	Dec/Feb
Amplitude					
$\bar{x}$	0.32	0.32	0.33	0.32	0.29
σ	0.04	0.06	0.06	0.08	0.03
Number of months in which mean NDVI					
>0.3	4	6	7	9	9
>0.4	2	4	4	6	7

*Mean NDVI values based on ten 3 × 3 pixel kernels in each land cover class.

Sudanian woodlands can be subdivided into three classes on the basis of vegetation responses which are controlled by the west African monsoon. They are subject to a summer wet season and a winter drought. The wet season is shortest in the north of the region and longest in the south. In addition, the mean monthly rainfall in the wettest months decreases in a northerly direction. If NDVI behaviour is related to green leaf activity, which over large areas can be considered a function of rainfall and soil moisture characteristics, it would be expected that the NDVI phenology for the three woodland types would be different. Lack of significant relief variations in these areas means that latitude will be the major predictor of woodland type.

To test this, ten 3 × 3 pixel kernels were chosen in each of the three woodland types. The mean NDVI for the nine pixels at each of the sites was calculated for each month. Statistics for each woodland type were derived (Table 3.8) and the mean NDVI curve for each of the woodland types was plotted (Fig. 3.6). The three NDVI profiles are very similar in shape but their relative position in terms of NDVI is important. The northernmost of the three classes – Dry Sudanian woodland – has lower NDVI values throughout the year than either Sudanian or Moist Sudanian woodland, whereas Moist Sudanian woodland has consistently higher values than either Sudanian or Dry Sudanian woodland (Fig. 3.6). The

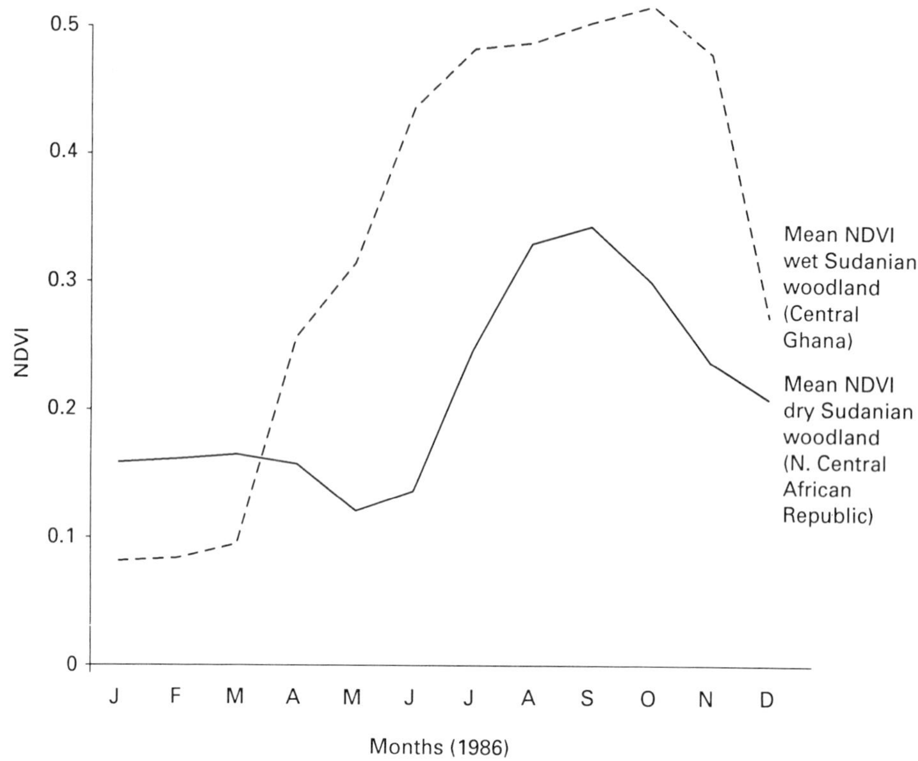

Sudanian Woodland type	Highest NDVI	Lowest NDVI	NDVI range
DRY (6)*	0.43	0.12	0.31
INTERMEDIATE (4)*	0.45	0.12	0.33
MOIST (WET) (7)*	0.50	0.15	0.37

* Numbers in parentheses indicates number of 3 x 3 pixel sites used to calculate NDVI statistics

Fig. 3.6 Mean NDVI profiles for Sudanian woodlands.

amplitude between the high, wet season, NDVI values (between August and September), and the low, dry season, (between December and May), only vary between 0.32 and 0.33 with low standard deviations between 0.04 and 0.06 (Table 3.8). Apart from their relative positions on the NDVI axis, attention needs to be drawn to two other differences in the NDVI behaviour of the woodlands. Firstly, the NDVI behaviour of the Dry Sudanian woodland in the dry season differs from that of the other two woodland types. Very low NDVI values are found from March to May in Dry Sudanian woodland, whereas in the other two woodland types this period is restricted to March. This woodland suffers from a longer dry season than the other two types of woodland because the rains start later and finish earlier. Consequently the period of soil moisture stress is relatively long and severe. Trees drop their leaves earlier than in the other two types of woodland and there is also a time lag in the response of green leaves to humid conditions before the rains begin.

Miombo woodlands were also subdivided on the basis of NDVI behaviour, and again this can be illustrated by reference to ten 3 × 3 pixels kernels in each class (Table 3.8). The division into 'seasonal' and 'wet' types is based upon the length of the time miombo woodland trees are in leaf, and is a modification of a four-fold division of Miombo woodlands in southern Africa mapped from 1984 AVHRR imagery (Millington *et al.*, 1989). The two-fold division correlates closely with observations made on Miombo woodlands in Malawi in August 1986 (Millington *et al.*, 1989) and is in general agreement with observation on the phenological characteristics of vegetation in southern Zaire (Malaisse, 1978). Seasonal Miombo woodland is most extensive in northern Angola, Malawi, Mozambique, Tanzania, southern Zaire, Zambia, and Zimbabwe. Wet Miombo woodland is best developed in northern and central Angola, Mozambique, south-eastern Tanzania, central Zaire, and northern Zambia. The division of the two types is based on the amplitude of the fluctuation of the NDVI curves (Fig. 3.6). For Wet Miombo woodland, NDVI values range from 0.45 to 0.51 during the wet season to 0.27 to 0.39 in the driest months. The amplitude of the mean seasonal NDVI profile is 0.29 with a standard deviation of 0.03. This compares to NDVI values for Seasonal Miombo woodlands of 0.41 to 0.51 in the wet season, 0.22 to 0.28 in the driest months and an amplitude of 0.32. The two NDVI profiles have the same shape with the Wet Miombo woodland profile being located higher on the NDVI axis.

The forest–savanna boundary and fuelwood energy planning

There is little evidence to suggest that the incursion of savanna vegetation into forested areas is a consequence of fuelwood collection alone. The main cause is land clearance for agriculture, timber and construction; any fuelwood offtake from cleared land is usually of secondary importance. In the woody biomass classes dominated by forest types, and the high woody biomass mosaics, there are only very localized fuelwood supply problems. For the most part, wood collection and the fuelwood trade relies on the collection of unburnt wood during burning of farmland prior to cultivation, or young trees in the fallow regrowth which are cut specifically for fuelwood. The fuelwood situation is more acute in some of the savanna woodlands where there are woody biomass resource shortages and savanna woodland degradation is occurring at the present time. These phenomena have been documented in Southern Africa (Munslow *et al.*, 1989; ETC, 1989) and the output of this research lies in the prediction of future trends in woody biomass supply in sub-Saharan Africa which will undoubtedly affect savanna woodland and forested areas.

Many of the drier savanna woodlands already show signs of severe degradation and in many cases have degraded into a mosaic of agricultural plots and shrubby fallow regrowth with only small areas of woodland remaining, e.g. west-central Senegal and the Lilongwe and Phalombe Plains in Malawi. The position of the moister savanna woodlands on the whole is rather better, but it is envisaged that, in the next decade, there will be significant pressure on such woodlands. This pressure is related to population growth and the movement of people from more marginal areas. Add to this the uncertainties of a changing climate and the outlook for such woodlands is bleak. Already there is significant population movement to the relatively intact seasonal and wet miombo woodland areas in Tanzania. In addition to agricultural activity, the demand for fuelwood and charcoal from urban centres will undoubtedly lead to an expansion of wood cutting in these areas. In the cities along the west African coast the main urban fuel is charcoal. In those countries which extend from the coast inland to reach the Sudanian woodlands (e.g. Ivory Coast, Togo, Benin), most of the charcoal to the coastal cities is supplied from the Sudanian woodlands. This long-distance transport to cities such as Abidjan and Accra, of charcoal produced in the Sudanian woodlands in the north, is due to the fact that the wood from many Sudanian woodland trees makes good quality charcoal which commands a high price in the urban markets. The Guinea woodlands in most west African countries have trees with wood having similar properties.

In most places Guinea woodlands are under-exploited in comparison to Sudanian woodlands, but with expanding urban populations the Guinea woodlands will come under severe pressure and their degradation will certainly affect the forest–savanna boundary. When such a switch occurs in west Africa, the forest adjacent to the Guinea woodlands will be affected and will impact upon the forest–savanna transition.

CONCLUSIONS

The value of analysing 8 × 8 km AVHRR data, in the context of this work, lies in its ability to map land cover at continental and sub-continental scales using a synoptic data set, acquired over a short period of time. In addition, the NDVI has been shown to be closely correlated with green biomass production; and the NDVI phenology is closely correlated with rainfall, the main constraint on green biomass production over most of Africa. However, the data cannot be used to estimate woody biomass directly, and has been achieved by interfacing secondary data on woody biomass stocks and yields with the land cover maps.

The land cover maps represent a situational over view for 1986 and, as such, probably represent one of the most accurate and recent maps of the status of forests, woodlands and forest–savanna woodland transition zones in Africa. However, the coarse resolution of these data means that monitoring changes in the distribution of these three types of land cover would be of limited value, only major changes caused by the destruction of vast areas of vegetation would be detected in the short-term. Nevertheless, the ability to monitor forest destruction has been demonstrated in Amazonia using 1.1 km AVHRR data, and this type of monitoring could be extended to other land cover types. AVHRR data at 1.1 km resolution probably represents the only accessible data set available to monitor land cover changes at the continental and sub-continental scale at the present time. This is because of its acquisition schedule which provides global coverage on a daily basis, and its proven ability to map land cover, and monitor changes that have been verified on the ground.

ACKNOWLEDGEMENTS

This work was carried out under contract to ETC(UK) from the World Bank/UNDP Energy Sector Management Assistance Programme, and the support of Paul Ryan (ESMAP) and the offices of ETC(UK) are gratefully acknowledged. Many institutions helped during the project; particularly FAO, The Museum of Middle Africa, International Forest Science Consultancy (Edinburgh), and the Oxford Forestry Institute. We would also like to thank the following for their help: R. Bevan, A.E. Critchley, T.D. Douglas, N.A. Drake, P. Kennedy, J.S. Kirkby, F.P. Mallaise, A. Mellor, P. O'Keefe, J.R.G. Townshend, C.J. Tucker, C.O. Justice, R.J. Saull, K.D. Singh. The greatest feedback came from the foresters and land-use planners that we met in Abidjan and Nairobi, to these people we owe a great deal.

REFERENCES

ETC (1989) *Woody Biomass Mapping: Sub-Saharan Africa*. Report submitted to UNDP/World Bank ESMAP. ETC(UK), North Shields.

Gatlin, J.A., Sullivan, R.J. and Tucker, C.J. (1983) Monitoring global vegetation using NOAA AVHRR data. *Proc. IGARRS Symp. San Francisco* I, PF2, 7.1.

Goward, S.N., Dye, D.G. and Tucker, C.J. (1985) North American vegetation patterns observed by NOAA-7 AVHRR. *Vegetatio* **64**, 3–14.

Gray, T.I. and McCrary, D.G. (1981) Meteorological satellite data – a tool to describe the health of the world's agriculture. *AgRISTARS Rept. EW-NI-04042*. NASA/JSC, Houston, Texas.

Henderson-Sellers, A., Seze, G., Drake, F. and Debois, M. (1987) Surface observed and satellite-received cloudiness prepared for the 1983 ISCCP special study in Europe. *Journal of Geophysical Research* **22**, D4, 4019.

Holben, B.N. (1986) Characteristics of maximum-value composite images from temporal AVHRR data. *International Journal of Remote Sensing* **7**(11), 1383–622.

Justice, C.O. (ed.) (1986) Monitoring the grassland of semi-arid Africa using NOAA-AVHRR data. *International Journal of Remote Sensing* **7**(11), 1328–622.

Kennedy, P.J. (1989a) Monitoring the vegetation of Tunisian grazing lands using the normalised difference vegetation index. *Ambio*, **18**(2) 119–23.

Kennedy, P.J. (1989b) Monitoring the phenology of Tunisian grazing lands, *International Journal of Remote Sensing* **10**(4 & 5), 835–45.

Lamotte, M. and Bourlière, F. (1978) Energy flow and nutrient cycling in tropical savannas. In (ed. F. Bourlière), *Ecosystems of the World 13: Tropical Savannas*, Elsevier, Amsterdam.

Malaisse, F.P. (1978) Phenology of the Zambezian woodland area with emphasis on the Miombo ecosystem. In (ed. H. Leith), *Phenology and Seasonal Modelling*, Springer-Verlag, Berlin.

Malingreau, J.P. (1986) Monitoring tropical wetland rice production systems: a test for orbital remote sensing. In (eds M.J. Eden and J.T. Parry), *Remote Sensing and Tropical Land Management*, John Wiley, Chichester, 279–305.

Malingreau, J.P., Stephens, G. and Fellows, L. (1985) Remote sensing of forest fires: Kalimantan and North Borneo in 1982–83. *Ambio*, **14**, 314–21.

Malingreau, J.P. and Tucker, C.J. (1988) Large scale deforestation in the Southeastern Amazon Basin of Brazil. *Ambio* **17**, 49–55.

Malingreau, J.P., Tucker, C.J. and Laporte, N. (1989) ACHRR for monitoring global tropical deforestation. *International Journal of Remote Sensing*, **10**(4 & 5), 855–67.

Menaut, J.C. and César, J. (1979) Structure and primary productivity of Lamto Savannas, Ivory Coast. *Ecology* **60**(6), 1197–210.

Menaut, J.C. and César, J. (1982) The structure and dynamics of a west

African savanna. In (eds B.J. Huntley and B.H. Walker), *Ecology of Tropical Savannas* 81–100, Berlin, Springer-Verlag.

Millington, A.C., Townshend, J.R.G., Kennedy *et al.* (1989) *Biomass Assessment: Woody biomass in the SADCC region*. Earthscan, London.

Millington, A.C., Critchley, R.W., Douglas, T.D. and Ryan, P. (in press). *Assessment Woody Biomass Resources in sub-Saharan Africa*, 2 vols. World Bank Publications, Washington DC.

Munslow, B., Katerere, Y., Ferf, A. and O'Keefe, P. (1989) *The Fuelwood Trap: A study of the SADCC region*. Earthscan, London.

Norwine, J. and Greegor, D.H. (1983) Vegetation classification based on AVHRR satellite imagery. *Remote Sensing of Environment*, **13**, 69.

Perry, J.R. and Lautenschlager, L.F. (1984) Functional equivalence of spectral vegetation indices. *Remote Sensing of Environment*, **14**, 169.

Prince, S.D. (1986) Monitoring the vegetation of semi-arid tropical rangelands with the NOAA-7 advanced very high resolution radiometer. In (eds M.J. Eden and J.T. Parry), *Remote Sensing and Tropical Land Management*, John Wiley, Chichester 307–34.

Rutherford, M.C. (1978) Primary production ecology in southern Africa. In (ed. M.J.A. Werger), *Biogeography and Ecology in Southern Africa*. Dr W. Junk, The Hague.

Sarmiento, G. and Monasterio, M. (1978) Life forms and phenology. In (ed. F. Bourlière), *Ecosystems of the World 13. Tropical Savannas*. Elsevier, Amsterdam.

Schneider, S.R., McGinnis, S.R. Jr and Gatlin, J.A. (1981) Use of NOAA/AVHRR visible and near-infra red data for land remote sensing. *NOAA Tech. Report, NESS84*. USDC, Washington DC.

Schneider, S.R., McGinnis, S.R. Jr and Gatlin, J.A. (1985) Monitoring Africa's Lake Chad Basin with Landsat and NOAA satellite data. *International Journal of Remote Sensing*, **6**, 59.

Sellers, P. (1985) Canopy reflectance, photosynthesis, and transpiration. *International Journal of Remote Sensing* **6**, 1335.

Sellers, P. (1986) Canopy reflectance, photosynthesis, and transpiration II. The role of biophysics and their interdependence. *Remote Sensing of Environment* **18**, 205.

Tarpley, J.D., Schneider, S.R. and Money, R.L. (1984) Global vegetation indices from NOAA-7 meteorological satellite. *Journal of Climate and Applied Metereology* **23**, 491.

Townshend, J.R.G., Justice, C.O. and Kalb, V. (1987) Characterization and classification of South American land cover types using satellite data. *International Journal of Remote Sensing* **8**, 1189–207.

Tucker, C.J., Townshend, J.R.G. and Goff, T.E. (1985) African land cover classification using satellite data. *Science* **227**, 110.

Townshend, J.R.G. and Tucker, C.J. (1984) Objective assessment of

AVHRR data for land cover mapping. *International Journal of Remote Sensing* **6**, 127.

Tucker, C.J. and Sellers, P. (1986) Satellite remote sensing of primary production. *International Journal of Remote Sensing* **7**, 1395.

Tucker, C.J., Gatlin, J.A., Schneider, S.R. and Kuchinos, A. (1984a). Monitoring vegetation in the Nile Valley with NOAA-6 and NOAA-7 AVHRR. *Photogram. Eng. & Remote Sensing* **50**, 53.

Tucker, C.J., Holben, B.N. and Goff, T.E. (1984b). Intensive forest clearing in Rondonia, Brazil as detected by satellite remote sensing. *Remote Sensing of Environment* **15**, 255.

Tucker, C.J., Townshend, J.R.G. and Goff, T.E. (1985) African land cover classification using satellite data. *Science* **227**, 110.

Tucker, C.J., Vanpraet, C.L., Sharman, M.J. and Van Ittersum, G. (1985) Satellite remote sensing of total herbaceous biomass production in the Senegalese Sahel: 1980–1984. *Remote Sensing of Environment* **17**, 233.

Woodward, F.I. (1987) *Climate and Plant Distribution*. Cambridge University Press, Cambridge.

White, F. (1983) *The Vegetation of Africa* (Natural Resources Research Series, 20), Paris, UNESCO.

4

Influence of physical factors on the nature and dynamics of forest–savanna boundaries

MONICA M. COLE

SUMMARY

1. This chapter identifies major differences in morphological features and physiological responses between the species that make up forest types of vegetation and those that make up savanna types of vegetation. It focuses attention on the outstanding characteristics of the forest–savanna boundaries in selected areas.
2. The abrupt nature of the boundaries with regard to both physiognomy and species composition is related to fundamental changes in edaphic conditions associated with changes of topography and/or lithology and structure that stem from geomorphological processes and underlying geology.

INTRODUCTION

The species that make up forest and savanna types of vegetation exhibit distinctive morphological features and physiological responses that are related to plant–soil moisture relationships, which are dependent on climatic and edaphic conditions.

A critical distinction is the dominance in all savanna areas of grasses exhibiting the C_4 photosynthetic pathway, that is favoured by high irradiance, high temperature and high evaporation, whereas the grasses of the semi-deciduous tropical forests belong to the C_3 group (see Huntley and Walker 1982; Medina, 1982; Cole, 1986). Within the savannas, aspartic acid-forming C_4 grasses predominate in areas of grey and brown loams of heavy texture whereas the malate-forming C_4 grasses prevail in areas of lateritic and ferrallitic soils. The trees and shrubs, as well as the grasses of the savannas, all display morphological features as well as physiological responses that enable them to survive in those soils which are inimical to the species of the forest (Cole, 1986).

The outstanding feature of the forest–savanna boundaries is abruptness with regard to changes in both the physiognomy and species composition of the vegetation. In the writer's view, this results from fundamental changes in the edaphic conditions usually associated with changes of topography and/or lithology and structure that stem from geomorphological processes and underlying geology (Figs 4.1–4.3).

Around the major areas of savanna vegetation the boundary may be between savanna woodland and forest, or savanna grassland and forest, while in vegetational tension zones subject to both long-term and short-term climatic changes and particularly active geomorphological processes and geological events (Figs 4.2 and 4.3) (Cole, 1982, 1986) areas of tropical forest occur in mosaic distributions of savanna woodlands, savanna grasslands, low tree and shrub savannas and savanna parklands (Cole, 1960, 1963a,b). The boundaries between each of these types are also abrupt.

THE BOUNDARIES BETWEEN FOREST AND DIFFERENT CATEGORIES OF SAVANNA

Over the great peneplaned or pediplaned plateaux of Brazil and south-central Africa, notably Zambia, where the savanna woodlands – the *campos cerrados* of Brazil and *miombo* of Africa – attain their most extensive distributions, the boundary is that between forest and savanna woodland (Cole, 1960, 1963b). In Brazil the savanna woodland grades into low tree and shrub savanna – the *campos sujos* – and thence to savanna grassland – *campos limpos* – over the poorest soils characterized by near-surface indurated laterite or plinthite over the more central parts of the plateaux. Near São José de Serra on the Chapada de Guimarães in Brazil, however, corridors of forest follow the headstreams of the Amazon and Paraguay rivers as they cut back into the watershed area of the plateaux (Fig. 4.4) (Cole, 1960).

In Zambia the forest–savanna boundary is again between forest and savanna woodland – the *miombo* – but the boundaries and the relationships to physical factors are more complex than in Brazil, notably on the Copperbelt. There the interplay of geomorphological processes and crustal warping has resulted in more varied edaphic conditions over relatively short distances. These are reflected mainly in different tree associations within the savanna woodlands; the forest boundary may

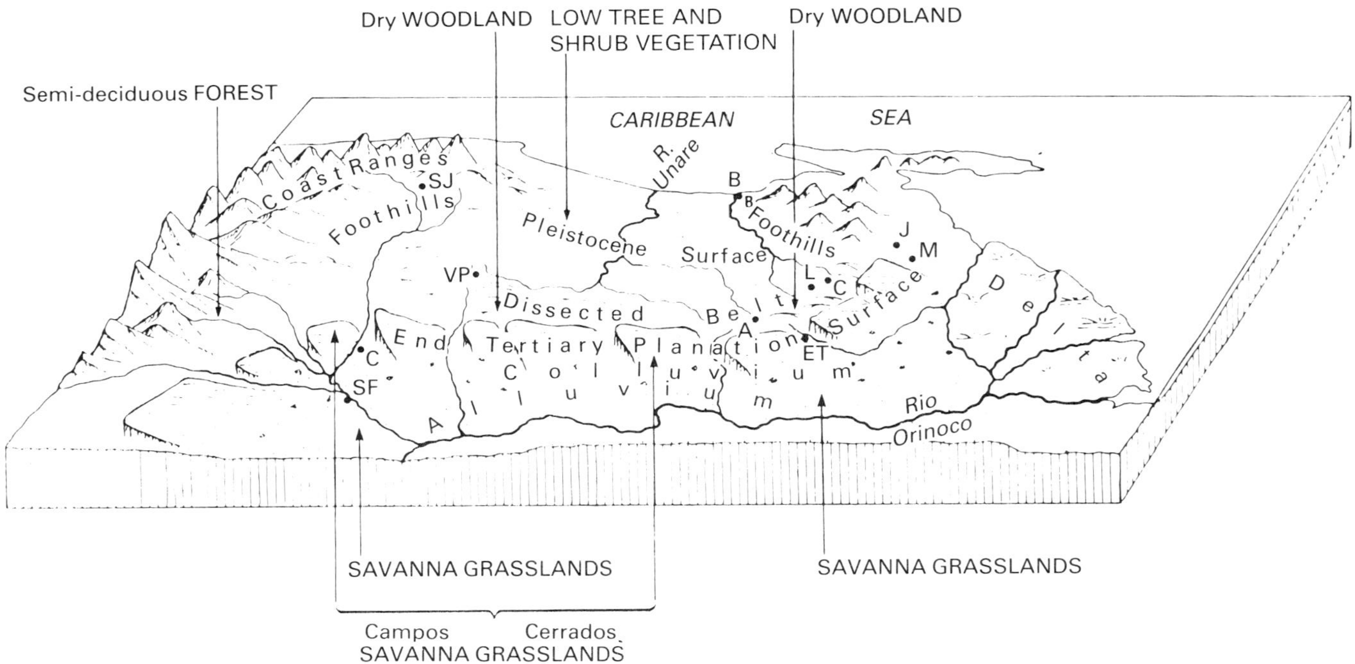

Fig. 4.1 Relationship between vegetation and geomorphology in northern Venezuela (from Cole, 1986, courtesy Academic Press).

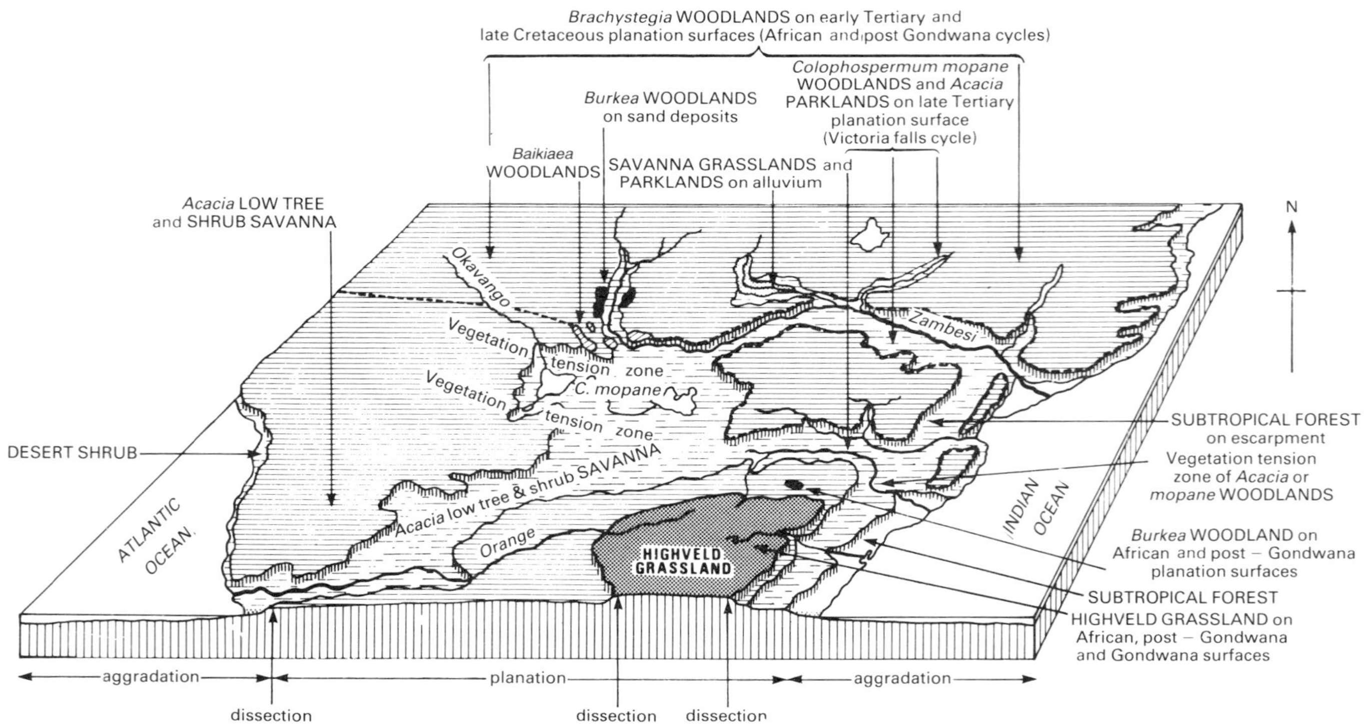

Fig. 4.2 Relationships between vegetation and geomorphology in south Central Africa (from Cole, 1986, courtesy Academic Press).

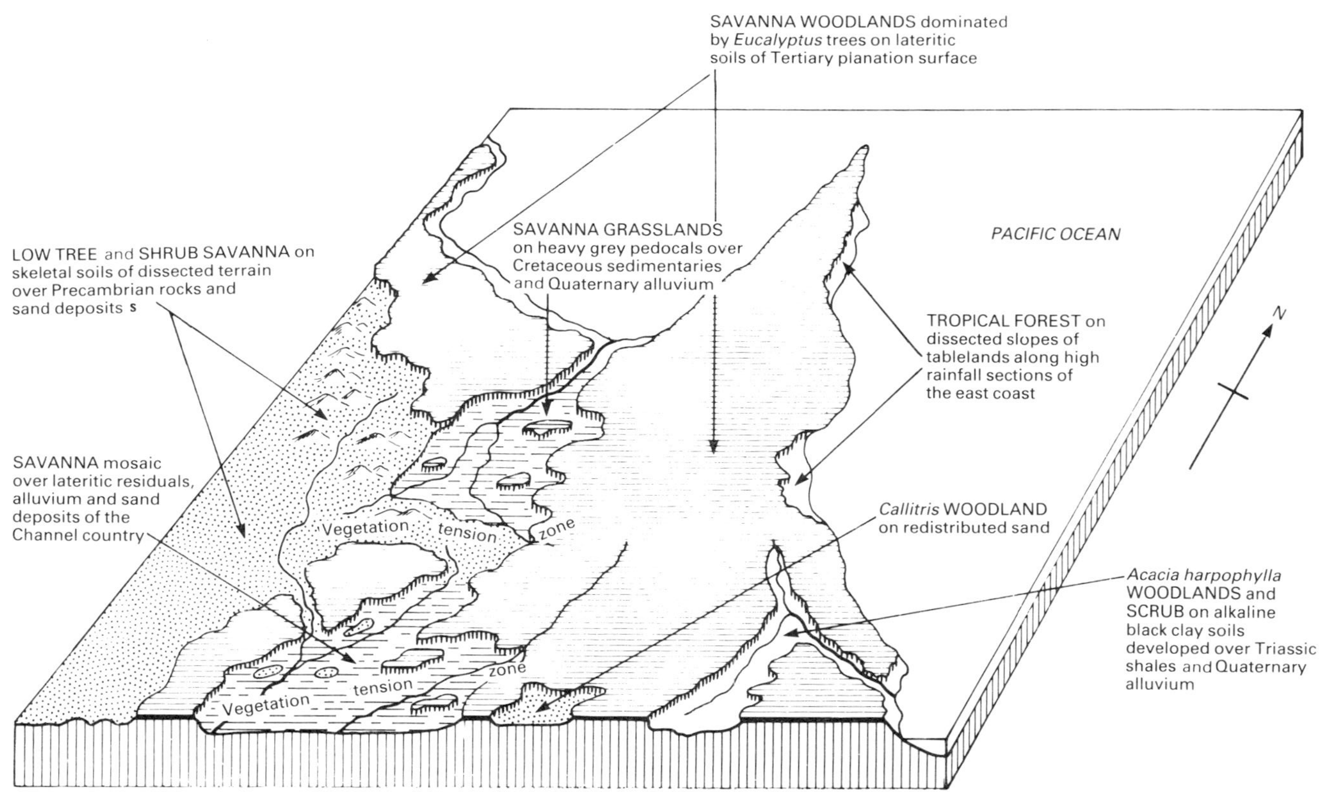

Fig. 4.3 Relationships between vegetation and geomorphology in northern Australia (from Cole, 1986, courtesy Academic Press).

Fig. 4.4 Campos cerrados (foreground) grading through campos sujos to campos limpos on the Chapada de Guimãraes, east of Cuiabá Mato Grosso, Brazil. The road crosses the watershed between drainage to the R. Araguaia which flows to the R. Tocantins and the Amazon and that to the R. Paraguay which joins the Paraná system. Gallery forest follows the R. Araguaia on the left. Similar forest follows the headstreams of the Paraguay, right of the road but not shown in the photograph.

be with savanna woodlands of differing physiognomy and species composition, while in some areas mosaic distributions of forest, savanna woodland and savanna grassland occur (Cole, 1963b, 1986).

Where tectonic forces have accelerated erosion and dismembered the plateaux surfaces, as in northern Australia and east Africa, the boundary may be between forest and savanna woodland or savanna grassland or savanna parkland. This is evident in the Gulf Fall area around the Gulf of Carpentaria, along the Roper, MacArthur, Cloncurry and Leichardt rivers in Australia, and in the Rift Valley zone of East Africa, notably where the volcanic peaks of Meru and Kilimanjaro rise abruptly from the Serengeti Plains (Cole, 1986). In these areas the edaphic controls over vegetation distribution are indisputable.

Fig. 4.5 Relationships between vegetation and geomorphology between Goiânia and Anapolis, Goiás, Brazil. *Kielmeyera* sp. tree frames the picture (from Cole, 1960, courtesy *Geographical Journal*, and Cole, 1986, courtesy Academic Press).

THE FOREST–SAVANNA BOUNDARIES AROUND CORE SAVANNA AREAS

Around the core savannas of Brazil, central, east and Southern Africa and Australia the boundaries with forest are usually sharply defined and frequently linear. Most commonly they are between savanna woodland and forest and occur where there is an abrupt change from the peneplaned or pediplaned plateaux, characterized by old, exhausted, non-renewable lateritic soils of low nutrient status with a well defined laterite or plinthite horizon within the profile, to dissected terrain with actively renewable free-draining soils of higher nutrient status. Such boundaries are evident at many localities at the periphery of the Brazilian plateau, notably between Goiánia and Anapolis in Goiás (Fig. 4.5) and along the São Vicente escarpment at the edge of the Chapada de Guimarães in the Mato Grosso where, in each case, the Gondwana surface of Jurassic age is being consumed in a later erosion cycle. In Goiais the surface is cut across highly inclined Precambrian rocks, whereas the Chapada de Guimarães is cut across Devonian and later rocks that are exposed in the São Vicente

Fig. 4.6 Contact between campos cerrados on ill-drained level taboleiro with sandy soils derived from Tertiary sandstone (foreground) and forest along river valley cutting into the plateau. Near Olinda, north of Recife, Brazil (from Cole, 1960, courtesy *Geographical Journal*, and Cole, 1986, courtesy Academic Press).

escarpment. The Cuiabá surface below this escarpment truncates highly inclined Precambrian rocks and represents an exhumed pre-Devonian surface exposed by erosion in post-Jurassic and later erosion cycles. In all cases the plateaux surfaces carry savanna woodland – campos cerrados grading either to campos limpos or to *cerradão* – whereas the dissected terrain is forested.

Equally sharp forest–savanna boundaries related to similar changes of relief and soil occur at the peripheries of the *tabuleiros* in the coastal zone of Bahia, Alagoas and Pernambuco in Brazil (Fig. 4.6). While in the Cuidad Bolivar – Cerro Bolivar area of Venezuela, the change from savanna to forest occurs on the slopes of the *cerros* which rise above the level peneplaned surface over which savanna grasslands prevail (Fig. 4.7). In this area the control of edaphic conditions over the form and composition of the vegetation is further exemplified over the iron orebodies contained in the Precambrian sedimentary rocks of Cerro Bolivar, Cerro Altimara and San Isidro where, owing to edaphic aridity over collapsed iron-rich crust called *canga* and skeletal soils with high

Fig. 4.7 Savanna grasslands with scattered *Curatella americana* trees over the level 300 m planation surface below Cerro Bolivar (right) whose slopes carry tropical forest (from Cole, 1986, courtesy Academic Press).

concentrations of iron, highly specialized communities of drought-resistant plants characterized by *Clusia* spp. and *Myrcia* spp. prevail, notwithstanding the constantly hot, wet climate (Cole, 1980, 1986).

In east, central and Southern Africa the forest–savanna boundaries are likewise abrupt (Fig. 4.8) but the savanna vegetation is floristically richer and more varied than that of South America and the geomorphological history has been more complex. Periods of continental uplift, regional uplift around the periphery of the subcontinent, local warping and major faulting have interrupted cycles of erosion dominated by planation processes since the close of the Mesozoic, while processes associated with drainage superimposition have played an increasingly important role since the Cretaceous. The latter processes have exposed rocks of varying age and lithology. Since soils are dependent on the rocks exposed by these processes, they are more varied than in South America and there are many discrete vegetation associations within the savanna woodlands. The change from savanna to forest occurs where the level plateaux with lateritic soils give way to dissected terrain with better drained soils of higher nutrient status, but different savanna woodland associations, savanna grasslands and parklands abut the forest in different localities.

Fig. 4.8 View westwards between Mpika and Shiwa Ngandu, Zambia, showing evergreen tropical forest along the valley, a pure stand of *Brachystegia tamarindoides* on the hilltop and *Brachystegia–Julbernardia* woodland on the end-Tertiary peneplain which forms the horizon (from Cole, 1963b, courtesy *Geographical Journal* and Cole, 1986, courtesy Academic Press).

The precise relationships can be understood only by detailed studies of individual areas (Cole, 1963b, 1986) which are beyond the scope of this chapter. As in Venezuela, the importance of edaphic influences over the form and composition of the vegetation is emphasized by the unique communities of metal-tolerant indicator plant species occupying toxic ground associated with copper and cobalt orebodies within the savanna woodlands of Katanga, Zambia and Zimbabwe (Duvigneaud, 1958, 1959; Duvigneaud and Denaeyer de Smet, 1960, 1963; Cole, 1965, 1986).

In Australia the savanna woodland–forest boundaries likewise coincide with the change from level plateaux to dissected terrain, notably in the Mareeba area of eastern Queensland, but in the Gulf Fall area of northern Australia the legacies of geomorphological processes affected by warpings and differential land movements in recent geological time and by major climatic changes have resulted in complicated, often mosaic, distributions of different types of savanna each of which may give way to forest. An understanding of these relationships requires detailed studies of individual areas (Cole, 1986).

A feature of the Australian savannas is the vast extent of savanna grasslands that occupy the grey and brown soils of heavy texture and the montmorillonitic black clays of the Barkly Tableland and the plains of the Great Artesian Basin. The extent of these grasslands narrows northwards where, over small areas of heavy soils, they alternate with areas of savanna parkland in the river valleys along which ribbons of forest extend from the Gulf of Carpentaria. Savanna woodlands cover the relict outliers of lateritic plateaux that stud these valleys, so that all types of savanna occur with forest in complex mosaic distributions related to the geomorphological evolution of the landscape referred to above.

THE FOREST–SAVANNA BOUNDARIES IN AREAS OF TECTONIC ACTIVITY AND IN TENSION ZONES

Over the east African plateau where the major faulting of the Rift Valley is associated with volcanic activity, the interplay of geological events and geomorphological processes has resulted in abrupt changes of elevation and relief, of climate and soils and, hence, of vegetation. Several categories of savanna occur in complex distributions related to these environmental conditions and give way to tropical forests on the east-facing escarpments and on the lower moisture-clad slopes of those volcanoes such as Meru and Kilimanjaro where the basaltic rocks have weathered to fertile soils. Here, the change from the savanna grasslands of the Serengeti Plains is particularly abrupt and is clearly related to climatic and edaphic conditions.

The best examples of tension zones within the savanna zone occur in the Okavango delta–Lake Ngami area of Botswana, the Pantanal of Brazil and the Channel country of Australia. Each of these areas is subject to periodic inundations which introduce moist conditions and renewable riverine soils, which in Brazil and Botswana are capable of supporting forest trees. The most complex vegetation patterns occur in the Okavango delta–Lake Ngami area where, in an area subject to tectonic faulting and both long-term and short-term fluctuations of rainfall, the south-eastward-flowing flood waters transgress the geological grain of level terrain covered mainly by low tree and shrub savannas south of Maun and by *Colophospermum mopane* woodlands to its north. The interplay of climatic oscillations and of geomorphological processes, however, has resulted in a great variety of edaphic conditions, each favoured by different types of vegetation with *Colophospermum mopane* woodland occupying the poorly drained alkaline riverine clays, stands of *Terminalia sericea* and *Acacia giraffae* trees studding the sand dunes, successional sequences of grassland communities occupying the inundated areas, and ribbons of forest trees following the elevated levées along the distributaries of the Okavango. Interpretation of sequential Landsat imagery reveals changes in vegetation patterns while dead trees along the Savuti channel in the north and around Lake Ngami reflect the effects of flooding and of alternating flooding and drought. Detailed studies of this area (Cole, 1986) emphasize the dynamic nature of the vegetation changes and pose reflections on the major regional changes of vegetation that have occurred over Southern Africa.

The mosaic vegetation patterns of the Pantanal (Cole, 1960) reflect similar edaphic influences, while those of the Channel country of Australia show the legacies of climatic changes and geomorphological processes. In the Channel country the occurrence of savanna woodland over lateritic plateau relicts and of stands of *Acacia combagei* trees over redistributed lateritic gravel reflect the importance of edaphic influences, as do the successive sequences of grassland communities in the inundated areas and the ribbons of trees over the more elevated levées.

CONCLUSION

Both savanna and forest vegetation exhibit distinctive morphological features and physiological responses that are related to plant–soil moisture relationships which, in turn, are dependent upon climatic and edaphic conditions. Characteristically, the forest–savanna boundaries are abrupt and occur where there are marked changes in topography and soils. For selected areas, these are shown to be the legacies of the interplay of geological events and geomorphological processes and of long-term climatic changes and short-term fluctuations in rainfall.

REFERENCES

Cole, M.M. (1960) Cerrado, caatinga and pantanal; distribution and origin of the savanna vegetation of Brazil. *Geographical Journal*, **126**, 168–79.

Cole, M.M. (1963a) Vegetation nomenclature and classification with particular reference to the savannas. *South African Geographical Journal*, **55**, 3–14.

Cole, M.M. (1963b) Vegetation and geomorphology in Northern Rhodesia; an aspect of the distribution of the savanna of central Africa. *Geographical Journal*, **129**, 290–310.

Cole, M.M. (1965) *Biogeography in the Service of Man*, Inaugural lecture, Bedford College, University of London, 1964.

Cole, M.M. (1980) Geobotanical expression of orebodies. *Transactions of the Institution of Mining and Metallurgy (Section B: Applied Earth Science)*, **89**, B73–143.

Cole, M.M. (1982) The influence of soils, geomorphology and geology on the distribution of plant communities in savanna ecosystems, in *Ecology of Tropical Savannas*, (eds B.J. Huntley and B.H. Walker), Springer-Verlag, Berlin, pp. 145–74.

Cole, M.M. (1986) *The Savannas: Biogeography and Geobotany*, Academic Press, London.

Duvigneaud, P. (1958) La végétation du Katanga et de ses sols metallifères. *Bullétin Royale Societé Botânique Belge*, **90**, 127–86.

Duvigneaud, P. (1959) Plantes cobaltophytes dans le Haut-Katanga. *Bullétin Royale Societé Botânique Belge*, **91**, 111–34.

Duvigneaud, P. and Denaeyer de Smet, S. (1960) Influence des sols toxiques sur la végétation. Action de certaines metaux lourds du sol (cuivre, cobalt, manganèse, uranium) sur la vegetation dans le Haut Katanga, in *Rapport de sol et la vegétation*, Masson, pp. 121–39.

Duvigneaud, P. and Denaeyer de Smet, S. (1963) Cuivre et vegétation au Katanga. *Bullétin Royal Societé Botânique Belge*, **93**, 93–231.

Huntley, B.J. and Walker, B.H. (eds) (1982) *Ecology of Tropical Savannas*, Springer-Verlag, Berlin.

Medina, E. (1982) Physiological ecology of neotropical savanna plants, in *Ecology of Tropical Savannas*, (eds B.J. Huntley and B.H. Walker), Springer-Verlag, Berlin, pp. 305–55.

5

Dynamics of the forest–savanna boundary in the Rio Branco–Rupununi region of northern Amazonia

MICHAEL J. EDEN and DUNCAN F.M. McGREGOR

SUMMARY

1. Earlier work in the southern Rupununi region of Guyana shows that, in spite of persistent shifting cultivation, forest regrowth generally occurs along the forest–savanna boundary.
2. Local evidence also exists in the southern Rupununi of extension of forest into savanna in an area where cultivation and settlement are currently lacking and fire incidence apparently reduced.
3. The present study shows similar conditions near Ilha de Maracá in Brazil, where forest clearance for cattle pasture is taking place at the forest–savanna boundary. Woody regeneration occurs in regularly burnt derived pastures, and no indication as yet exists of their deflection to savanna.
4. It is suggested that nutrient retention within the pasture system is mainly responsible for the recurrent regeneration, although, with continuing burning, long-term net nutrient loss will probably occur.
5. Local forest advance into savanna is also evident on Ilha de Maracá itself in an area where the incidence of savanna fires has apparently declined.

INTRODUCTION

In many parts of the tropics, there is evidence of past changes in the forest–savanna boundary. They include large-scale, long-term changes associated with climatic fluctuations as well as recent smaller-scale changes associated with anthropic intervention. Anthropic changes mostly involve a deflection of forest to savanna following clearance and burning for agricultural or other purposes. Evidence of such deflection is reported from tropical Africa (Hopkins, 1983), south-east Asia (Taylor, 1968; Eden, 1974; Seavoy, 1975) and parts of central and South America (Budowski, 1956; Scott, 1977). However, deflection of forest to savanna is not ubiquitous. In places, forest that has been subject to persistent clearance may repeatedly regenerate, and even subsequently advance into existing savanna. One area where this is occurring is the southern Rupununi region of Guyana, where indigenous shifting cultivation has long been undertaken in forest areas marginal to the savanna (Eden, 1986). The savanna forms part of the extensive Rio Branco–Rupununi savanna that straddles the Brazil–Guyana frontier in northern Amazonia (Fig. 5.1).

In 1986–87, the authors visited the Brazilian part of the Rio Branco–Rupununi savanna during the Royal Geographical Society Maracá Rainforest Project and studied the effects of anthropic clearance along a sector of the forest–savanna boundary. The study area lies adjacent to the Uraricoera river, immediately south of Ilha de Maracá, where forest is currently being cleared for cattle ranching (Eden *et al.*, 1990, 1991). In this chapter, aspects of the ecological dynamics of clearance in the area are examined, the impact on the local forest–savanna boundary is discussed and a comparison is made with boundary conditions in the adjacent southern Rupununi region.

RIO BRANCO–RUPUNUNI SAVANNA

Environmental conditions

The Rio Branco–Rupununi savanna covers about 5.5×10^6 ha, and is largely flanked by semi-evergreen to evergreen seasonal forest. The savanna occupies mostly an undulating land surface at 100–180 m above sea level, marginally extending onto adjoining higher and more dissected terrain. Where free-draining soils are present, savanna with low woody species exists mostly, although there are tracts of poorly drained savanna without woody species. The forest–savanna boundary in the area is usually abrupt, with a transition from closed-canopy forest to open savanna commonly occurring over a few metres.

The regional climate is of seasonal tropical type, with mean monthly temperatures in the range 26–30 °C. Rainfall seasonality is characteristic of the area. Boa Vista, in the southern savanna, has an annual rainfall of

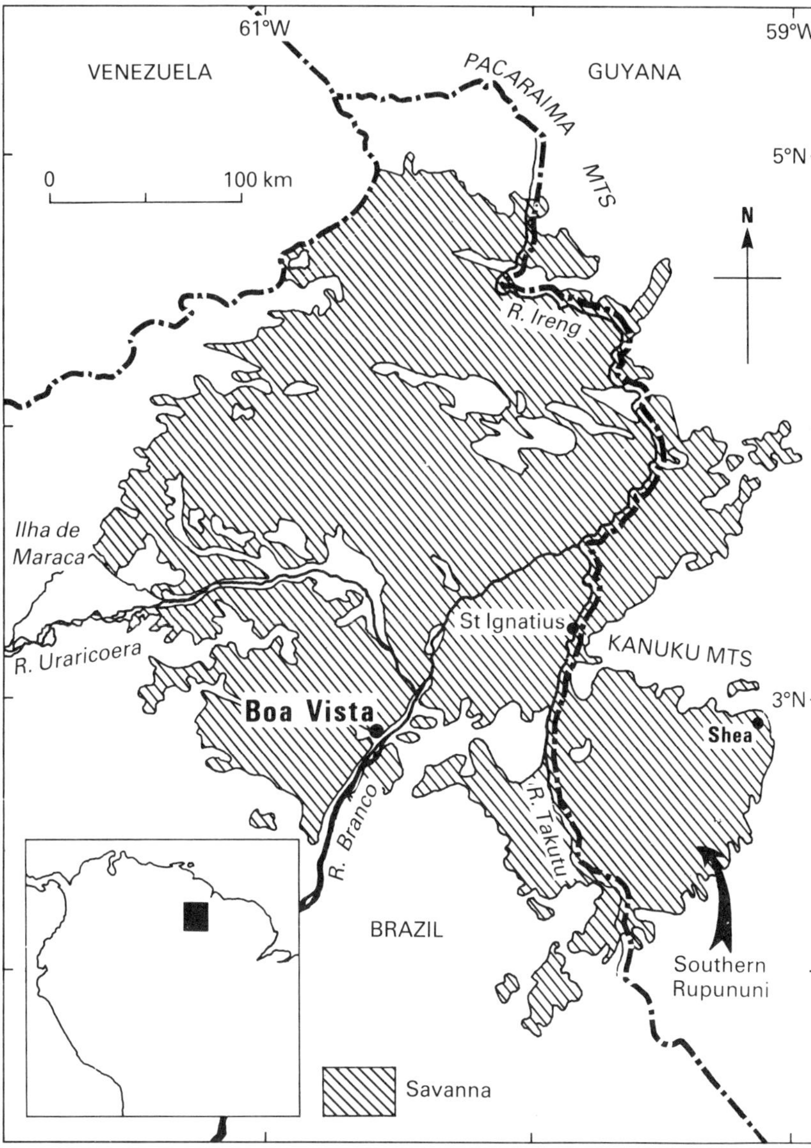

Fig. 5.1 Rio Branco – Rupunumi Savanna.

1751 mm, of which 75–80% commonly falls from April to August. As elsewhere in Amazonia, rainfall intensity is generally high. In the dry season, occasional rain falls, but its effectiveness is minimized by high rates of evapo-transpiration. Widespread burning of the savanna occurs in the dry season (Eden, 1964; Brasil, 1975).

Savanna soils are mostly developed over highly weathered parent materials, and are mainly Orthic Ferralsols and Plinthic Acrisols (FAO-Unesco, 1971). Local occurrences of Eutric Gleysols are reported, but the majority of soils are acid and inherently infertile.

Forest–savanna boundary in the southern Rupununi

Most of the forest–savanna boundary zone in the southern Rupununi is occupied by Wapisiana Indians, who live in villages in the savanna and undertake shifting cultivation in the adjacent forest. In recent decades, Wapisiana numbers have risen and their cultivation impact on the forest has increased. It was previously assumed that the savanna was extending as a result of the shifting cultivation (Waddell, 1963; Hills, 1964; Potter, 1978) but examination of sequential aerial photography and Landsat imagery has failed to confirm this (Eden, 1986). In most areas, the forest–savanna boundary has been stable of late, with no evidence of fire-deflection of abandoned forest fields to savanna. Woody regrowth generally occurs in abandoned fields, apparently reflecting a low incidence of herbaceous weeds and fire (Eden, 1986).

Some evidence exists of localized forest advance (Hills, 1964). In the Bunawau basin northeast of Shea village, aerial photographs show the forest–savanna boundary as diffuse and irregular and that secondary forest is extending. Notably, this is an area where indigenous settlement and cultivation is currently lacking and fire-incidence presumably reduced (Eden, 1986).

Forest–savanna boundary south of Ilha de Maracá

Methods

Investigations were undertaken on four adjacent cattle ranches, 1000–2000 ha in extent, lying to the south of Ilha de Maracá. Soil profiles were widely examined and sampled in the study area; also topsoil samples (0–10 cm) were collected in adjacent forest, derived pasture and savanna sites. At each site, four topsoil samples were collected; all were single samples obtained from a central 30 × 30 m plot at the site. In total, topsoil samples were collected at 7 sites in forest, 19 in derived pasture, and 6 in savanna.

Laboratory analysis of soil samples was undertaken as follows: organic carbon was determined by the Walkley–Black method; exchangeable sodium, calcium, magnesium, hydrogen and aluminium were extracted with 1 M KCl and exchangeable potassium with 0.02 M HCl, and were measured by atomic absorption spectrophotometer or flame photometer; available phosphorus was determined using the 'Bray No. 1 method'; total nitrogen was determined using a micro-Kjeldahl procedure; pH was measured in a 1:2.5 suspension in H_2O. Effective cation exchange capacity (ECEC) is expressed as the sum of exchangeable bases, aluminium and hydrogen. Particle size analysis was undertaken using the pipette method (Black *et al.*, 1965).

On one cleared forest site, data were obtained on residual timber and other forest debris remaining after cutting and burning. Surface forest debris (twigs, bark, charcoal, etc.) was collected in 20 quadrats, 0.5 × 0.5 m, at the site, and weighed after drying to constant weight. Volumetric sampling of large timber debris was undertaken along four 30 × 1 m transects, and total dry weight estimated on the basis of representative bulk density wood samples. The total phosphorus, potassium, sodium, calcium and magnesium content of representative ashed samples of the above were determined by atomic absorption spectrophotometry (calcium, magnesium), flame photometry (potassium, sodium) and UV spectrophotometry (phosphorus) after digestion in 1 M HNO_3. Total nitrogen was obtained by micro-Kjeldahl procedure (Black *et al.*, 1965).

Pasture establishment

Investigations of the forest–savanna boundary zone focused on an area of forest clearance south of Ilha de Maracá, where small-scale cattle ranching has recently extended. The area became accessible in the early 1950s through a dirt road from Boa Vista, and forest clearance for pasture began on a small scale. However, most of the pasture has been developed over the last 15 years.

The study area is gently to moderately dissected, with a local elevation of 100–140 m above sea level. Soils are mainly developed over weathered micaceous schists with some granitic and gneissic materials. Both residual and colluvial parent materials occur, giving soils with sandy loam to sandy clay loam topsoils over sandy clay to clay at depth. In places, plinthite or ironstone gravel may be present. Most soils are Acrisols, with local occurrence of Eutric Nitosols (FAO-Unesco, 1971; Brasil, 1975; Nortcliff and Robison, 1988). In general, the soils of adjacent forest and savanna sites have developed from similar rock materials, although the savanna soils are relatively degraded. Analytical data for forest soils are given in Table 5.1.

In mature forest, taller trees attain 30–35 m in height, with occasional emergents reaching 40 m. Characteristic large species include *Anacardium*

Table 5.1 Analytical data for forest soils to the south of Ilha de Maracá

	Sand	*Texture silt %*	*Clay*	*Organic carbon*	*pH*	*Ca*	*Mg*	*Exchangeable Na (cmol kg^{-1})*	*K*	*Al*	*H*	*ECEC (cmol kg^{-1})*	*Total N (%)*	*Available P (ppm)*
(a) Fazenda Patchuli[a]														
0–25 cm	55	12	33	0.40	4.3	0.10	0.08	0.06	0.06	0.03	0.93	1.26	0.08	2.2
25–50 cm	50	12	38	0.35	4.3	0.06	0.08	0.06	0.04	0.03	0.70	0.97	0.06	2.1
50–75 cm	45	14	41	0.43	4.2	0.06	0.08	0.07	0.05	0.04	0.58	0.88	0.06	1.5
75–100 cm	36	16	48	0.35	4.3	0.10	0.10	0.08	0.05	0.04	0.58	0.95	0.07	0.0
(b) Fazenda Pau Roxo[b]														
0–25 cm	67	11	22	0.47	5.6	0.90	0.30	0.06	0.10	tr	0.27	1.63	0.09	1.6
25–50 cm	59	10	31	0.68	5.3	0.64	0.32	0.06	0.10	tr	0.25	1.37	0.04	2.2
50–75 cm	55	11	34	0.51	5.2	0.10	0.60	0.07	0.29	tr	0.34	1.40	0.04	1.9
75–100 cm	47	9	44	0.40	5.4	0.04	0.84	0.10	0.53	0.02	0.39	1.92	0.03	2.3

[a] Free-draining mid-slope profile, developed over *in situ* schistose parent material.
[b] Free-draining local profile, developed over fine-grained metamorphic parent material.

giganteum Hancock ex Engl., *Jacaranda copaia* (Aubl.) D. Don and *Micropholis melinoniana* Pierre. Most trees are evergreen, but some species, such as *Tabebuia uleana* (Kranzl.) A. Gentry and *Lueheopsis duckeana* Bussett, lose their leaves during the dry season (Milliken and Ratter, 1989; Eden *et al.*, 1991). Adjacent to the forest lie areas of savanna with low woody species, containing a characteristic *Trachypogon–Curatella* community, with associated panicoid grasses and sedges. Local areas of poorly drained savanna without woody species also occur.

Pasture establishment follows forest cutting during the dry season from September to March and firing of the debris. Foliage and small branches are generally consumed, but larger branches and trunks are only superficially charred. Thereafter, maize and dry rice are planted to coincide with the rains in April or May. Grass cuttings are usually planted amid the food crops. In earlier times, *colonião* (*Panicum maximum* Jacq.) was the commonest pasture grass in use, together with *jaraguá* (*Hyparrhenia rufa* (Nees) Stapf), but since about 1980 *kikuyu* grass (*Brachiaria humidicola* (Rendle) Schweik.) has been planted widely.

During the initial cropping period, hand weeding is undertaken, but increasing shrubby invasion occurs. In the subsequent dry season, the accumulated woody weeds are generally cut by hand and the site reburnt. This disposes of the weed debris and depletes the remaining timber debris. After the burn, the grass cuttings re-establish themselves and a pasture is created onto which cattle are introduced. This normally occurs in the second rainy season after clearance.

Once established, the introduced forage grasses develop a more or less continuous ground cover. New kikuyu growth provides a well developed ground layer, which is relatively effective in suppressing early weeds. Erect colonião also grows quickly, often up to 2 m, but the pasture is more weedy. In all cases, however, pioneer woody species soon establish and would dominate the community as the initial stage of forest regeneration.

Among the more common regenerating woody plants are the palms *Astrocaryum aculeatum* G.F.W. Meyer and *Maximiliana maripa* (Correa) Drude, which are common understorey species in local forest stands, various leguminous shrubs including *Mimosa* spp., and sundry forest and other trees (Eden *et al.*, 1990). In the face of woody regrowth, periodic hand cutting of weeds is necessary. Pastures are usually burned each dry season to clear the weed debris and renew the grass cover. The recurrent burning also removes the residual timber, which is largely lost from pastures by 4–6 years from clearance.

Although the above sequence is widespread, variation exists in older pastures. Some colonião pastures are still well established on land that was cleared ten years or more ago. Elsewhere colonião pastures of similar age are overrun with woody weeds, or, in a few cases, only support a low cover of native grasses and herbaceous and shrubby weeds. However, most older pastures contain a mixture of grasses and woody weeds, with no general early deflection of cleared forest land to savanna. In places, scattered examples of *Curatella americana* occur in older weedy pastures, but the typical *Trachypogon–Curatella* association of the savanna does not occur on cleared forest land.

Soil and pasture conditions

Soil contrasts exist between adjacent older derived pastures and savanna in the vicinity of Ilha de Maracá. The soils are developed over similar rock materials and, in sites examined, are physically comparable, but the pasture soils contain more nutrients (Table 5.2). This fact, plus the increased biomass that generally characterizes the pastures, suggests that their total nutrient capital is higher than that of the savanna. The increased nutrients in the pasture system presumably derive from the initial clearing and burning of the forest and from subsequent decomposition of residual timber debris (Hecht, 1981; Uhl, 1987). In one first-year cultivation site at Fazenda Nova Olinda, the dead wood remaining after burning was about 117 t ha^{-1} (dry weight). The material, sampled about seven months after the burn, contained a substantial nutrient capital, which is gradually released to the pasture (Table 5.3).

Table 5.2 Topsoil (0–10 cm) analytical data for eight older derived pastures (6–25 years from clearance) and six savanna sites to the south of Ilha de Maracá. Bold values in column 2 are significantly different from those in column 1 at the 95% level (Mann-Whitney *U*-test – two-tailed probabilities)

	Old pastures (6–25 years) (n = 32)	*Savanna* (n = 24)
Clay (%)	17	21
Organic carbon (%)	1.65	1.49
pH	5.9	**5.1**
Exchangeable Ca (cmol kg^{-1})	3.64	**0.47**
Exchangeable Mg (cmol kg^{-1})	0.59	**0.20**
Exchangeable Na (cmol kg^{-1})	0.08	0.06
Exchangeable K (cmol kg^{-1})	0.16	0.13
Exchangeable Al (cmol kg^{-1})	0.03	**0.53**
Exchangeable H (cmol kg^{-1})	0.17	**0.97**
ECEC (cmol kg^{-1})	4.67	**2.36**
Total N (%)	0.13	**0.08**
Available P (ppm)	5.8	2.6

Table 5.3 Total nutrient supply in forest debris in a recently cleared field area to the south of Ilha de Maracá. The area was burnt in January/February 1987, and sampled in September 1987

	Timber	*Surface organic debris (twigs, bark, charcoal, etc.)*	*Total*
Total dry weight ($t\,ha^{-1}$)	110.1	7.2	117.3
Total N ($kg\,ha^{-1}$)	509.4	52.3	561.7
Total P ($kg\,ha^{-1}$)	45.9	3.2	49.1
Total K ($kg\,ha^{-1}$)	129.1	4.9	134.0
Total Na ($kg\,ha^{-1}$)	2.2	0.3	2.5
Total Ca ($kg\,ha^{-1}$)	171.4	121.1	292.5
Total Mg ($kg\,ha^{-1}$)	55.0	18.7	73.7

In respect of maintaining the pasture nutrient supply, the situation is less clear. There is considerable soil nutrient variability among older pastures, but no general nutrient decline is apparent by comparison with younger pastures (Eden *et al.*, 1990). It is assumed that this reflects efficient nutrient cycling in the pastures and low nutrient loss after recurrent fires. However, there is no positive correlation between topsoil nutrient content and total above-ground biomass in older pastures. Instead, it appears that soil nutrient levels are more a function of differential sequestration by the plant biomass (Buschbacher *et al.*, 1988) than an indicator of the overall nutrient status of the soil–plant system. The influence of woody weeds is no doubt important in view of their assumed efficiency at taking up soil nutrients in short supply, particularly phosphorus (Jordan, 1985). Nitrogen fixation may also occur among the leguminous shrubs that exist in derived pastures. Admittedly, the woody weeds do nothing to improve pasture quality directly, but the woody regrowth increases nutrient retention in the soil–plant system. In the present case, however, nutrient accumulation in the biomass is reversed by dry-season burning, which probably causes a slow net nutrient loss from the pasture system. This is indicated by the degraded status of the adjacent savanna over similar rock materials, and apparently occurs in spite of assumed nutrient inputs in rainwater.

DISCUSSION

A feature of cleared forest land in the vicinity of Ilha de Maracá is that woody regrowth occurs in annually burnt derived pastures. This contrasts with adjacent savanna, also regularly burnt, which is not so

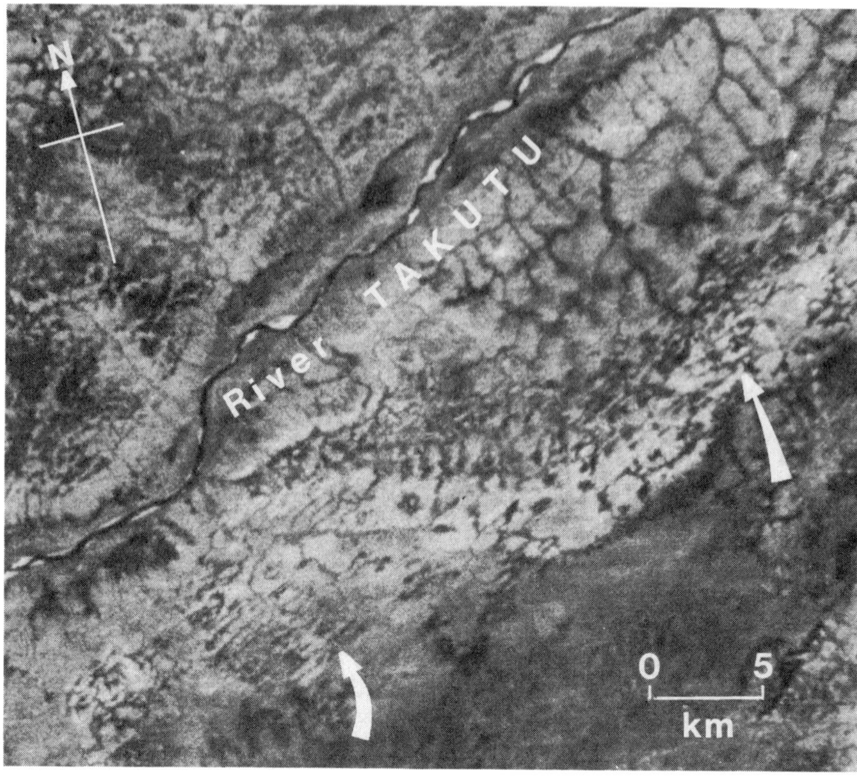

Fig. 5.2 Relict linear dunes (arrowed) south of the Takutu river, Rio Branco Savanna, Brazil. LANDSAT MSS band 6 subscene, March 1973, ID zn. 1224 13472 (reproduced by courtesy of NOAA)

invaded except by scattered savanna species such as *Byrsonima crassifolia* and *Curatella americana*. No clear soil physical differences exist between the older derived pasture and savanna sites examined, but there are significant nutrient contrasts that may explain the differential plant response (Table 5.2). The contrasts reflect the fact that, although many nutrients are lost from the pasture system following initial burning of forest debris (Eden *et al.*, 1990), an important supply is retained in and recycled through the system over subsequent years. The role of woody weeds in this process appears critical, with the nutrients in question currently sustaining woody regrowth to an extent that is lacking in the more degraded savanna (Nye and Greenland, 1960; Jordan, 1985; Uhl, 1987).

Forest–savanna boundary conditions in the vicinity of Ilha de Maracá are similar to those reported in the southern Rupununi where persistent woody regrowth occurs in areas of shifting cultivation. According to Eden (1986), forest regrowth in the area is primarily a response to a relatively low incidence of burning in newly abandoned fields, but may also be encouraged, as at Maracá, by the retention of nutrients from the original forest cover.

In addition, in one part of the southern Rupununi, it is reported that local forest advance into the savanna has occurred in an area where there is currently no indigenous settlement or cultivation (Hills, 1964; Eden, 1986). A comparable situation is also reported on eastern Ilha de Maracá, where small, isolated savannas within the main forest are being invaded by forest trees. The process may again be associated with a recent reduction in human impact, which in this case results from the designation of Ilha de Maracá as an ecological reserve and from possible associated cessation of burning (Milliken and Ratter, 1989). Such forest advance is rare, but not exceptional, in Amazonia or other tropical areas, and appears primarily to be associated with a reduced incidence of savanna fires (Hopkins, 1983; Milliken and Ratter, 1989). Under such conditions, low nutrient supply does not preclude woody regrowth, with sufficient nutrients being available over time to permit gradual forest encroachment on the savanna (Kellman, 1979).

The local dynamics of the forest–savanna boundary zone are thus primarily a function of fire incidence, which determines the survival or otherwise of invading forest species. However, potential rates of woody invasion, which are rapid in derived pastures and slower in savanna, appear to reflect nutrient levels in the system. That recent derived pastures and old-established savanna coexist at different nutrient levels over similar rock materials suggests that gradual net nutrient loss from the pastures occurs over time. This may eventually deflect the pastures to the compositional status of the savanna, although no evidence of this yet exists in the study area.

ACKNOWLEDGEMENTS

Grant assistance was received from the Ford Foundation/Royal Geographical Society (MJE/DFMM), the Carnegie Trust for the Universities of Scotland (DFMM) and the Central Research Fund of the University of London (MJE). Grateful acknowledgement is made of logistical support provided by the Royal Geographical Society Maracá Rainforest Project and by Brazilian government agencies, notably the Secretaria Especial do Meio Ambiente, the Instituto Nacional de Pesquisas da Amazônia and the Empresa Brasileira de Pesquisa Agropecurária. Field assistance was provided by Nelson Vieira. Laboratory analyses were undertaken by Margaret Onwu and cartographic work by Justin Jacyno.

REFERENCES

Black, C.A., Evans, D.D., White, J.L. *et al.* (1965) *Methods of Soil Analysis*, American Society of Agronomy, Madison.

Brasil (1975) *Levantamento de Recursos Naturais. Vol. 8. Folha NA.20 Boa Vista e Parte das Folhas NA.21 Tumucumaque, NB.20 Roraima e NA.21*, Departmento Nacional de Produção Mineral, Ministério das Minas e Energia, Rio de Janeiro.

Budowski, G. (1956) Tropical savannas, a sequence of forest felling and repeated burnings. *Turrialba*, **6**, 22–33.

Buschbacher, R., Uhl, C. and Serrão, E.A.S. (1988) Abandoned pastures in eastern Amazonia. II. Nutrient stocks in the soil and vegetation. *Journal of Ecology*, **76**, 682–99.

Eden, M.J. (1964) The savanna ecosystem – northern Rupununi, British Guiana. *Savanna Research Series, McGill University*, **1**, 1–216.

Eden, M.J. (1974) The origin and status of savanna and grassland in southern Papua. *Transactions of Institute of British Geographers*, **63**, 97–110.

Eden, M.J. (1986) Monitoring indigenous shifting cultivation in forest areas of southwest Guyana using aerial photography and Landsat, in *Remote Sensing and Tropical Land Management* (eds M.J. Eden and J.T. Parry), John Wiley, Chichester, pp. 255–77.

Eden, M.J., McGregor, D.F.M. and Vieira, N.A.Q. (1990) Pasture development on cleared forest land in northern Amazonia. *Geographical Journal*, **156**, 251–66.

Eden, M.J., Furley, P.A., McGregor, D.F.M. *et al.* (1991) The impact of forest clearance and burning on soil properties in northern Roraima, Brazil. *Forest Ecology and Management*, **38**, 283–90.

FAO-Unesco (1971) *Soil Map of the World. Vol. 4. South America*, Unesco, Paris.

Hecht, S.B. (1981) Deforestation in the Amazon basin: magnitude, dynamics and soil resource effects. *Studies in Third World Societies*, **13**, 61–108.

Hills, T.L. (1964) *Progress Report, Savanna Research Project*, Unpublished report, Geography Department, McGill University, Montreal.

Hopkins, B. (1983) Successional processes, in *Ecosystems of the World. 13. Tropical Savannas*, (ed. F. Bourlière), Elsevier, Amsterdam, pp. 605–16.

Jordan, C.F. (1985) *Nutrient Cycling in Tropical Forest Ecosystems*, John Wiley, Chichester.

Kellman, M. (1979) Soil enrichment by neotropical savanna trees. *Journal of Ecology*, **67**, 565–77.

Milliken, W. and Ratter, J.A. (1989) *The Vegetation of the Ilha de Maracá. First Report of the Vegetation Survey of the Maracá Rainforest Project (INPA/*

RGS/SEMA), Royal Botanical Garden, Edinburgh.
Nortcliff, S. and Robison, D. (1988) *The Soils and Geomorphology of the Ilha de Maracá, Roraima. The Second Approximation*, Department of Soil Science, University of Reading, Reading.
Nye, P.H. and Greenland, D.J. (1960) *The Soil Under Shifting Cultivation*, Commonwealth Agricultural Bureau, Harpenden.
Potter, L. (1978) *The Amerindians of Guyana and their environment*, paper presented at 10th Conference of Caribbean Historians, St Thomas, Virgin Islands, March–April 1978.
Scott, G.A.J. (1977) The role of fire in the creation and maintenance of savanna in the Montaña of Peru. *Journal of Biogeography*, **4**, 143–167.
Seavoy, R.E. (1975) The origin of tropical grasslands in Kalimantan, Indonesia. *Journal of Tropical Geography*, **40**, 48–52.
Taylor, B.W. (1968) Changes in the forest/savanna boundary in the perhumid lowlands of New Guinea. *Savanna Research Series, McGill University*, **13**, 62–5.
Uhl, C. (1987) Factors controlling succession following slash-and-burn agriculture in Amazonia. *Journal of Ecology*, **75**, 377–407.
Waddell, E.W. (1963) *The Anthropic Factor in a Savanna Environment*, MA thesis, McGill University, Montreal.

6

Edaphic changes at the forest–savanna boundary with particular reference to the neotropics

PETER A. FURLEY

SUMMARY

1. Five different boundaries have been studied, within three dispersed regions ranging from Central America, the northern Amazon Basin to the central Brazilian plateau.
2. The edaphic influence can be seen to be significant at each of the localities examined. Soil moisture is the determining factor in forest–savanna transitions located over toposequences, such as those in Belize or the Federal District of Brazil, where there is a downslope increase in water content and where growth at the upslope forest edge is constrained by seasonal water deficiencies.
3. Variations in soil water also determine the boundary between forest and hyperseasonal savannas, as demonstrated on Maracá Island in the northern Brazilian Amazon. Where underlying geology affects topography and surface soils, as in southern Mato Grosso, the soils possess marked nutrient variations, especially of calcium. This distinguishes the richer soils of the mesotrophic forests and *cerradão* from the poorer, dystrophic and coarse-textured *cerrado* soils.
4. In each case study, evidence has been presented for forest or thicket advance upon savanna, whether arboreal or open grassland.
5. Climate acts as the major regional determinant, whilst topography and nutrient levels provide the subregional determinant and the local boundary, at any given site, is strongly determined by the soil-water regime.

INTRODUCTION

The evidence for soil properties having a determining role in vegetation change at the forest edge is varied and, at first sight, contradictory. At one extreme, there is the view that vegetation is so closely related to soil that pedological differences inevitably condition the nature of the plant community. Brown (1987) for instance, expresses this as 'under a given climatic regime and without human interference, each major soil class tends to support a distinct and recognizable type of vegetation'. At the other extreme, there is the argument that vegetation changes occur so rapidly in comparison to those of soil and, for a number of different reasons, that the two processes proceed at different spatial and temporal rates and are therefore unrelated. If theories of savanna advance during Pleistocene ice expansion are correct, then at a broad scale the forest–savanna boundaries must have fluctuated greatly over the recent geological past approximately coincident with known oscillations in glacial advance and retreat (Brown and Ab 'Saber, 1979). If it is argued that soil properties and plant communities have a co-evolutionary relationship, then this association must have been constantly changing and could only have stabilized during longer phases of climatic and vegetational equilibrium. Evidence for such past equilibrium between soils and plant communities has been put forward to substantiate savanna episodes now found beneath contemporary forest (Ab 'Saber, 1977). However, surface soils, which are likely to react to these changes more rapidly than deeper horizons, may be out of phase with their more slowly evolving profiles.

There is ambiguous evidence for the 'savannization' of soils during dry periods associated with forest retreat. On the one hand it is suggested that weathering, pedogenic and geomorphological processes were sufficiently distinctive to generate recognizable semi-arid products beneath widespread areas currently covered by tropical forest. It is argued that conditions favoured the rapid removal of unprotected soil (Bigarella *et al.*, 1969; Bigarella and Andrade-Lima, 1965, 1982) and that the erodate created characteristic downstream and downslope deposits (Beigbeder, 1959; Ab 'Saber, 1977). In particular, widespread stone-lines and stretches of infertile white sand have been interpreted as results of landscape evolution under dry climates (Journaux, 1975; Ab 'Saber, 1982). On the other hand, the locations where sufficient evidence is available are scattered and few in number, whilst sites which ought to demonstrate laterization and ironstone formation are not widespread under forest, although sesquioxide enrichment with various degrees of hardening is abundantly obvious in savanna areas. Despite the lack of convincing evidence at a regional scale, it is not difficult to demonstrate the key role of soils in determining the forest–savanna boundary at a local scale or specific site, although the vegetation distribution often forms a much more complicated mosaic pattern than is generally recognized.

The aim of this chapter is to examine a number of localities in the neotropics where a sharp boundary occurs between forest and savanna formations, in order to assess the role of soil properties in determining vegetation change. Several catenary sites are analysed in central America and central Brazil respectively to examine the gallery forest edge. An additional toposequence in western Brazil illustrates the determining influence of parent material upon soil properties and consequently upon vegetation zonation, whilst two further locations in the Brazilian Amazon provide evidence for transitions over near-level topography with more complex soil–plant community interactions.

SOIL VARIATIONS OVER THE FOREST–SAVANNA BOUNDARY

Maya Mountains, Belize, Central America

Three catenary sequences have been examined over different parent materials in areas where a sharp zonation of vegetation was evident (Furley, 1974a,b, 1976). In theoretical terms, a convex–concave slope with an open system develops a characteristic sequence of soils as illustrated in Fig. 6.1. Length of slope and to a lesser extent slope angle, combine to produce a pattern of erosion and deposition, though this can frequently be masked by transportation processes over ancient weathered landscapes. Furthermore, groundwater and drainage tend to be related to slope form.

Despite slope processes which can obscure soil and vegetation transitions, a clear zonation of plant communities was evident at the three transects examined (Fig. 6.2). The zonation was related to a number of soil properties but with increasing complexity, from the upper to the lower parts of each transect. For example, calcium and potassium increased strikingly towards the footslope. This was reflected in an increase in the herbaceous plant diversity and abundance. At the same time, pH rose steadily over the transect although levels dropped away in the wettest sites in the valley bottom (Fig. 6.3). On the other hand, the clay fraction, which might have been expected to mirror the erosion–deposition sequence, showed no statistically significant trends. The change in soil properties coincided almost exactly with the change from predominantly herbaceous to predominantly tree vegetation in gallery forest. In the case of Granite Cairn, there was virtually no footslope, and

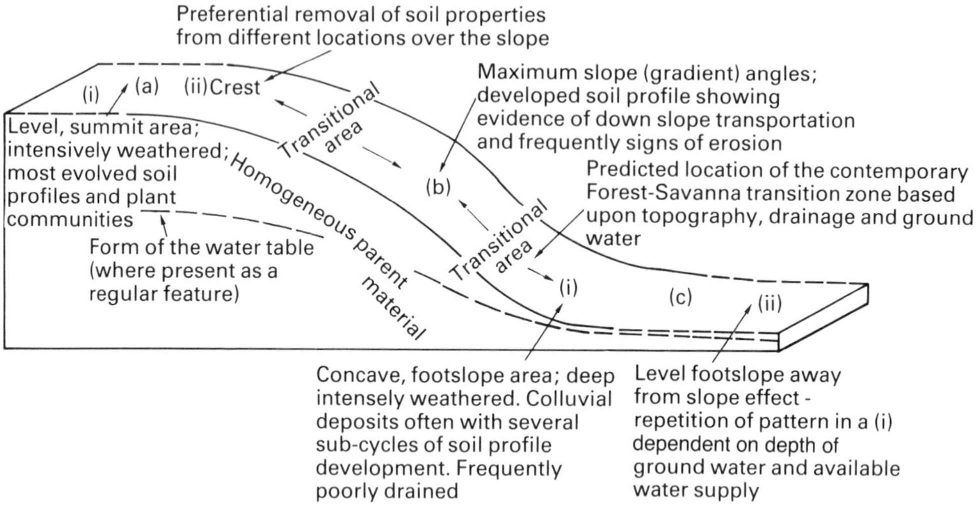

Fig. 6.1 Predicted forest–savanna boundaries over catenary slope forms within an open system.

this accounts for the absence of downslope accumulation of nutrient elements and the lack of a clear change in soil properties. Even in this case, a rise in organic carbon, exchangeable calcium, magnesium and potassium is discernible over the diffuse transition zone where the plant communities changed. Further, at the Despair and Cooma Cairn sites, there was a marked buried stone-line, increasingly broken up and buried to greater depths with distance downslope. This evidence for downslope movement is paralleled by the increasing mottling in lower horizons downslope and the appearance of a water table over the footslope profiles (Fig. 6.4). Although surface soil properties reflect the changes in vegetation to a greater extent than subsurface horizons, the profile analyses complemented those of the A-horizon depicted here and this may signify a relatively long period of soil–plant equilibrium.

Consequently, from these examples, it may be summarized that forest–savanna transitions can be sharply distinguished over toposequences, especially where they take place over convex–concave slopes and are emphasized by marked zones of soil erosion and deposition. Changes in soil properties and soil-water status coincided with the position of the forest edge over the slope and are interpreted as having a determining role in locating the boundary.

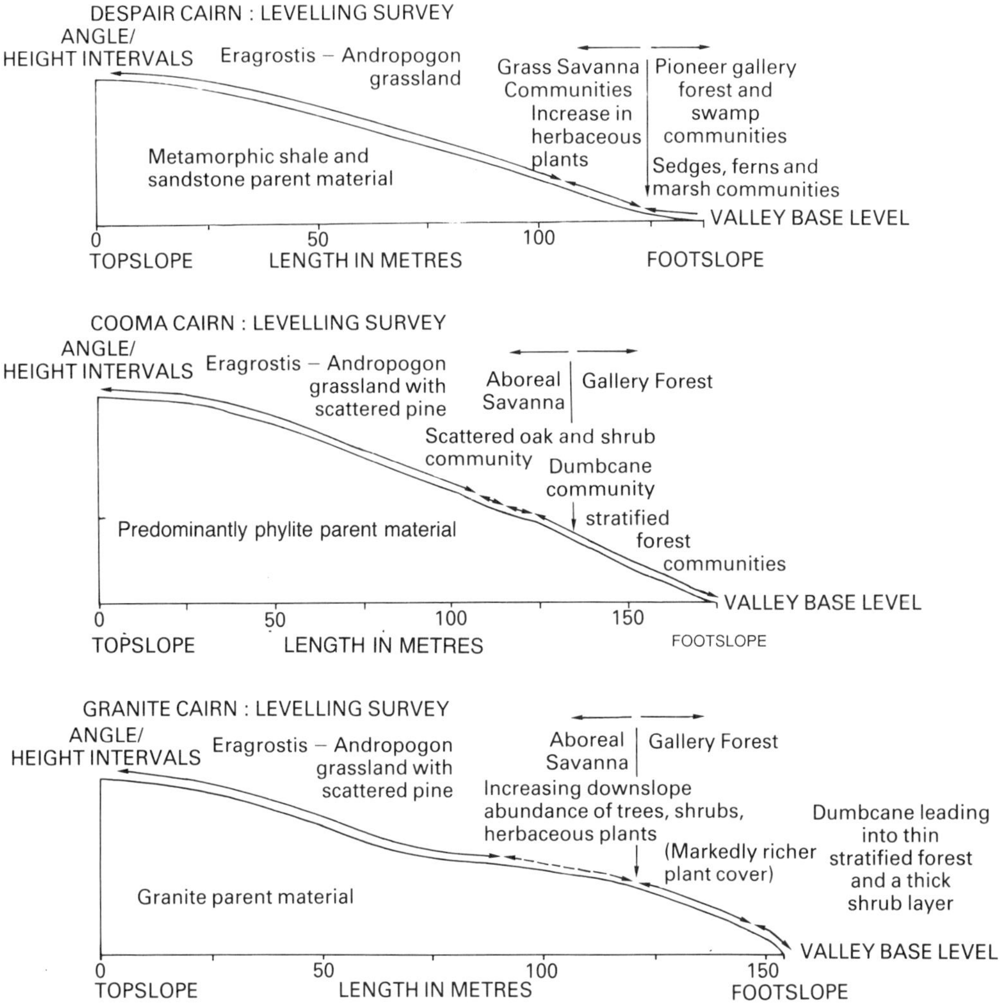

Fig. 6.2 The zonation of plant communities in the Maya Mountains, Belize (modified from Furley, 1974a, b).

Fazenda Água Limpa, Central Brazil

Whereas relatively straightforward soil–vegetation relationships can be demonstrated over some toposequences, there are numerous examples of forest–savanna boundaries occurring over sloping land with less direct concordance between geomorphology and soil properties (Askew *et al.*,

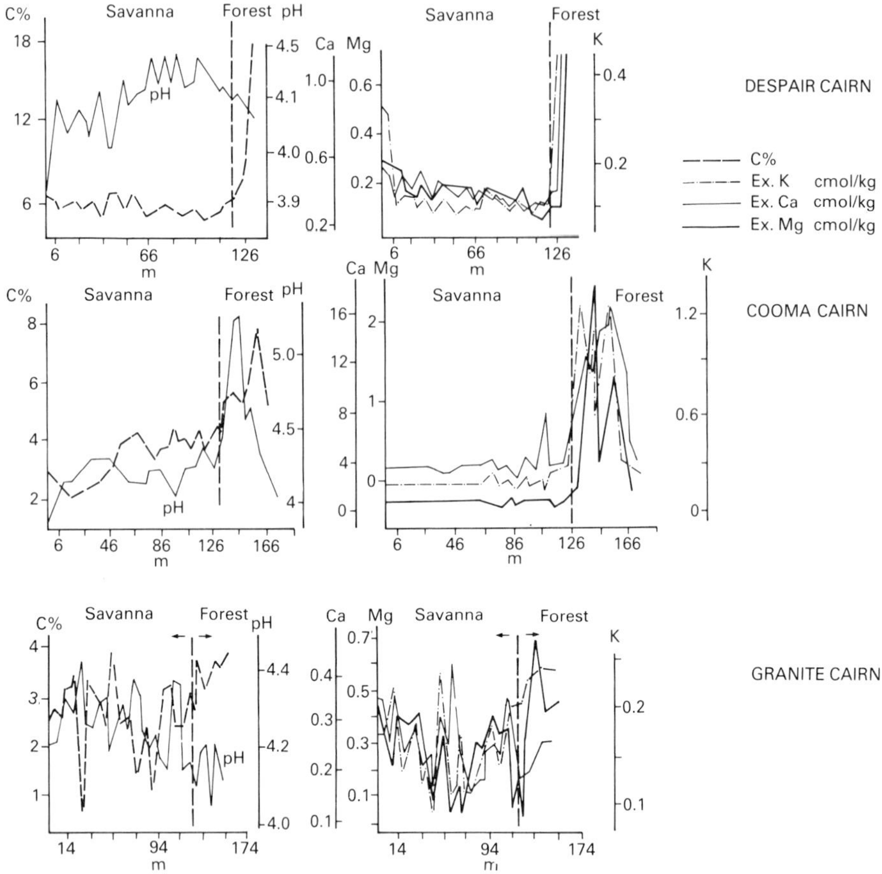

Fig. 6.3 Changes in soil properties over the three Maya Mountain transects (modified from Furley, 1974, 1976a, b).

1970). Two transects were sited over such topographic forms within the Fazenda Água Limpa ecological reserve at the University of Brasília.

In both cases, transect lines orthogonal to the basal stream passed downslope from open woody savanna (*campo cerrado*) to open shrubby grassland (*campo sujo* and *campo limpo*) with occasional patches of woody cerrado distributed on earth mounds raised above the seasonally wet grassland (*campo de murundus*; Diniz *et al.*, 1986, Oliveira-Filho and Furley, 1990) before passing into dense gallery forest or *mata ciliar* (Ratter, 1985, Eiten, 1972; Fig. 6.5). Transect 2 is more complex, having a second woody savanna tract between the wet grassland and gallery forest. The soils of this campo cerrado, which are shallow with depths varying with proximity

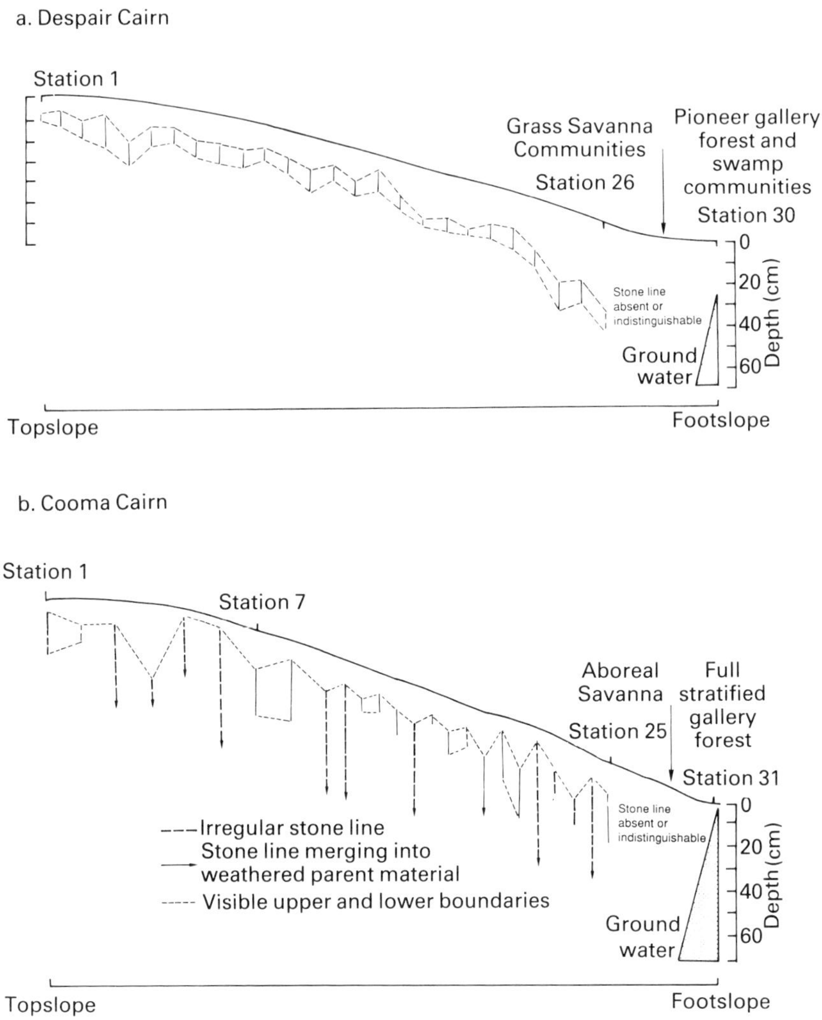

Fig. 6.4 The positions of the stone-line and groundwater at Despair and Cooma Cairns, Maya Mountains, Belize (modified rom Furley, 1974a, b, 1976).

to concretions close to the surface, differ in soil and drainage characteristics, from the campo cerrados upslope.

On the assumption that it is the surface soils that most closely relate to vegetation change and correspondingly plant growth, the present discussion will concentrate upon spatial changes of the A-horizon. The full data sets for the two sites are published in Freire (1979) and Furley (1985). The clearest trends in soil properties are those associated with soil organic

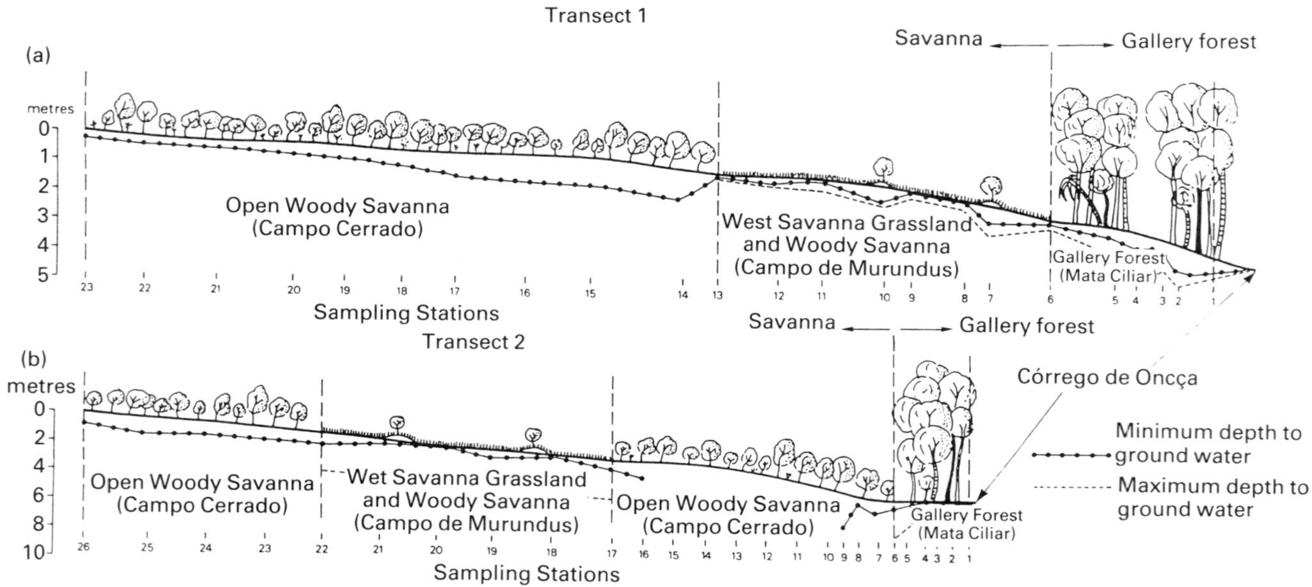

Fig. 6.5 The sequence of plant communities over two catenary transects at Fazenda Água Limpa, DF, Brazil (modified from Furley, 1985).

matter, which tends to respond rapidly to changes in plant biomass input and which, in turn, provide the most likely source of recycled nutrients for plant growth. The distribution of organic carbon, total nitrogen and available phosphorus is shown in Fig. 6.6. At both transect sites there is a striking change between forest and the upslope vegetation formation – either campo cerrado or campo de murundus. At transect 1, the forest edge is very clearly demarcated by all three soil parameters. At transect 2, although the means for soil properties at forest and non-forest sites are different, the vegetation boundary, which is evident in the field, is less closely matched to the soil properties. This is likely to result from two factors: the first is the short but sharp break of slope above profile 7. It is likely therefore that profiles 7 and 8 are affected by downslope seepage to the water table which reaches close to the surface in the wet season at these points (refer to Fig. 6.5). The second factor is that profiles 1 and 2 are extremely wet, and 1 in particular is likely to be affected by permanent high water table plus downstream movement from the small catchment.

The distribution of total exchangeable acidity parallels the pattern of the organic properties with high levels of exchange hydrogen over the gallery forest section and significant but less spectacular increases in exchange aluminium (Fig. 6.6b). As might be expected, the pH values follow the same pattern inversely, such that the forest areas are characterized by low pH (except for the high profile 1, transect 1 figures explained earlier) (Fig. 6.6c). Despite oscillations in values over the savanna (campo) zones and the overall acidity of all the soils, pH and total exchangeable acidity pick out the forest–savanna boundary clearly and are closely correlated with organic matter (in the correlation matrix not shown). Turning to the nutrient cations, Transect 2 illustrates an evident drop upslope over the forest–savanna body. This is less clear in Transect 2 where the soil boundary could be fixed either within the forest or at the two profiles immediately upslope (Fig. 6.6). This may be at least partly explained by the seepage and groundwater distribution but may simply reflect the natural movement of the boundary edge – which will not necessarily have a close relationship to soil properties unless a considerable period of vegetation stability has occurred.

A further way of exploring the extent of the 'savannization' process is to examine the sesquioxide variations and the degree of iron induration (see Chapter 11). Enrichment of less mobile aluminium and iron oxides should accompany exposure to an oxidizing drying environment characteristic of the markedly seasonal, open vegetation of the well drained cerrado sites. This may be observed in surface horizons, where the effect should be most clearly manifest (Fig. 6.6). Both Al_2O_3 and Fe_2O_3 increase from the forest to the non-forest areas, although the precise boundary is

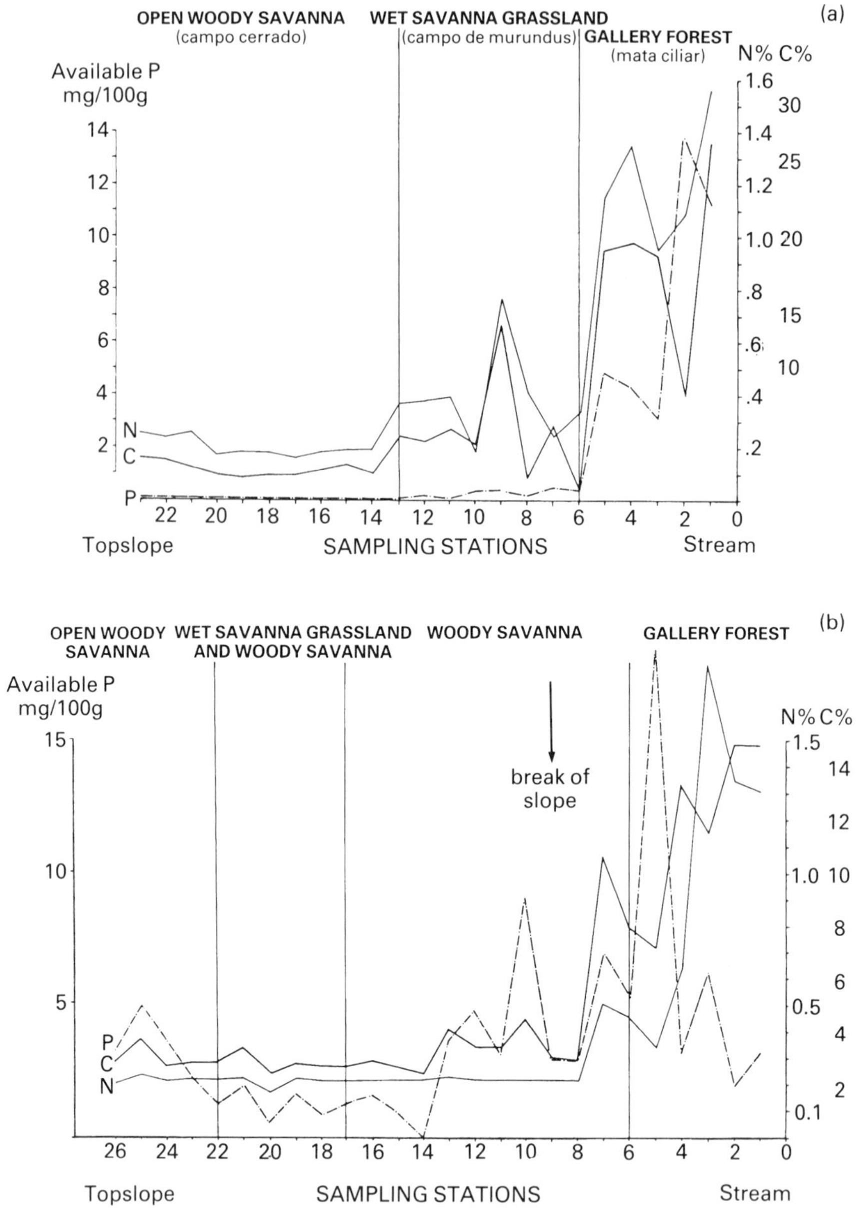

Fig. 6.6 Variations in surface soil properties over the forest–savanna boundary, Fazenda Água Limpa (Transect 1, upper; Transect 2, lower).

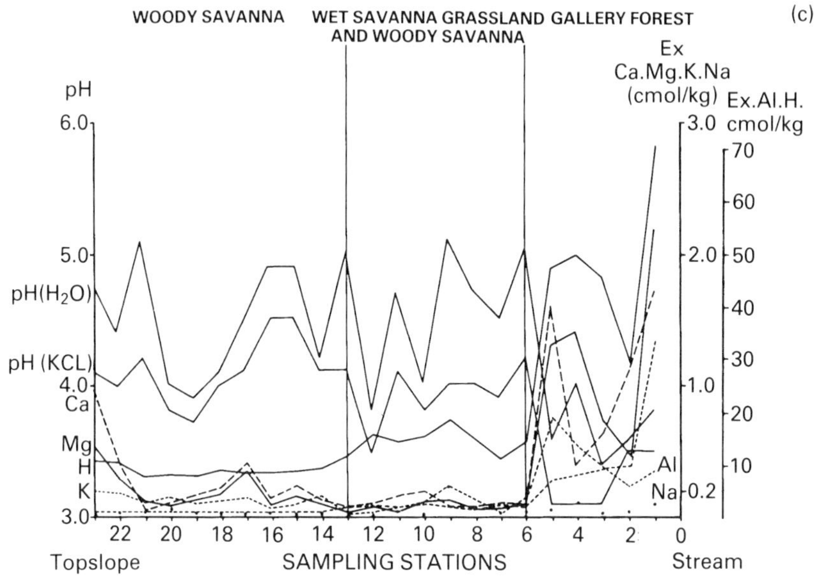

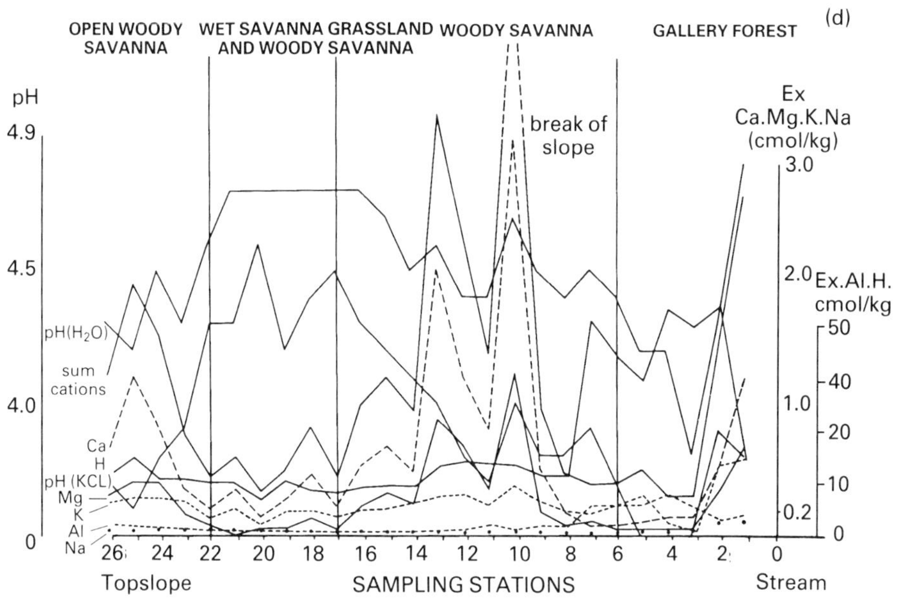

Fig. 6.6 *Continued*

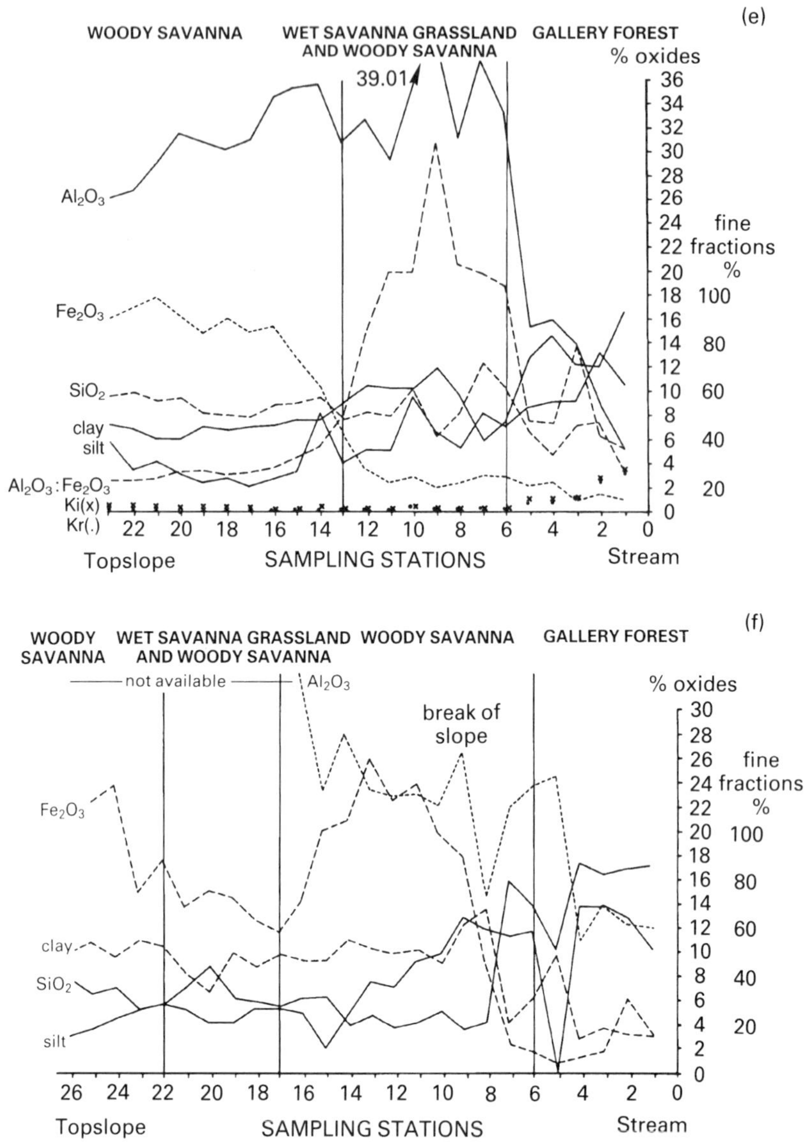

Fig. 6.6 *Continued*

less apparent in Transect 2 (similar to the distribution of other soil properties described above); however, the many seasonally damp sites of the campo de murundus in Transect 1 have depressed the Fe_2O_3 figures. The high temperature and more continuously moist environments may also account for the lower SiO_2 mean values in the forest as opposed to non-forest locations. These distributions are nevertheless unlikely to be related to vegetation so much as with changes in water regime over the transects. Indurated iron, in nodular or block form, was also observed only in areas of savanna. If, as suggested in this chapter and Chapter 20, the forest margin is advancing upslope at the expense of savanna, then it would be interesting to investigate possible changes in oxide state with redox variations. In the present case, the savanna formation is likely to be of considerable age and, therefore, the 'savannization' processes are well advanced. The pedogenesis of these savanna versus forest sites is insufficiently understood.

Variations in the soil particle sizes would not be expected to match vegetation changes although they might be expected to relate to slope in an old geomorphic landscape. The development of clay and colloidal fractions is known to take far longer, in pedogenic terms, than most of the soil properties considered so far. They would not therefore be likely to relate to shorter-term changes in the vegetation cover, assuming similar drainage conditions, unless those zones persisted for a very long time when weathering rates might conceivably have been higher in the intense decomposition environments of humid forest. At Fazenda Água Limpa, there are higher mean values for the silt fraction and lower for the clay fraction within the forest at both transects (Fig. 6.6). The degree of flocculation, illustrated for Transect 1, did not vary significantly. Although there are visible differences in the fine fraction distribution, they are difficult to interpret in relation to the forest–savanna boundary.

Thus soil properties influence both the possible shift of the forest–savanna boundary (more or less nutrient availability) and are in turn affected by the vegetation changes (as shown in the surface properties). In particular, soil-water availability over the course of the year seems to be a crucial factor in limiting the upslope migration of forest over savanna.

So far, the discussion has concentrated on the contemporary vegetation and soil boundaries and has not considered dynamic change. The question is whether the vegetation formations are mobile and have, in the past, shifted position. Equally, it would be valid to query whether soil properties may influence such changes. Unfortunately, there is little convincing palaeoecological evidence for the central cerrado regions of Brazil and various assumptions on the forest–savanna boundary fluctuations are usually based on tropical lowland forest (Graham, 1973, 1982), or fluctuations of the forest edge in the mountain regions as in the north

Andean area (Van der Hammen, 1982) with cooler temperatures and drier conditions attached to Pleistocene and humid, warmer conditions in Holocene (see, for example, Salgado-Labouriau, 1982). As a broad generalization, it might be assumed that arboreal and grass cerrado expanded at the expense of forest during the ice maxima in Pleistocene and that the Holocene, despite climatic fluctuations, has witnessed a gradual re-colonization by forest over the past 12 500 years. If this is the case and if such change is occurring today, we should find:

1. soil properties more characteristic of cerrado under contemporary forest; such properties could arguably be demonstrated near to and within the current edge in the examples described earlier; and
2. plants typical of the forest edge and cerrado found today within the forest and there are in fact indicator plants of this nature; the Buriti palm (*Mauritia flexuosa*) has mature specimens some 30–100 m within the gallery forest where they could not have become established (Ratter, 1985). This suggests that, at least over recent time, the forest edge has moved upslope by 30–100 m during the age of the trees (unlikely to be more than 200 years). Ratter *et al.* (1978b) also found old, giant trees of cerrado species deep within Amazonian forest in Mato Grosso (Chapter 20, p. 424).

South-eastern Mato Grosso

The forest–savanna boundary in the area around Torixoreu is located over a series of low hills, with the upper slopes clothed in seasonal deciduous forest and the lower ground consisting of open savanna (campo cerrado). Between these formations there is a narrow zone of characteristic dense arboreal savanna (*cerradão*). A toposequence was investigated in detail at Morro de Fumaça (Furley *et al.*, 1988) and represented a succession of vegetation formations typical of the region. The soil distribution patterns showed a clear association with changes in the plant communities (Fig. 6.7).

The forest zone on the Morro de Fumaça is dominated by *Anadenanthera peregrina*, a species known to be associated with soils of higher base status. The forest reaches to around 20 m with the taller trees being deciduous at the height of the dry season but retaining some understorey evergreen species such as *Casearia sylvestris*. Patches of dense bamboo and Marantaceae carpet the floor. Much of the forest floor vegetation is characterized as calciphilous and has been identified elsewhere in Brazil as being associated with calcareous soils (Ratter *et al.*, 1978a). The narrow band of floristically distinct arboreal savanna has been termed *Magonia pubescens – Callisthene fasciculata* cerradão (Ratter *et al.*, 1973, 1977) and represents a mesotrophic facies. Most of the species lose their leaves in

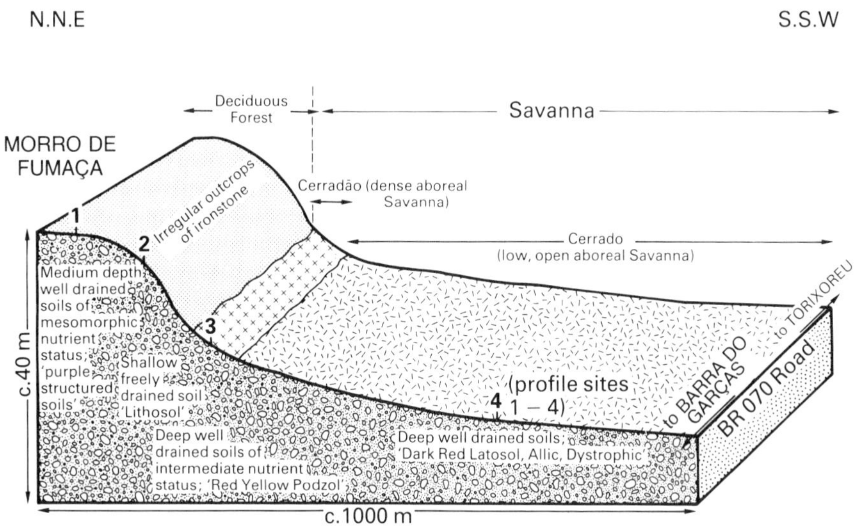

Fig. 6.7 The forest-savanna boundary over the Morro de Fumaça, eastern Mato Grosso (after Furley *et al.*, 1988).

the dry season. The cerrado vegetation found over the lower slopes was dominated by *Qualea parviflora* and *Lafoensia pacari*, etc. Most species in this formation are indifferent to base status but are usually associated with acidic soils, although there is some indication that the topsoils are ameliorated, possibly as a result of colluviation from the better soils upslope.

Unlike the examples discussed so far, the soils at Morro de Fumaça are all well drained across the toposequence. The convex, predominantly sandstone hills in the region give rise to shallow calcareous lithosols and concretionary soils. Plinthitic soils develop ironstone crusts locally but can also develop deeper profiles with argillic horizons. They have base saturation values over 50% at near neutral pH and high (17 cmol kg^{-1}) topsoil levels of exchangeable calcium. The forest–savanna boundary strictly speaking, lies at the junction of the forest and mesotrophic cerradão but the greatest pedological contrast occurs between the cerradão and the more open savanna (campo cerrado). The cerrado zone occurs over sandy plains with dystrophic (base saturation levels below 50%) and allic or aluminium-rich soils which are low in organic matter (Fig. 6.8).

In summary, the richer hilltop soils carry a thick cover of deciduous or semi-deciduous forest, whilst the poorer dystrophic soils downslope are able to support only an open arboreal savanna. The junction between the two can be clearly distinguished and relates to soil properties which are

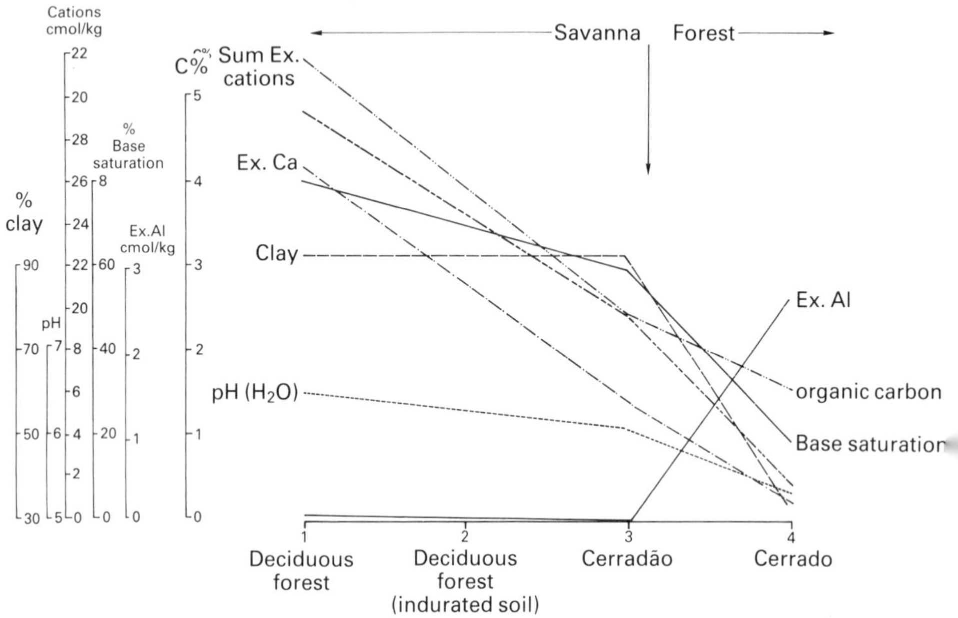

Fig. 6.8 Variations in soil properties over the forest–savanna boundary at the Morro de Fumaça (after Furley *et al.*, 1988).

partly a result of topography, which is in turn dominated by the change in parent material.

Northern Brazilian Amazon, Roraima

A number of studies were made of the forest–savanna boundary as part of the Maracá Rainforest Project (Furley, 1989; Milliken and Ratter, 1989). The most intensive studies were carried out at the eastern end of Maracá Island. These may be divided into two principal locations: forest-wet savanna transitions and forest-dry savanna island sites.

Forest-wet savanna transitions

Duplicate transects were sited across a grass and arboreal savanna depression (Santa Rosa) stretching from forest to savanna and back into forest (Fig. 6.9: sampling stations MS 3 to MS 13).

Some of the results of the forest transect, worked on by several researchers, are indicated in Chapter 18, whilst a replicate transect is being analysed by a team from the Instituto Nacional de Pesquisas da Amazônia (INPA) and are referred to in Chapter 8. A more complete interpretation of the edaphic transition at these sites will be presented at a later date but

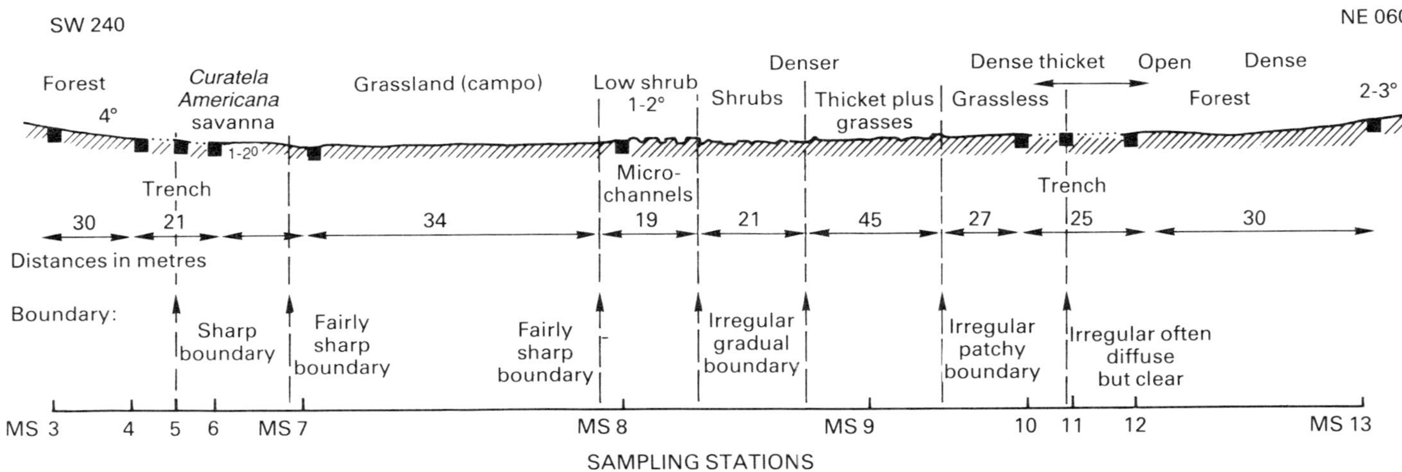

Fig. 6.9 The forest–savanna boundary at Santa Rosa, Maracá Island, Roraima (after Furley, 1989).

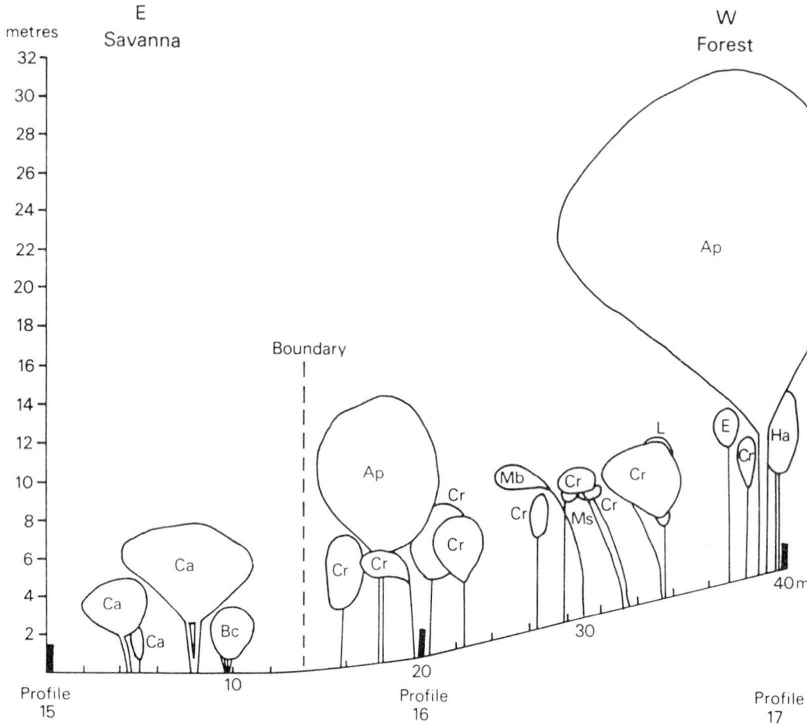

Fig. 6.10 The forest–savanna boundary at the 'Angico' transect (after Furley, 1989).

it would appear that seasonal flooding in the lower parts of the basin is responsible for inhibiting forest tree growth and conversely encouraging arboreal and grassy savanna. The forest–savanna boundary has been shown to relate to the upper limits of the seasonal water table (Furley and Ratter, 1990).

A second transect, known as the Angico transect after the dominance of *Anadenanthera peregrina* (Ap) in the forest section, provides a short but clear-cut boundary (Fig. 6.10). The site characterizes many of the features of the regional forest–savanna boundary in terms of its relation to slope and drainage. The boundary is exceptionally sharp with a wall of forest, fronting a short zone of *Curatella americana* arboreal savanna which in turn passes downslope into a wet grassland savanna (campo). The slope ranges from 23° under forest (profile 17) to 6° under arboreal cerrado at profile 15. A number of plants typify this boundary such as *Cupania rubiginosa*, the spiny *Randia armata*, *Bauhinia ungulata* and the climbing *Davilla kunthii*. The dense, low forest behind (reaching to around 10 m) is characterized by the Angico, *Anadenanthera peregrina*, *Cochlospermum orinocense* and *Didymopanax morototoni*. The composition of the forest seems to relate to the mesotrophic nature of the soil which in turn is associated with slightly higher gound and the likelihood of a different parent material from that of the coarse-textured deposits of the lower, wet grassland. The forest zone is characterized by high nutrient cation levels (especially calcium and magnesium) and high cation exchange capacities, high organic carbon and slightly elevated phosphorus levels, an increase in pH and depressed exchangeable acidity (Fig. 6.11a). It is interesting to note that there is a slight increase in iron content from savanna to forest – the reverse of the situation described for Fazenda Água Limpa. This may be readily explained by the fact that the Angico forest is well drained and situated above seasonally wet savanna, whereas at Fazenda Água Limpa the reverse vegetation and moisture patterns prevail. This Angico site is published in more detail elsewhere (Furley and Ratter, 1990) and will not therefore be discussed in detail here except to comment on the role of soil moisture. The contemporary forest edge matches exactly the observed maximum height of the water table which reaches the tree root zone during the one-year field monitoring and also the zone of gleying, as indicated by mottling, within the soil profile (Fig. 6.11b). The current slope form together with increasingly impeded drainage makes it unlikely that the forest will migrate downslope since the campo areas are too seasonally wet for semi-deciduous forest growth. However, with the onset of possible drier conditions (since this is also the regional forest–savanna boundary), it is possible to envisage an upslope colonization by the savanna.

Clearly in this example, the edaphic nutrient controls are significant but soil water status appear to be the predominant factor.

'Island savannas' and forest-dry savanna boundaries

During the course of digital image processing of Landsat TM wavebands 3, 4 and 5 on Maracá (Dargie and Furley, in press), it became apparent that there were numerous small isolated patches of savanna within the forest and also a number of linear connected patches of non-forest vegetation related to periodic streamflows (*vazantes*) which have been included in a broad definition of savannas (Milliken and Ratter, 1989).

Many of these savanna patches were too remote for practical access but the two major groups were examined along one of the cross-island trails. For the present, soil variation across these two type locations will be considered.

1. A traverse was made across two small campos (around 20 × 50 m in area) separated by dense thick/low forest (Fig. 6.12; key numbers 8 and 7).

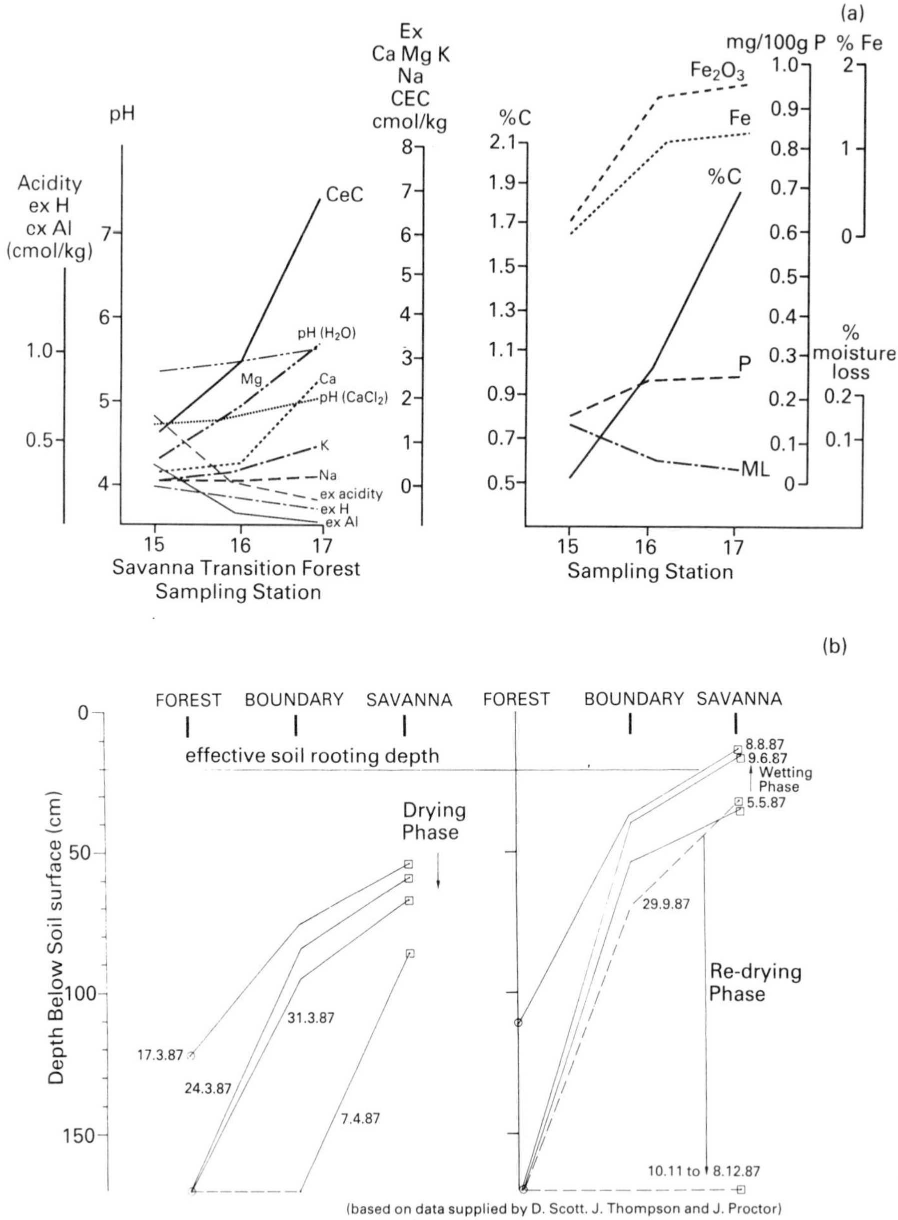

Fig. 6.11 Surface soil and moisture variations across the 'Angico' transect (after Furley, 1989).

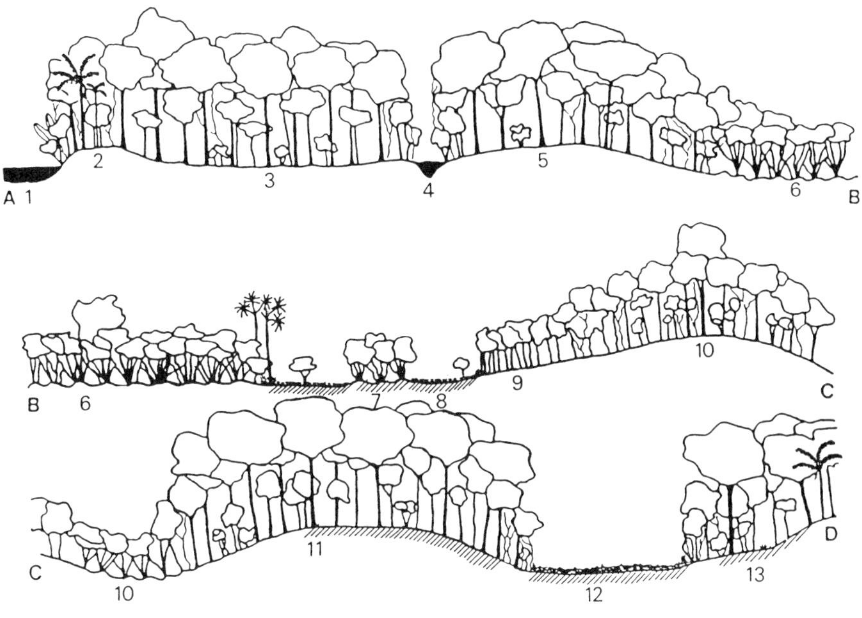

Fig. 6.12 The forest–savanna boundary along the southern section of the Preguiça Trail, Maracá Island (after Milliken and Ratter, 1989).

2. A further traverse cut across a vazante lying between tall forest with different dominants to either side (Fig. 6.12: key numbers 11, 12 and 13).

The data for these sites have not been fully analysed (notably the striking differences in soil texture) but the preliminary results confirm the distinct zonations of vegetation and soils. The pattern of surface soil variation is more clearly exemplified in the distribution of soil organic matter (Fig. 6.13 top). The two campo areas have extremely low organic carbon values and zero or nearly negligible nitrogen values. The intervening thicket/low forest has a very slightly greater level – nowhere matching the tall forest which in turn, has distinctly lower organic values than the vazante areas. Turning to the pH and exchangeable acidity (Fig. 6.13 bottom) it is obvious that the vazante stands out in terms of exchangeable aluminium and this influences the total exchangeable acidity. Exchange hydrogen is very similar over campo, thicket and tall forest, and not much different for the vazante sites. The pH values did not vary significantly, which is curious, and at present inexplicable in view of the high nutrient values (mostly calcium and magnesium) at the

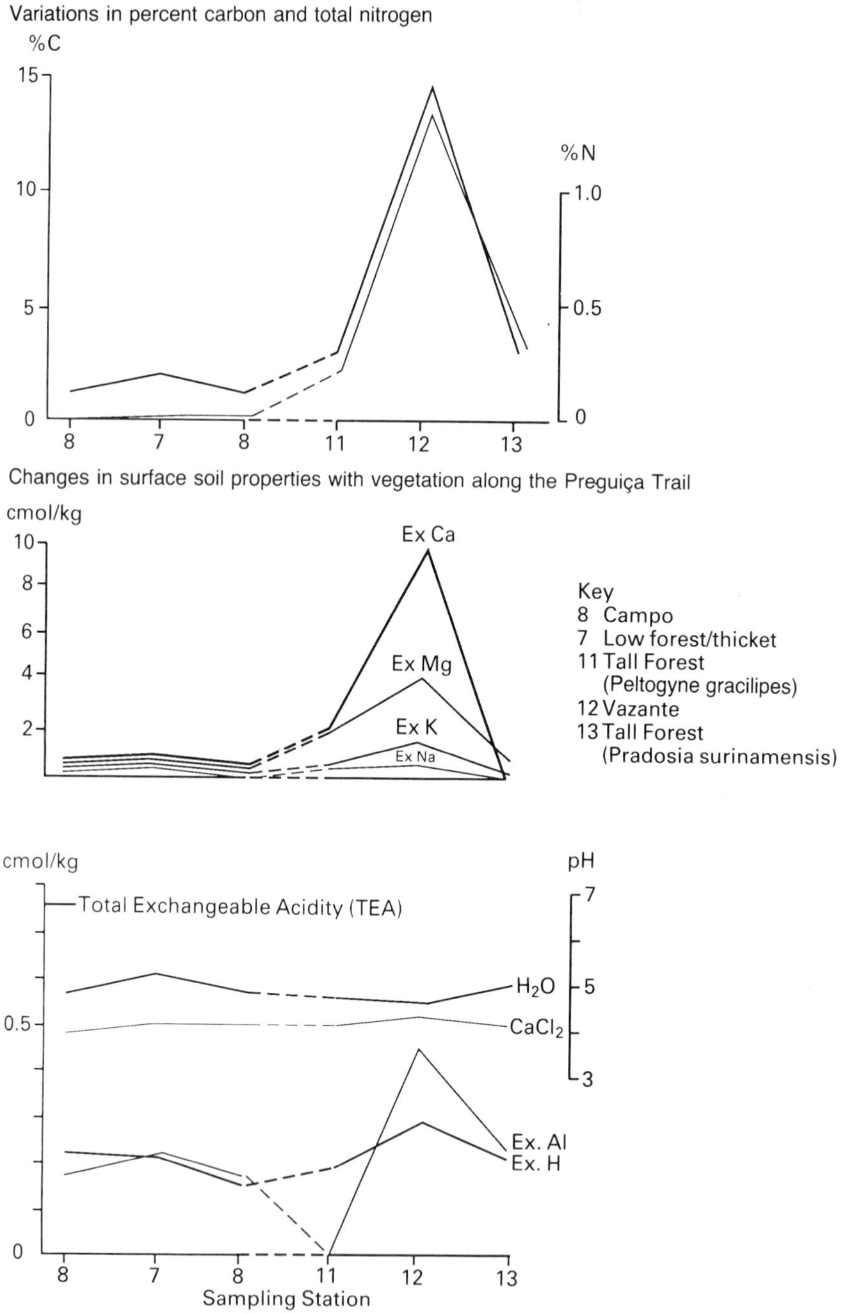

Fig. 6.13 Surface soil variations across the forest–savanna boundaries on the Preguiça Trail.

forest vazante sites (Fig. 6.13 middle), contrasting with exceptionally low figures for the campo areas and marginally better values for the intervening thicket.

In summary, these surface distributions pick out the mesotrophic nature of vazante soils, the intermediate soil status of the tall forest sites and the poor oligotrophic nature of the campo and thicket sites.

The two savanna locations are quite different both in soil, vegetation and origin. The vazante formation results from seasonally wet conditions, with large volumes of water being shed down the stream-line which dries out completely at the surface during the rainless period of the year. Such soils are widespread over the island and are dark and clay-rich (from field profile descriptions) supporting a dense herbaceous and low shrub vegetation 60–200 cm tall, made up predominantly of *Mimosa pellita*, and several other shrubs such as *Senna alata* with grasses such as *Eriochloa punctata*. As indicated in Milliken and Ratter (1989), these appear to be two plant components which reflect the origin of the formation: semi-aquatic species characteristic of the flooded run-off conditions of the wet season (e.g. *Thalia geniculata* and *Canna glauca*) alternating with dry-season tall grasses and herbaceous leguminous and convolvulaceous vines. The tall forest to either side is different in floristic composition and also to a lesser extent, in soil composition. All three sites are, however, distinctly richer than the three campo and thicket sites. The campo sites form strips partly consisting of open grassland and partly open arboreal cerrado (dominated by *Curatella americana* and *Byrsonima crassifolia*), and are extremely damp and sandy, with little surface organic matter. The intervening thicket is interlaced by a network of microchannels 30 cm deep which must be mostly flooded in the wet season. The thicket is very dense but low in height over damp ground and consists of species characteristic of the nearby campo areas.

Since the soils and plant species composition are essentially similar, it is suggested that the thicket may be evolving at the expense of the neighbouring campos which are therefore being closed, despite having inherently poor soils. It has also been observed elsewhere on the island that there are many strips of thicket either lying close to more open campo or where campo is not observed and cannot be detected on satellite images. It is believed that many of these localities may represent older savanna areas which have now become overgrown.

CONCLUSIONS

Five different localities have been studied in three dispersed regions ranging from Central America, the northern Amazon Basin to the central Brazilian plateau.

The toposequences investigated in Belize and in the Federal District of Brazil, demonstrate that distinctive surface soil properties occur in conjunction with particular vegetation zones. Sometimes properties precisely mirror the changes in plant communities, predictably those connected with soil organic matter, but also nutrient cations and exchangeable acidity. This correspondence is generally clear but not always exact. It may therefore be interpreted that these soil properties are related to, but cannot necessarily determine, the forest–savanna boundary. However, soil moisture variations, drainage and movements of the groundwater do frequently match the upslope limit of the gallery forest edge and may be considered as a dominant controlling factor.

The toposequence examined in the Mato Grosso illustrates a different edaphic control. In this case, changes in the underlying geology have affected topography and soil formation, which has resulted in variations in nutrient status, especially calcium. This distinguishes mesotrophic forest and cerradão from dystrophic cerrado.

The predominant form of the forest edge at Maracá Island has a number of pedogenic characteristics in common with those illustrated in the drainage *catenas*. Where the slope gradients are pronounced and where they are associated with seasonal drainage differences, then the soil and plant community relationships are very closely allied and the boundaries are extremely sharp. In other cases a more diffuse transition is apparent, usually allied to a more complicated pattern of soil property distribution.

Finally, the savanna islands within the forest of Maracá exhibit two distinct types of margin. The vazante soils stand out as mesotrophic and are reflected in a distinctive plant association, whilst the adjoining tall forest zones have an intermediate soil character. The campo savannas, on the other hand, are uniformly dystrophic, lacking organic matter and nutrient content and possessing very sandy textures.

The edaphic influence can be seen to be significant in each of the examples examined. Soil moisture is the determining factor in forest–savanna boundaries located on toposequences where there is a downslope increase in water content and where growth at the upslope forest margins is constrained by seasonal water deficiencies. Variations in soil water also determine the boundary between forest and hyperseasonal savannas. Higher base status accounts for the deciduous forest distribution where the cerrado margin is characterized by a drop in available bases, an increase in acidity and is frequently associated with coarse soil textures. Similar pedological characteristics accompany the occurrence of savanna islands within forests.

In each of these examples, evidence has been presented for forest or thicket advance upon savanna, whether arboreal or open grassland. If in the central Brazilian cerrado, rates of forest advance upslope of 30–100 m over some 200 years are postulated, then over relatively level ground in lowland tropical forest, characterized by a more diffuse and complex transition or mosaic (Ackerly *et al.*, 1985; Projecto Radam, 1980, 1981, 1982; Ratter *et al.*, 1973, 1978b) the boundary may be moving much more rapidly. In any case, outside these broad patterns, the local boundary at any given site seems to be stronly conditioned by the soil-water regime.

REFERENCES

Ab 'Saber, A.N. (1977) Espaços ocupados pela expansão dos climas seces na America do Sul, por ocasião dos períodos glaciais quaternários., *Paleoclimas*, **3**, 1–19.

Ab 'Saber, A.N. (1982) The paleoclimate and paleoecology of Brazilian Amazonia, in *Biological Diversity in the Tropics*, (ed. G.T. Prance), Columbia University Press, New York, pp. 41–59.

Ackerly, D.D., Thomas, W.W., Cid Ferreira, C.A. and Pirani, J.R. (1989) The rainforest–cerrado transition in southern Amazonia: Results of the 1985 Projecto Flora Amazonica Expedition in Mato Grosso. *Brittonia*, **41**, 113–28.

Askew, G.P., Moffat, D.J., Montgomery, R.F. and Searl, P.L. (1970) The interrelationships of soils and vegetation in the savanna–forest boundary of north east Mato Grosso. *Geographical Journal*, **136**, 371–6.

Beigbeder, Y. (1959) La région moyennes du Haut Rio Branco (Brésil), étude geomorphologique, *'Travaux et Mémoires de l'Institut des Hautes Ètudes de l'Amerique Latine*, Vol. 10, Institute of Advanced Studies in Latin America, Paris.

Bigarella, J.J. and de Andrade-Lima, D. (1965) Contribution to the study of Brazilian Quaternary. *Geological Society of America, Special Paper*, **84**, 433–51.

Bigarella, J.J. and de Andrade-Lima, D. (1982) Palaeoenvironmental changes in Brazil, in *Biological Diversity in the Tropics*, (ed. G.T. Prance), Columbia University Press, New York, pp. 27–41.

Bigarella, J.J., Mousinho, M.R. and da Silva, J.X. (1969) Processes and environments of the Brazilian Quaternary, in *The Periglacial Environment*, (ed. T.L. Pewe), McGill University Press, Montreal, 417–87.

Brown, K.S. (1987) Soils and vegetation, in *Biogeography and Quaternary History in Tropical America*, (eds T.C. Whitmore and G.T. Prance), Clarendon Press, Oxford, pp. 19–45.

Brown, K.S. and Ab 'Saber, A.N. (1979) Ice age refuges and evolution in the Neotropics. *Paleoclimas*, **5**, 1–30.

Dargie, T.C.D. and Furley, P.A. (in press) The pace of land cover change, in *The Rainforest Frontier; Settlement and Change in Brazilian Roraima*, (ed. P.A. Furley), Routledge, London.

Diniz, M. de Araújo Neto, Furley, P.A., Haridasan, M. and Johnson, C.E. (1986) The murundus of the cerrado region of central Brazil. *Journal of Tropical Ecology*, **2**, 17–35.

Eiten, G. (1972) The cerrado vegetation of Brazil. *Botanical Review*, **38**(2), 201–341.

Freire, E., Maues da Serra (1979) *Influência das propriedades do solo na distribuição de comunidades de vegetação em uma topossequência em área da 2ª superfície de erosão do planalto central Brasileiro, na Fazenda Água Limpa, DF*, unpublished Master's thesis, University of Brasília.

Furley, P.A. (1974a) Soil–slope–plant relationships in the northern Maya Mountains, Belize, Central America, 1. The sequence over metamorphic sandstones and shales. *Journal of Biogeography*, **1**, 176–86.

Furley, P.A. (1974b) Soil–slope–plant relationships over the northern Maya Mountains, Belize, Central America, 2. The sequence over phyllites and granites. *Journal of Biogeography*, **1**, 263–79.

Furley, P.A. (1976) Soil–slope–plant relationships over the northern Maya Mountains, Belize, Central America, 3. Variations in the properties of soil profiles. *Journal of Biogeography*, **3**, 313–19.

Furley, P.A. (1985) *Notes on the soils and plant communities of Fazenda Água Limpa, Brásilia DF*, Occasional Publications 5, Department of Geography, University of Edinburgh.

Furley, P.A. (1989) The soils and soil–plant relationships of the eastern sector of Maracá Island, in *The Vegetation of the Ilha de Maracá, First Report of the Vegetation Survey of the Maracá Rainforest Project* (INPA/RGS/SEMA), (eds W. Milliken and J.A. Ratter), Royal Botanic Garden, Appendix 5, pp. 229–76.

Furley, P.A. and Ratter, J.A. (1990) Pedological and botanical variations across the forest–savanna transition on Maracá Island. *Geographical Journal*, **156**, 251–66.

Furley, P.A., Ratter, J.A. and Gifford, D.R. (1988) Observations on the vegetation of eastern Mato Grosso, III. Woody vegetation and soils of Morro de Fumaça, Torixoreu, Brazil. *Proceedings of the Royal Society of London*, **B203**, 191–208.

Graham, A. (1973) *Vegetation and Vegetation History in Northern Latin America*, Elsevier, Amsterdam.

Graham, A. (1982) Diversification beyond the Amazon Basin, in *Biological Diversity in the Tropics*, (ed. G.T. Prance), Columbia University Press, New York, pp. 78–92.

Journaux, A. (1975) Recherches géomorphologiques en Amazonie bresilienne. *Bulletin Centre Géomorphologique*, Caen (CNRS), **20**, 1–67.

Milliken, W. and Ratter, J.A. (1989) *The Vegetation of the Ilha de Maracá,*

First Report of the Vegetation Survey of the Maracá Rainforest Project INPA/RGS/SEMA, Royal Botanic Garden, Edinburgh.

Oliveira-Filho, A.T. de and Furley, P.A. (1990) Monchão, Cocurutu, Murundu. *Ciência Hoje*, **11**(61), 30–7.

Projeto Radam (1980, 1981, 1982) *Levantamento de Recursos Naturais*, Volumes 20 (*Juruena*), 22 (*Tocantins*), 25 (*Goiás*), 26 (*Cuiabá*), DNPM, Rio de Janeiro.

Ratter, J.A. (1985) *Notas sobre a vegetação de Fazenda Água Limpa (Brásilia, DF, Brasil)*, Texto Universitário, Universidade de Brásilia, Brásilia.

Ratter, J.A., Richards, P.W., Argent, G. and Gifford, D.R. (1973) Observations on the vegetation of north-eastern Mato Grosso 1. The woody vegetation types of the Xavantina–Cachimbo Expedition area. *Philosophical Transactions of the Royal Society, London*, **B266**, 449–92.

Ratter, J.A., Askew, G.P., Montgomery, R.F. and Gifford, D.R. (1977) Observações adicionais sobre o cerradão de solos mesotróficos no Brasil Central, in *IV Simpósio Sobre o Cerrado*, (ed. M.G. Ferri), Universidade de São Paulo, Brazil, pp. 303–16.

Ratter, J.A., Richards, P.W., Argent, G. and Gifford, D.R. (1978a) Observations on forests of some mesotrophic soils in central Brazil. *Revista de Botânica*, **1**, 47–58.

Ratter, J.A., Askew, G.P., Montgomery, R.F. and Gifford, D.R. (1978b) Observations on the vegetation of north-eastern Mato Grosso 2. Forest and soils of the Rio Suiá-Missu area. *Proceedings of the Royal Society, London*, **B203**, 191–208.

Salgado-Labouriau, M.L. (1982) Climatic change at the Pleistocene–Holocene boundary, in *Biological Diversity in the Tropics*, (ed. G.T. Prance), Columbia University Press, New York, pp. 74–7.

Van der Hammen, T. (1982) Palaeoecology of tropical South America, in *Biological Diversity in the Tropics*, (ed. G.T. Prance), Columbia University Press, New York, pp. 60–6.

7

Soil and litter nutrient losses in forest clearings close to a forest–savanna boundary on Maracá Island, Roraima, Brazil

S.M. ROSS

SUMMARY

1. This chapter reports the results of an experiment to examine litter decomposition and the soil and nutrient losses under fully cleared, partially cleared and undisturbed terra firme forest close to the boundary between forest and savanna on Maracá Island in northern Brazil.
2. Rates of soil erosion after clearance were up to eight times higher than those after partial clearance and under undisturbed forest. Rates of leaf litter decomposition were not significantly different between treatments, but the rate of litter removal from the soil surface by erosion processes was significantly higher after full forest clearance.
3. During soil erosion, nitrogen, calcium and potassium were all lost in significantly higher quantities than sodium, phosphorus and magnesium. The order of nutrient loss from totally cleared sites was: nitrogen > potassium > calcium > phosphorus > magnesium > sodium.
4. Nitrogen losses of 90% and between 25 and 60% of the phosphorus losses occur in organic form in eroded leaf litter and organic matter fragments. Only sodium was lost predominantly in soil mineral material.
5. The rapidity and magnitude of soil degradation on fully cleared slopes, compared to undisturbed forest, over very short time periods and individual storm events, show how quickly soil changes can occur in these areas.

6. The significance of leaf litter as a nutrient reserve and as protection against soil erosion is indicated, and the importance is stressed of the need to retain leaf litter and soil organic matter in the soil after deforestation if soil degradation is to be avoided.

INTRODUCTION

The objectives of the study reported in this chapter are to assess the effects of forest canopy removal on soil erosion and nutrient losses under terra firme forest. A subsidiary objective, although the time period of this study was short, was to assess the likelihood of soil degradation, caused by forest clearance, close to the forest–savanna boundary, tipping the balance between forest regeneration and the invasion of savanna species.

Perhaps the clearest approach to understanding the location and form of the transition zone between forest and savanna grassland is to examine the different scales at which various controlling factors operate. Three hierarchical scales for consideration (in decreasing strength of importance) are climatic, edaphic and anthropogenic. Climate operates primarily through severe soil moisture deficits which limit tree growth in the dry season; the edaphic scale is controlled in part by regional hydrological characteristics, and in part by local geology and geomorphology, and affects the physical and chemical properties of soil; and, finally, the anthropogenic scale, primarily through the agents of fire and cultivation, affects tree seedling regeneration. Hopkins (1983) views these controls as operating, in the case of climate and soils, on *primary* succession. This type of succession would be rare at the forest–savanna boundary. Fire and cultivation operate on *secondary* succession, which is a much more common form of man-induced change, often in association with the hierarchical controls of climate and soils.

The balance between survival of forest and survival of savanna can be viewed from different points of view. Firstly, from the point of view of savanna survival, forest regeneration and encroachment is adversely affected by fire, cultivation and drought. From the point of view of forest survival, a sufficiency of soil water and nutrients is required, although a lack of waterlogging is also important. Forest clearance may lead to soil organic matter and nutrient losses, followed by soil degradation, with poorer soil conditions for supporting rainforest. The low stature and species diversity of campina forest on extremely nutrient-poor quartzitic soils (e.g. Anderson *et al.*, 1975) compared to terra firme forest on adjacent latosols is testimony to the effect that soil might have on forest growth. This chapter is concerned with the soil degradation due to tree removal at the forest–savanna boundary. An important question to ask is whether soil and nutrient loss due to forest disturbance depletes the soil to such an extent that regenerating tree seedlings could be out-competed by influx of less nutrient-demanding savanna species. In parts of the forest–savanna transition zones on Maracá Island, Northern Roraima, Furley and Ratter (1990) have suggested that the opposite migration has been in operation: that of forest advance into the savanna, but that current vegetation boundaries are defined by soil hydrological conditions, not soil infertility. They note the presence of savanna trees about 25 m inside the forest, suggesting that, without fire, the forest is extending and invading the savanna. Current halting of the migration of the boundary, they suggest, is due to soil saturation of the hyperseasonal Maracá savanna, which inhibits further forest advance.

Two component elements of soil degradation after forest disturbance are physical soil loss through overland flow and erosion, and the change in availability and rates of leaf litter decomposition at the soil surface after canopy removal. Soil erosion after forest clearance is a well documented problem. It is also considered to be one of the main reasons for soil impoverishment after forest clearance. Two main factors control the rate of soil and nutrient loss: the availability of erodable material and the power of the transporting agency. In coarse and poorly consolidated soils, the major limitations to soil loss are rainfall and runoff energy. Both are significantly altered by vegetation removal which reduces the extent to which raindrops are intercepted and their energy dissipated before they reach the soil surface. Elwell and Stocking (1976) found that after 70% of the vegetation cover was removed, soil erosion was dramatically increased. The presence of an understory canopy and leaf litter at the soil surface reduces erosion due to raindrop splash (Wiersum, 1985), so their removal plays a vital role in enhancing soil erosion (Brandt, 1988).

The geometric arrangement of fallen leaves on the forest floor acts both as protection of the soil surface and as a nutrient reserve, or nutrient conservation mechanism (Herrera, 1978). With the loss of vegetation and litter after forest disturbance, nutrients are no longer returned to the soil through decomposition processes. Brinkmann (1985), working in terra firme forest near Manaus, showed that the amount of nutrients returned to the soil in litterfall was higher than from any other input source, amounting to 72.4% for nitrogen, 58.3% for phosphorus, 52.8% for calcium and 62.2% for magnesium.

Several methods are available for assessing leaf litter decomposition and nutrient release rates. The two most generally used techniques are *k*-values (litterfall/standing crop quotients) (e.g. Anderson *et al.*, 1983; Heaney and Proctor, 1989) and weight loss in litter bags (e.g. Ewel, 1976; Lowman, 1988). Other techniques have included the use of individually tethered leaves (Gong and Ong, 1983), or the laying out of discs of leaf material (Heath and Arnold, 1966). In a critical review of several techniques for assessing decomposition processes and rates, Anderson and Swift (1983) warn of the over-simplification of decomposition processes that each individual method represents. It is rarely possible to calculate k-values for total litter in terrestrial systems and usually a k_L-value for aboveground litter is measured instead (Swift *et al.*, 1979). Turnover rates calculated in this way represent decay under 'steady-state' conditions. Shorter-term measures of litter decomposition have frequently been sought, either to characterize different seasonal periods thought to deviate from 'steady-state' conditions, or to provide more rapid comparisons of soil biological activity between different habitat or treatment sites. In this study, the litter bag technique was chosen as a relatively simple, although limited, comparative index of decomposition rates after different kinds of forest clearance. Weight and nutrient loss from litter bags is due to the sum of three component processes: physical litter comminution, microbial catabolism and leaching of water-soluble materials (Swift *et al.*, 1979). In the case of comminution and leaching, the nutrients contained in particulate matter and solutes enter the soil directly. In the case of catabolism, faecal materials and decay of bodily tissues provide a readily mineralizable source of nutrients, particularly nitrogen, for root uptake and recycling to vegetation. Singh *et al.* (1989) has shown that microbially immobilized nitrogen and phosphorus provide important, readily mineralizable inputs of these nutrients to both tropical forest soils and adjacent savanna soils. Several authors (e.g. Herrera *et al.*, 1978) have suggested that some of these decomposition and mineralization processes are short-circuited in the litter layer by the growth and ramification of feeding roots onto decaying leaf surfaces where direct absorption or mycorrhizal transfers operate.

Three main types of variable control the rate of decomposition processes: the physico-chemical environment, the resource quality and the number and type of decomposer organisms (Swift *et al.*, 1979). Anderson and Swift (1983) suggest that these variables operate in a hierarchical way, in decreasing rank order: macroclimate > microclimate > resource quality > organisms. After forest clearance, there is likely to be a change in microclimate, resource quality (and quantity) and decomposer micro-organisms.

The aim of this study was to examine rates of nutrient loss from both decomposing leaf litter on the forest floor and in eroded soil materials after tree and vegetation removal. Soil erosion has been targeted as the prime reason for soil degradation after deforestation. This study aimed to quantify soil mass and nutrient losses resulting from different degrees of forest removal on slopes of 13°–16°. Leaf litter decomposition studies were used initially to assess change in soil surface protection after tree canopy removal, but also to allow quantification of relative rates of nutrient release after different types of forest clearance.

The location used for this study, the Brazilian Environment Secretariat (SEMA) Ecological Reserve of the Ilha de Maracá in northern Roraima, lies close to the northern limit of tropical rainforest in the Amazon Basin. In north-eastern Roraima, close to the Guyanan and Venezuelan borders, there is an interdigitation of forest and savanna, with *Curatella americana*/*Byrsonima crassifolia* – type *campo* extending north and eastwards over a wide area into the Rio Branco–Rupununi region, Sarmiento (1983). The countryside surrounding Maracá, and Maracá Island itself, has areas of tall and short grass savanna, as well as wooded savanna dominated by *Curatella americana*, of the type described by Sarmiento (1984) as *campo sujo*. Many of these areas are typified by coarse-textured quartzitic soils with either thin layers of surface organic matter, or no obvious surface organic matter at all. In some areas, plinthitic layers are present at depth, becoming exposed at the surface where soil erosion has occurred (Nortcliff and Robison, 1989).

MATERIALS AND METHODS

The predominant soil type of the forest in the eastern part of Maracá Island is coarse textured and quartzitic, described by Nortcliff and Robison (1989) as a *Grossarenic plinthic Paleudult* (USDA classification), with a very thin layer of surface organic matter, only a small amount of surface litter and rarely any identifiable root mat at the soil surface. The forest vegetation at the eastern end of the Island was described generally by Milliken and Ratter (1989) and in more detail for the Santa Rosa savanna region in chapter 18. Near the experimental plots, the tree canopy was about 25–30 m tall, with a significant number of *Astrocaryum* palms. The ground vegetation was made up of *Calathea*, other Marantaceae, and Bromeliaceae. About 25–30% of the ground surface was unvegetated and covered by leaf litter.

Three plots, each 15 × 20 m, were set up on each of three slope transects, representing undisturbed forest, partially cleared forest and fully cleared forest. On each transect, the three plots represented top of slope, mid-slope and foot of slope locations (Fig. 7.1), with maximum mid-slope angles of 12–13°. During full clearance, all trees, vegetation and leaf litter at the soil surface were removed. During partial clearance, all trees were cut and vegetation removed above 1.5 m, but below this, all vegetation and leaf litter were left *in situ*. Treatment plots were set up in May 1987, at the beginning of the wet season.

Weight loss and release of nitrogen, phosphorus and potassium from

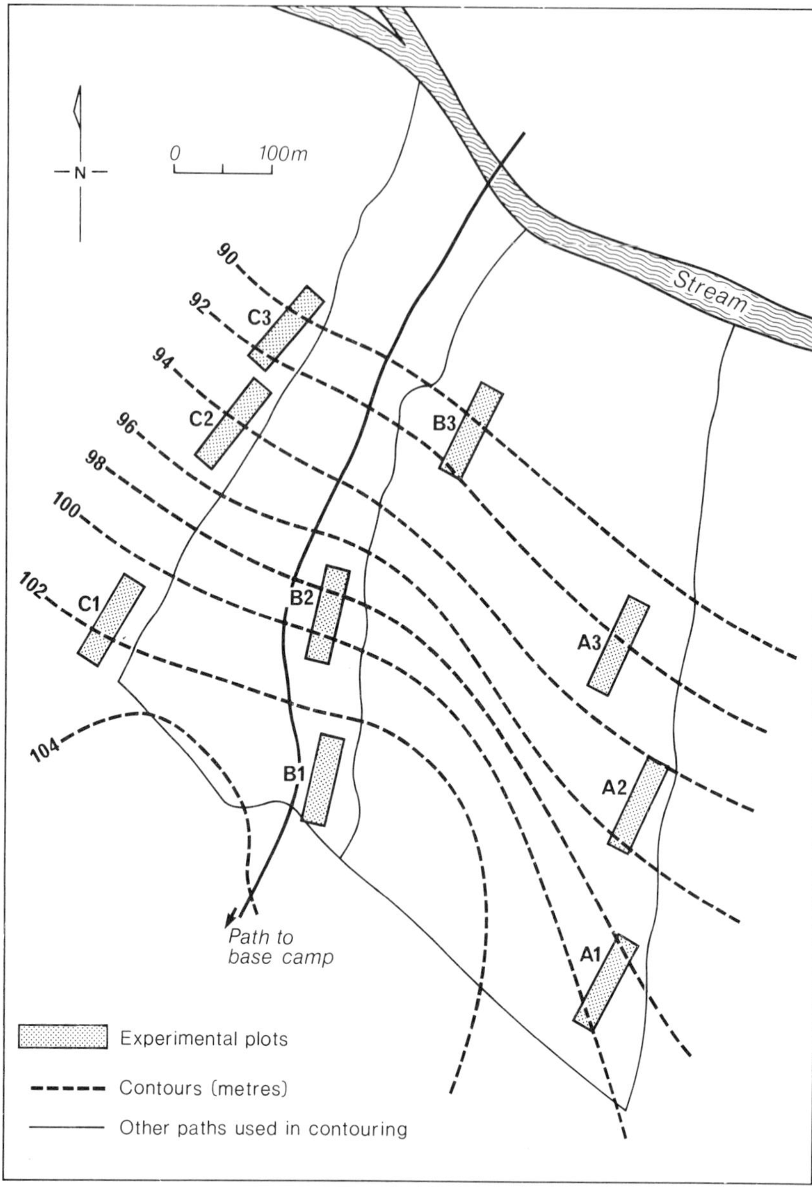

Fig. 7.1 Plan of field plots on Maracá Island, showing three downslope transects of three plots each, on terra firme forest (a), partial clearance (b) and total clearance (c).

leaf litter in nylon mesh bags were determined for two common forest tree species, *Tetragastris* sp. and *Ecclinusa guiaensis*, over a period of three months. Leaves were collected in June before abscission, their surfaces very gently moistened to remove dust particles, then air-dried for three days. Nylon mesh bags, approximately 200 × 150 mm in size, with a 2 mm mesh size, were made and filled with about 4.5 g of air-dried leaf material. Four rows of larger holes, 4 mm in diameter, were pierced in the top surface of each bag to ease entry of larger soil mesofauna. The litter bags were placed in random locations in each of the forest, partially cleared and fully cleared plots, so that their lower surfaces touched the soil and their upper surfaces were level with the top of the litter layer. Each bag was tethered using one wire pin in one corner and flagged for identification purposes. Litter bags were sampled after 30, 60 and 90 days, and their weight loss and total nitrogen, phosphorus and potassium contents determined after microdigestion (Allen, 1989) using the techniques described below for soil chemical analysis.

A series of instruments for monitoring surface runoff and soil hydrology (Nortcliff *et al.*, 1989), and a series of experiments for studying soil erosion, leaf litter decomposition and soil nutrient status, were located within each of the nine plots. Slope/soil conditions, rainfall, soil hydrology and soil erosion results for the period of March to November 1987 are reported in Nortcliff *et al.* (1989). The losses of soil nutrients in eroded sediments from each treatment are reported in the present study for a period during the wet season, from July to September 1987.

A gerlach trough, 2 × 5 m, was constructed on each treatment plot to collect runoff waters and eroded sediments. This is a bounded plot, running perpendicular to the contours of the slope, with a collection channel along the lower end for intercepting runoff waters and sediments and conveying them to a collection vessel. Volumes of runoff waters were measured every two days during May to October 1987. Eroded mineral sediments, leaf litter and particulate organic matter was collected at the outfall from each gerlach trough using a nylon net bag of mesh size around 100–250 μm. Three fractions of accumulated eroded sediments (two mineral fractions: >500 μm, <500 μm and particulate organic matter >500 μm) were measured every two weeks for three months during July to September. Mineral fractions were separated using wet sieving, while particulate organic matter was separated by hand from all samples.

Laboratory methods

Soil and leaf litter samples were oven dried at 105 °C and returned to the UK for chemical analysis. Total nitrogen, phosphorus, potassium, sodium, calcium and magnesium were determined in each of the sediment fractions after microdigestion using concentrated sulphuric acid and hydrogen peroxide (Allen, 1989). Nitrogen was determined colorimetrically as ammonium nitrogen using an automated version of the method described by Crooke and Simpson (1971). Phosphorus was determined colorimetrically using an automated version of the technique described by Murphy and Riley (1962). Potassium and sodium were determined directly by atomic emission, while calcium and magnesium were determined by atomic absorption spectrophotometry after the addition of lanthanum chloride releasing agent.

In situ extractable ammonium and nitrate nitrogen in the surface 5 cm of soil was determined on site in Brazil using a DREL portable environmental laboratory manufactured by the HACH Chemical Company, USA, which employs the Nessler and cadmium reduction colorimetric methods for NH_4-N and NO_3-N determination respectively. These methods were subsequently calibrated using the standard laboratory colorimetric methods of Crook and Simpson (1972) for NH_4-N and Henriksen and Selmer-Olsen (1970) for NO_3-N.

RESULTS AND DISCUSSION

Leaf litter decomposition

Inspection of litter bags returned to the laboratory after 30, 60 and 90 days in the field plots showed that litter from the two tree species behaved differently. The highly sclerophyllous and brittle *Tetragastris* litter was very fragmented, with virtually no intact leaves after 30 days. *Ecclinusa* litter, on the other hand, retained an intact leaf shape for longer, but showed much evidence of fungal and microbial attack. One surprising overall result of the litter bag experiment was that while significant differences in decomposition and nutrient loss were observed between the two leaf species, there were no significant differences between slope locations and between virgin forest and different clearance methods. The percent weight loss measured over a three month period from June to September, averaged from all slope locations for each treatment during the wet season, is shown in Fig. 7.2. Although none of the results are significantly different ($p < 0.05$), it is worth describing the trends in litter breakdown. *Ecclinusa* litter shows slightly faster weight loss than *Tetragastris,* so that, after three months, 37–38% of the *Tetragastris* litter had disappeared, while 41–43% of the *Ecclinusa* litter had disappeared, giving organic matter half-lives for the two species of greater than 100 days for *Tetragastris* and around 100 days for *Ecclinusa*. For both litter types, fully cleared plots showed slightly slower rates of weight loss than forest or partially cleared

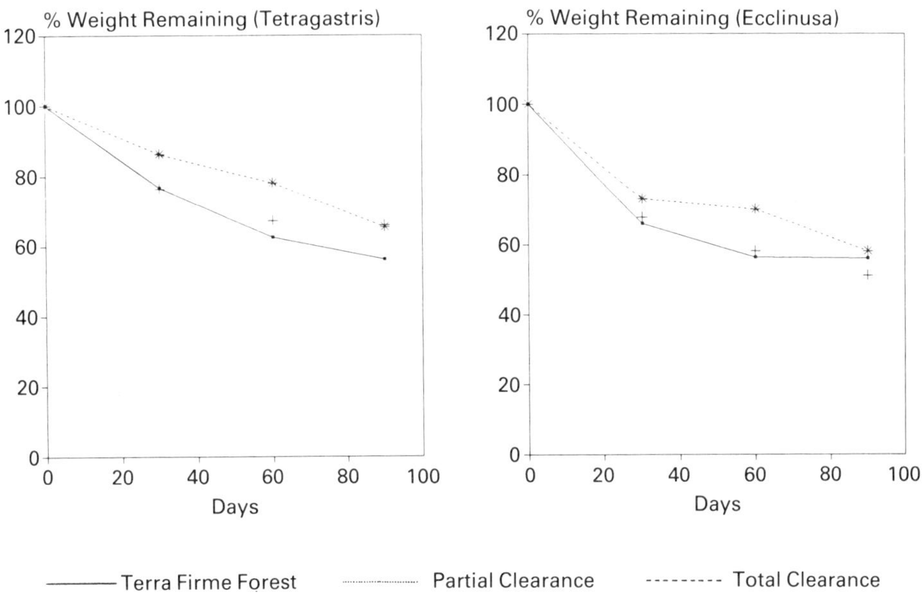

Fig. 7.2 Percent weight remaining in litter bags at 30, 60 and 90 days for *Tetragastris* and *Ecclinusa* leaves, in terra firme forest, partial clearance and total clearance.

sites. This result could be due to three factors. Firstly, since all leaf litter at the soil surface was removed regularly, this probably caused the soil decomposer organisms to abandon the site, resulting in slower decomposition rates once the litter bags were placed in the field. Secondly, since the enhanced raindrop splash and overland flow observed on fully cleared plots caused leaf material in litter bags to become plastered with sediment, slight errors in both mass and chemical determinations may have arisen where it was impossible to brush off all sediment particles prior to litter analysis. A third contributing factor could have been reduced palatability of sediment-plastered litter on fully cleared sites. Other studies have reported slower litter decomposition rates in clearings compared to under forest (e.g. Ewel, 1976). As in Ewel's study in Guatemala, decomposition appears to be retarded by vegetation removal. The intense daytime drying and wide diurnal temperature range provide a harsh environment for soil organism activity. In this respect, forest clearings are similar to the adjacent savanna grassland. Madge (1969) also found slower rates of litter decay in savanna compared to forest sites.

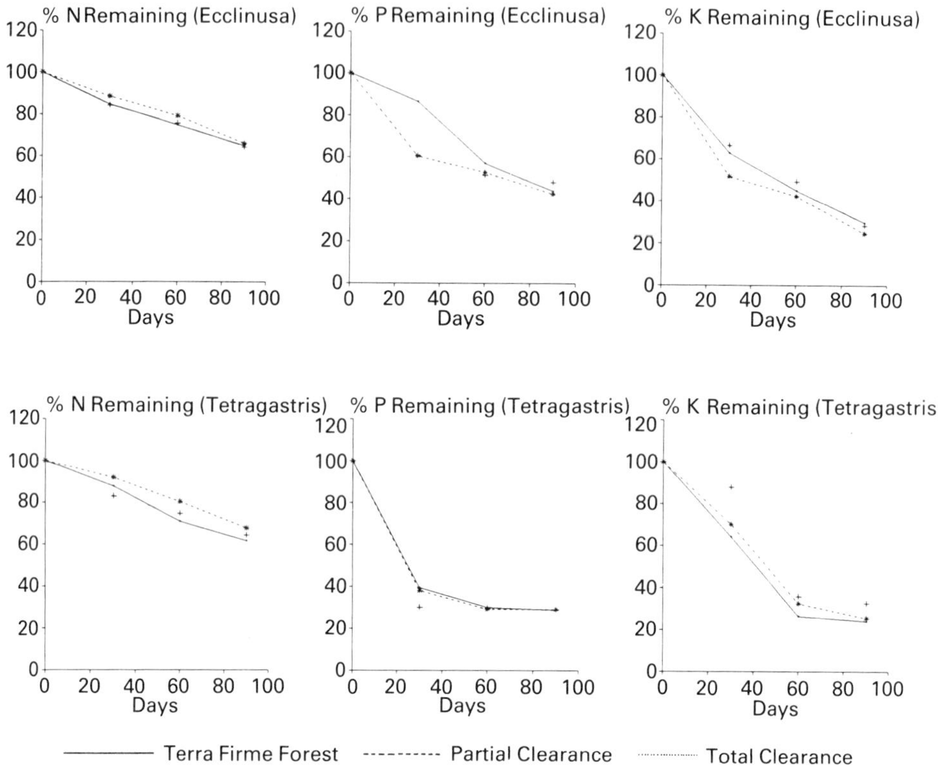

Fig. 7.3 Percent N, P and K remaining in litter bags at 30, 60 and 90 days for *Tetragastris* and *Ecclinusa* leaves, in terra firme forest, partial clearance and total clearance.

Leaf litter nutrient loss

Nitrogen, phosphorus and potassium losses from litter bags are illustrated in Fig. 7.3, using means of all treatment plots, irrespective of slope location. Identical, linear rates of nitrogen loss, of around 10% per month, were observed in both litter types (Figs 7.3a and d), resulting in around 30% nitrogen loss from all leaves after three months. The pattern for phosphorus loss from *Tetragastris* litter is quite different from that of *Ecclinusa* litter. In the first 30 days, *Tetragastris* leaves show a sharp loss of as much as 75% of the initial litter phosphorus. There is very little subsequent phosphorus loss from the litter. *Ecclinusa* litter shows a slightly more gradual loss of around 45–55% initial phosphorus after 60 days, falling to around 50–60% loss after 90 days. *Ecclinusa* litter on the virgin forest plots shows a significant retention of phosphorus after 30 days compared to litter under both types of forest clearance. Potassium loss from both litter types is more variable than for nitrogen and phosphorus, with slightly faster loss of potassium from *Ecclinusa* litter in the first 30 days. Both litter types show more gradual potassium losses after 60 days, to end with around 70% of the original litter potassium lost from both species after 90 days. Comparing nutrient loss half-lives, both leaf types show nitrogen half-lives longer than 100 days. The half-lives for phosphorus are quite different, at only 20 days for *Tetragastris* and 90 days for *Ecclinusa*. Potassium half-lives were 40 days for *Tetragastris* and 50 days for *Ecclinusa*. These decay rates are similar to those reported for litter decomposition in Guatemalan rain forest: nitrogen half-lives of around 180–200 days, phosphorus half-lives of around 90 days and potassium half-lives of around 60 days (Ewel, 1976).

If litter nutrient loss rates are compared to litter weight loss rates, both species show a relative retention of nitrogen (faster weight loss than nitrogen loss) and relative loss of phosphorus and potassium (faster nutrient loss than weight loss). Faster nutrient losses than weight losses suggest that nutrients are being leached from leaf litter rather than mineralized. The largest nutrient losses are of phosphorus and potassium from *Tetragastris* litter (75% in 30 and 60 days respectively), and for potassium from *Ecclinusa* litter (75% in 90 days). The relative retention of nitrogen in decomposing leaf litter, the slightly faster loss of phosphorus and the rapid loss of potassium were also noted by Medina and Cuevas (1989), working in different nutrient-poor forest types at San Carlos in southern Venezuela and similarly by Luizão and Schubart (1987) in decomposing litter of tropical forest on yellow latosols near Manaus. Using k_L-value estimates, Heaney and Proctor (1989) found rather similar nutrient release patterns, with nitrogen retained in litter and phosphorus and potassium lost more rapidly. Although much attention has been focused on nitrogen in tropical forests and on nitrogen deficiencies after forest clearance, the loss of soil phosphorus and the likelihood of phosphorus deficiency is now attracting attention.

Faster nutrient losses in the earlier stages of litter decomposition, illustrated in this study for phosphorus and potassium, and illustrated elsewhere in Amazonia for calcium and magnesium loss from decaying litter (Medina and Cuevas, 1989; Luizão and Schubart, 1987), indicate rapid nutrient losses from cleared sites immediately after tree removal, when the litter which was put in the sites just before or during disturbance decomposes. It would also appear that nutrient depletion on a longer time-scale after clearance is more likely associated with a curtailment of litter influx rather than with slower nutrient mobilization due to significantly reduced rates of litter decomposition.

Several authors have described the invasion of roots into leaf litter bags during decomposition experiments and have suggested that this illustrates one aspect of the very rapid cycling and conservation of nutrients that take place in the litter layer (e.g. Herrera *et al.*, 1978; Luizão and Schubart, 1987). Despite the general lack of a root layer in Maracá surface soils, litter bags in the present study did contain roots.

Nutrient losses in eroded sediments

Field observations showed greater volumes of surface runoff and larger amounts of eroded sediments on fully cleared plots, particularly in mid-slope positions. While Ross *et al.* (1990) showed that rates of runoff and erosion under virgin forest and under partial clearance were relatively small and not significantly different from each other, total forest clearance on mid-slope positions showed runoff rates about five times higher and sediment erosion rates about seven times higher. Masses of fragmented leaf litter and particulate organic matter eroded from all plots under all treatments were much lower than masses of mineral sediments (Fig. 7.4). At the top of slope and at mid-slope locations, significantly greater quantities of organic matter were eroded from virgin forest and partially cleared plots than from totally cleared plots. Over time after clearance this is an expected result since there should be no further litter influx to fully

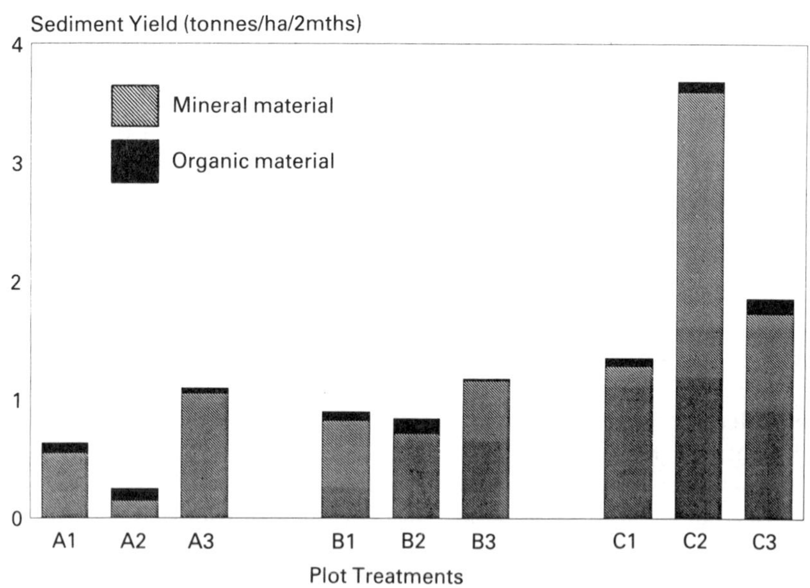

Fig. 7.4 Amounts of mineral and organic soil material eroded from top (1), mid (2) and foot (3) slope locations of terra firme forest (a), partial clearance (b) and total clearance (c), over a two month period from July to September.

cleared plots, while forest and partially cleared plots maintain litterfall. Over the two-month period from July to September, the proportion of soil material eroded in organic form ranged from 4 to 20% of the total mass of eroded sediments in virgin forest, from 5 to 10% on partially cleared plots and from 0.5 to 2.5% on totally cleared plots.

Total amounts of nitrogen, phosphorus, potassium, sodium, calcium and magnesium eroded from all plots are calculated in $kg\,ha^{-1}$ in Fig. 7.5 for the two month period from July to September. Not surprisingly, highest amounts of all nutrients were lost from fully cleared plots at mid-slope and foot of slope locations. All undisturbed forest and partially cleared plots show relatively small losses of nitrogen, phosphorus, potassium and magnesium, even at steepest (mid-slope) sites, compared to totally cleared plots. Partially cleared plots show higher losses of sodium and calcium which are more similar to undisturbed forest than to fully cleared treatments. As much as 4 kg nitrogen ha^{-1}, 0.7 kg phosphorus ha^{-1} and 6.5 kg potassium ha^{-1} can be washed from totally cleared sites in just two months of the wet season. Nitrogen is predominantly lost in eroded organic matter. Under undisturbed forest and partial clearance, 90% of nitrogen losses are in organic form (Fig. 7.6). After full clearance, as little as 30% of the nitrogen loss is in organic form. This result confirms the laboratory observation that sediments eroded from fully cleared plots contain large amounts of very finely fragmented organic matter, smaller than 500 μm, which was impossible to separate by hand from the mineral fraction prior to digestion. Clearly a more efficient mineral soil/organic matter separation technique is required. Lower, more variable, amounts of eroded phosphorus are in organic form, ranging from 50% under forest and partial clearance, and falling to 20% after total clearance. Only 25–30% of potassium lost from forest and from partially and totally cleared sites is in organic form. Very small amounts of sodium are lost in organic form from any site, but a surprisingly high proportion of the magnesium and calcium losses are in organic form, up to 80% for magnesium from forest and partially cleared sites. These results indicate the importance of employing site management practices after forest clearance to retain organic leaf litter at the soil surface and thus help to maintain soil fertility.

Summing the total elemental contents of all eroded sediments and organic matter indicates that nitrogen, calcium and potassium are lost in higher quantities than sodium, phosphorus and magnesium from all treatment plots (Table 7.1). On both virgin forest and partially cleared plots, the nutrient lost in highest amounts is calcium, with nitrogen as a close second. This order is reversed on totally cleared plots, where potassium is lost in greatest quantities, with nitrogen and calcium losses next highest. On totally cleared plots, the high potassium losses reflect

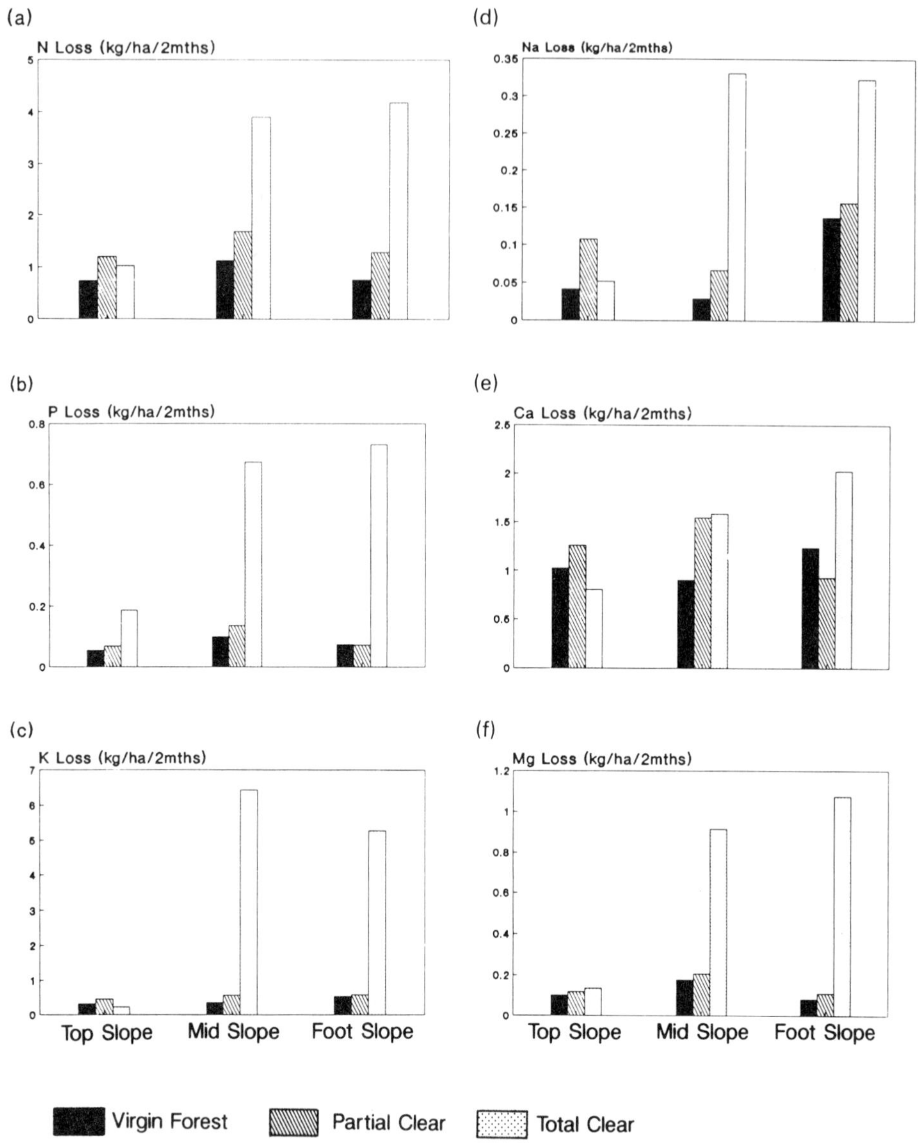

Fig. 7.5 Amounts of N, P, K, Na, Ca and Mg lost in eroded soil material over a two month period from July to September, for top, mid-and foot of slope locations of terra firme forest, partial clearance and total clearance.

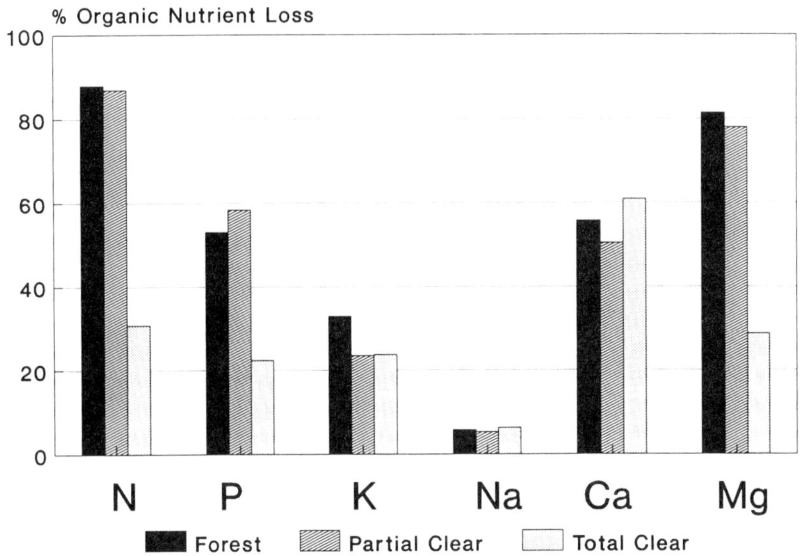

Fig. 7.6 Proportions of N, P, K, Na, Ca and Mg eroded in organic soil material from plots under forest, partial clearance and total clearance; mean values over a two month period from July to September.

Table 7.1 Order of nutrient loss in eroded soil materials and in decomposing leaf litter

(a) Decreasing order of nutrient loss in eroded sediments:

Forest: Ca > N > K > P = Na > Mg

Partial clearance: Ca > N > K > Na > P > Mg

Total clearance: N > K > Ca > P > Mg > Na

(b) Decreasing order of nutrient loss from decomposing leaf litter:

	Tetragastris	*Ecclinusa*
After 30 days		
Forest	P > K > N	K ≫ N = P
Partial clearance	P ≫ N = K	P = K > N
Total clearance	P > K > N	K > P > N
After 90 days		
Forest	K = P ≫ N	K > P > N
Partial clearance	P = K ≫ N	K > P > N
Total clearance	K = P ≫ N	K > P > N

high mineral sediment loss, probably containing substantial amounts of secondary clay minerals, such as micaceous minerals, of which potassium is a major component. The calcium concentration of woody organic litter in this study is around 3.5–4 times higher than leaf litter, while the nitrogen concentration of woody litter is around 2 times higher than in leaf litter. This allows the suggestion that while particulate woody organic debris is the main component of organic matter loss under virgin forest and partial clearance, fragmented leaf litter is the main form of organic matter loss after total clearance. This probably indicates that the experimental plots were too small, allowing continued leaf litter influx after clearance. It also indicates that leaf litter is more likely than twigs and branches to fall into totally cleared plots.

Sediment removal by soil erosion depletes the site of surface soil nutrients. For this reason, we would expect the nutrient status of sediments eroded under undisturbed forest, and from partially and fully cleared plots, (a) to differ from each other because of differential nutrient inputs and uptakes, and (b) to change through time as the supplies of new, erodable mineral soil and organic litter change (decline) after tree and vegetation removal. We would thus expect soil surface lowering by soil erosion and progressive erosion of soil nutrients to result in reduced concentrations of nutrients in eroded material with time after clearance. Figure 7.7 shows the total concentration of nitrogen, phosphorus, potassium, sodium, calcium and magnesium in eroded sediments from the top-slope locations of undisturbed forest, partially cleared and fully cleared transects for four sampling periods from July to October. Firstly, it can be seen that sediments eroded from the fully cleared top-slope plot consistently show higher nitrogen and phosphorus concentrations compared to forest and partial clearance. Sediments eroded from the totally cleared top-slope plot show lower potassium and magnesium concentrations. There is no clear increase or decrease in concentration of any of the nutrients in eroded material with time after forest clearance. The proportional organic matter content of eroded sediments over the same time period was also examined. The percent loss-on-ignition (LOI) values for sediments on six consecutive sampling dates from July to October are given in Fig. 7.8. Again, if organic matter is being progressively eroded with time after clearance, we would expect LOI values for eroded sediments to decline with time after clearance. Apart from one sampling date, organic matter contents are smaller in both partially cleared and fully cleared sites, compared with forest. Only on fully cleared sites do organic matter contents of eroded sediments fall through time. This is perhaps an expected result, since the fully cleared sites are being depleted of organic matter through erosion and decomposition while receiving considerably reduced litter influx. Both forest and partially cleared sites are still

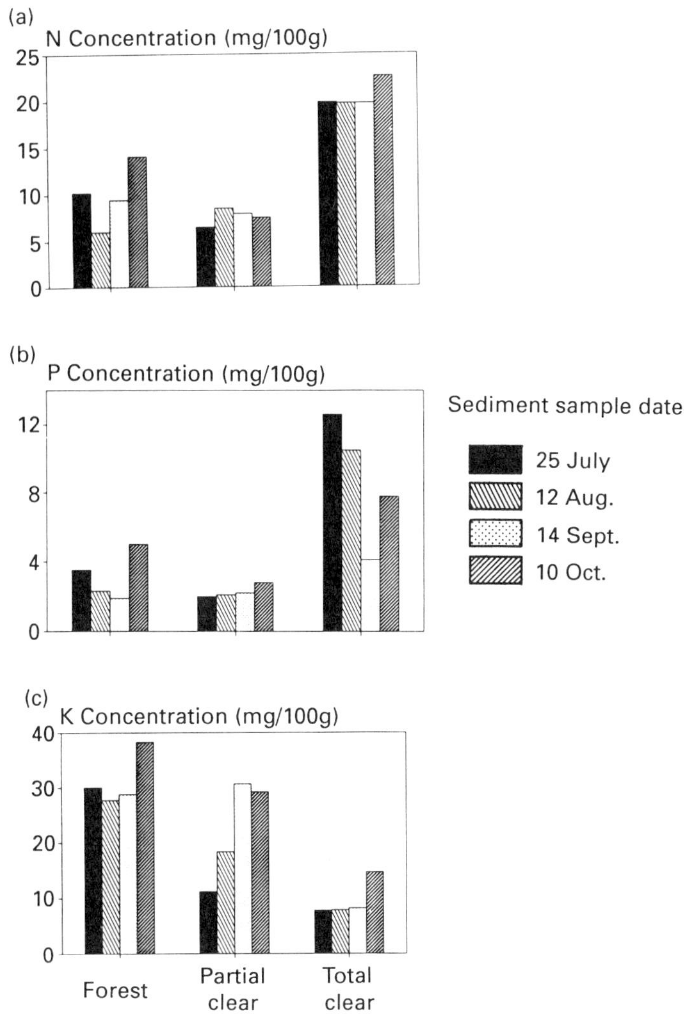

Fig. 7.7 Concentrations of N, P, K, Na, Ca and Mg in eroded soil material from mid-slope plots under forest, partial clearance and total clearance on four consecutive dates from July to September.

receiving some replacement for any lost soil organic matter, so we might expect their organic matter contents to remain fairly constant. The fact that they do not may simply reflect spatial variability of litter influx. Deforested plots appear to be losing organic matter and nutrients at a variable, but not decreasing, rate over the 4–6 sampling dates illustrated in Figs 7.7 and 7.8. This probably indicates that soil erosion has not yet progressed far enough on these sites for severe soil degradation to be identifiable.

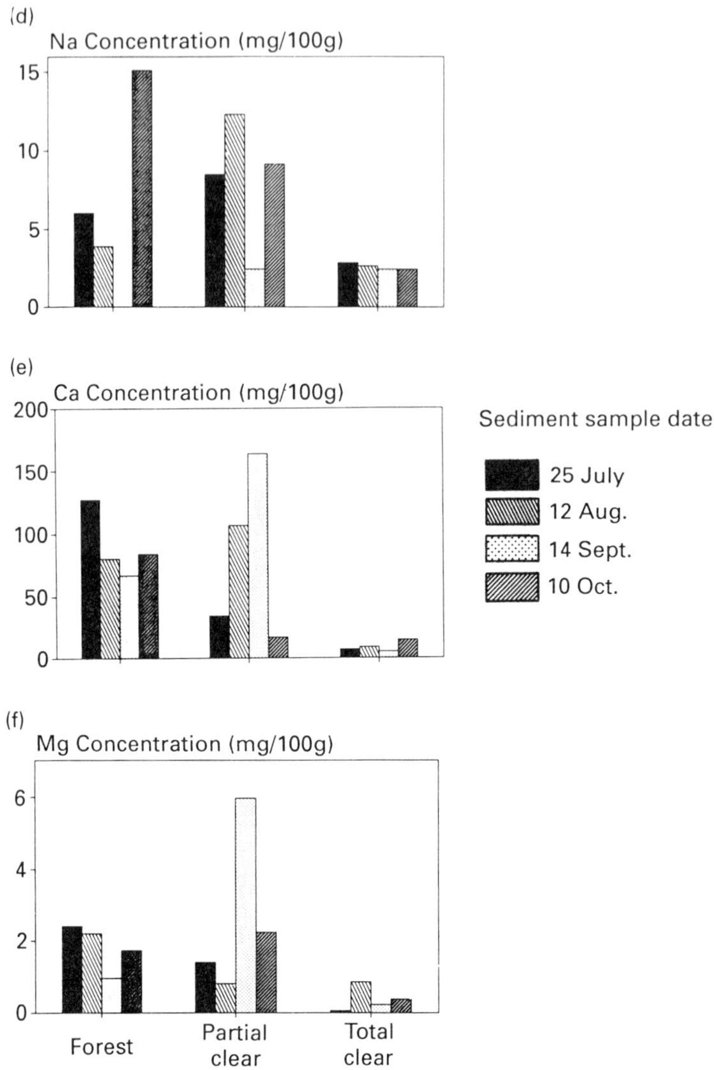

Fig. 7.7 *Continued*

FINAL DISCUSSION

The short time period of this study allows only relative process rate comparisons between sites and makes difficult any more comprehensive estimates of longer-term nutrient losses and soil degradation. Nevertheless, patterns of litter decomposition and nutrient release, as well as nutrient losses in soil erosion, agree well with studies of these processes

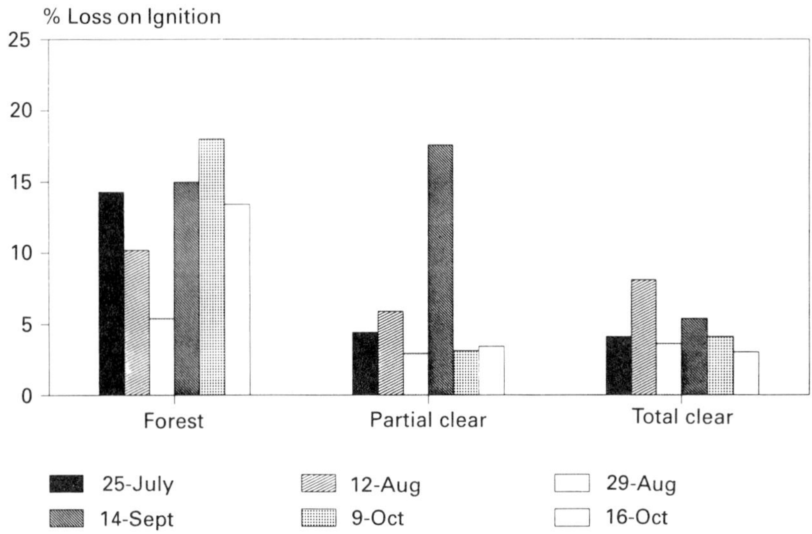

Fig. 7.8 Loss on ignition values for eroded soil material from mid-slope plots under forest, partial clearance and total clearance on six consecutive dates from July to October.

in other tropical forests and forest clearings. To assess the impact of these soil nutrient depletions on the relative colonizing abilities of forest and savanna vegetation, it is necessary to compare soil conditions after forest disturbance with soil conditions in adjacent savanna.

Soil nitrogen stocks in terra firme forest and after partial and total clearance (Fig. 7.9a) are compared with those in savanna grassland, dry campo and adjacent terra firme forest (Fig. 7.9b) around 4 km away, described in chapter 8. The rather surprising result is that total soil nitrogen increases after complete forest clearance, perhaps because of incorporation of fragments of organic matter at the time of disturbance which will decompose or be eroded through time. Total soil nitrogen in savanna, dry campo, terra firme forest and after partial forest clearance are similar, but KCl-extractable soil NH_4-N and NO_3-N are higher in the forest and cleared plots than in any of the habitats at the forest–savanna boundary only 4 km away, including forest. These comparisons support the above suggestion that the cleared plots do not yet show soil degradation, despite the large nutrient losses (Fig. 7.5). Reviewing the subject of soil nitrogen loss after forest clearance, Sanchez (1982) notes that felling and burning practices result in only modest soil nitrogen losses. More intensive techniques, such as bulldozing, cause very dramatic declines in soil nitrogen content in the first six months after forest clearance, as shown by Seubert *et al.* (1977) at Yurimaguas in Peruvian Amazonia. The relatively small differences, not only in soil nitrogen, but

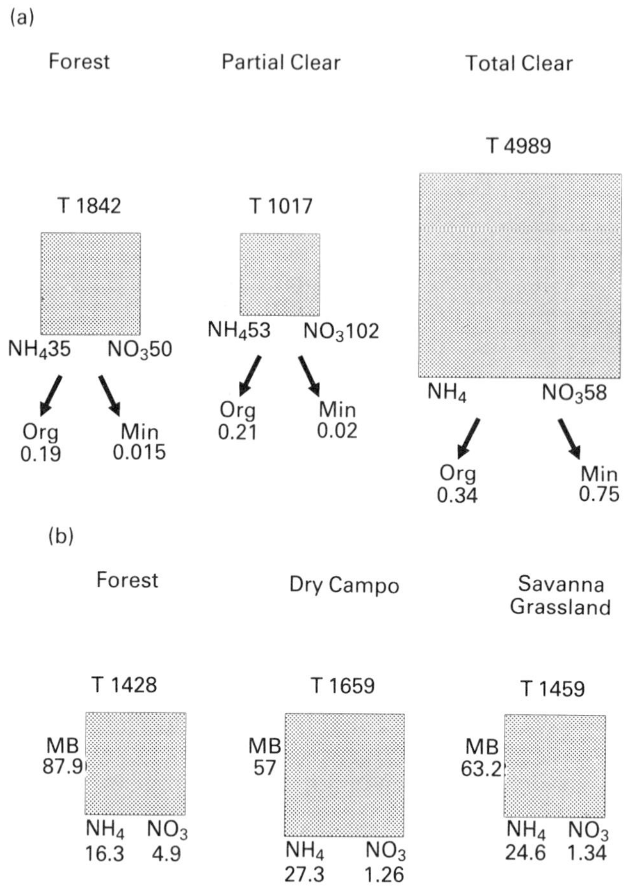

T = Total soil nitrogen (kg ha^{-1})
NH_4 = KCl – extractable soil NH_4 (kg ha^{-1})
Org = Eroded soil N in organic matter (kg ha^{-1} 2 $months^{-1}$)
Min = Eroded soil in N in 'mineral' matter (kg ha^{-1} 2 $months^{-1}$)
(All soil particles < 500μ m)
MB = Microbial biomass nitrogen (kg ha^{-1})
NO_3 = KCl – extractable soil NO_3 (kg ha^{-1})

Fig. 7.9 Soil nitrogen compartments in forest, partial clearance and total clearance plots, compared to habitats in adjacent savanna (all units kg ha^{-1}).

also in other soil nutrients, after clearance in the present study are likely to be due to the fact that vegetation and litter were removed from the plots carefully, with the intention of causing as little disturbance as possible. It is probably because both forest and savanna soils are relatively nutrient poor that differences in their organic matter and nutrient contents are very small.

Ramakrishnan (1989) reported generally lower soil nutrient contents after deforestation, with further soil nutrient reduction with time after clearance. This pattern of soil impoverishment is due only in part to soil erosion and litter removal, since in most cases, nutrients are also volatilized during the burning which generally accompanies deforestation in the tropics. Since Maracá Island is a SEMA Ecological Reserve, experimental burning treatments were not permitted. Thompson *et al.* in chapter 18 identify charcoal in both forest and savanna soils on Maracá Island. Salas and Fölster (1976) found large volatile losses of carbon and nitrogen: 25 kg carbon ha^{-1} and 673 kg nitrogen ha^{-1}, when tropical rainforest was burnt in Colombia. Clearly there have been nutrient losses due to burning from both forest and savanna soils on Maracá Island in the past, although some of the losses in smoke would fall back on the island, but the magnitude of losses due to fire are unknown and undetectable in contemporary soil chemistry.

In a study in central Amazonia which examined nutrient inputs and losses to terra firme forest other than litter decomposition and soil erosion, Brinkmann (1985) found net losses of nitrogen, phosphorus, calcium, magnesium and iron in soil leachates, compared to the amounts of these nutrients received in rainfall. Although this result agrees with studies in other tropical rainforests (e.g. Turvey, 1974; Golley *et al.*, 1975), Jordan (1982) found the opposite result: that the infertile *caatinga* forest at San Carlos in Venzuela showed net gains in rainfall over leachate losses for all nutrients apart from phosphorus. The magnitude of leaching losses is generally around 10–100 times lower than the soil-eroded nutrient losses reported for completely cleared sites in the present study (e.g. Maass *et al.*, 1988). Rainfall inputs of nutrients may thus be able to offset leaching losses, but not the combined losses of leaching and erosion after forest removal.

Few studies have reported nutrient losses in eroded sediments after rainforest removal. Maass *et al.* (1988) found high rates of erosion of soil nutrients when forest was converted to agriculture on steep slopes on the Pacific coast of Mexico. Eight times more nitrogen and ten times more phosphorus were lost from cleared plots growing buffel grass than from virgin forest. In both the study by Maas *et al.* (1988), and on Maracá Island, nutrient losses are associated with the loss of soil organic matter. These results highlight the need to retain leaf litter and soil organic matter on site after forest clearance to prevent nutrient impoverishment.

The difference in humidity between wet and dry seasons in central Amazonia is cited by Luizão and Schubart (1987) as the main reason for significantly faster leaf litter decomposition rates under terra firme forest in the wet season than in the dry season. They suggest that more intense microarthropod activity and greater penetration of litter by fine roots in the wetter conditions are the main agencies causing leaf litter weight loss. Analogous differences in humidity at the forest–savanna boundary are likely to be the cause of slower decomposition rates in savanna grassland than in the forest. Theoretically it is not the rate of soil and litter nutrient mineralization *per se* that cause soil degradation, but the asynchrony between nutrient 'release' and the ability of the vegetation to take up available nutrients. Thus slower decomposition rates in savanna may act as a nutrient conservation mechanism in vegetation systems which have generally lower nutrient uptake requirements.

The results of this study indicate that nutrient losses in soil erosion can be high if full forest clearance occurs on sloping sites of even 13°–16°. The rate of supply of mobilized nutrients through litter decomposition in this study is little different after clearance from that in undisturbed forest, so it is likely to be the curtailment of litter influx after clearance that is the main control on nutrient inputs to, and resultant soil fertility of, cleared soils. In this study, the small size of plots may have allowed root input to the soil and canopy overlap from surrounding trees. Several authors have noted the role of trees in savannas in enriching the soil. Soils under tree canopies tend to show higher nitrogen and organic matter, as well as other nutrients (e.g. Bernhard-Reversat, 1982; Kellman, 1979). The prime reason given for this effect is that trees transport nutrients from surrounding soils to their canopy and drop the nutrients in leaf and stem litter. Bird and animal droppings in the vicinity of the tree may also contribute significantly to nutrient inputs. Belsky *et al.* (1989) noted differences in floristic composition and herbage biomass under tree canopies in the savanna, compared to open savanna grassland. The relationship between herbage biomass and soil fertility was significant, but that between species diversity and soil chemical or physical properties was not. These kinds of herbage changes under tree canopies in savanna presumably illustrate the way in which forest edges could advance into savanna if soil nutrition limitations were relieved.

The degree to which the results of this study can be used to understand vegetation competition and the location of boundaries or transition types at the forest–savanna boundary is limited. On Maracá Island, other evidence (Furley and Ratter, 1990) suggests that soil hydrology rather than soil fertility is the contemporary determinant of the location and form of the forest–savanna boundary. Further work on comparative decomposition indices should improve our knowledge of nutrient cycling processes in these environments.

ACKNOWLEDGEMENTS

The author thanks the Brazilian Environment Secretariat (SEMA) for the use of field sites on Maracá Island, the Royal Geographical Society for support, particularly the help of Mr S. Bowles, Miss F. Watson and Miss S. Latham, and the field assistance of Dr J. Thompson and Mr D. Scott of Stirling University. Professor J. Thornes, Dr S. Nortcliff, Miss B. Gregory, Miss S. Aitade, Miss K. Fawcet, Miss D. Adams and Mr S. Wild all provided scientific help.

REFERENCES

Allen, S.E. (1989) *Chemical Analysis of Ecological Materials* 2nd edn, Blackwell Scientific, Oxford.

Anderson, A.B., Prance, G.T. and de Albuquerque, B.P. (1975) Estudos sobre a vegetação das Campinas Amazônicas. III A vegetação lenhosa da Campina da Reserva Biológica INPA–SUFRAMA (Manaus–Caracarai, km 62). *Acta Amazônica*, **5**, 225–46.

Anderson, J.M., Proctor, J. and Vallack, H.W. (1983) Ecological studies in four contrasting lowland rain forests in Gunung Mulu National Park, Sarawak. III Decomposition processes and nutrient losses from leaf litter. *Journal of Ecology*, **71**, 503–27.

Anderson, J.M. and Swift, M.J. (1983) Decomposition in tropical forests. in *Tropical Rain Forest: Ecology and Management*, (eds S.L. Sutton, T.C. Whitmore and A.C. Chadwick), Blackwell Scientific, Oxford, pp. 287–309.

Belsky, A.J., Amundson, R.G., Duxbury, J.M. *et al.* (1989) The effects of trees on their chemical and biological environments in a semi-arid savanna in Kenya. *Journal of Applied Ecology*, **26**, 1005–24.

Bernard-Reversat, F. (1982) Biogeochemical cycles of nitrogen in a semi-arid savanna. *Oikos*, **38**, 321–32.

Brandt, J. (1988) The transformation of rainfall energy by a tropical rain-forest canopy in relation to soil erosion in *Biogegraphy and Development in the Humid Tropics* (ed P.A. Furley), Special Issue. *Journal of Biogeography*, **15**, 41–8.

Brinkmann, W.L.F. (1985) Studies on hydrobiogeochemistry of a tropical lowland forest system. *GeoJournal*, **11**, 89–101.

Crooke, W.M. and Simpson, W.E. (1971) Determination of ammonium in kjeldahl digests of crops by an automated procedure. *Journal of the Science of Food and Agriculture*, **22**, 9–10.

Elwell, H.A. and Stocking, M.A. (1976) Vegetative cover to estimate soil erosion hazard in Rhodesia. *Geoderma*, **15**, 61–70.

Ewel, J.J. (1976) Litter fall and leaf decomposition in a tropical forest succession in eastern Guatemala. *Journal of Ecology*, **64**, 293–308.

Furley, P.A. and Ratter, J.A. (1990) Pedological and botanical variations across the forest–savanna transition on Maracá Island. *Geographical Journal*, **156**, 251–66.

Golley, F.B., McGinnis, J.T., Clements, R.G. *et al.* (1975) *Mineral Cycling in a Tropical Moist Forest Ecosystem*, University of Georgia Press, Athens, Georgia, USA.

Gong, W.K. and Ong, J.E. (1983) Litter production and decomposition in a coastal hill dipterocarp forest, in *Tropical Rain Forest: Ecology and Management*, (eds S.L. Sutton, T.C. Whitmore and A.C. Chadwick), Blackwell Scientific, Oxford, pp. 275–87.

Heaney, A. and Proctor, J. (1989) Chemical elements in litter in forests on Volcán Barva, Costa Rica, in *Mineral Nutrients in Tropical Forest and Savanna Ecosystems*, (ed. J. Proctor), Blackwell Scientific, Oxford, pp. 251–71.

Heath, G.W. and Arnold, M.K. (1966) Studies in leaf-litter breakdown. II Breakdown rates of 'sun' and 'shade' leaves. *Pedobiologia*, **6**, 238–43.

Henriksen, S. and Selmer-Olsen, A.R. (1970) Automatic methods for determining nitrite and nitrate in water and soil extracts. *Analyst*, **95**, 514–18.

Herrera, R. (1978) *Nutrient Distribution and Cycling in an Amazonian Caatinga Forest on Spodosols in Southern Venezuela*, Unpublished PhD thesis, University of Reading, UK.

Herrera, R., Jordan, C.F., Klinge, H. and Medina, E. (1978) Amazon ecosystems: their structure and functioning with particular emphasis on nutrients. *Interciência*, **3**, 223–32.

Hopkins, B. (1983) Successional processes, in *Ecosystems of the World*, 13: Tropical Savanna, (ed. F. Bourlière), Elsevier, Amsterdam, pp. 605–15.

Jordan, C.F. (1982) The nutrient balance of an Amazonian rain forest. *Ecology*, **63**, 647–54.

Kellman, M. (1979) Soil enrichment by neotropical savanna trees. *Journal of Ecology*, **67**, 565–77.

Lowman, M.D. (1988) Litterfall and leaf decay in three Australian rain-forest formations. *Journal of Ecology*, **76**, 451–65.

Luizão, F.J. and Schubart, H.O.R. (1987) Litter production and decomposition in a terra-firme forest of central Amazonia. *Experiêntia*, **43**, 259–65.

Maass, J.M., Jordan, C.F. and Sarukhán, J. (1988) Soil erosion and nutrient losses in seasonal tropical agroecosystems under various management techniques. *Journal of Applied Ecology*, **25**, 595–607.

Madge, D.S. (1969) Litter disappearance in forest and savanna. *Pedobiologia*, **9**, 288–99.

Medina, E. and Cuevas, E. (1989) Patterns of nutrient accumulation and release in Amazonian forests of the upper Rio Negro basin, in *Mineral Nutrients in Tropical Forest and Savanna Ecosystems*, (ed. J. Proctor), Blackwell Scientific, Oxford, pp. 217–40.

Milliken, W. and Ratter, J.A. (1989) *Vegetation of the Ilha de Maracá,*

Roraima, Northern Brazil. Royal Botanic Garden, Edinburgh.

Murphy, J. and Riley, J.P. (1962) A modified single solution method for the determination of phosphate in natural waters. *Analytica Chimica Acta*, **27**, 31–6.

Nortcliff, S. and Robison, D. (1989) *The Soils and Geomorphology of the Ilha de Maracá, Roraima, The Second Approximation*, Royal Botanic Garden, Edinburgh.

Nortcliff, S., Ross, S.M. and Thornes, J.B. (1989) Soil moisture, runoff and sediment yield from differentially cleared tropical rainforest plots, in *Vegetation and Erosion*, (ed. J.B. Thornes), John Wiley, Chichester, pp. 419–36.

Ramakrishnan, P.S. (1989) Nutrient cycling in forest fallows in north eastern India, in *Mineral Nutrients in Tropical Forest and Savanna Ecosystems*, (ed. J. Proctor), Blackwell Scientific, Oxford, pp. 337–52.

Ross, S.M., Thornes, J.B. and Nortcliff, S. (1990) Soil hydrology, nutrient and erosional response to the clearance of terra firme forest, Maracá Island, Roraima, Northern Brazil. *Geographical Journal*, **156**, 267–82.

Salas, G. de las, and Fölster, H. (1976) Bio-element loss on clearing tropical rainforest. *Turrialba*, **26**, 179–86.

Sanchez, P.A. (1982) Nitrogen in shifting cultivation systems of Latin America. *Plant and Soil*, **67**, 91–103.

Sarmiento, G. (1983) The savannas of Tropical America, in *Ecosystems of the World, 13: Tropical Savannas*, (ed. F. Bourlière), Elsevier, Amsterdam, pp. 245–88.

Sarmiento, G. (1984) *The Ecology of Neotropical Savannas*, Harvard University Press, Cambridge, Mass.

Seubert, C.E., Sanchez, P.A. and Valverde, C. (1977) Effects of land clearing methods on soil properties and crop performance in an ultisol of the Amazon jungle of Peru. *Tropical Agriculture, Trinidad*, **54**, 307–21.

Singh, J.S., Raghubanshi, A.S., Singh, R.S. and Srivastava, S.C. (1989) Microbial biomass acts as a source of plant nutrients in a dry tropical forest and savanna. *Nature*, **338**, 499–500.

Swift, M.J., Heal, O.W. and Anderson, J.M. (1979) *Decomposition in Terrestrial Ecosystems*, Blackwell Scientific, Oxford.

Turvey, N.D. (1974) Water in the nutrient cycle of a Papuan rain forest. *Nature*, **251**, 414–15.

Wiersum, F. (1985) Effects of various vegetation layers of an *Acacia auriculiformis* forest plantation on surface erosion in Java, Indonesia, in *Soil Erosion and Conservation*, (eds S.A. El-Swaify, W.C. Moldenhauer and A. Lo), Soil Conservation Society of America, pp. 79–89.

8

Soil conditions and soil biology in different habitats across a forest–savanna boundary on Maracá Island, Roraima, Brazil

S.M. ROSS, F.J. LUIZÃO and R.C.C. LUIZÃO

SUMMARY

1. A series of soil biological attributes, including soil microbial biomass, numbers of individual soil fauna groups and soil respiration, plus a series of soil chemical characteristics, including organic carbon, total soil nitrogen and three indices of 'available' soil nitrogen, namely: initial KCl-extractable NH_4-N and NO_3-N, 'potentially available' NH_4-N and NO_3-N after incubation, and microbially immobilized mineral N, released as NH_4-N and NO_3-N after fumigation, were measured in five habitats at the forest–savanna boundary on Maracá Island in northern Roraima, Brazil.
2. Duplicate field plots in terra firme forest, dry grass savanna, wet and dry wooded savanna, and the boundary zone between forest and savanna were studied in the early part of the wet season. The soil biology of grass tussock hummock and hollow microsites in grass and wooded savanna were also compared.
3. Significant differences in soil moisture content between habitats were paralleled by significant differences in soil nitrogen processes, soil microbial biomass, respiration and numbers of individual soil faunal groups, such as Collembola, Acari and Formicidae. The forest sites showed significantly different numbers of all soil faunal groups studied, apart from the very mobile Formicidae.
4. Forest and boundary sites showed both higher available and higher potentially available nitrate nitrogen compared with grass and wooded savanna sites. The pattern for ammonium nitrogen was the inverse,

with higher available and potentially available amounts in the savanna sites. These trends were shown to be due to distinct differences in soil physical conditions, induced by higher soil moisture levels which inhibit nitrification in savanna and *campo* sites. It is suggested that microbial immobilization of nitrogen may be an important nutrient conservation mechanism, providing sources of immediately mineralizable N for plant uptake in otherwise fairly nutrient-poor soils.

INTRODUCTION

Edaphic and climatic conditions are generally invoked as prime controls, together with fire, on the location and form of the boundary between tropical rainforest and savanna. Soil physical characteristics, particularly soil moisture and aeration status, are important controls on the activity of the soil microbial population and hence they are important controls on soil organic matter dynamics, nutrient cycling and soil fertility. Soil moisture status and fertility are potentially the most important controls on colonization and survival of seedlings, and thus the ability of forest species to migrate into the savanna or of savanna species to penetrate the forest. Since numbers and activity of decomposer soil organisms show a very distinct seasonal pattern, it is clear that the seasonality of climate and soil hydrology in both tropical savanna and forest systems imposes regular pulses of activity and inactivity on soil processes such as organic matter decomposition and nutrient cycling (e.g. Lavelle, 1983; Levings and Windsor, 1982). Most studies divide the activities of the soil organism population into soil *faunal* processes, relating to breakdown and decomposition of litter, and soil *microbial* processes, relating primarily to bacterial mineralization and chemical transformation of nutrients. Both show some seasonality, but their activities have rarely been studied together.

Treeless savannas and their soils have been described as oligotrophic and 'highly conservative of their small nutrient capital' (Kellman, 1989). We might also expect to find a rather small soil organism population, perhaps of low diversity, with a seasonal activity which peaks at the beginning of the wet season. It is clear from the work of Jung (1967) and Kellman (1979), for example, that the presence of trees in the savanna enriches the soil immediately below the canopy. In the present study, we might thus expect to find more fertile conditions, with larger soil organism populations, firstly in *campo sujo* (wooded savanna) compared with open *campo limpo* (grass savanna), and secondly in terra firme forest compared with wooded savanna.

In this study at the forest–savanna boundary on Maracá Island, our intention was to examine soil chemical and biological conditions in a series of forest–savanna 'boundary' habitats. The typical soil of the Santa Rosa savanna region of the eastern part of Maracá Island is described by Nortcliff and Robison (1989) as an *aquic haplorthox* (USDA classification), of sandy and silty texture. More detail on the textural and chemical characteristics of soils of the region is provided by Thompson *et al.* (Chapter 18), working in an area of savanna 1.5 km from the present study. Their work shows that soil texture varies from sites dominated by the sand-sized fraction (around 70%), to soils with substantial silt (up to 50%) and clay (up to 60%) fractions. The most common textural type contains around 75% sand and has the appearance of a pure white sand with flecks of mica and differing amounts of finely divided organic matter. Discussion of the soil–vegetation relationships across a number of savanna–forest boundaries in the region is supplied by Furley and Ratter (1990) and Furley (in Milliken and Ratter, 1989). Some of the savanna sites at Santa Rosa are semi-seasonal, described by Sarmiento (1984) as hyperseasonal, having a distinct wet season, in which the water level rises above the soil surface, despite the high permeability of the soil, and a very distinct dry season, in which the water table drops to around 40–50 cm depth below the soil surface. During the dry season, active evapo-transpiration causes the surface 5–10 cm of soil to become excessively dry. The savanna habitats range from areas of low, open grass/sedge sward, to areas of taller mixed rank grasses up to 1 m high with scattered low stature trees of *Curatella americana* and *Byrsonima crassifolia*, rarely exceeding 5 m in height. The vegetation of the region is described in detail by Milliken and Ratter (1989) who confirm that these areas are frequently flooded to around 20–30 cm depth during the rainy season, but are dry for the rest of the year. The low grass savanna is of the type described as campo limpo by Sarmiento (1983), while the tree/shrub savanna areas are similar to the campo sujo of Sarmiento (1983).

In this study, five habitat types were chosen in the eastern part of Santa Rosa savanna (Fig. 8.1), according to their vegetation (Milliken and Ratter, 1989), above-ground biomass (J. Ferraz, personal communication) and soil moisture conditions, as indicated by seasonal water-table level (J. Thompson and D.A. Scott, personal communication). These habitats were: terra firme forest (TF); *dry* savanna (DS), with low stature sward and low diversity of grasses and sedges; *dry* campo (DC), with taller sward and short stature trees; *wet* campo (WC), with wet soils and lush tall sward with short stature trees, and the boundary zone (B) between forest and savanna. *Dry* and *wet* descriptors for savanna and campo are used in a relative sense, in relation to the degree of saturation observed in the wet season, and the depth of the water table below the soil surface in the dry season.

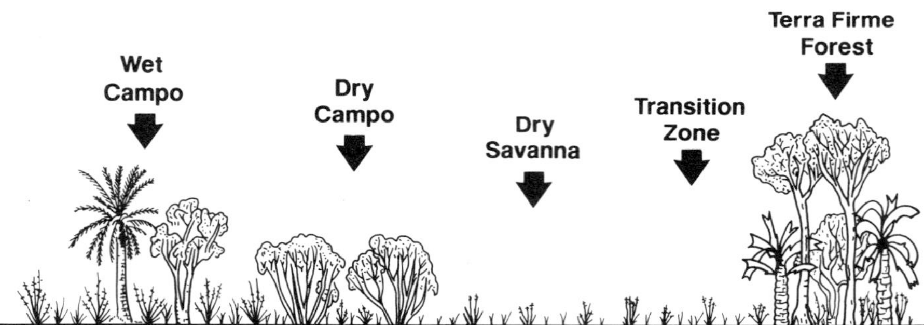

	AOV	WC	DC	DS	B	TF
Aboveground Litter plus Herbage Biomass (gm^{-2})	$F = 39.27$ $p = 0.000$	347 (97.28)	246.5 (35.81)	104.5 (3.66)	215 (40.62)	417 (79.97)
Root Biomass (gm^{-2}/8cm depth)	$F = 8.91$ $P = 0.000$	7.47 (2.11)	5.83 (2.88)	8.46 (6.03)	17.98 (7.15)	18.85 (8.33)
Water Table Depth (wet season) (cm)		20 - 0	40 - 25	60 - 35	150	200 - 180

Fig. 8.1 Diagrammatic representation of five vegetation habitat types at the forest–savanna boundary on Maracá Island, with data on herbage and litter biomass, root biomass and water table depth. Water table data supplied by Thompson and Scott (personal communication). Figures in parentheses represent standard deviations. Results of a one-way analysis of variance (AOV) are given.

Soil organic matter in seasonal savanna grasslands tends to be concentrated in a shallow surface layer, due in part to the physical and chemical inhospitability of lower soil horizons for soil microbial activity and plant root growth. This means that litter arriving at the soil surface is not translocated, but is decomposed *in situ*. The general lack of roots at depth reduces the organic matter input below the soil surface. The current study is limited to an examination of soil chemistry and biology in the surface 10 cm of soil in each of the five habitats designated above.

Litter decomposition experiments and soil organic matter and decomposition indices were also studied in the same habitats noted above and will be reported elsewhere.

MATERIALS AND METHODS

A transect, laid down by J. Ferraz in 1987, across the forest–savanna boundary in the eastern part of the Santa Rosa region of Maracá Island, about 1.5 km east of the study reported by Thompson *et al.* (Chapter 18), was used to locate sampling sites for the five habitats. At each site, two randomly located plots, each 10 × 10 m, were marked out with x and y axes, from which random coordinate sampling locations could be measured. In each of the ten plots, a series of soil organism and soil chemistry studies were carried out. The soil data reported in this paper are for May 1989, during the early rainy season on Maracá, when conditions were optimal for microbial and soil faunal activity.

Soil physical and chemical characteristics

In each plot, six randomly located soil samples were taken from the surface 10 cm. In savanna, dry campo and wet campo sites, the tussock-forming grasses produce a hummock and hollow microtopography with a maximum height of 5–10 cm and hummock diameter of 10–30 cm, with both hummock height and diameter greatest in wet campo. Soil samples were taken from both hummock and hollow locations in these habitats. All surface soil samples were used for the determination of loss-on-ignition, organic carbon and total nitrogen. Soil samples were oven-dried at 105 °C and sieved to pass through a 500 μm mesh. Samples weighing 1 g were then digested, using a modified micro-kjeldhal technique, with concentrated sulphuric acid and hydrogen peroxide (Allen, 1989) and total nitrogen was determined using the continuous flow auto-analyser colorimetric method of Crooke and Simpson (1972). Organic carbon content was determined using a modification of the Walkley-Black technique (Black, 1965). Soil C:N ratios were calculated for all habitats. Loss-on-ignition was determined at 450 °C for 18 h.

In each plot, at least six randomly located samples were taken of the surface 5 cm of soil. These soil samples, together with samples taken from hummock and hollow locations in savanna and campo sites, were then analysed for soil microbial biomass, soil respiration, gravimetric moisture content, organic carbon, pH and root biomass. Several indices of nitrogen availability and immobilization were also measured: immediately extractable ammonium and nitrate, potentially available ammonium and nitrate, and extractable ammonium and nitrate nitrogen after fumigation. Soil samples used for the determination of soil microbial biomass, respiration and available nitrogen were taken back to the laboratory in their fresh condition, packed in an insulated polystyrene box and analyses carried out immediately. Microbial biomass was determined by the chloroform fumigation-incubation technique of Jenkinson and Poulson (1976), modified for acid soils (Vance *et al.*, 1987), with 2 M KCl-extractable NH_4-N and NO_3-N after fumigation representing microbially immobilized N (Brookes *et al.*, 1985). Incubations of the type described by Stanford and Smith (1972) were used to assess and compare potentially available inorganic nitrogen in the five habitats. This type of incubation technique has been shown by Keeney (1980) to provide useful estimates of potential nitrogen mineralization and nitrification in forestry systems and has subsequently been used by several authors to assess potential nitrogen availability in different types of tropical forest systems (e.g. Robertson and Vitousek, 1981; Vitousek and Denslow, 1986; Vitousek and Matson, 1988). Soil samples to be determined for potentially available nitrogen were incubated for ten days at 24 °C after adjusting the moisture content to 50% water-holding capacity. The carbon dioxide produced during the ten-day incubation was considered as the soil respiration in that period. Evolved carbon dioxide was trapped in a 1 M solution of sodium hydroxide and determined by titration (Hesse, 1971). After incubation, soil inorganic nitrogen was extracted in 25 g samples with 60 ml of 2 M KCl by shaking for one hour, followed by vacuum filtration. Extractable ammonium and nitrate before and after incubation were determined colorimetrically by flow injection using modifications of, respectively, the indophenol blue method (Allen, 1989) and the cadmium reduction method (Henriksen and Selmer-Olsen, 1972). Soil organic carbon was determined in the oven-dried 0.5 mm fraction using a modification of the Walkley-Black technique (Black, 1965). Soil pH was determined in a 1:2 vol.:vol. paste of soil:distilled-deionised water. Root biomass was determined in soil cores after water washing.

In each plot, the above-ground herbage biomass and litter were collected in five quadrats, each of 22 cm edge, and oven-dried at 60 °C.

Microfauna present in soil and leaf litter

In May 1989, three randomly located sets of three surface soil cores, each 3.5 cm diameter by 8 cm long, were bulked for each plot and the soil fauna extracted by a Berlese–Tulgren funnel system. These extracts were taken to represent the characteristic wet-season surface soil fauna population. In a second experiment, to be reported elsewhere, in which leaf litter bags were laid out on the soil surface in each habitat to examine litter decomposition rates over periods of one, three and six months, soil fauna were extracted from the bags immediately after sampling using the Berlese–Tulgren funnel. These litter fauna extractions represented the colonizing fauna of June (mid wet season), August (late wet season) and November (early dry season). Typical seasonal rainfall conditions on Maracá are given in Chapter 18.

For statistical comparisons between sites for soil pH, soil moisture, soil carbon and soil nitrogen, a simple, one-way analysis of variance was used. In all cases, this was run on eight replicates from each of five habitats. For all simple correlation analyses reported, the same number (40) of samples was used.

RESULTS AND DISCUSSION

Vegetation

Figure 8.1 shows that there are highly significant differences in above-ground litter and herbage biomass in the five habitats. Highest values were recorded under forest and in the wet campo, with low, but less variable amounts in savanna grassland. These differences were clearly visible on the ground, with, for example, bare soil patches between grass tussocks in the savanna grassland (DS). The mean herbage heights in savanna grassland, dry campo and wet campo were 13, 42 and 57 cm respectively. Only in the forest was a significant proportion of the above-ground litter and herbage biomass made up of undecomposed leaf litter. Root biomass in the surface 5 cm was also different between sites (Fig. 8.1) with highest values under forest vegetation.

Soil properties

Soil moisture and chemical characteristics

The wet-season depth of the water table is deepest under terra firme forest and closest to the soil surface in the hyperseasonal savanna (wet campo in this study) (Fig. 8.1). In this study, our single rainy season

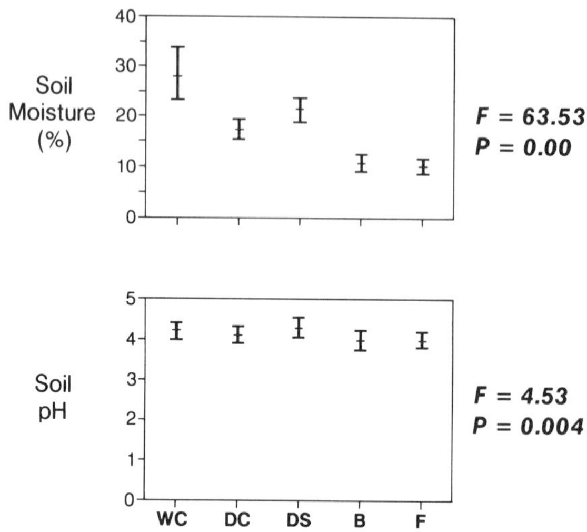

Fig. 8.2 Soil moisture and soil pH in five vegetation habitats at the forest–savanna boundary. WC = wet campo, DC = dry campo, DS = savanna grassland, B = boundary between savanna grassland and terra firme forest, F = terra firme forest. Standard deviations are given in bars. Statistical results are for one-way analysis of variance, showing degrees of freedom, *F*-statistic and probability level.

measurement of gravimetric moisture content in the surface 10 cm of soil showed this trend very well: that of much wetter conditions in the surface rooting zone of the hyperseasonal savanna (wet campo in this study), tending towards drier conditions in the surface rooting zone of terra firme forest (Fig. 8.2). Differences in gravimetric soil moisture between the five habitats are highly significant, and are likely to be the principal cause of differences in soil microbial activity, in decomposition rates and in soil nutrient status. Soil pH along the transect ranges from 4 to 4.5, with very little difference between the habitats (Fig. 8.2). Soils under terra firme forest and on the forest–savanna boundary zone show very slightly more acidic pH conditions.

Soil organic carbon was highest in wet campo and in the forest (Fig. 8.3). The dry campo and savanna sites have the lowest organic carbon values, reflecting smaller litter inputs than in the other habitats, particularly the forest. Despite this, there is no relationship across the five habitats between above-ground herbage biomass and soil organic carbon, with a correlation coefficient of 0.503 (n = 40). The trend in soil organic carbon from lowest in the dry campo to highest in terra firme forest, is, however, paralleled by an identical trend in root biomass (Fig. 8.1), but,

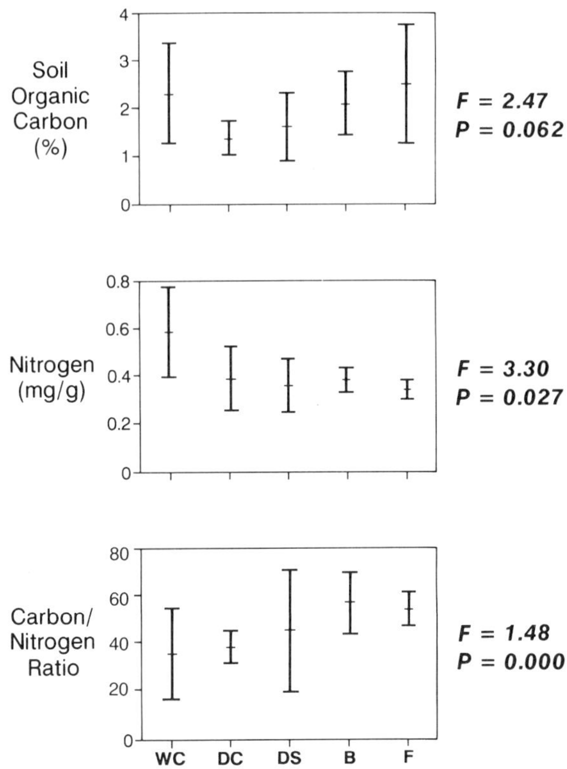

Fig. 8.3 Soil organic carbon, total nitrogen and C/N ratio in five vegetation habitats at the forest–savanna boundary. Habitat key as in Fig. 8.2. Standard deviations are given in bars. Statistical results are for one way analysis of variance, showing the *F*-statistic and probability level.

again, no significant relationship was found. Soil organic carbon values, ranging quite widely from 0.63% to 2.88% in savanna and from 0.91% to 1.65% in dry campo sites, lie in the same range of values quoted by Askew *et al.* (1970) and Eden (1974) for other savanna regions in Amazonia. The high spatial variability of soil organic carbon in all five habitats may be due to an irregular spatial distribution of litter input, particularly in the savanna where there is less than 100% vegetation cover, with bare soil patches between grass and sedge tussocks. In the savanna, black algal mats formed in patches of bare soil during the wet season when water ponded in the microtopographic hollows at the soil surface. These mats then dried and fragmented in the dry season to form an input of particulate organic matter to the soil. In the wet campo, higher amounts of both herbage biomass and root biomass, combined with much wetter soil conditions, account for higher amounts of litter input, slower organic matter decomposition rates and the longer retention of organic matter in the soil.

Soil nitrogen

Total soil nitrogen in the top 10 cm of all five habitats is relatively low (Fig. 8.3) compared with savanna soils elsewhere (Table 8.1). Eden (1974) found, in southern Venezuela, sandy latosolic soils similar to those on Maracá Island. He reported total nitrogen for savanna topsoils of 0.08%, rising to 0.22% in poor-draining savanna sites and 0.06–0.09% under forest. Values for total soil nitrogen on Maracá Island range from 0.021–0.049% in the savanna to 0.028–0.079% in the wet campo and 0.03–0.04% in forest. These low values emphasize the nutrient poverty of the Maracá soils. This is a little surprising, since other authors have reported nitrogen fixation in tropical savanna grasslands by free-living bacteria, particularly *Beijerinckia*, in the rhizosphere of tropical grasses (e.g. Dobereiner and Day, 1975), and by mats of blue-green algae in sites with discontinuous grass cover (e.g. Raud, 1977).

All soil C:N ratios are high (Fig. 8.3c). This appears to be due to the extremely low total soil nitrogen contents. Results from other savanna soils show lower C:N ratios and the more expected trend of higher C:N in savanna soils compared to forest soils (Table 8.1). Eden (1974), for example, found that C:N ratios in terra firme forest were 11–15, while those in adjacent savanna were 15–17.5. His higher values were found in the more poorly drained sites. Along the Maracá savanna–forest transition, the lowest C:N ratios occurred in the wet campo. This is an unexpected result, since soil waterlogging is likely to reduce decomposition rates and to encourage denitrification of the system. Both of these processes would act to maintain high C:N ratios. The generally higher C:N ratios in the topsoil of forest is again unexpected, because the lower soil moisture and larger inputs of easily decomposable leaf litter, combined with conservation of nitrogen by immobilization and root uptake, would act to reduce C:N values. Bernhard-Reversat (1982), working in Senegal, found higher soil organic carbon and total nitrogen, combined with lower C:N ratios, in soils immediately underneath tree canopies in the savanna. Both Bernhard-Reversat (1982) and Kellman (1979) have illustrated the relative soil nutrient enrichment that occurs under individual trees in the savanna. In the Maracá savanna, although we should expect to see a relative nutrient enrichment in dry campo compared to savanna soils, this is not the case for carbon and nitrogen. Generally higher than normal C:N ratios in topsoils from both savanna and adjacent forest sites were reported by Singh (1989) in northern India. He described his forest and savanna field sites as dry and particularly poor in

Table 8.1 Soil organic carbon and total nitrogen contents of savanna and forest soils

Location	*Vegetation*	*% organic carbon*	*% total nitrogen*	*Mean C:N ratio**	*Author*
Northern India	Forest	1.6–2.1	0.06–0.09	25	Singh (1989)
	Savanna	1.6–1.8	0.05–0.07	28.9	
Nicaragua	Forest	5.3	0.53	10	Alexander (1973)
	Savanna	2.1	0.14	15	
Southern Venezuela	Forest	0.7–1.4	0.06–0.09	13.6	Eden (1974)
	Savanna	1.2–1.4	0.08–0.09	16.3	
Northern Senegal	Savanna with trees	4.2–8.1	0.3–0.6	11.9	Bernhard–Reversat (1982)
Northern Brazil	Forest	1.7–2.0	0.03–0.04	52.7	This study
(Maracá)	Savanna	0.6–2.9	0.02–0.05	31.0	

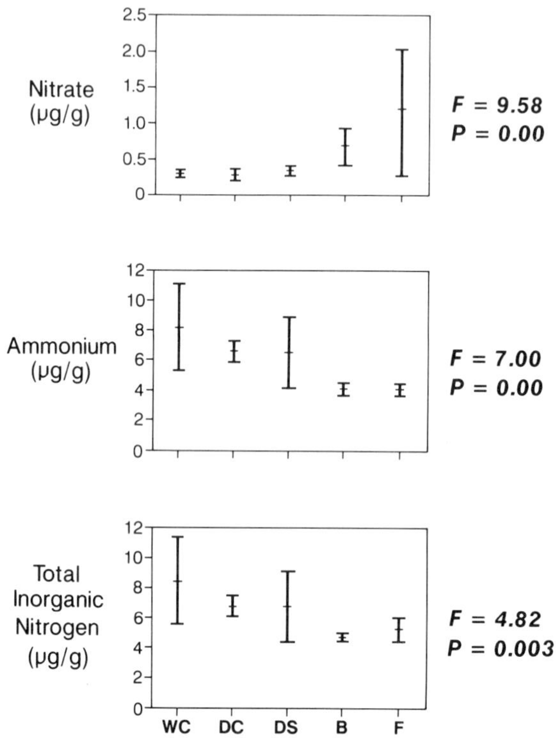

Fig. 8.4 Initial KCl-extractable soil nitrate and ammonium nitrogen in five vegetation habitats at the forest–savanna boundary. Habitat key as in Fig. 8.2. Statistical results are for one way analysis of variance, showing the *F*-statistic and probability level.

both soil carbon and nitrogen (see Table 8.1), and recorded C:N ratios ranging from 26.7–23.3 in the forest to 25.7–32.0 in the savanna.

Nitrogen mineralization

Early in the rainy season, ammonium and nitrate nitrogen concentrations in the surface-soil rooting zone show an inverse relationship with distance along the forest–savanna transect (Fig. 8.4). Ammonium concentrations are high and nitrate concentrations low under wet and dry campo, with low ammonium and high nitrate concentrations under forest. This trend can be well explained in terms of water-table level and soil moisture trends. The very much wetter conditions under wet campo allow the development of anaerobic conditions and prevent nitrification. Under these conditions, any nitrate which is produced would be quickly denitrified. The result is a build-up of ammonium nitrogen and a deficit

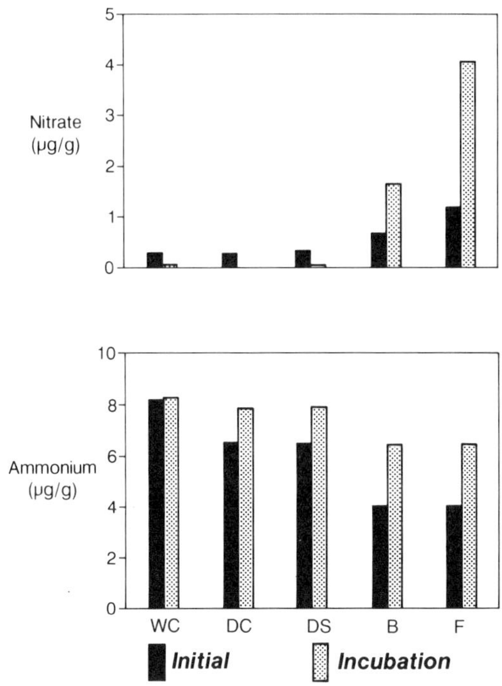

Fig. 8.5 Comparisons of (a) initial KCl-extractable NO_3-N, NO_3-N after incubation and NO_3-N after fumigation, and (b) initial KCl-extractable NH_4-N, NH_4-N after incubation and NH_4-N after fumigation for five vegetation habitats at the forest–savanna boundary. Habitat key as in Fig. 8.2.

of nitrate. The relationship between soil moisture content and KCl-extractable ammonium shows a significant positive correlation, with a correlation coefficient of 0.722 (n = 40). The relationship between soil moisture content and KCl-extractable nitrate is less strong and is negative, with a correlation coefficient of −0.505 (n = 40). Soil nitrate values under forest are highly variable. This may be due to a more heterogeneous distribution under the tree canopy of soil organic matter, soil organism activity, decomposition and nitrification processes. Since concentrations of ammonium are around six to eight times higher than those of nitrate, the overall trend for total inorganic nitrogen parallels that of ammonium.

The above values for 'available' inorganic nitrogen simply illustrate one point in time during the early wet season. Potentially mineralizable nitrogen in Maracá soils is illustrated in Fig. 8.5 as incubation NO_3-N and as incubation NH_4-N. In the case of ammonium nitrogen, potentially available amounts are higher than immediately available amounts in soils from all five habitats. The potential for nitrogen mineralization (i.e. the difference between initial and incubated N amounts) is highest under the forest and the forest–savanna boundary zone. Exactly the same pattern is seen for potential nitrification. Potentially available nitrate is nearly four times as high as immediately available nitrate under forest. This result demonstrates a larger amount of more easily mineralized organic nitrogen in forest sites, perhaps because of the different leaf-litter type under forest.

The main value of estimates of potentially available ammonium and nitrogen in these soil systems is to assess the magnitude of relative mobilization and immobilization processes. During the wet season, Marrs *et al.* (1991), working on the same forest/savanna transect as Thompson *et al.* (Chapter 18), found net immobilization of nitrogen in savanna sites, but net mobilization in adjacent forest sites. Although very low nitrification rates were observed during the wet season in the present study, no immobilization was found. Marrs *et al.* also reported much higher mineralization and nitrification rates for the savanna, shrub and forest sites than were found in the present study. It is rather difficult to assess why there is a difference between the two studies since Marrs *et al.* (in press) do not report information on total soil nitrogen, organic carbon or indices of soil microbial activity. Good correlations between total soil nitrogen and mean annual nitrogen mineralization (NH_4-N + NO_3-N) were reported by Bernhard-Reversat (1982) in savanna with trees in the Ivory Coast.

Other authors have reported a similar trend in soil nitrogen mineralization as was found on Maracá Island, with highest mineralization and nitrification rates in forest, declining with loss of tree cover, to lowest rates in treeless savanna or open pasture. A secondary rainforest succession in Costa Rica, extending from pasture to 60-year-old forest, showed lowest nitrogen mineralization rates in the pasture and gradually increasing nitrogen mineralization rates with age and forest maturity up to more than 60-years-old (Robertson, 1984). A fairly sharp increase in mineralization and nitrification rates under tree canopies, compared to open grass savanna, was measured by Bernhard-Reversat (1982). This type of result, namely, higher total soil nitrogen and higher nitrogen mineralization rates in dry campo compared to treeless savanna grassland, was not observed in the Maracá sites.

Apart from low total soil nitrogen, a second possible reason for low nitrogen mineralization and nitrification rates in this Maracá study, may be phosphorus deficiency. Purchase (1974) suggested that nitrification in Zimbabwean savanna soils was limited by low levels of available phosphorus ($<4\,\mu g\,g^{-1}$). All Maracá phosphorus availability results reported by Thompson *et al.* (Chapter 18) (extracted by Melich solution of hydrochloric and sulphuric acids) lie in the lower ranges of the Purchase (1974) index, at between 1–14 µg phosphorus g^{-1}.

Soil biology

The soil microbial biomass, soil respiration and mean numbers of extracted soil mesofauna are illustrated in Fig. 8.6 for replicate surface soil samples from each of the five habitats. Soil mesofauna data are not available for the forest–savanna boundary (B in Fig. 8.6). The different habitat conditions of wet and dry campos are picked out by microbial biomass and numbers of soil mesofauna. Mesofauna show a clear preference for the dryer surface soil of dry campo, while soil microbial biomass is higher in the wet campo, with higher soil organic carbon. For the three indices of biological activity illustrated in Fig. 8.6, the patterns for dry campo and terra firme forest are most alike. There is a strong relationship between microbial biomass in all habitats and soil moisture content (Fig. 8.7), giving a correlation coefficient of 0.822 (n = 40). The positive relationships between soil respiration and microbial biomass (r = 0.642; n = 40) and between soil respiration and soil moisture (r = 0.535; n = 40) content are not particularly strong. It is likely that soil microbial activity is strongly seasonal, depending on moisture influx during the wet season, but that the wet campo probably has a more persistent soil population, owing to generally wetter conditions during the dry season.

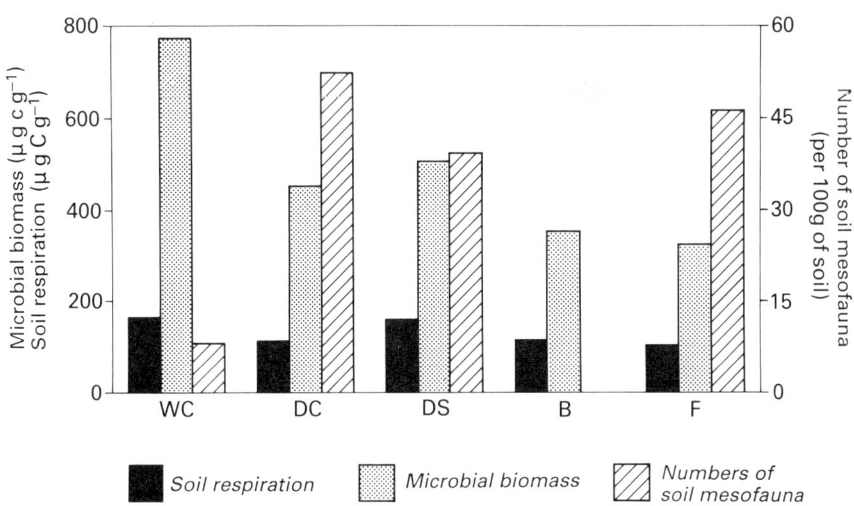

Fig. 8.6 Soil respiration, microbial biomass and mean extracted total soil mesofauna for five vegetation habitats at the forest–savanna boundary. Habitat key as in Fig. 8.2.

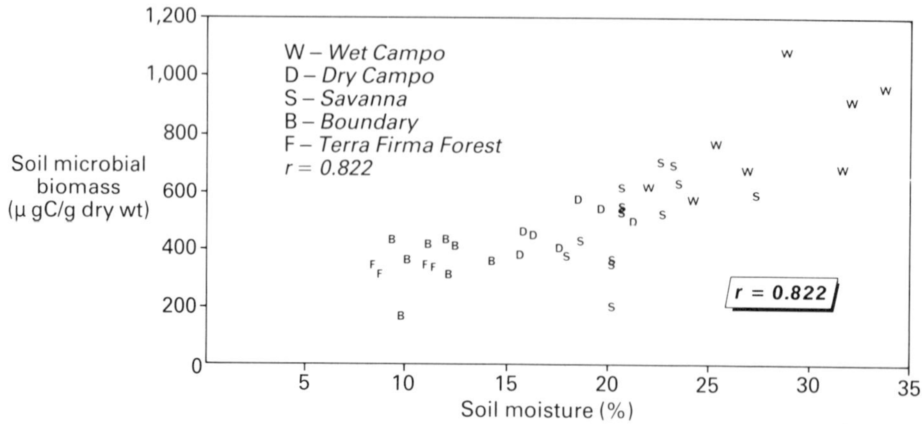

Fig. 8.7 Relationship between soil microbial biomass and gravimetric soil moisture content (wet season) for five vegetation habitats at the forest/savanna boundary, showing the correlation coefficient.

Soil microbial biomass

KCl-extractable ammonium and nitrate nitrogen after fumigation–incubation represent the flush of decomposition which occurs as organisms killed by the fumigant are decomposed by the recolonizing population. The flush thus represents microbially immobilized nitrogen, minus that immobilized by the recolonizing population. The first mineralization flush after fumigation commonly produces as much as five times as much mineral nitrogen in the ten days after fumigation as does unfumigated soil (Shen *et al.*, 1984). The production of ammonium nitrogen is usually rapid immediately after fumigation, with nitrification peaking as much as 10–20 days after fumigation. In Fig. 8.5 it is noteworthy that, for all five habitats studied, amounts of microbially immobilized N which is mineralized as NH_4 are all higher than that potentially mineralizable from soil organic matter. Microbial NH_4 is more than twice that which could potentially be mobilized from soil organic matter (incubation NH_4 in Fig. 8.5b) in dry campo, savanna, boundary and forest soils. Amounts of NO_3-N released on fumigation are much lower than NH_4-N (Fig. 8.5a), although in wet campo, dry campo and savanna soils, even these small amounts are greater than that potentially nitrified from the soil (incubation NO_3 in Fig. 8.5a). These results emphasize the value, particularly in nutrient-poor soils such as those of the Maracá, of the soil microbial biomass as a potential source of readily mineralized nitrogen (Jenkinson and Ladd, 1981). Singh *et al.* (1989) suggested that the inverse relationship reported in several savanna ecosystem studies between microbial biomass and plant growth probably indicates that nutrients, immobilized by microbes, are a major pool for plants and that these processes are valuable methods of conserving soil nutrients. In Maracá soils, microbial nitrogen represented 3.24–8.1% of the total soil nitrogen in the savanna and 4.7–6.27% of the total soil nitrogen in the forest. These proportions are similar to those reported for temperate soils (e.g. Paul, 1984; Bonde *et al.*, 1988). The amount of microbially immobilized nitrogen depends on the availability of readily decomposable soil organic matter. Thus variable amounts of microbial nitrogen might be expected, depending on litter amount and quality. Singh *et al.* (1989) found nitrogen immobilized in Indian soils to be much higher than in Maracá soils. They reported microbially immobilized nitrogen in forest soils which was around 10–20 times higher than initially extractable mineral nitrogen, and microbial nitrogen in savanna soils which was around 8–10 times higher than initially extractable mineral nitrogen.

Fungi are probably a major component of the microbial biomass in Maracá soils, because of the soil acidity (around pH 4.5 on all sites) and the existence in some experimental mesh litter bags of fungal hyphae, particularly samples taken from the field after six months. Only in the wet campo site, under waterlogged conditions, are fungi probably replaced by anaerobic bacteria as the dominant group.

In our studies of the soil biology of hummock and hollow microtopographic locations in the campo and savanna sites, only soil microbial biomass and fumigation (microbially immobilized) mineral-nitrogen were found to be significantly different in hummocks compared with hollows, giving Student's *t*-test values of $t = 2.3$, $p = 0.033$ and $t = 2.44$, $p = 0.027$ respectively for a total of 16 samples in each microtopographic location. No significant differences were found for soil moisture, soil organic carbon or total soil nitrogen. This appears to indicate that increased microbial activity beneath grass tussocks is not directly due to soil physical and chemical differences, but may be related to root distribution and density. Root biomass and soil microbial biomass are, however, not significantly correlated.

Soil mesofauna

Soil organisms extracted from soil cores in May (wet season) show that the three drier sites, dry campo, dry savanna and forest, have relatively high animal numbers, up to 70 animals per core, while there is only around one-seventh this number in the wet campo (Fig. 8.8). Soil fauna extractions from leaf litter bags tethered in the field for up to six months show that at the beginning of the dry season (December), in the dry campo, organism numbers per litter bag rose to an average of 213, although there was much variation between replicates. Results for the three main soil faunal groups, Acari (mites) and Collembola (springtails)

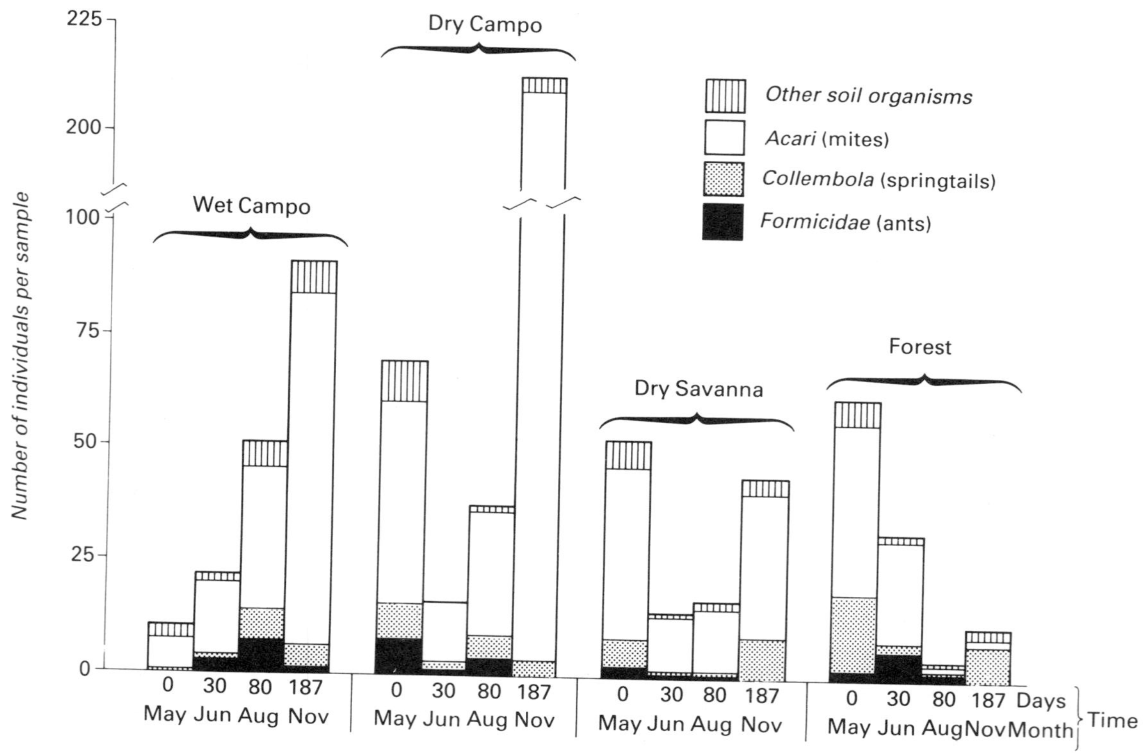

Fig. 8.8 Mean numbers and composition of soil mesofauna extracted by Berlese–Tullgren apparatus per soil core (0 days, May), and per leaf litter bag (at 30, 80 and 187 days, representing June, August and November, 8 replicates per date).

and Formicidae (ants), are shown in Fig. 8.8, and for the less frequent groups in Table 8.2. All numbers of soil organisms are reported as means per soil core (eight replicates per site) or per leaf litter bag (four replicates per site). This allows direct comparisons to be made between sites. Numbers of Acari reflect the trend in total soil organisms, of very low numbers in wet campo and around eight times as many individuals in the other four habitats. In contrast, Collembola numbers in both dry campo and dry savanna are low, perhaps because these two sites also experience wet and ponded water conditions at certain times of the year. In common with studies in other tropical savanna soils (e.g. Lavelle, 1983), the Acari were the most abundant group in Maracá soils. Beck (1971) reported microarthropod numbers of around 70 000 individuals m^{-2} in terra firme forest in Amazonia. If the Maracá data are calculated in this way, the soil core data (0 days in Fig. 8.8) give maximum numbers of soil organisms in the dry campo, with around 4200 Acari m^{-2} and around 1000 Collembola m^{-2} in the top 8 cm of soil. Populations of organisms (around 4000 Acari m^{-2} and 2000 Collembola m^{-2}) were recorded in soil cores rom the forest. Of the mesofauna and macrofauna, only termites are commonly found at depth in the soil. Most mesofauna occupy the surface 10–20 cm of soil, in the zones predominantly exploited by plant roots.

Collembola and Acari generally live on litter surfaces and in air-filled micropores in the soil. Smaller species feed more selectively on fungi (Anderson, 1975). Acari can have more general feeding habits during certain times of the year, but may have 'feeding refuges' during periods of limited food availability due to unfavourable environmental conditions such as flooding or drought when they are forced from their characteristic surface soil habitat (Anderson, 1975).

Levings and Windsor (1982) noted a distinct pattern in soil microarthropod numbers in forest soils of Barro Colorado Island, Panama. Numbers of soil fauna groups and of microarthropods decreased sharply at the onset of the dry season, but rose sharply at the beginning of the wet season. Numbers of most soil fauna groups were strongly associated with litter mass and litter moisture. In the Maracá habitats, wet-season soil fauna numbers were not significantly correlated with litter mass. In the Lamto savanna of the Ivory Coast, soil microarthropods are always more numerous under grass tussocks than in bare areas between clumps of grass (Malcevschi, 1978). We tested for this in Maracá savanna and campo sites by extracting soil fauna from 12 replicate hummock and 16 replicate hollow sites. The total numbers of soil fauna extracted from the different sites were found to be significantly different at $p = 0.037$. Despite the lack of significant difference in this study between soil moisture content under hummocks and hollows; it may be that grass tussocks act to channel rainfall into the soil, so that tussock rooting zones

Table 8.2 Numbers of soil mesofauna extracted from soil cores* and leaf litter** samples (Data are means of 8 replicates; standard deviations in brackets)

Faunal Group	*Site*	*Initial* (May, 0 days)*		*30 days*** (June)*		*80 days*** (August)*		*187 days*** (November)*	
	WC	0.4	(0.52)	1.0	(1.60)	6.9	(7.30)	5.3	(3.8)
	DC	8.0	(6.3)	1.9	(1.73)	4.6	(3.71)	3.1	(3.6)
Collembola	DS	6.1	(5.5)	0.5	(0.75)	1.0	(2.07)	8.5	(6.6)
	F	16.8	(9.4)	2.3	(3.6)	0.4	(1.05)	7.0	(5.8)
	WC	0.5	(0.35)	0.37	(0.52)	0.8	(0.63)	0.4	(0.52)
	DC	0.6	(0.57)		0		0		0
Coleoptera	DS	1.0	(0.61)		0	0.3	(0.46)	1.3	(0.40)
	F	1.5	(0.98)	0.12	(0.35)	0.1	(0.35)	0.3	(0.46)
	WC	0.3	(0.71)	0.37	(0.74)	0.5	(0.92)	3.4	(1.85)
	DC	0.4	(0.52)		0		0	0.5	(1.07)
Diptera	DS	0.4	(0.52)	1.1	(1.88)	0.9	(1.36)	1.8	(1.28)
	F	1.0	(0.89)	0.12	(0.35)	0.4	(0.52)	0.5	(0.53)
	WC	0.1	(0.35)		0	2.3	(4.10)	0.8	(2.87)
	DC	5.6	(1.42)		0	0.1	(0.35)	2.8	(5.87)
Homoptera	DS*	0.5	(12.4)		0		0		0
	F	0.5	(0.84)	0.12	(0.35)		0		0
	WC	0.3	(0.46)	3.0	(8.09)	7.0	(18.1)	0.8	(1.16)
	DC	7.6	(17.9)	1.0	(1.60)	3.8	(10.2)	0.1	(0.35)
Formicidae	DS	2.0	(2.67)	0.3	(0.46)	0.1	(0.35)		0
	F	1.3	(1.21)	5.5	(13.6)	0.4	(1.05)		0
	WC	6.8	(3.5)	16.1	(25.2)	31.9	(27.6)	78.5	(74.9)
	DC	45.1	(24.3)	13.1	(15.5)	28.3	(15.1)	205.8	(426.2)
Acari	DS	38.1	(20.3)	12.3	(19.0)	13.8	(10.7)	31.8	(13.2)
	F	37.8	(18.1)	22.3	(15.4)	1.6	(1.77)	1.8	(1.4)
	WC		0		0		0		0
	DC	1.6	(1.50)		0		0		0

Symphyla	DS	0.5	(0.75)		0		0		0
	F	0.7	(1.03)	0.1	(0.35)		0		0
	WC	0.1	(0.35)	0.4	(1.06)		0	0.4	(0.52)
	DC		0	0.1	(0.35)	0.1	(0.35)		0
Earthworms	DS	0.1	(0.35)		0	0.1	(0.35)		0
	F	0.2	(0.41)	0.1	(0.35)	0.1	(0.35)	0.8	(1.16)
	WC	10.9	(4.05)	22.0	(28.2)	51.8	(35.4)	91.8	(79.1)
	DC	70.0	(46.0)	16.3	(17.0)	37.3	(22.3)	213.0	(424.1)
Total fauna	DS	52.5	(31.1)	14.1	(20.1)	16.6	(12.3)	43.9	(18.3)
	F	61.7	(26.9)	31.6	(34.6)	3.1	(3.44)	10.6	(7.71)
	WC	3.8	(1.28)	2.6	(2.56)	4.5	(3.62)	5.8	(1.60)
	DC	5.9	(1.81)	2.1	(1.04)	2.8	(0.89)	3.1	(1.46)
No. of faunal groups	DS	6.3	(3.01)	1.7	(1.19)	2.9	(1.46)	4.1	(0.10)
	F	6.5	(1.37)	2.6	(2.39)	1.6	(1.06)	3.0	(1.93)

WC = Wet campo; DC = Dry campo; DS = Savanna; F = Terra firme forest become better supplied with water than surrounding soil and hence provide a more suitable environment for soil organisms. If roots of grass tussocks do influence soil organism numbers, several factors could be involved: root exudates, root litter, root architecture or root density distribution, since root biomass and soil fauna numbers are not significantly correlated.

CONCLUDING DISCUSSION

Five habitats at the boundary between forest and savanna on Maracá Island showed significantly different soil chemical and soil biological characteristics. The differences are shown to be closely linked to the level of the water table and its effect on soil moisture conditions in the surface 10 cm of soil which is the zone predominantly occupied by soil fauna and soil micro-organisms. Several questions have been raised concerning the availability of soil nitrogen, such as the possibility that all Maracá sites may be phosphorus-deficient as far as N-mineralization and nitrification processes are concerned (Purchase, 1974). To what extent the vegetation of these ecosystems is nutrient-stressed remains unknown, although Thompson *et al.* (Chapter 18) suggest that both nitrogen and phosphorus deficiency may be limiting. In many cases, it is the synergistic effect of phosphorus deficiency on nitrogen availability which promotes the nutrition imbalance.

Quantification of microbially immobilized nitrogen in this study shows that the microbial biomass is an important source of easily mineralized nitrogen for plant uptake. However, environmental conditions causing autolysis of microbial cells, such as waterlogging, anaerobism or desiccation, are also likely to inhibit root growth and nutrient uptake. The synchronization of microbial release of nutrients and the plants' uptake abilities are critical for good nutrient conservation. In savanna and campos with periodic innundation, the possibilities for nutrient losses from surface soil through leaching when the watertable is lower in the profile, and, in the case of nitrogen, denitrification, are high. Despite the view of Singh *et al.* (1989) that microbially immobilized nitrogen may be a major source for plant uptake, Schnurer and Rosswall (1987) found that only 10% of nitrogen immobilized in the soil microbial biomass was taken up by barley plants during the growing season.

Significant differences in above-ground herbage biomass and root biomass between the five vegetation habitats in Maracá savanna probably indicate differences in the magnitude of nutrient cycling in these habitats.

The results reported in this chapter for soil habitats near to the forest–savanna boundary on Maracá Island relate to a short-term study which provides only an indication of on-going biological soil processes. The findings require verification over wet- and dry-season conditions, together with the wetting-up and drying-down periods between seasons. It would be useful to extend the present study to cover soil chemical and biological dynamics in relation to soil hydrology over at least a one-year period, to confirm whether the trends reported in this paper relate to the wet season only, and to examine the relative dry-season microbial and soil fauna activity and nitrogen dynamics operating in the different habitats studied. There is a need for integrated studies of microbial immobilization and release of nutrients, particularly nitrogen and phosphorus in conjunction with (a) studies of relative nutrient uptake rates and amounts in forest and savanna, and (b) studies of *in situ* soil solute concentrations of the major plant nutrients, under varying conditions of soil moisture and redox conditions.

ACKNOWLEDGEMENTS

The authors wish to thank the Brazilian Environment Secretariat (SEMA) for the use of field sites and the field station on Maracá Island and the National Amazonian Research Institute (INPA) for the use of laboratory facilities. They thank Miss S. Aitade who assisted in collection and preparation of field samples, Dr J. Thompson and Mr D.A. Scott (Stirling University) who collected seasonal water-table data along the forest–savanna transect of this study area in 1987, and Dr J. Ferraz who described soil profiles and vegetation along the forest–savanna transect. The Royal Society is thanked for financial support for S. Ross.

REFERENCES

Alexander, E.B. (1973) A comparison of forest and savanna soils in north-eastern Nicaragua. *Turrialba*, **23**, 181–91.

Allen, S.E. (1989) *Chemical Analysis of Ecological Materials*, 2nd edn, Blackwell Scientific, Oxford.

Anderson, J.M. (1975) Succession diversity and trophic relationships of some soil animals in decomposing leaf litter. *Journal of Animal Ecology*, **44**, 475–95.

Askew, G.P., Moffatt, D.J., Montgomery, R.F. and Searl, P.L. (1970) Soil landscapes in north-eastern Mato Grosso. *Geographical Journal*, **136**, 211–27.

Beck, L. (1971) Bodenzoologische Gliederung und Charakteriserung des amazonischen Regenwaldes. *Amazoniana*, **3**, 69–132.

Bernhard-Reversat, F. (1982) Biogeochemical cycle of nitrogen in a semi-arid savanna. *Oikos*, **38**, 321–32.

Black, C.A. (1965) *Methods of Soil Analysis. Part 2: Chemical and Microbiological Properties*, American Society of Agronomy, Madison, Wisconsin.

Bonde, T.A., Schnurer, J. and Rosswall, T. (1988) Microbial biomass as a fraction of potentially mineralizable nitrogen in soils from long-term field experiments. *Soil Biology and Biochemistry*, **17**, 837–42.

Brookes, P.C., Landman, A., Pruden, G. and Jenkinson, D.S. (1985) Chloroform fumigation and the release of soil nitrogen: a rapid direct extraction method to measure microbial biomass nitrogen in soil. *Soil Biology and Biochemistry*, **17**, 837–42.

Dobereiner, J. and Day, J.M. (1975) Nitrogen fixation in the rhizosphere of tropical grasses, in *Nitrogen Fixation by Free Living Microorganisms*, (ed. W.D.P. Stewart), Cambridge University Press, Cambridge, pp. 39–56.

Eden, M.J. (1974) Palaeoclimatic influences and the development of savanna in southern Venezuela. *Journal of Biogeography*, **1**, 95–109.

Furley, P.A. and Ratter, J.A. (1990) Pedological and botanical variations across the forest–savanna transition on Maracá Island. *Geographical Journal*, **156**, 251–66.

Henriksen, S. and Selmer-Olsen, A.R. (1970) Automatic methods for determining nitrite and nitrate in water and soil extracts. *Analyst*, **95**, 514–18.

Hesse, P.R. (1971) *A Textbook of Soil Chemical Analysis*, John Murray, London.

Jenkinson, D.S. and Ladd, J.N. (1981) Microbial biomass in soil: measurement and turnover, in *Soil Biochemistry*, Vol. 5, (eds E.A. Paul and J.N. Ladd), Dekker, New York, pp. 415–71.

Jenkinson, S. and Poulson, D.S. (1976) The effects of biocidal treatments on metabolism in soil. 5. A method for measuring soil biomass. *Soil Biology and Biochemistry*, **8**, 209–13.

Jung, G. (1967) *Influence de* l'Acacia albida *(Del.) sur la biologie des sols*, ORSTROM, Hann-Dakar.

Kellman, M. (1979) Soil enrichment by neotropical savanna trees. *Journal of Ecology*, **67**, 565–77.

Kellman, M. (1989) Mineral nutrient dynamics during savanna–forest transformation in Central America, in *Mineral Nutrients in Tropical Forest and Savanna Ecosystems*, (ed. J. Proctor), Blackwell Scientific, Oxford, pp. 137–51.

Lavelle, P. (1983) The soil fauna of tropical savannas. 1. The community structure, in *Ecosystems of the World, 13: Tropical Savannas*, (ed. F. Bourlière), Elsevier, Amsterdam, pp. 477–84.

Levings, S.C. and Windsor, D.M. (1982) Seasonal and annual variations in litter arthropod populations, in *The Ecology of a Tropical Forest. Seasonal Rhythms and Long-term Changes*, (eds E.G. Leigh, A.S. and D.M. Windsor), Smithsonian Institution Press, Washington DC, pp. 355–87.

Malcevschi, S. (1978) Considerations quantitatives sur la nematofaune d'une savane herbeuse à Andropogonées de Lamto (Cote D'Ivoire). *Revue d'Ecologie et Biologie du Sol*, **15**, 487–96.

Marrs, R.H., Thompson, J., Scott, D. and Proctor, J. (1991) Nitrogen mineralisation and nitrification in terra firme forest and savanna soils on Ilha de Maracá, Roraima, Brazil. *Journal of Tropical Ecology*, **7**, 123–37.

Milliken, W. and Ratter, J.A. (1989) *Vegetation of the Ilha de Maracá, Roraima, Northern Brazil*, Royal Botanic Garden, Edinburgh.

Nortcliff, S. and Robison, D. (1989) *The Soils and Geomorphology of the Ilha de Maracá, Roraima. The Second Approximation*, Royal Geographical Society, London.

Paul, E.A. (1984) Dynamics of organic matter in soils. *Plant and Soil*, **76**, 275–85.

Purchase, B.S. (1974) The influence of phosphate deficiency on nitrification. *Plant and Soil*, **41**, 541–7.

Raud, G. (1977) Données preliminaires sur les cyanophycées du sol fixatrices d'azote dans la savane de Lamto. *Revue d'Ecologie et Biologie du Sol*, **14**, 311–19.

Robertson, G.P. (1984) Nitrification and nitrogen mineralization in a lowland rainforest succession in Costa Rica, Central America. *Oecologia*, **61**, 99–104.

Robertson, G.P. and Vitousek, P.M. (1981) Nitrification potentials in primary and secondary succession. *Ecology*, **62**, 376–86.

Sarmiento, G. (1983) The Savannas of Tropical America, in *Ecosystems of the World, 3: Tropical Savannas*, (ed. F. Bourlière), Elsevier, Amsterdam, pp. 245–87.

Sarmiento, G. (1984) *The Ecology of Neotropical Savannas*, Harvard University Press, Harvard.

Schnurer, J. and Rosswall, T. (1987) Mineralisation of nitrogen from ^{15}N labelled fungi, soil microbial biomass and roots and its uptake by barley plants. *Plant and Soil*, **102**, 71–8.

Shen, S.M., Pruden, G. and Jenkinson, D.S. (1984) Mineralisation and immobilisation of nitrogen in fumigated soil and the measurement of microbial biomass nitrogen. *Soil Biology and Biochemistry*, **16**, 437–44.

Singh, J.S., Raghubanshi, A.S., Singh, R.S. and Srivastava, S.C. (1989) Microbial biomass acts as a source of plant nutrients in a dry tropical forest and savanna. *Nature*, **338**, 499–500.

Singh, K.P. (1989) Mineral nutrients in tropical dry deciduous forest and savanna ecosystems in India, in *Mineral Nutrients in Tropical Forest and Savanna Ecosystems,* (ed. J. Proctor), Blackwell Scientific, Oxford, pp. 153–68.

Stanford, G. and Smith, S.J. (1972) Nitrogen mineralisation potentials of soil. *Proceedings, Soil Science Society America,* **36**, 465–72.

Vance, E.D., Brokes, P.C. and Jenkinson, D.S. (1987) Microbial biomass measurements in forest soils: the use of a chloroform fumigation–incubation method in strongly acid soils. *Soil Biology and Biochemistry,* **19**, 697–702.

Vitousek, P.M. and Denslow, J.S. (1986) Nitrogen and phosphorus availability in treefall gaps of a lowland tropical rainforest. *Journal of Ecology,* **74**, 1167–78.

Vitousek, P.M. and Matson, A. (1988) Nitrogen transformations in tropical forest soils. *Soil Biology and Biochemistry,* **20**, 361–7.

9

Observations on soils, foliar nutrient concentrations and floristic composition of cerrado *sensu stricto* and *cerradão* communities in central Brazil

M. HARIDASAN

SUMMARY

1. Soil chemistry, comparative floristic composition and foliar concentrations of calcium, magnesium and potassium of savanna woodland (cerrado *sensu stricto*) and xeromorphic forest (*cerradão*) on both dystrophic and mesotrophic soils are considered in relation to the savanna–forest transition.
2. Occurrence of various characteristic tree species indicating the higher calcium levels of the mesotrophic soils is discussed.
3. Species of cerrado *s.s.* and cerradão communities on dystrophic (nutrient-poor) soils have lower foliar concentrations of calcium, magnesium and potassium than the corresponding communities on mesotrophic soils (of medium nutrient status).
4. The cerrado *s.s.* and cerradão communities on dystrophic soils do not differ in foliar concentrations of calcium, magnesium and potassium, although their floristic composition is dissimilar. The same applies to the equivalent communities on mesotrophic soils.
5. Cerrado *s.s.*–cerradão transitions on dystrophic soils are usually not accompanied by an increase in soil fertility or a reduction in exchangeable aluminium. Aluminium toxicity is not a limiting factor preventing savanna woodland developing into forest.
6. It seems possible that cerradão on dystrophic soils has established when the soils were richer, and since then they have become

impoverished. Some workers consider that cerradão may be a natural succession of cerrado *s.s.* on dystrophic soils when there is protection from fire. This, however, seems unlikely since nutrients in the dystrophic soil would be insufficient to support such an increase in biomass.

INTRODUCTION

The cerrado (savanna) region of central Brazil extends over 180 million ha and comprises widely varying physiognomic types of vegetation. The most extensive of these is an open tree and scrub woodland (Eiten, 1972) (savanna woodland/*campo cerrado* according to Cole, 1986, Chapter 5). This savanna woodland is known as cerrado *sensu stricto* and originally occupied more than 65% of the region. Xeromorphic forest, on upland soils, known as cerradão (the augmentative of cerrado), occupied an estimated area of nearly 20 million ha and accounted for almost 11% of the geographic area.

The physiognomic form of cerrado *s.s.* seems to be a consequence of the dystrophic nature of the soil in combination with climatic type and the effect of intermittent fires. Cerrado *s.s.* occurs most commonly on deep, well drained, dystrophic, acid latosols. When it occurs on mesotrophic soils with higher calcium and magnesium levels (see Table 9.1), the floristic composition is different from that on dystrophic soils, although the physiognomy is similar (Ratter *et al.*, 1973, 1977). Sandy soils (Typic Quartzipsamments) also support cerrado *s.s.* over large areas.

Cerradões (plural of cerradão) occur on dystrophic as well as mesotrophic soils, and their floristic composition varies according to soil base status (Ratter, 1971; Ratter *et al.*, 1973, 1977, 1978a,b). In many situations, cerrado *s.s.* and cerradão vegetation occur side by side on very similar upland dystrophic soils. The occurrence of cerradão in such cases has often been attributed to the presence of a more fertile soil (Goodland, 1971a,b; Goodland and Pollard, 1974). However, a number of workers have demonstrated that cerrado *s.s.* and cerradão can occur on dystrophic soils of similar nutrient status (Askew *et al.*, 1970; Ratter *et al.*, 1973; Haridasan, 1982; Ribeiro, 1983, etc.).

The aim of this chapter is to consider the information available from foliar nutrient data, soil characteristics and comparison of floristic composition in relation to differentiation of cerrado *s.s.* and cerradão on both dystrophic and mesotrophic soils in central Brazil.

MATERIALS AND METHODS

Phytosociological surveys of selected native cerrado *s.s.* and cerradão communities on mesotrophic and dystrophic soils were conducted to compare their floristic composition (Haridasan, 1987; Haridasan and Araújo, 1988; Ribeiro, 1983). All tree and shrub species occurring in native vegetation on a deep, well drained dark red latosol at Fazenda Água Limpa, Brasília, were sampled once and leaf samples analysed for phosphorus, potassium, calcium, magnesium and aluminium (Haridasan, 1982, 1987). All tree species in a small patch of cerrado *s.s.* on a calcium-rich soil in the same area were sampled to determine their leaf nutrient concentrations. These data on cerrado *s.s.* communities were then compared with those from two cerradão communities, one on a dystrophic soil at Brasília Botanic Garden and the other on a mesotrophic soil in the Município of Padre Bernardo, Goiás (Haridasan and Araújo, 1988).

RESULTS

Soil chemistry

The chemistry of the dystrophic and mesotrophic soils which support the cerrado *s.s.* and cerradão communities are compared in Table 9.1. There were no significant differences between the dystrophic soils which support cerrado *s.s.* and cerradão. On the other hand, there was an important difference between the calcium-rich soil which supports a cerrado *s.s.* at the Fazenda Água Limpa site in the Federal District and the mesotrophic soil which supports the cerradão in Padre Bernardo. The calcium levels, as well as those of available potassium and magnesium, were high only in the surface horizon of the soil of the mesotrophic cerrado *s.s.*, dropping off rapidly in the sub-surface horizons. On the other hand, the levels of calcium and magnesium in the mesotrophic cerradão soil fell much less rapidly with depth in the profile. The mesotrophic soils had significantly lower levels of available aluminium in the surface layers than the dystrophic soils.

Floristic composition

Cerrado *s.s.* on dystrophic and mesotrophic soils

Comparison of the floristic composition of cerrado *s.s.* communities on dystrophic and mesotrophic soils was presented in Haridasan (1987). Among the 20 species which were encountered in the mesotrophic soil cerrado, all except six also occurred in the dystrophic soil cerrado. The six

Table 9.1 The chemical characteristics of dystrophic and mesotrophic soils which support cerrado *sensu stricto* and cerradão

Soil horizon (depth, cm)	*Soil pH in water*	*Soil pH in 1 N KCl*	*Organic carbon (%)*	*Total N ($mg\,g^{-1}$)*	*Exchangeable cations ($mg\,kg^{-1}$ soil) Ca*	*Mg*	*K*	*Al*
			Dystrophic soil					
Cerrado *s.s.* (Fazenda Água Limpa, Brasília)								
Al (0–12)	4.9	3.7	3.3	2.3	30	16	64	108
A3 (12–28)	5.0	4.1	1.8	1.5	5	2	31	45
B21 (28–56)	5.2	4.4	1.1	0.8	2	2	11	13
B22 (56–100+)	5.5	5.0	0.7	0.5	1	2	6	5
Cerradão (Jardim Botânico de Brasília, Brasília)								
Al (0–5)	4.9	3.7	3.9	1.5	22	10	70	243
A3 (5–25)	5.0	4.1	2.4	0.9	4	4	20	54
B21 (25–60)	5.2	4.4	2.0	0.6	2	2	12	18
B22 (60–120+)	5.5	5.0	1.7	0.5	1	1	4	9
			Mesotrophic soil					
Cerrado *s.s.* (Fazenda Água Limpa, Brasília)[a]								
Al (0–15)	6.2	5.3	5.3	3.3	3837	312	150	5
B21 (15–30)	5.2	4.3	1.5	1.5	29	10	27	45
B22 (30–45)	5.1	4.6	1.2	1.0	12	8	19	18
Cerradão (Fazenda Lagoa Santa, Padre Bernardo)								
Al (0–10)	6.0	4.3	3.2	1.8	2260	166	187	54
B1 (10–33)	5.9	4.3	2.2	1.0	1000	41	27	36
B21 (33–56)	5.8	4.2	1.6	0.7	1280	43	16	27
B22 (56–120+)	5.8	4.4	1.2	0.06	720	71	20	18

[a] Soil samples collected by Dr Peter A. Furley (see Furley, 1985).

species restricted to the mesotrophic cerrado were *Terminalia argentea* Mart. and Zucc., *Luehea paniculata* Mart., *Platypodium grandiflorum* Benth., *Pseudobombax tomentosum* (Mart. and Zucc.) A. Robyns, *Pseudobombax longiflorum* (Mart. and Zucc.) A. Robyns and *Aegiphila lhotskyana* Cham. Among these six, only the first three were among the five most phytosociologically important species of the community (in terms of importance value index). However, four of these species are regarded as characteristic of cerrado soils with higher calcium levels (Ratter *et al.*, 1977), while *P. longiflorum* is often associated with such soils and only *A. lhotskyana* can be regarded as 'calcium-indifferent' in its soil requirements (J.A. Ratter, personal communication).

Cerradão on dystrophic and mesotrophic soils

Comparison of the floristic composition of cerradão communities on dystrophic and mesotrophic soils were presented in Haridasan and Araújo

(1988). Thirty-nine species belonging to 24 families were encountered on the dystrophic soil and the same number of species belonging to 22 families on the mesotrophic soil. Only six species were common to both communities, all of which generally occur with greater frequencies on dystrophic soils of the cerrado region. Several species among the more important ones of the mesotrophic cerradão were classified by Ratter *et al.* (1978a) as exclusively calciphilous: *Callisthene fasciculata* (Spr.) Mart., *Astronium urundeuva* (Fr. All.) Engl., *Dilodendron bipinnatum* Radlk. and *Celtis iguanaea* (Jacq.) Sargent.

Cerrado *s.s.* and cerradão on dystrophic soils

Data on the floristic composition of a cerrado *s.s.* and a cerradão occurring at Fazenda Água Limpa on very similar dystrophic soils were presented by Haridasan (1987). Thirty-eight species belonging to 28 families were found in the cerrado and 51 species belonging to 33 families in the cerradâo. Twenty-eight species were common to both communities. Only one of the phytosociologically more important species of the cerradão, *Ocotea spixiana* (Nees) Mez, was not encountered in the cerrado community. Similarly, only one of the more important species of the cerrado *s.s.* was not found in the cerradão: *Ouratea hexasperma* (St. Hil.) Baill.

Cerrado *s.s* and cerradão on mesotrophic soils

Comparison of the floristic composition of the cerrado *s.s.* and cerradão communities on mesotrophic soils is also interesting. Among the 20 species belonging to 16 families occurring in the cerrado *s.s.* on the calcium-rich soil, only six were found in the cerradão on the mesotrophic soil in Padre Bernardo. None of the nine most important species of the Padre Bernardo community occurred in the cerrado *s.s.* on calcium-rich soil of Fazenda Água Limpa. However, it is extremely important to note that four of the species in common, *Terminalia argentea*, *Luehea paniculata*, *Platypodium grandiflorum* and *Pseudobombax tomentosum*, are indicators of soils richer in calcium in the cerrado region, and it was their presence that alerted us to the existence of this small patch of mesotrophic soil (Furley *et al.*, 1984; Ratter, 1980, 1985, 1986). Thus the recognition of floristic clues of this type can be very important, even when the percentage similarity of the vegetation is low.

Foliar concentrations of nutrients

Detailed data by species of foliar concentrations of nutrients in the cerrado *s.s.* community on the dystrophic soil at Fazenda Água Limpa were presented by Haridasan (1987) and those for the cerradão communities by Haridasan and Araújo (1988). Since there were no major differences among the soils in nitrogen and phosphorus levels, foliar concentrations of these nutrients are not discussed in the present paper.

The frequency distribution of calcium, magnesium and potassium for the four communities are presented in Figs 9.1–9.3. The means and standard deviations for each community are summarized in Table 9.2.

The calcium concentrations in the leaves of trees varied from less than $1\,mg\,g^{-1}$ to about $10\,mg\,g^{-1}$ in the plant communities on dystrophic soils, with mean values of $2.33\,mg\,g^{-1}$ for the cerradão and $3.48\,mg\,g^{-1}$ for the cerrado *s.s.* The corresponding mean concentrations for the mesotrophic soil communities were more than three times these values: $10.73\,mg\,g^{-1}$ for the cerradão and $7.22\,mg\,g^{-1}$ for the cerrado *s.s.* The mean concentrations of magnesium on mesotrophic soil communities (cerrado: 2.83; cerradão: $2.71\,mg\,g^{-1}$) were also higher than those for dystrophic soil communities (cerrado: 1.95; cerradão; $1.78\,mg\,g^{-1}$). Leaf concentrations of potassium also varied similarly between the two soils. The order of mean foliar concentrations was $K > Ca > Mg$ in dystrophic soils and $Ca > K > Mg$ in mesotrophic soils.

DISCUSSION

It is clear that the plant communities on the mesotrophic soils (cerrado *s.s.* or cerradão) possess higher foliar concentrations of potassium, calcium and magnesium than those on dystrophic soils (Figs 9.1–9.3, Table 9.2). Analyses also show that when soil fertility does not vary, cerrado *s.s.* and

Table 9.2 Mean foliar concentrations of Ca, Mg and K in the tree species of cerrado *sensu stricto* and cerradão communities on dystrophic and mesotrophic soils in central Brazil (number in parentheses represents standard deviation)

Nutrient	*Soil*	*Concentration ($mg\,g^{-1}$ dry matter)* Cerrado *s.s.*	*Cerradão*
Ca	Dystrophic	3.48 (1.95)	2.33 (1.13)
	Mesotrophic	7.22 (4.68)	10.73 (6.86)
Mg	Dystrophic	1.95 (0.74)	1.78 (0.78)
	Mesotrophic	2.83 (1.65)	2.71 (1.30)
K	Dystrophic	4.98 (3.56)	4.81 (2.05)
	Mesotrophic	6.68 (3.56)	7.53 (3.17)

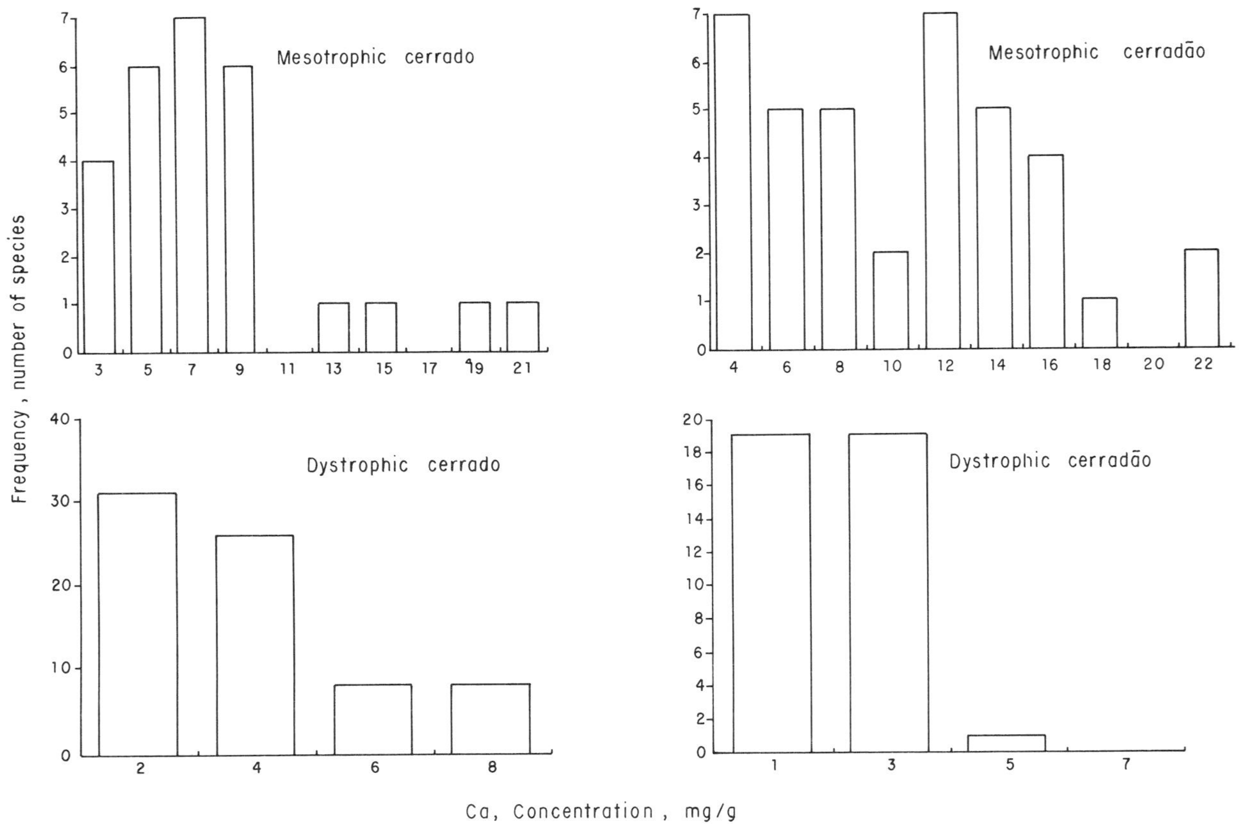

Fig. 9.1 Distribution of foliar concentrations of calcium in the tree species of cerrado *s.s.* and cerradão communities on dystrophic and mesotrophic soils in central Brazil.

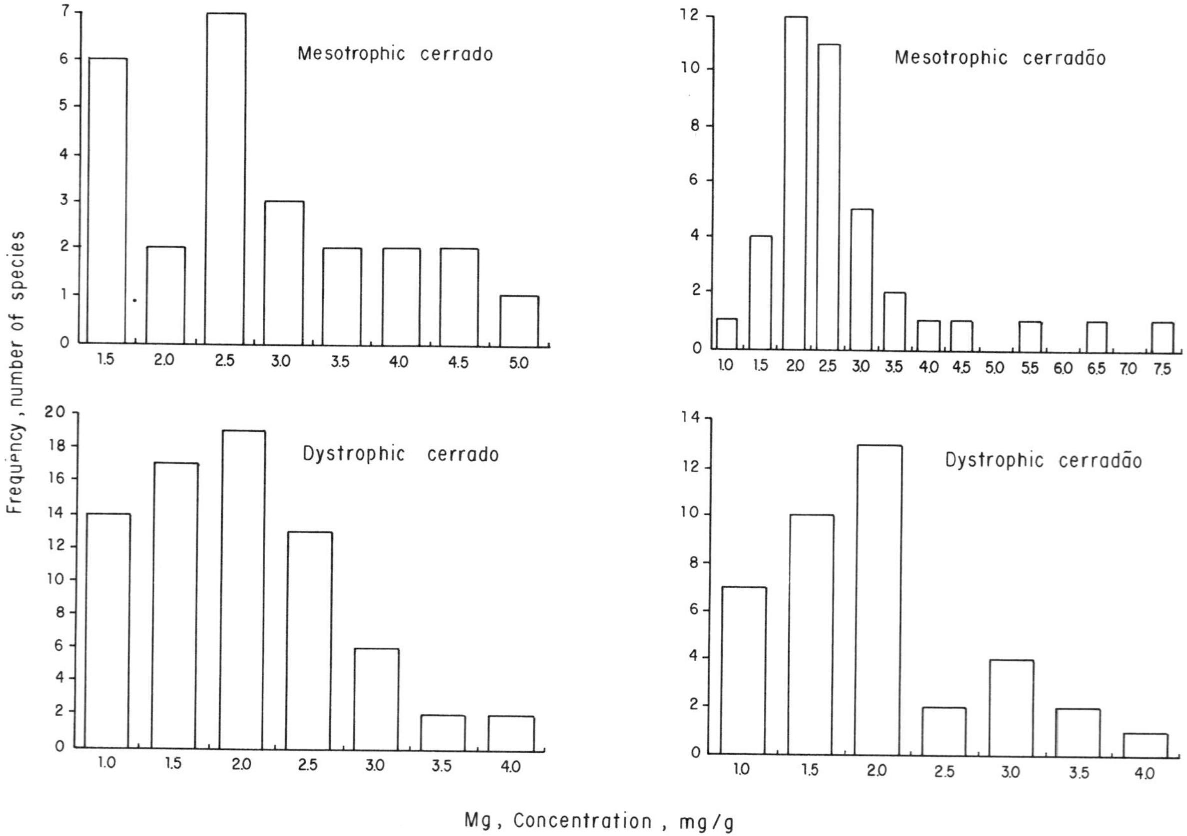

Fig. 9.2 Distribution of foliar concentrations of magnesium in the tree species of cerrado *s.s.* and cerradão communities on dystrophic and mesotrophic soils in central Brazil.

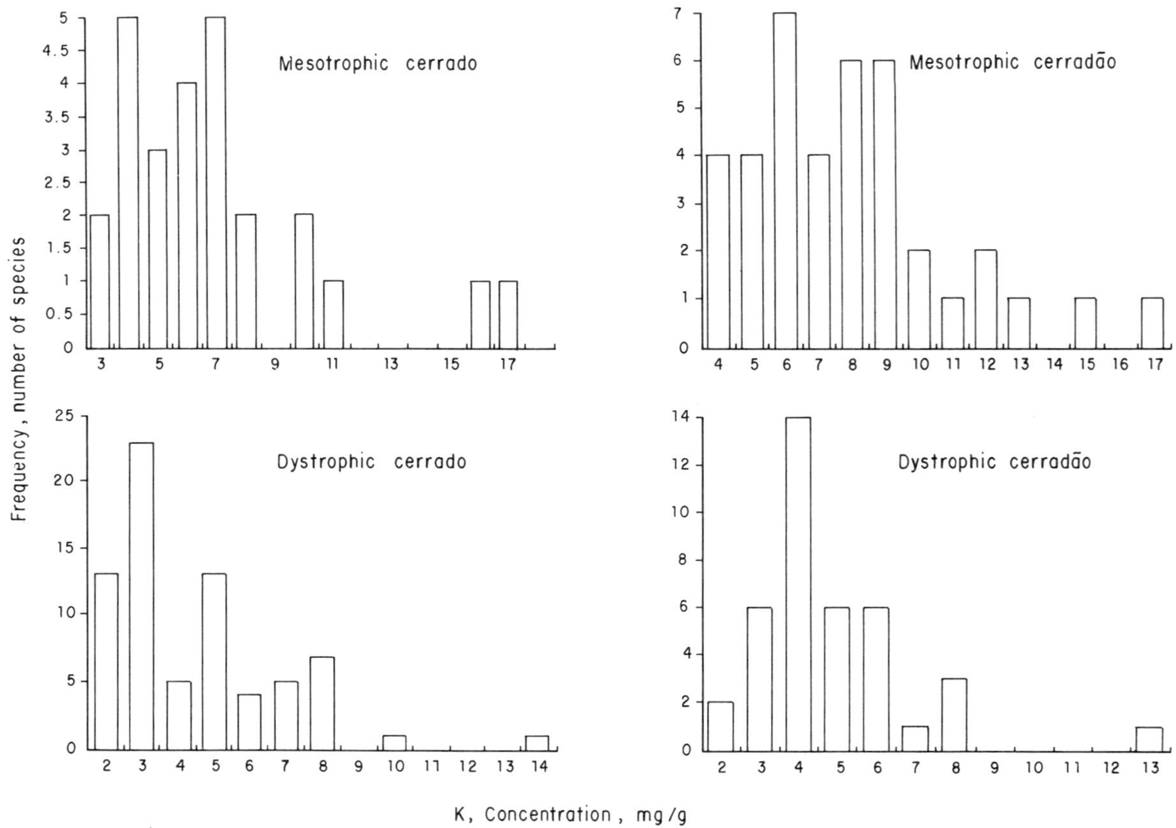

Fig. 9.3 Distribution of foliar concentrations of potassium in the tree species of cerrado *s.s.* and cerradão communities on dystrophic and mesotrophic soils in central Brazil.

cerradão communities do not differ greatly in foliar concentrations of calcium, magnesium and potassium. Thus differences in foliar concentrations between plant communities reflect soil fertility rather than any difference associated with the contrasting physiognomy of the vegetation. This suggests that soil fertility, or at least the availability of calcium, magnesium and potassium, is not primarily responsible for the transition from cerrado *s.s.* to cerradão on upland soils in the cerrado region.

Such a savanna–forest (i.e. cerrado–cerradão) transition, then, could be a consequence of other favourable edaphic factors such as better availability of water. It could also reflect a higher availability of nutrients during soil development in the past which enabled a higher plant biomass to establish at the cerradão sites. Such vegetation, after reaching an edaphic climax, could continue to survive under the present dystrophic soil conditions through closed nutrient cycling. This would also explain the difference in floristic composition between the cerradão and cerrado communities, if the cerradão species were more competitive under higher availability of nutrients in the past. This seems possible since several cerradão species also occur in gallery forests in the cerrado region where soils are more fertile.

One of the questions that should be addressed as far as savanna–forest transition dynamics are concerned in the cerrado region is the possibility of cerrado *s.s.* developing into cerradão. There are authors who believe that cerrado *s.s.* is a result of frequent fires, or of human interference, or simply that it represents a climax which is now being replaced under changing climatic conditions. They argue that in many cases cerrado *s.s.* on a dystrophic soil would develop in due course into cerradão if protected from fires (e.g. Ratter, 1980, 1985, 1986, Chapter 20). However, there are two arguments showing why this is unlikely.

1. The reserve of nutrients in the biomass and the soil of the cerrado *s.s.* is very low (Silva, 1990; Batmanian and Haridasan, 1985; Villela, 1990; Haridasan, unpublished data), and part of this is in the ground layer biomass dominated by grasses. The difference in tree biomass between cerrado *s.s.* and cerradão is substantial. Any increase in tree biomass accompanying a cerrado *s.s.*–cerradão transition would be possible only if large quantities of nutrients were transferred from the ground layer biomass to the tree biomass since soil reserves are low. This seems unlikely with only protection from fire.
2. The reasons for the difference in floristic composition between the cerrado *s.s.* and cerradão communities on dystrophic soils are not very clear even when they occur side by side – although Ratter (Chapter 20) claims that the cerradão species are much more fire-sensitive. Cerrado *s.s.* and cerradão communities with similar floristic compositions have

been described as 'undifferentiated cerradão' by Ratter *et al.* (1973) in Mato Grosso, but in general seem rare. If their rarity reflects a general inability of the species of cerrado *s.s.* to develop a higher biomass under the existing edaphic conditions, then protection from fire alone is unlikely to transform cerrado *s.s.* into cerradão, without a change in floristic composition.

The differences in floristic composition between the cerrado *s.s.* and cerradão are generally greater than encountered by Haridasan (1987). Ribeiro (1983) compared the phytosociological parameters of a cerrado *s.s.* and a cerradão which occur side by side on very similar dystrophic soils in Brasília. Fifty species belonging to 26 families were found in the cerrado and 61 species belonging to 37 families in the cerradão. None of the eight phytosociologically most important species of the cerradão were encountered in the cerrado community: *Emmotum nitens* (Benth.) Miers, *Ocotea spixiana* (Nees) Mez, *Sclerolobium paniculatum* Vog. var. *subvelutinum*, *Siparuna guianensis* Aubl., *Tapirira guianensis* Aubl., *Xylopia grandiflora* A. St. Hil., *Diospyros sericea* A. DC., and *Simarouba versicolor* St. Hil. On the other hand, five of the eight most important species of the cerrado *s.s* also occurred in the cerradão: *Qualea parviflora* Mart., *Q. grandiflora* Mart., *Annona crassifolia* Mart., *Caryocar brasiliense* Camb., and *Vochysia rufa* Mart.

Araújo and Haridasan (1988) discussed the difficulties in comparing leaf concentrations of nutrients between plant communities. To elucidate the differences between the cerradão communities on dystrophic and mesotrophic soils in leaf nutrient concentrations, they compared the species exclusive to each community and those common to both separately. This avoided confusion of the differences characteristic of species with those that could be attributed to variations in soil fertility. The important conclusions were: (a) species restricted to the mesotrophic soil had higher foliar concentrations of phosphorus, potassium, calcium and magnesium, but lower concentrations of nitrogen and manganese, than species restricted to the dystrophic soil, and (b) species common to both soils showed higher concentrations of calcium, magnesium and manganese in the mesotrophic soil.

The last important aspect discussed here is whether aluminium toxicity plays a decisive role in preventing the savanna formation of cerrado *s.s.* from developing into the forest formation of cerradão on dystrophic soils. This seems unlikely for two reasons. Firstly, a comparison of soil properties shows that cerradão soils are equally or more acid with higher aluminium saturation than soils of cerrado *s.s.* Soils under the gallery forests in the region are also more acid with higher exchangeable aluminium levels, which does not prevent the vegetation from being a closed-canopy forest. Secondly, and related to this, the native species which are more important in the plant communities on dystrophic soils overcome aluminium toxicity in one of two ways: most of them exclude aluminium from uptake but some of the most frequent species absorb, transport and accumulate aluminium in different organs without any detrimental effect on the uptake and transport of other essential elements (Haridasan, 1982; Haridasan *et al.*, 1986, 1987). It has been suggested that aluminium might even have some beneficial effects in some of the native species (Haridasan, 1988). Aluminium-tolerant and aluminium-accumulating species seem just as abundant amongst the larger trees of the cerradão and dystrophic forest as they are in the smaller species of the cerrado *s.s.* and therefore this characteristic does not seem to be associated with differences in the stature of the community.

CONCLUSIONS

Three major conclusions may be drawn from the data on soil fertility, floristic composition and leaf nutrient concentrations of the tree species of cerrado *s.s.* and cerradão communities on dystrophic and mesotrophic soils.

1. There are no large differences between the cerrado *s.s.* (savanna) and cerradão (xeromorphic forest) communities in leaf concentrations of nutrient cations, either on dystrophic or on mesotrophic soils.
2. Native plant communities on mesotrophic soils possess higher foliar concentrations of major essential nutrient cations.
3. The cerrado *s.s.*–cerradão transition on upland soils in the cerrado region is not caused by differences in present day soil fertility conditions. There are no data available to compare the total reserves of nutrient elements in different components of the biomass of these plant communities. Past differences in soil fertility could have contributed to the build-up of higher biomass and reserves of nutrients in the cerradão than in the cerrado *s.s.* on dystrophic soils.

REFERENCES

Araújo, G.M. and Haridasan, M. (1988) A comparison of the nutritional status of two forests on dystrophic and mesotrophic soils in the cerrado region of Brazil. *Communications in Soil Science and Plant Analysis*, **19**, 1075–89.

Askew, G.P., Moffat, D.J., Montgomery, R.F. and Searl, P.L. (1970) Soil landscapes in north eastern Mato Grosso. *Geographical Journal*, **136**, 211–27.

Batmanian, G.J. and Haridasan, M. (1985) Primary production and accumulation of nutrients by the ground layer community of a cerrado vegetation of central Brazil. *Plant and Soil*, **88**, 437–40.

Cole, M.M. (1986) *The Savannas – Biogeography and Geobotany*, Academic Press, London.

Eiten, G. (1972) The cerrado vegetation of central Brazil. *Botanical Review*, **38**, 201–341.

Furley, P.A. (1985) *Notes on the Soils and Plant Communities of Fazenda Água Limpa, Brasília, DF, Brasil*, Dept of Geography, Occasional Publications, NS, No. 5, University of Edinburgh, Edinburgh.

Furley, P.A., Haridasan, M. and Ratter, J.A. (1984) Uso de espécies indicadoras de solo mesotrófico no cerrado, in *Resumos XXXV Congresso Nacional de Botânica*, p. 152, São Paulo, Brazil.

Goodland, R. (1971a) A physiognomic analysis of the cerrado vegetation of central Brazil. *Journal of Ecology*, **59**, 411–19.

Goodland, R. (1971b) Oligotrofismo e alumínio no cerrado, in *III simpósio sobre o cerrado*, (ed. M.G. Ferri), Universidade de São Paulo, Brazil, pp. 44–60.

Goodland, R. and Pollard, R. (1974) The Brazilian cerrado vegetation: a fertility gradient. *Journal of Ecology*, **61**, 219–23.

Haridasan, M. (1982) Aluminium accumulation by some cerrado native species of central Brazil. *Plant and Soil*, **65**, 265–73.

Haridasan, M. (1987) Distribution and mineral nutrition of aluminium-accumulating species in different plant communities of the *cerrado* region of central Brazil, in *La Capacidad Bioproductiva de Sabanas*, (eds J.J. San José and R. Montes), IVIC, Caracas, Venezuela, pp. 309–48.

Haridasan, M. (1988) Performance of *Miconia albicans* (Sw.) Triana, an aluminium-accumulating species in acidic and calcareous soils. *Communications in Soil Science and Plant Analysis*, **10**, 1091–103.

Haridasan, M. and Araújo, G.M. (1988) Aluminium-accumulating species in two forest communities in the cerrado region of central Brazil. *Forest Ecology and Management*, **24**, 15–26.

Haridasan, M., Hill, P.G. and Russell, D.G. (1987) Semiquantitative estimates of aluminium and other cations in the leaf tissues of some Al-accumulating species using electron probe microanalysis. *Plant and Soil*, **104**, 99–2.

Haridasan, M., Paviani, T.I. and Schiavini, I. (1986) Localization of aluminium in the leaves of some aluminium-accumulating species. *Plant and Soil*, **94**, 435–7.

Ratter, J.A. (1971) Some notes on two types of cerradão occurring in North Eastern Mato Grosso, in *III Simpósio sobre o cerrado*, (ed. M.G. Ferri), Universidade de São Paulo, Brazil, pp. 100–102.

Ratter, J.A. (1980) *Notes on the vegetation of Fazenda Água Limpa (Brasília,*

DF, Brasil). Including a key to the woody genera of dicotyledons of the cerrado, Special Series No. 1, Royal Botanic Garden, Edinburgh.

Ratter, J.A. (1985) *Notas sobre a vegetação da Fazenda Água Limpa (Brasília, DF); com uma chave para os gêneros lenhosos de dicotyledôneas do cerrado*, Royal Botanic Garden, Edinburgh.

Ratter, J.A. (1986) *Notas sobre a vegetação da Fazenda Água Limpa (Brasília, DF); com uma chave para os gêneros lenhosos de dicotyledôneas do cerrado*, Publ. University of Brasília as Texto universitário no. 003, Brasília, Brazil.

Ratter, J.A., Richards, P.W., Argent, G. and Gifford, D.R. (1973) Observations on the vegetation of north eastern Mato Grosso, 1. The woody vegetation types of the Xavantina–Cachimbo Expedition Area. *Philosophical Transaction of the Royal Society, London (B)*, **226**, 449–92.

Ratter, J.A., Askew, G.P., Montgomery, R.F. and Gifford, D.R. (1977) Observações adicionais sobre o cerradão de solos mesotróficos no Brasil central, in *IV Simpósio sobre o cerrado*, (ed. M.G. Ferri), Universidade de São Paulo, Brazil, pp. 306–16.

Ratter, J.A., Askew, G.P., Montgomery, R.F. and Gifford, D.R. (1978a) Observations on the vegetation of northeastern Mato Grosso. II. Forests and soils of the Rio Suiá–Missú area. *Proceedings of the Royal Society, London (B)*, **203**, 191–208.

Ratter, J.A., Askew, G.P., Montgomery, R.F. and Gifford, D.R. (1978b) Observations on the forests of some mesotrophic soils in central Brazil. *Revista brasileira de Botânica*, **1**, 47–58.

Ribeiro, J.F. (1983) *Comparação da concentração de nutrientes na vegetação arbórea e nos solos de um cerrado e um cerradão no Distrito Federal, Brasil*, Unpublished M.Sc. thesis, Departamento de Ecologia, Universidade de Brasília, Brazil.

Silva, F.C. (1990) *Compartilhamento de nutrientes em differentes componentes da biomassa aérea em espécies arbóreas de um cerrado*, Unpublished M.Sc. thesis, Departamento de Ecologia, Universidade de Brasília, Brazil.

Villela, D.M. (1990) *Resposta do estrato rasteiro de um cerrado à irrigação e à calagem*, Unpublished M.Sc. thesis, Departamento de Ecologia, Universidade de Brasília, Brazil.

10

Soil formation and nutrient dynamics at the woodland–savanna boundary in East Africa

JULIA ALLEN JONES

SUMMARY

1. Located within the transition zone from forest to savanna in East Africa, the vegetation of the 240 km^2 study area is a mosaic of savanna, savanna–woodland (including miombo woodland and *Acacia–Commiphora* bushland) and riparian forests (Brenan, 1978; Cole, 1982; Huntley, 1982). Distinct boundaries also occur between woody plant communities which are (a) dominated by ectomycorrhizal species (e.g. miombo woodland), (b) dominated by N_2-fixing species (e.g. *Acacia* bushland and thicket) and (c) on termite mounds. This chapter examines landform and parent material influences on soils, which affect the distribution of soil biota (especially termites) and thereby influence soil nutrient availability, which interacts with fire, clearing, cultivation and grazing to produce boundaries between vegetation types.
2. Most vegetation boundaries are weakly related to soil chemistry, which is strongly associated with landform and parent material. Woodland, bushland, and savanna have different concentrations of soil organic carbon, total nitrogen, acid-extractable phosphorus, total phosphorus, exchangeable calcium and exchangeable magnesium ($p < 0.02$). These differences are partly due to landform and parent material which are significantly associated with soil organic carbon, total nitrogen, acid-extractable phosphorus, total phosphorus, exchangeable calcium, magnesium, potassium and sodium ($p < 0.0001$) and exchangeable magnesium ($p < 0.02$).
3. Edaphic savanna and riparian woodland occur only on soils lacking colonies of large mound-building, fungus-cultivating termites (Macrotermitinae), whereas secondary savanna, miombo woodland,

and *Acacia–Commiphora* bushland occur on soils with and without Macrotermitinae. Boundaries between dry-deciduous miombo woodland and *Acacia–Commiphora* bushland, between *Acacia–Commiphora* bushland and riparian woodland, and between *Acacia–Commiphora* bushland and savanna occur at junctions between mountains and plains, plains and channels, and plains and depressions. Soil drainage and mineralogy on these landforms affect the distribution of termite colonies. Macrotermitinae occur only on >1 m deep, well drained, medium to coarse-textured soils formed on granitic sediments. Colonies are absent from the mountains, channels and depressions. Soils which lack Macrotermitinae have significantly higher organic carbon, total nitrogen, acid-extractable phosphorus and exchangeable calcium in the intermound surface horizons than soils with Macrotermitinae.

4. Termite mounds are occupied by woody species different from those found on surrounding soils. Termite mounds have significantly greater amounts of total nitrogen, exchangeable cations, total phosphorus and acid-extractable phosphorus than soils surrounding mounds. This situation appears to result from termites' removal of plant litter from foraging zones between mounds, transport of this litter to the mound, and its efficient breakdown by termite–fungal symbioses.
5. Boundaries between communities with and without root symbioses are related to soil nitrogen and phosphorus availability. However, boundaries between communities with different root symbioses cannot be explained by differential soil nitrogen and phosphorus concentrations as proposed by Högberg (1986b). N_2-fixing *Acacia* species dominate on soils rich in nitrogen and phosphorus, poor in nitrogen and phosphorus, and rich in total phosphorus but poor in total nitrogen and acid-extractable phosphorus. Ectomycorrhizal miombo woodland also occurs on soils rich and poor in both nitrogen and phosphorus. Determination of foliar nitrogen and phosphorus and mineralizable soil nitrogen and phosphorus fractions is needed to examine this pattern further.
6. Fire, grazing, vegetation clearance and cultivation have created and altered boundaries between vegetation types, visible in aerial photographs from 1956 and 1978. *Acacia–Commiphora* bushland appears to be replacing miombo woodland on the soils of the plains, which have supported domestic cattle for at least 100 years and were cleared and cultivated in the last four decades. Present boundaries between bushland and miombo woodland therefore may be due to the greater fire-tolerance of bushland species, or to the enhancement of *Acacia* spp. germination when seeds are ingested by cattle. However, *Acacia–Commiphora* bushland species are not replacing miombo woodland in

the mountains where fire also occurs. The impact of anthropogenic factors on vegetation boundaries will vary according to the intensity and duration of human activity as well as the underlying patterns of landform and parent material.

INTRODUCTION

The forest–savanna boundary stretches over several hundred kilometres in East Africa. In central Tanzania, the vegetation makes a striking transition from predominantly grassland in the north to predominantly woodland in the south-west (Bell, 1982). The most important aspect of the forest–savanna boundary here is that it is neither distinct nor uniform in its characteristics. Rather this transition zone is made up of patterns of different vegetation communities, both natural and managed, whose characteristics can be defined at spatial scales ranging from metres to tens of kilometres and whose dynamics operate over temporal scales ranging from seasons to centuries.

Vegetation in the study site is a changing mosaic of distinct vegetation types overlying distinctly bounded landforms and soil parent materials, and has been modified by fire, grazing, fuelwood cutting and cultivation. The vegetation includes examples of savanna grasslands, both edaphic and secondary (White, 1983), and savanna woodlands, namely miombo woodland and *Acacia–Commiphora* bushland (Huntley, 1982; Cole, 1982) as well as termite mound thicket and riparian forest (White, 1983). Miombo woodland is dominated by woody species with ectomycorrhizal root symbioses, and *Acacia–Commiphora* bushland has rhizobia and endomycorrhizae (Högberg and Nylund, 1981; Högberg, 1982, 1986a). The distinct boundaries among these formations have been attributed to geomorphology and soil drainage (Cole, 1982), soil nutrient status (Bell, 1982) and rainfall (Huntley, 1982). This study examines the boundaries among six vegetation formations, including four savanna vegetation types, in a 240 km^2 site in central Tanzania and relates them to landforms, soil drainage, soil chemistry, soil biology and land use.

STUDY SITE

Climate

The 240 m^2 study area in the Dodoma region, central Tanzania, lies at 7°S, 36°E and 1000–2000 m elevation in a climatic zone transitional between the wetter highlands to the south and west, and the drier lowlands to the north and east. The site has a single five-month rainy season with 98% of mean annual precipitation (MAP) falling between November and April, and a 50-year MAP of about 550 mm (East African Meteorological Department [EAMD], 1975; Rodhe and Virgi, 1976). Mean monthly temperatures range between a minimum of 20 °C and a maximum of 25 °C. The climate is semi-arid with potential evaporation exceeding rainfall in every month (Rapp *et al.*, 1972; Christiansson, 1981).

Geology and Geomorphology

The study site consists of three major landforms: hills and mountains with slopes of 35–60%; a surrounding plain with slopes of 3–9%; and an internally-drained depression or 'mbuga' 20 km to the south-west (Wade and Oates, 1938). The plain, which has a mantle of sediments 1 to >3 m thick, is dissected by seasonal stream channels.

The mountains are composed predominantly of Precambrian synorogenic granite and gneiss, but a central range of hills is formed from intruded amphibole schist and ferruginous quartzite schist (Government of Tanganyika Territory, 1936; Government of Tanzania, 1963, 1964). The deep red sediments which surround these hills are formed from the schist, and hereafter are called 'schist sediments' (Government of Tanzania, 1963, 1964). The remainder of the pediment is covered by granitic and gneissose sediments whose mineralogy is primarily quartz, feldspars and biotite. Hereafter they are referred to as 'granitic sediments'.

Soil classification and soil fauna

Soils consist of Lithic Ustropepts, Lithic Dystropepts, Oxic Ustropepts, Ustoxic Dystropepts, Ustic Dystropepts, Typic Ustipsamments, Ultic Paleustalfs, Plinthustalfs, Fluventic Ustropepts, Fluventic Dystropepts, Typic Haplustalfs, and Typic Pellusterts following the USDA Soil Taxonomy (Soil Survey Staff, 1975; Jones, 1989). The plains on granitic sediments are occupied by mounds of Macrotermitinae (*Macrotermes* and *Odontotermes* spp.); other termites (such as *Microcerotermes* spp., *Amitermes unidentatus* [Wasmann] and *Hodotermes mossambicus* (J. Darlington, personal communication)) occupy the plains on amphibole schist sediments. No termite mounds or dwellings were observed on soils on mountains, in seasonal drainages or in the internally drained depression.

Vegetation

The study site is located at the margins of two regional centres of endemism: the Zambezian to the south and the Somalia–Masai to the north. The Zanzibar–Inhambane regional mosaic lies to the east, the Lake Victoria regional mosaic to the west (White, 1983). Following White (1983), the vegetation types (shown in Fig. 10.1) are:

(a) Zambezian dry deciduous miombo woodland;
(b) Zambezian semi-deciduous to evergreen miombo woodland;
(c) Somalia–Masai *Acacia–Commiphora* deciduous bushland and thicket;
(d) Zambezian termite mound thicket;
(e) northern Zambezian riparian woodland;
(f) Somalia–Masai/Zambezian secondary grassland; and
(g) Somalia–Masai edaphic grassland.

At this site the miombo woodlands are dominated by *Brachystegia spiciformis* and *Julbernardia globiflora* with occasional *Pterocarpus angolensis* (Table 10.1) (J.A. Jones, unpublished). Ectomycorrhizae were reported on *B. spiciformis* and *J. globiflora* in Tanzania (Högberg and Nylund, 1981; Högberg, 1982, 1986a) while *P. angolensis* had nodulation, nitrogenase activity and endomycorrhizae (Högberg, 1986a). Miombo has been interpreted as a recent vegetation type resulting from the destruction of evergreen forest (Fanshawe, 1971), i.e. the midpoint of a progression from closed-canopy dry woodland to savanna caused by fire and clearing (Trapnell, 1959; Strang, 1974; Malaisse, 1978; Stromgaard, 1986). Because miombo often lacks shrubs, but has some cover of grasses 0.6–1.2 m tall (White, 1983), it has been called a wooded savanna (Morris, 1970) or savanna-woodland (Brenan, 1978).

In the study site, dry, deciduous miombo woodland occupies <0.5 m deep, excessively well drained, sandy loam soils on mountain slopes (Lithic Ustropepts and Lithic Dystropepts) over at least 100 km^2 (Unit 1, Fig. 10.1). The canopy of *B. spiciformis* and *J. globiflora* reaches only 3–5 m height with 40–60% cover; herb cover is 40–80%, and there is virtually no shrub canopy (J.A. Jones, unpublished; Ackerson, 1985; Allen, 1986). This formation is considered a wooded savanna (Morris, 1970; Brenan, 1978). In contrast, semi-deciduous to evergreen miombo woodland occurs in two patches covering an estimated area of 20–25 km^2 on >1.0 m deep, well drained sands and sandy clay loam soils on the plains (Typic Ustipsamments, Ustoxic Dystropepts and Plinthustalfs) (Fig. 10.1). This woodland is also dominated by *B. spiciformis* and *J. globiflora*, but the canopy reaches 20–25 m in height. Tree cover is 80–100%, shrub cover is 40–80%, and herb cover is 40–60% (P.S. Jones and J.A. Jones, unpublished). This vegetation type may be considered a forest or woodland.

Somalia–Masai *Acacia–Commiphora* deciduous bushland and thicket has a dense canopy 3–5 m high, dominated by spinous species with some succulents. Grasses contribute little to the phytomass and shrubs are physiognomically dominant. *Acacia*, *Commiphora* and *Grewia* spp. are

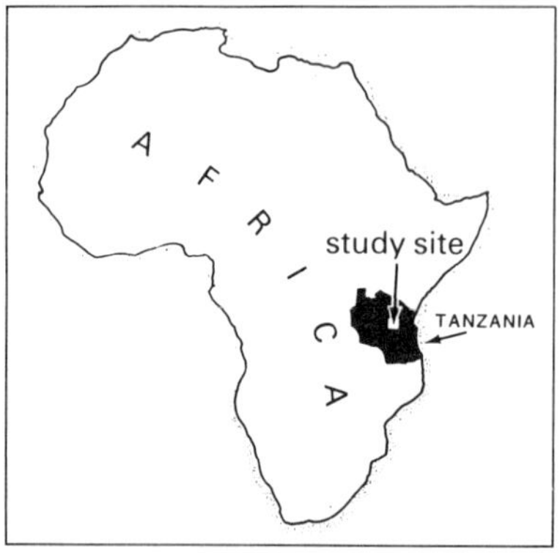
A F R I C A
study site
TANZANIA

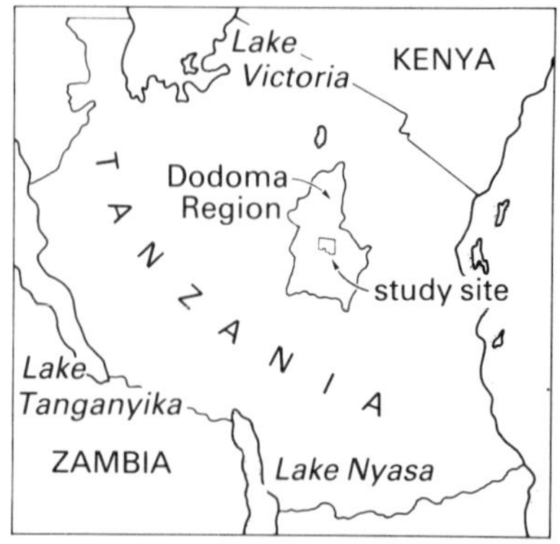
Lake Victoria
KENYA
T A N Z A N I A
Dodoma Region
study site
Lake Tanganyika
ZAMBIA
Lake Nyasa

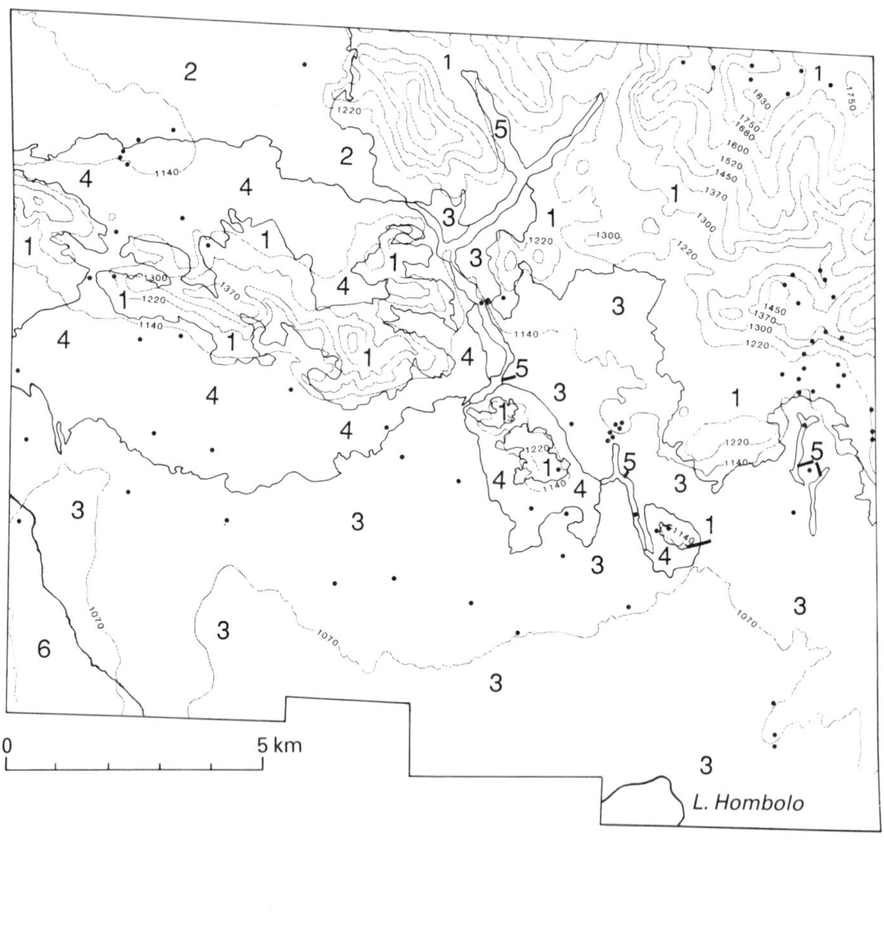
0
5 km
L. Hombolo

nearly always present (White, 1983). *Acacia* spp., *Dalbergia melanoxylon*, *Dichrostachys cinerea* and *Xeroderris stuhlmannii* in central Tanzania had nodulation, nitrogenase activity and endomycorrhizae, while *Combretum zeyheri* had endomycorrhizae but was not N_2-fixing (Högberg, 1986a). Brenan (1978) notes a 'tree savanna' dominated by *Acacia* and *Combretum* spp. at lower elevations and lower rainfall in Takhtajan's (1986) Zambezian subregion, which includes most of Tanzania. Huntley (1982) refers to *Acacia–Commiphora* communities as 'arid savanna'.

In this study site, *Acacia–Commiphora* deciduous bushland and thicket occupies three soil units over nearly 100 km^2. It occurs on >1.0 m deep, well drained sands, sandy loams and sandy clay loams on the plains (Ustoxic Dystropepts, Ustic Dystropepts, Typic Ustipsamments, Ultic Paleustalfs, and Plinthustalfs) (Unit 3, Fig. 10.1), deep clays on the plains (Oxic Ustropepts) (Unit 4, Fig. 10.1), and <0.5 m deep, excessively well drained sandy loam soils on a few mountain slopes (Lithic Ustropepts and Lithic Dystropepts) (Unit 1, Fig. 10.1). It is dominated by *Acacia tortilis* subsp. *spirocarpa*, and also includes *A. mellifera* subsp. *detinens*, *Adansonia digitata*, *Commiphora* spp., *Delonix elata*, *Grewia bicolor*, *G. flavescens*, *Maerua* spp., *Rhus natalensis*, *Sclerocarya birrea* subsp. *multifoliolata*, *Strophanthus eminii*, *Vitex iringensis*, *Xeroderris stuhlmanni*, *Ximenia caffra* var. *caffra* and *X. caffra* var. *natalensis* (Table 10.1) (White, 1983). Species included in White's (1983) Zambezian undifferentiated woodland and scrub thicket (*Combretum apiculatam*, *C. zeyheri*, *Dalbergia melanoxylon*, *Dichrostachys cinerea*, *Euphorbia candelabrum* and *Haplocoelum foliolosum*) also occur. This vegetation type has a canopy height of 3–5 m with 40–60% cover by small trees and shrubs, and 40–60% herb cover (P.S. Jones and J.A. Jones, unpublished). It can be considered a savanna-woodland (Huntley, 1982; Cole, 1982).

Species composition of Zambezian termite mound thicket varies from place to place but is completely different from that on surrounding soil (White, 1983). In this site, termite mounds are found only on >1.0 m deep

Fig. 10.1 Soil and vegetation map units for a 240 km^2 area north-east of Dodoma in central Tanzania. Dots indicate locations of sites where soil and vegetation were sampled; these data were used to construct mapping units. Solid lines indicate map unit boundaries; half-tone lines indicate elevation in metres. Unit 1 = dry deciduous miombo woodland with some patches of *Acacia–Commiphora* deciduous bushland and thicket on small hills in centre of figure; unit 2 = semi-deciduous to evergreen miombo woodland; unit 3 = *Acacia–Commiphora* deciduous bushland and thicket on granitic sediments with many small patches of termite mound thicket and secondary grassland; unit 4 = *Acacia–Commiphora* deciduous bushland and thicket on schist sediments with patches of secondary grassland; unit 5 = riparian woodland; unit 6 = edaphic grassland. Numbered units are described in detail in the text.

Table 10.1 Woody species in Dodoma District, Tanzania, according to their occurrence in units 1 to 6 from Fig. 10.1

Woody plant species[a]	*Units* 1	2	3	4	5	6
Abrus schimperi Hochst. ex Bak. subsp. *africanus* (Vatke) Verdc.		X				
Acacia lockii de Willd.	X					
Acacia mellifera (Vahl) Benth. subsp. *detinens* (Burch.) Brenan			X			X
Acacia polyacantha Willd. subsp. *campylacantha* (Hochst. ex A. Rich.) Brenan[b]					X	
Acacia senegal (L.) Willd.	X			X		
Acacia seyal Delile subsp. *seyal*						X
Acacia tortilis (Forssk.) Hayne? subsp. *spirocarpa* (Hochst. ex A. Rich.) Brenan			X	X		
Acacia spp.			X			
Adansonia digitata		X	X	X	X	
Albizia anthelmintica (A. Rich.) Brenan[e]			X			X
Albizia spp.[g]				X		
Balanites pedicellaris Mildbr. and Schlecht			X			X
Boscia mossambicensis Klotzsch[d,f,g]		X	X			
Brachystegia? spiciformis (Benth.)[c]	X	X				
Bussea massaiensis (Taub.) Harms		X				
Cassia siamea		X		X		
Cassia singueana[g]				X		
Combretum apiculatum Sond.[d]	X	X	X	X		
Combretum zeyheri	X		X			
Combretum spp.			X			
Commiphora spp.[e,g,h]	X					
Dalbergia arbutifolia Bak. subsp. *aberrans* Polhill[b,g]	X					
Dalbergia melanoxylon Guill. & Perr.	X			X		
Delonix elata (L.) Gamble		X		X		
Dichrostachys cinerea (L.) Wright & Arn.[b,e]			X			
Dombeya aff. *pedunculata* K. Schum.		X		X		
Euphorbia candelabrum Trém.[e,f,g,h]			X			
Ficus sycamorus L.					X	
Grewia bicolor Juss.[d,e]			X			
Grewia flavescens Juss.[d,e]			X			
Grewia? mollis Juss.[d,e]	X		X			
Haplocoelum foliolosum (Heim) Bullock			X			
Indigofera rhynchocarpa Welw. ex Bak. var. *rhynchocarpa*		X				
Julbernardia globiflora (Benth.) Troupin[c]	X	X				
Kigelia spp.					X	
Lannea humilis (Oliv.) Engl.[e,f]		X				
Lannea? schimperi (Hochst. ex A. Rich.) Engl.[e,f]			X	X		

Table 10.1 *Continued*

Woody plant species[a]	*Units* 1	2	3	4	5	6
Maerua crassifolia Forsk.						X
Maerua edulis (Gilg. & Bened.) De Wolf			X			
Maerua eminii Pax[e]		X	X			
Maerua parvifolia Pax			X			
Maerua triphylla sens. lat.				X		
Margaritaria discoidea (Baill.) Webster		X				
Mundulea sericea (Willd.) S. Chev.				X		
Pterocarpus angolensis D.C.[b]	X	X				
Rhus natalensis Bernh. ex Krauss[i]			X			
Sclerocarya birrea (A. Rich.) Hochst. subsp. *multifoliolata* (Engl.) Kokwero[i]			X			
Strophanthus eminii Asch. & Pax			X			
Tamarindus indica L.					X	
Veclis glomerata (F. Hoffm.) Engl. var. *glomerata*		X				
Vitex iringensis Guerke		X	X			
Xeroderris stuhlmannii (Taub.) Mendonça & E.P. Sousa			X			
Ximenia caffra var. *caffra*, var. *natalensis*			X			
Zanthoxylum chalybeum Engl. var. *molle* Kokwaro		X				

[a] Identifications according to Ms Diane M. Bridson of the Royal Botanic Garden, Kew.
[b] N_2-fixing and endomycorrhizal (Högberg, 1986a).
[c] Ectomycorrhizal (Högberg, 1986a).
[d] Termite mound genus (Morison *et al.*, 1948; Burtt, 1942; Wild, 1952; Hesse, 1955).
[e] Exclusive to *Odontotermes* mounds, Kafue flats, Zambia (Ellenbroek, 1987).
[f] Termite mound genus, miombo woodland, Zaire (Aubreville, 1957).
[g] Mound-specific, central Tanzania (Howard, 1959).
[h] On termite mounds, west central Tanzania (Boaler, 1966).
[i] On termite mounds in wooded grassland of Serengeti National Park (White, 1983).

well drained sands, sandy loams and sandy clay loams on the plains (Ustoxic Dystropepts, Ustic Dystropepts, Typic Ustipsamaments, Ultic Paleustalfs and Plinthustalfs) (Unit 3, Fig. 10.1). Woody species found on termite mounds in the study site are shown in Table 10.1, along with those genera which have been noted on termite mounds elsewhere in east and Southern Africa (Burtt, 1942; Morison, *et al.*, 1948; Wild, 1952; Hesse, 1955; Aubreville, 1957; Howard, 1959; Boaler, 1966; White, 1983; Ellenbroek, 1987).

Northern Zambezian riparian woodland occupies an estimated area of $<10\,km^2$ in the study site. It is represented by *Acacia polyacantha* subsp. *campylacantha*, *Adansonia digitata*, *Ficus sycamorus*, *Kigelia* spp. and *Tamarindus indica* (White, 1983) (Table 10.1). *A. polyacantha* in central Tanzania had nodulation, nitrogenase activity and endomycorrhizas

(Högberg, 1986a). Riparian woodland occurs on >1.0 m deep seasonally flooded loamy sands (Fluventic Ustropepts and Fluventic Dystropepts) in stream channels dissecting the plain (Unit 5, Fig. 10.1).

White (1983) describes two types of Somalia–Masai edaphic grassland, both of which occur in the study site (Unit 6, Fig. 10.1). In mbugas, edaphic grassland occurs on hydromorphic soils with an ecotone of gall Acacias (White, 1983). In the study site, the mbuga is a grassland on >2.0 m deep seasonally flooded alkaline clays (Typic Pellusterts); the only tree present is *Acacia seyal* subsp. *seyal*. The other type of edaphic grassland described by White (1983) occurs on non-cracking calcimorphic clays with dwarf grasses and scattered trees. One area of calcimorphic clays (Typic Haplustalfs) was surveyed in the study site but the herb canopy was sparse. Trees or shrubs occurring on this site included *Acacia mellifera, Albizia anthelmintica, Balanites pedicellaris* and *Maerua crassifolia*.

Somalia–Masai/Zambezian secondary grassland occurs as patches throughout the plains (Units 3 and 4, Fig. 10.1) and occasionally on mountain slopes where woodland or bushland has been cleared. In the study site these patches have only a sparse herb cover and spinescent or unpalatable shrubs are common.

METHODS

Field sampling and photograph interpretation

Soil and vegetation were sampled at 85 points located within the study site on the basis of features distinguishable on 1978 1:25 000 black and white aerial photographs (Government of Tanzania, 1978). Nineteen units were outlined on a photomosaic of the aerial photographs based on vegetation canopy cover (0–10, 11–40, 41–80 and 80–100%); soil colour (dark, light); topographic position (mountain/hill, plain, depression, channel); and presence/absence of termite mounds (Goldschmidt, 1987). Sampling locations were selected at random points within each unit.

At each sampling location (Fig. 10.1), the soil profile was described using standard methods (Birkeland, 1984) and sampled using a bucket auger. Soil samples were also taken from the 0–10 cm depth of termite mounds within 15 m of ten soil sampling locations. Three parallel 25 m transects were drawn perpendicular to the hillslope gradient with mid-points at the soil sampling site, 20 m upslope, and 20 m downslope. Along these transects specimens of woody perennials were collected and percentage cover was estimated by line-intercept (Barbour *et al.*, 1987). Plant specimen collection was not exhaustive because it occurred in June, July and August when some species were leafless.

In order to determine vegetation change, percent cover at 383 points on the plains was compared between 1956 1:30 000 black and white aerial photographs and the 1978 photographs (Government of Tanganyika Territory, 1956; Government of Tanzania, 1978) using a Bausch and Lomb zoom transferscope. For each point in both years one of four vegetation cover classes was designated: 0–20, 21–40, 41–80 or 80–100% cover. Two hundred and forty eight of the points were on granitic sediments (pale soil surface colour) and 125 were on schist sediments (dark surface colour).

Laboratory analysis

Chemical and physical analyses were performed on air-dried <2 mm soil fractions. Organic carbon was determined by the Walkley-Black method (Walkley, 1947; Olsen and Sommers, 1982). Total nitrogen and total phosphorus were determined in acid digests (H_2SO_4 and H_2O_2 with $LiSO_4$ catalyst) using a modified Indophenol assay for nitrogen (Scheiner 1976; Dorich and Nelson, 1983; Setaro and Jones, 1990). 'Available' (acid-extractable) phosphorus was extracted with 0.05 M HCl + 0.025 M H_2SO_4 (Nelson *et al.*, 1953; Olsen and Sommers, 1982). Phosphorus adsorption and desorption were measured from 3 g soil samples equilibrated in a 0.01 M $CaCl_2$ solution for six days with end-over-end shaking for two 30 min periods (Fox and Searle, 1978). Phosphorus in extracts was determined by the ammonium molybdate–ascorbic acid assay. Exchangeable cations were extracted with 1 M NH_4COOCH (Thomas, 1982). Iron was extracted from selected samples using the dithionite-citrate-bicarbonate (DCB) method of Mehra and Jackson (1960). Calcium, magnesium and iron were determined by atomic absorption spectrophotometry (AAS), and sodium and potassium were determined by flame photometry (Avery and Bascomb, 1974). Soil pH was determined using a glass electrode in 1:2 soil:water and 1:2 soil 1 M KCl. Aluminium, extracted by 1 M KCl (Barnhisel and Bertsch 1982), was determined by AAS.

Soil particle size was determined by hydrometer (Gee and Bauder, 1986). Mineralogical analysis using X-ray diffractometry (XRD) was conducted on a subset of 40 whole soil samples (including gravel) from representative soil profiles following Whittig and Allardice (1986), Carroll (1970) and Bigham *et al.* (1978).

Soil chemical, physical and mineralogical properties were used to classify the soils following the USDA Soil Taxonomy (Soil Survey Staff, 1975). These classes, combined with vegetation information, were used to construct the map units shown in Fig. 10.1.

Statistical design

The relationships between vegetation boundaries, landforms (mountains, plains, channels and the depression) and parent materials (granite- and schist-derived soils) were examined by statistical analysis of soil chemical properties. Following logarithmic transformations to correct for unequal variances, soil chemical data were grouped according to (a) landform and parent material and (b) vegetation type (woodland, bushland, secondary grassland) and analysed using analyses of variance (ANOVA). *Post hoc* multiple comparison tests (i.e. contrast statements) were performed using the Bonferroni procedure for multiple comparisons (Neter *et al.*, 1989). Data were analysed on a Macintosh SE/30 using Systat statistical software (Wilkinson, 1989). Data included organic carbon, total nitrogen, exchangeable cations, exchangeable aluminium, acid-extractable phosphorus and total phosphorus data from A-horizons (128 samples), B-horizons (51 samples) and C-horizons (55 samples). In a separate analysis, ten samples from termite mounds were compared to samples from surface horizons and lithic contacts of adjacent soil profiles using a difference of means test (Bhattacharyya and Johnson, 1977).

RESULTS

Many of the boundaries between vegetation types coincide with boundaries between soil types. For example, the dry-deciduous miombo woodland/*Acacia–Commiphora* bushland boundary coincides with an abrupt transition from shallow mountain soils to deep sandy loams of the plains (Units 1 and 3, Fig. 10.1). The semi-deciduous miombo woodland/*Acacia-Commiphora* bushland boundary coincides with a transition from sands and sandy loams to clays (Units 2 and 4, Fig. 10.1). The riparian woodland/*Acacia–Commiphora* bushland boundary coincides with a transition from seasonally flooded sands to well drained sandy loams (Units 3 and 5, Fig. 10.1). The *Acacia–Commiphora* bushland/edaphic grassland boundary coincides with a transition from sandy loams to clays (Units 3 and 6, Fig. 10.1). In addition, the boundary between termite mound thicket and surrounding vegetation coincides with a transition from the cation- and phosphorus-rich, cemented soils of the mounds to the surroundings with low base status.

However, some vegetation boundaries do not coincide with soil boundaries. For example, small patches of miombo woodland species occur in large areas of *Acacia–Commiphora* bushland and thicket on deep sandy loams on the plains. Patches of secondary grassland occur on all soil classes on the plains and the mountains. *Acacia–Commiphora* bushland and thicket also occurs on mountain soils, on deep clays of the plains and on deep sandy loams of the plains.

Soil chemical properties are weakly related to vegetation type. The soil sampling design did not permit the examination of individual tree effects as in Kellman (1979). Instead, data from pedons in Units 1 to 4 were grouped into three vegetation types: miombo woodland (Units 1 and 2), *Acacia–Commiphora* bushland (Units 3 and 4) and secondary grassland sites (small areas within Units 2, 3 and 4). Too few soil samples were available to include Units 5 and 6. For A-horizons, there were weakly significant differences ($p < 0.02$) between soil properties under these three vegetation types (Table 10.2). Soils under miombo woodland have significantly greater concentrations of acid-extractable phosphorus, exchangeable calcium and magnesium, organic carbon, total nitrogen and total phosphorus than soils under *Acacia–Commiphora* bushland. Soils under secondary grassland also have significantly higher concentrations of exchangeable calcium, exchangeable magnesium, and total phosphorus than those under *Acacia–Commiphora* bushland. There are higher concentrations of exchangeable aluminium in B- and C-horizons of soils under *Acacia–Commiphora* bushland, but there are no significant differences for B- and C-horizons according to vegetation type for other soil chemical properties in B- and C-horizons (data not shown).

Soil chemical properties are closely related to landform and soil parent material. Soils of mountains and channels have significantly greater ($p < 0.0001$) concentrations of organic carbon, total nitrogen, total phosphorus, acid-extractable phosphorus and exchangeable calcium, than the A-horizons of soils of the plains. Moreover, A-horizons of soils developed on schist sediments have significantly greater concentrations of total phosphorus than those on granitic sediments (Table 10.3). The B-horizons

Table 10.2 Means of A-horizon chemical properties, grouped by vegetation type, from soils in Dodoma District, Tanzania. Means followed by the same letter are not significantly different from others in that column based on contrast statements with $p < 0.02$

Class	n	*Organic C* ($g\,kg^{-1}$)	*Total N* ($g\,kg^{-1}$)	*Total P* ($mg\,kg^{-1}$)	*Acid extractable P* ($mg\,kg^{-1}$)	*Exchangeable* K	Na	Ca	Mg
						cmol[+]kg^{-1}			
Miombo woodland	48	15.4[a]	1.7[a]	285[a]	11.6[a]	0.5	0.1	6.9[a]	2.1[a]
Acacia–Commiphora bushland and thicket	43	3.6[b]	0.6[b]	150[b]	6.8[b]	0.5	0.4	2.9[b]	1.1[b]
Fields	17	3.8[b]	0.8[b]	237[b]	5.4[b]	0.5	0.2	4.5[a]	1.6[a]

Table 10.3 Means of A-horizon chemical properties, grouped by landform and parent material, from soils in Dodoma District, Tanzania. Means followed by the same letter are not significantly different from others in that column based on contrast statements with $p < 0.0001$

Class	n	*Organic C* ($g\,kg^{-1}$)	*Total N* ($g\,kg^{-1}$)	*Total P* ($mg\,kg^{-1}$)	*Acid extractable P* ($mg\,kg^{-1}$)	*K*	*Exchangeable* *Na*	*Ca*	*Mg*
							cmol[+]kg^{-1}		
Mountains	36	19.3^{a}	2.1^{a}	314^{a}	15.0^{a}	0.5	0.1	8.8^{a}	2.5^{a}
Plain, schist sediments	16	4.9^{b}	0.8^{b}	371^{b}	3.1^{b}	0.7	0.1	3.7^{b}	1.6^{b}
Plain, granitic sediments	52	2.9^{b}	0.5^{b}	115^{b}	3.5^{b}	0.4	0.4	2.5^{b}	1.0^{a}
Channels	4	7.9^{a}	1.1^{a}	252^{a}	42.0^{a}	0.9	0.1	7.8^{a}	1.8^{ab}

Table 10.4 Means of B- and C-horizon total P and acid-extractable P concentrations, grouped by landform and parent material, in soils from Dodoma District, Tanzania. Means followed by the same letter are not significantly different from others in that column based on contrast statements with $p < 0.05$ (upper-case letters) and $p < 0.0001$ (lower-case letters)

Class	*B-horizons* n	*Total P* ($mg\,kg^{-1}$)	*Acid-extractable P* ($mg\,kg^{-1}$)	*C-horizons* n	*Total P* ($mg\,kg^{-1}$)	*Acid-extractable P* ($mg\,kg^{-1}$)
Plain, schist sediments	15	295^{a}	0.9^{A}	15	278^{a}	0.7^{a}
Plain, granitic sediments	27	92^{b}	1.0^{A}	27	56^{b}	4.2^{b}
Channels	9	90^{b}	4.7^{B}	9	119^{b}	2.0^{b}

of soils in the depression have significantly greater concentrations of acid-extractable phosphorus ($p < 0.05$) than soils on the plains. The C-horizons of schist-derived soils have significantly lower concentrations of acid-extractable phosphorus, compared with soils of the depression or granitic soils on the plains ($p < 0.0001$). The B- and C-horizons of soils developed on schist sediments on the plains also have significantly greater concentrations of total phosphorus than other soils (Table 10.4).

Large termite mounds do not occur on shallow soils or deep schist-derived soils, Units 1 and 4 (A, Fig. 10.2) or on deep seasonally flooded soils of channels or depressions, Units 5 and 6. Deep soils on schist sediments (Unit 4) have termites, but no large mounds. Large mounds occur only in vegetation Units 2 and 3, i.e. under semi-evergreen miombo woodland or *Acacia–Commiphora* bushland, on deep, well drained, medium-textured soils developed on granitic sediments. Soil profiles described from areas with termite mounds have almost no surface litter whereas mountain soils which lack termite mounds but have similar vegetation cover have up to 3 cm deep litter layers. Soils of units with termite mounds also have very low C:N ratios (1:1 to 10:1) compared to soils without termite mounds (12:1 to 20:1) (data not shown).

Based on counts from aerial photographs (Fig. 10.2), large mounds in Units 2 and 3 appear to cover 150 km^2 at densities of about two mounds ha^{-1}, with average spacing of 65–75 m between mounds. Three distinct types of large mounds occur in the study site: (a) bare, flattened mounds with vegetated intermound spaces (B, Fig. 10.2), (b) bare mounds with vegetated pediments and relatively bare intermound spaces (C, Fig. 10.2), and (c) completely vegetated mounds with distinct vegetation cover or cattle trails in intermound spaces (D and E, Fig. 10.2). The bare mounds with or without vegetated pediments are probably active *Macrotermes subhyalinus* or *M. falciger* mounds while the vegetated mounds are probably abandoned *Macrotermes* or active *Odontotermes* mounds (Hesse, 1955; Lind and Morrison, 1974; Darlington, 1984; Aloni *et al.*, 1990).

Soils of large termite mounds differ in their chemistry from surrounding soils according to previously documented trends summarized in Lal (1987): soils from mounds of Macrotermitinae had significantly greater concentrations of total nitrogen, acid-extractable phosphorus, total phosphorus and exchangeable cations, but similar concentrations of organic carbon, compared with surrounding soils. On a per hectare basis, termite mounds, which typically comprise less than 0.5% of the soil volume to 0.5 m depth, contain 3–20% of soil exchangeable cations and acid-extractable phosphorus (Table 10.5).

DISCUSSION

Descriptive evidence presented here suggests that the vegetation patterns of the study site interact in a complex manner with landforms, parent material, soil drainage, depth and texture, soil fauna, soil chemistry, root symbioses and land use. Briefly, the major factors affecting vegetation boundaries are as follows.

Landform and parent material influence soil depth, drainage and textures, which in turn influence the distribution of termites. Termite activity appears to alter soil chemistry, creating large zones of soil nutrient impoverishment in foraging areas and small patches of nutrient accumulation in mounds (Jones, 1989, 1990). The zones of soil nutrient impoverishment on the plains coincide with areas of miombo woodland and

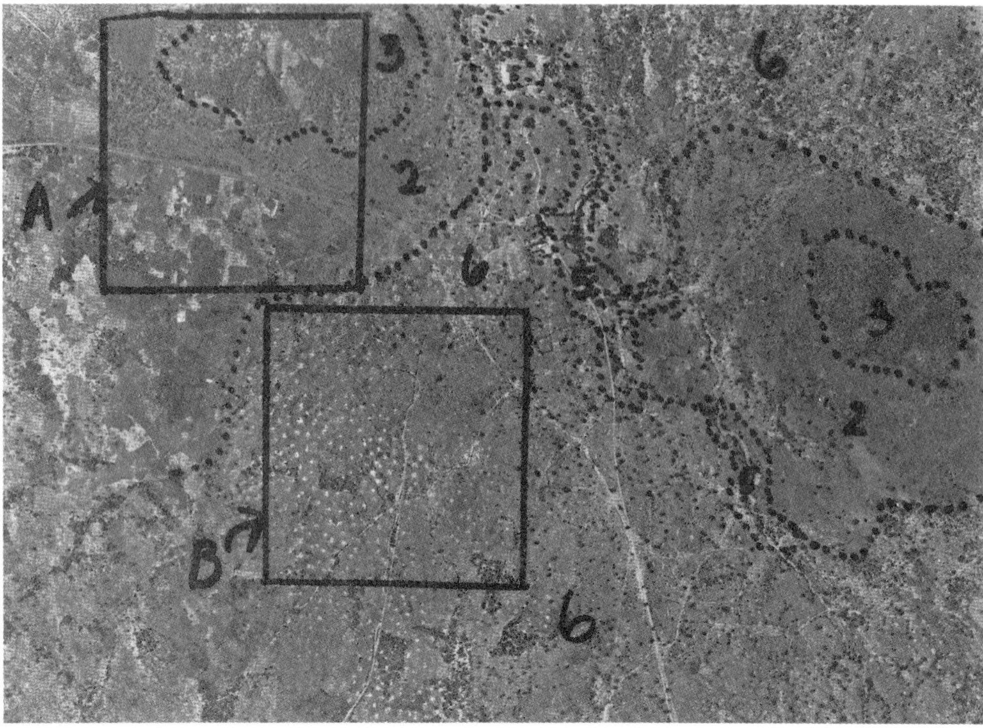
3
6
2
A
6
5
3
2
B
6

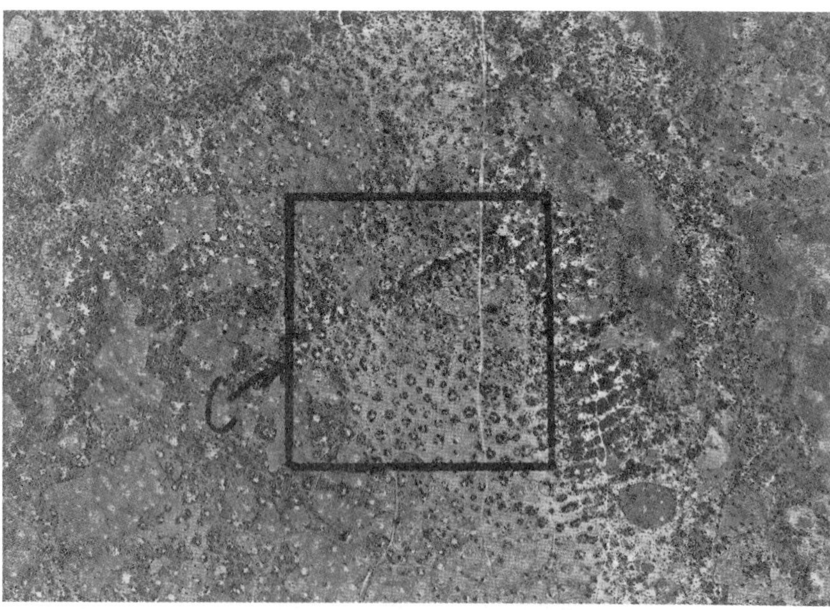
C

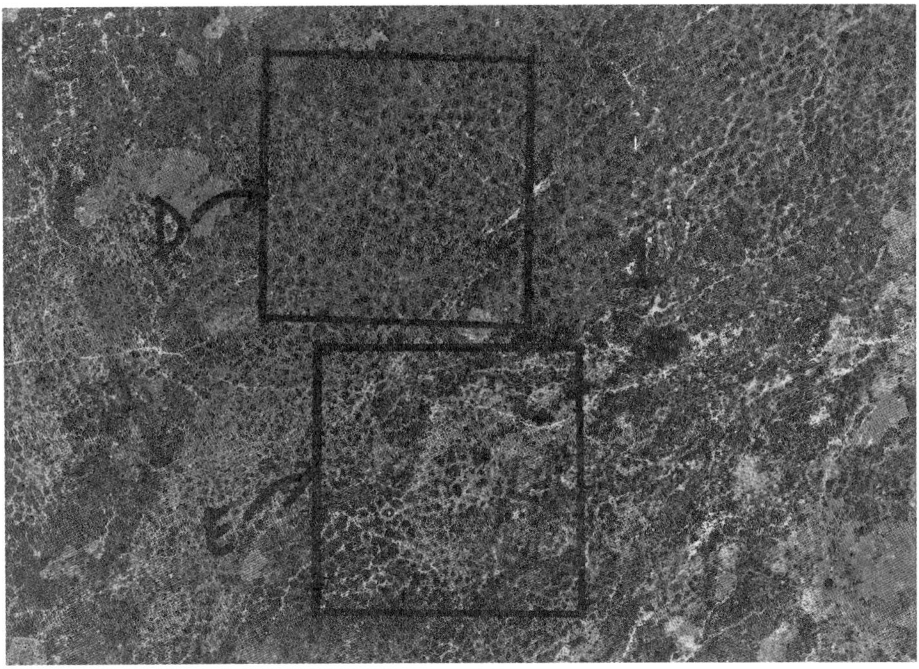

Fig. 10.2 Portions, 1 km^2 in size, of a 240 km^2 area north-east of Dodoma in central Tanzania from enlarged 1:25 000 black and white aerial photographs (Government of Tanzania, 1978). A = square kilometre of units 1 and 4 showing absence of large termite mounds from shallow mountain soils (marked with number 2) and deep schist-derived soils (marked with number 3); B = square kilometre of unit 3 showing bare, flattened termite mounds on deep granite-derived soils (marked with number 6); C = square kilometre of unit 3 showing bare mounds with rings of vegetation on their pediments; D and E = square kilometre of unit 2 showing mounds with vegetation cover that is distinct from surrounding vegetation.

Acacia–Commiphora bushland, whose dominant woody species have root symbioses. Nutrient-enriched termite mounds are apparently colonized by woody species that are rare in the nutrient-poor intermound soils. Riparian woodland occurs in the channels where there are no termite mounds and soils are rich in organic carbon and associated nutrients. Edaphic grassland occurs in the internally drained depression where there are no termite mounds and soils are deep, poorly drained, alkaline (pH 8.1–9.0) vertic clays. However, miombo woodland and *Acacia–Commiphora* bushland occur not only on nutrient-impoverished soils with termite mounds, but also on the shallow soils of the mountains where there are no termite mounds and soils are relatively rich in organic carbon and associated nutrients.

Table 10.5 Nutrient characteristics of soils, from Dodoma District, Tanzania, from mounds of Macrotermitinae (0–50 cm) compared to average contents of adjacent soil profiles ($n = 7$). Means in the same row with the following symbols are significantly different at $p < 0.05^a$, $p < 0.01^b$, $p < 0.001^c$, $p < 0.0001^d$ according to a one-sided, two sample t-test (Bhattacharyya and Johnson, 1977)

Soil property	*Termite mound* $(g\,kg^{-1})$	*Soil profile* $(g\,kg^{-1})$	*Termite mound*[e] $(kg\,ha^{-1})$	*Soil profile*[e] $(kg\,ha^{-1})$	*% in mounds*[f] *(%)*
Organic C	1.5	1.5	72	11 250	0.6
Total N	0.6[a]	0.4[a]	29	3 000	1.0
C:N ratio	2.4[a]	3.2[a]	–	–	–
Total P	0.249[b]	0.087[b]	12.0	653	1.8
Acid-extractable P	0.038[b]	0.001[b]	1.8	7.5	19.6
Exchangeable K	1.6[d]	0.3[d]	30	880	3.3
Exchangeable Na	1.7[c]	0.3[c]	19	515	3.6
Exchangeable Ca	9.7[d]	1.0[d]	93	1 500	5.8
Exchangeable Mg	3.1[c]	0.6[c]	18	547	3.2
Sum of cations	15.1[c]	2.3[c]	160	3 442	4.4

[e] Based on the following estimated volume and weight of mounds and soil. (1) Two mounds per hectare at $12\,m^3$ each $= 24\,m^3$ with bulk density of $2.0 = 48\,t\,ha^{-1}$. (2) Soil volume to $0.5\,m = 5000\,m^3$ per hectare with bulk density $1.5 = 7500\,t\,ha^{-1}$.

[f] Amount of soil element in mounds per hectare divided by total soil element content (mounds plus soil) per hectare.

Termites and vegetation boundaries

The Macrotermitinae may consume 50% of litterfall in dry African woodlands (Malaisse *et al.*, 1975; Lepage, 1978, 1979). Litter removal by termites in dry tropical Africa accelerates the decomposition of organic matter (Trapnell *et al.*, 1976; Lee and Butler, 1977; Buxton, 1981; Lee, 1982), especially in colonies of Macrotermitinae; whose symbiotic relation with fungi of *Termitomyces* spp. may achieve up to 80% effective carbon breakdown in litter (Noirot and Noirot-Timothée, 1969). The ability of termite colonies to regulate humidity and temperature (Singh and Singh, 1981) permits them to be active year-round in soils where the activity of other soil fauna may be restricted. Thus termites, especially Macrotermitinae, may short-circuit the food web and accelerate the decomposition of organic matter in soil over large areas. This hypothesis is supported by the lack of surface litter and the significantly lower concentrations of organic carbon and associated nitrogen, phosphorus and cations in soils of the plains, which have termite mounds, compared to copious soil litter and high carbon, nitrogen and phosphorus in soils which do not have termite mounds.

Depletion of organic carbon and associated nutrients in the intermound spaces is reflected by lower populations of bacteria and fungi in intermound soils compared to mounds (Meiklejohn, 1965; Arshad *et al.*, 1982; Keya *et al.*, 1982). Litter collected by termites in the foraging zones is carried to the mounds, where it is decomposed by the termite–fungal symbiosis. The fact that mounds are significantly enriched in nitrogen, phosphorus and cations but not in organic carbon with respect to surrounding soils (Table 10.5) suggests that plant litter is the source of the nutrients in mounds, rather than groundwater or other proposed mechanisms of mound nutrient enrichment (Jones, 1990).

In summary, termites appear to contribute to vegetation boundaries by partitioning soil nutrients at two spatial scales. Mountains, channels and depressions have soils which lack termite mounds, are relatively rich in organic carbon, total nitrogen, total phosphorus and associated cations, and have dry-deciduous miombo woodland, some *Acacia–Commiphora* bushland and thicket, riparian woodland or edaphic grassland. Within miombo woodland and *Acacia–Commiphora* bushland on deep soils with large termite mounds, the mounds are enriched in nitrogen, phosphorus and cations and are occupied by distinct woody species.

Root symbioses and vegetation boundaries

Högberg (1986a) noted relationships between soil and foliar nitrogen and phosphorus content under ectomycorrhizal *Brachystegia* and *Julbernardia* spp. versus N_2-fixing species in central Tanzania. He suggested (Högberg, 1986b) that trees with ectomycorrhizae are likely to have an advantage on soils that are phosphorus-poor but not nitrogen-poor, while trees which are N_2-fixing should have an advantage on soils which are nitrogen-poor but not phosphorus-poor. In the study site, Högberg's hypothesis would predict *Acacia–Commiphora* bushland on soils with high phosphorus and low nitrogen, and miombo woodland on soils with high nitrogen and low phosphorus. This is not the case.

Soils of the study site cover only three of the four possible combinations of nitrogen and phosphorus levels implied by Högberg's argument. Soils with relatively high nitrogen and high phosphorus ($>1\,g\,kg^{-1}$ total nitrogen, $>15\,mg\,kg^{-1}$ acid-extractable phosphorus and $>300\,mg\,kg^{-1}$ total phosphorus in A-horizons) occur on mountains and in channels; soils with low nitrogen and low phosphorus occur on granitic sediments on plains and in the depressions; and soils with low nitrogen and high total phosphorus but low acid-extractable phosphorus occur on schist sediments on the plains. No soil in the study site has high nitrogen and low phosphorus (Table 10.6).

Based on these data, woody species' distributions, while sharply demarcated, do not follow the patterns predicted by Högberg (1986b)

Table 10.6 Mean concentrations of N and P in A-horizons of soils in Dodoma District, Tanzania, under vegetation types whose dominant species have root symbioses

Vegetation	*Map unit (Fig. 10.1)*	*Total N ($g\,kg^{-1}$)*	*Total P ($mg\,kg^{-1}$)*	*Acid-extractable P ($mg\,kg^{-1}$)*
Miombo woodland, *Acacia–Commiphora* bushland	1	2.1	314	15.0
Miombo woodland	2	0.5	115	3.5
Acacia–Commiphora bushland	3	0.5	115	3.5
Acacia–Commiphora bushland	4	0.8	371	3.1
Riparian *Acacia* spp.	5	1.1	252	42.0
Acacia seyal subsp. *seyal*	6	0.4	90	5.0

(Table 10.6). Miombo woodland and *Acacia–Commiphora* bushland both occur on soils with high nitrogen and phosphorus and also on soils with low nitrogen and phosphorus. Boundaries between miombo woodland and *Acacia–Commiphora* bushland occur between soils which have no significant differences in soil nitrogen, phosphorus or other chemical properties, and have similar morphology. In addition, while shown by Högberg (1986a) to be N_2-fixing, *Acacia polyacantha* was observed only on high nitrogen soils in the study site.

The measures of nitrogen and phosphorus used by Högberg (1986a) and in this study (total nitrogen, total phosphorus and acid-extractable phosphorus) are not good indicators of soil nitrogen and phosphorus bioavailability (Harrison, 1982; Olsen and Khasawneh, 1980; Stevenson, 1986) and may mislead conclusions about soil–plant relations. This can be seen by examination of soil phosphorus chemistry across the boundary between *Acacia* spp. on schist-derived soils with high total phosphorus, and *Brachystegia* or *Julbernardia* spp. on granite sediment-derived soils with low total phosphorus, which appears to support Högberg's hypothesis. Despite high total phosphorus, schist-derived soils have very low acid-extractable phosphorus, because phosphorus is adsorbed by large amounts of iron oxide minerals (haematite and goethite) and free iron ($>20\,g\,kg^{-1}$ DCB extractable iron) (J.A. Jones, unpublished). The A-horizons of schist-derived soils adsorbed $1500\,\mu g$ phosphorus g^{-1}, at a soil concentration of 0.2 mg phosphorous per lit, while A-horizons of soils on granitic sediments absorbed only $200\,\mu g$ phosphorus g^{-1} soil. Schist-derived surface soils also desorbed only 1% of adsorbed phosphorus, while soils derived from granite and gneiss desorbed 20–30% of adsorbed phosphorus over six days (J.A. Jones, unpublished). High phosphorus-adsorption capacity in schist-derived soils is due to the high surface reactivity of mica, hematite and goethite, while low phosphorus adsorption in granite-derived soils is due to the low specific surface of quartz sand and hydrolysed silica (Uehara and Gillman, 1981).

These levels of phosphorus adsorption are among the highest reported (Juo and Fox, 1977; Fox, 1978). They probably account for the high concentrations of total phosphorus and low acid-extractable phosphorus in C-horizons (Table 10.4) and are suggestive of very low soil solution phosphorus in schist-derived soils. The large pool of phosphorus stored in iron complexes is apparently released very slowly to the soil solution in available form (as extracted by a dilute concentration of a strong acid). Consequently, the prevalence of *Acacia–Commiphora* bushland and absence of miombo woodland on deep schist-derived soils, provides at best equivocal support for the hypothesis that N_2-fixing species have an advantage on nitrogen-poor, phosphorus-rich soils.

According to these data, the distribution of vegetation types with root symbioses does not correspond to patterns of soil nitrogen and phosphorus as suggested by Högberg (1986b). Högberg's hypothesis deserves further examination using data on foliar nitrogen and phosphorus and mineralizable soil nitrogen and phosphorus fractions. The role of root symbioses in vegetation boundaries remains to be clarified because woody species with root symbioses are clearly dominant both on well drained soils that are poor in nitrogen and phosphorus and on soils that are rich in nitrogen and phosphorus but excessively well drained. Miombo woodland may be able to survive on shallow soils if ectomycorrhizae increase plants' access to soil moisture (J.A. Maghembe, personal communication). Moreover, *Acacia–Commiphora* bushland appears to be replacing miombo woodland on the deep soils of the plains. If *Acacia* spp. are both N_2-fixing and endomycorrhizal (Högberg, 1986a), they may be able to utilize not only nitrogen but also otherwise unavailable phosphorus bound up in iron complexes. A related explanation suggested by J. Wilson and P. Mason (personal communication) is that if endomycorrhizal fungi are generalists (few, common species) while ectomycorrhizae are specialists (numerous, relatively rare and possibly successional stage-specific species), the probability of successful infection might be higher for an *Acacia* than for a *Brachystegia* or *Julbernardia* spp. seedling.

Land use and vegetation boundaries

The mountains are part of the East Chenene Forest Reserve, which has been more or less protected from clearing and human settlement, whereas vegetation on the plains has been increasingly used by the local Wagogo people. Over the period from 1956 to 1978, 15% of the monitored points on granitic sediments on the plains changed from >41% cover to ≤40% cover. On schist sediments, 20% of the monitored points changed from >21% cover to ≤20% cover. Woodland clearance in the study site is associated with fire, cattle grazing, fuelwood cutting and cultivation of millet, sorghum and groundnuts (Allen, 1986; Goldschmidt and Jones, 1988). Land use may enhance the conversion of miombo woodland to *Acacia–Commiphora* bushland apparent on the plains and also seems to account for patches of secondary grassland within the woodland and bushland. Fires, fuelwood cutting and cultivation remove woody species, creating clearings with early successional herbs, while later succession on burned and cultivated plots may favour *Acacia–Commiphora* bushland spp. over miombo woodland species (Trapnell, 1959; Strang, 1974; Stromgaard, 1986). Grazing may modify species composition as well as cover; for example, seed germination of a common bushland species, *Acacia tortilis* subsp. *spirocarpa*, is favoured by passage through cattle (Lamprey *et al.*, 1974).

CONCLUSIONS

Factors operating to produce distinct vegetation boundaries in central Tanzania range from the localized foraging of termites and the spatial distribution of termite mounds, to soil morphology and regional landform characteristics. Many vegetation boundaries coincide with boundaries defined by landforms and by termite mound presence or absence. Spatial patterns of soil, carbon, nitrogen, phosphorus and cations and the presence or absence of termite mounds on certain landforms suggest that termites produce zones of soil carbon and nutrient impoverishment which apparently can be competitively exploited by savanna-woodland vegetation types with root symbioses. However, vegetation with root symbioses also dominate excessively drained carbon-, nitrogen- and phosphorus-rich mountain soils. Moreover, the boundaries between ectomycorrhizal vegetation (miombo woodland) and N_2-fixing vegetation (*Acacia–Commiphora* bushland) do not correspond to the relative concentrations of soil nitrogen and phosphorus as predicted by Högberg (1986b). In addition, all vegetation boundaries have been altered by fire, grazing, fuelwood cutting and cultivation, apparently favouring an increase in secondary grassland and *Acacia–Commiphora* bushland at the expense of miombo woodland.

ACKNOWLEDGEMENTS

This study was partially supported by the University of California, Santa Barbara, Academic Senate and conducted with the cooperation of J.A. Maghembe, the Tanzania National Scientific Research Council, G. Mashurano, J. Macha and other colleagues in Tanzania. I thank Ms Diane M. Bridsson of the Royal Botanic Gardens, Kew, for plant identifications; F. Davis, N. Dudley, A. Goldschmidt, P.S. Jones and F. Setaro for photograph interpretation, field and laboratory work; and J.P.E.C. Darlington, P.S. Jones, J. Proctor, W. Schlesinger, W.G. Whitford, P. Zimmerman and an anonymous reviewer for helpful discussions and comments on the manuscript.

REFERENCES

Ackerson, V. (1985) *Multi-stage Variable Probability Sampling of Tropical Dry Woodland for Fuelwood Resource Assessment*, M.A. thesis, Department of Geography, University of California, Santa Barbara, California, USA.

Allen, J.C. (1983) Deforestation, soil degradation and wood energy in developing countries. Ph.D. thesis (unpublished), The John Hopkins University, Baltimore, Maryland, USA.

Allen, J.C. (1986) Soil properties and fast-growing tree species in Tanzania. *Forest Ecology and Management*, **16**, 127–47.

Aloni, K., Malaisse, F. and Mbenza, M. (1990) Comportment et activité de récolte de *Macrotermes falciger* (Isoptera, Termitidae) dans une forêt claire zambeziénne du Shaba (Zaïre). *Bulletin des Séances de l'Académie Royale des Sciences d'Outre-mer*, **35**, 301–22.

Arshad, M.A., Murieria, N.K. and Keya, S.O. (1982) Effect of termite activities on the soil microflora. *Pedobiologia*, **24**, 161–7.

Aubreville, A. (1957) Muhulus, termitières fossiles géantes et forêts claire katanguiens. *Bois et forêts des tropiques*, **51**, 33–9.

Avery, B.W. and Bascomb, C.L. (1974) *Soil Survey Laboratory Methods*, Soil Survey of England and Wales, Harpenden, UK.

Barbour, M.G., Burk, J.H. and Pitts, W.D. (1987) *Terrestrial Plant Ecology*, Benjamin Cummings, Menlo Park, California, USA.

Barnhisel, R. and Bertsch, P.M. (1982) Aluminium in *Methods of Soil Analysis. Part 2. Chemical and Microbiological Methods*, (eds A.L. Page, R.H. Miller and D.R. Keeney), American Society of Agronomy, Madison, Wisconsin, USA, pp. 275–300.

Bell, R.H.V. (1982) The effect of soil nutrient availability on community structure in African ecosystems, in *Ecology of Tropical Savannas*, (eds B.J. Huntley and B.H. Walker), Springer-Verlag, Berlin, pp. 193–216.

Bhattacharyya, G.K. and Johnson, R.A. (1977) *Statistical Concepts and Methods*, John Wiley, New York.

Bigham, J.M., Golden, D.C., Bowen, L.H. *et al.* (1978) Iron oxide mineralogy of well-drained Ultisols and Oxisols: 1. Characterization of iron oxides in soil clays by Mossbauer spectroscopy, X-ray diffractometry, and selected chemical techniques. *Soil Science Society of America Journal*, **42**, 816–30.

Birkeland, P. (1984) *Soils and Geomorphology*, Oxford University Press, New York.

Boaler, S.B. (1966) Ecology of a miombo site, Lupa North Forest Reserve, Tanzania. II. Plant communities and seasonal variation in the vegetation. *Journal of Ecology*, **54**, 405–79.

Brenan, J.P.M. (1978) Some aspects of the phytogeography of tropical Africa. *Annals of the Missouri Botanic Gardens*, **65**, 437–74.

Burtt, R.D. (1942) Some East African vegetation communities. *Journal of Ecology*, **30**, 65–146.

Buxton, R.D. (1981) Termites and the turnover of dead wood in an arid tropical environment. *Oecologia (Berlin)*, **51**, 379–84.

Carroll, D. (1970) *Clay Minerals: A Guide to their Identification*, Geological Society of America, Boulder, Colorado.

Christiansson, C. (1981) *Soil Erosion and Sedimentation in Semi-arid Tanzania: Studies of Environmental Change and Ecological Imbalance*, Scandinavian Institute of African Studies and Department of Physical Geography, University of Stockholm, Stockholm, Sweden.

Cole, M.M. (1982) Influence of soils, geomorphology, and geology on the distribution of plant communities in savanna ecosystems in *Ecology of Tropical Savannas* (eds B.J. Huntley and B.H. Walker) Springer-Verlag, Berlin, pp. 145–74.

Darlington, J.P.E.C. (1984) Two types of mound built by the termite *Macrotermes subhyalinus* in Kenya. *Insect Science Applications*, **5**, 481–92.

Dorich, R.A. and Nelson, D.W. (1983) Direct colorimetric measurement of ammonium in potassium chloride extracts of soils. *Soil Science Society of America Journal*, **47**, 833–6.

East African Meteorological Department (1975) *Climatological Statistics for East Africa. Part III. Tanzania*, East African Meteorological Department, Nairobi, Kenya.

Ellenbroek, G.A. (1987) *Ecology and Productivity of an African Wetland System: The Kafue Flats, Zambia*, W. Junk, Boston.

Fanshawe, D.B. (1969) *The Vegetation of Zambia, Forest Research Bulletin* **7**, Government Printer, Lusaka.

Fox, R.L. (1978) *Studies on Phosphorus Nutrition in the Tropics*, CSIRO, Brisbane, Australia.

Fox, R.L. and Searle, P.G.E. (1978) Phosphate adsorption by soils in the tropics, in *Diversity of Soils in the Tropics*, (ed. M. Stelly), American Society of Agronomy, Madison, Wisconsin, USA, pp. 97–119.

Gee, G.W. and Bauder, J.W. (1986) Particle size analysis, in *Methods of Soil*

Analysis. Part I. Physical and mineralogical methods, (ed. A. Klute), American Society of Agronomy, Madison, Wisconsin, USA, pp. 383–411.

Goldschmidt, A.P. (1987) *Resource Mapping and Land Evaluation around Three Ujamaa Villages in Semi-arid Dodoma Region, Tanzania.* MA thesis, Department of Geography, University of California, Santa Barbara, California, USA.

Goldschmidt, A.P. and Jones, J.A. (1988) A land evaluation of three Ujamaa villages in Dodoma Region, Tanzania. *Soil Survey and Land Evaluation*, **8**, 183–201.

Government of Tanganyika Territory (1936) *Degree sheet No 52 (Dodoma) (1:250 000)*, Department of Lands and Mines, Dodoma, Tanzania.

Government of Tanganyika Territory (1956) *1:30 000 Black and White Aerial Photographs: 35°E to 37°E by 5°S to 7°S*, Ministry of Land Use Planning, Dar es Salaam, Tanzania.

Government of Tanzania (1963) *Quarter Degree Sheet No 143 (Meia Meia)*, Ministry of Industry, Mineral Resources and Power, Geological Survey Division, Dodoma, Tanzania.

Government of Tanzania (1964) *Quarter Degree Sheet No 144 (Zoissa)*, Ministry of Industry, Mineral Resources and Power, Geological Survey Division, Dodoma, Tanzania.

Government of Tanzania (1978) *1:25 000 Black and White Aerial Photographs: 35°E to 37°E by 5°S to 7°S*, Ministry of Land Use Planning, Dar es Salaam, Tanzania.

Harrison, A.F. (1982) Labile organic phosphorus mineralisation in relationship to soil properties. *Soil Biology and Biochemistry*, **14**, 343–51.

Hesse, P.R. (1955) A chemical and physical study of the soils of termite mounds in East Africa. *Journal of Ecology*, **43**, 449–61.

Högberg, P. (1982) Mycorrhizal associations in some woodland and forest trees and shrubs in Tanzania. *New Phytologist*, **92**, 407–15.

Högberg, P. (1986a) Nitrogen-fixation and nutrient relations in savanna woodland trees (Tanzania). *Journal of Applied Ecology*, **23**, 675–88.

Högberg, P. (1986b) Soil nutrient availability, root symbioses, and tree species composition in tropical Africa: A review. *Journal of Tropical Ecology*, **2**, 359–72.

Högberg, P. and Nylund, J.E. (1981) Ectomycorrhizae are coastal miombo woodland of Tanzania. *Plant and Soil*, **63**, 283–9.

Högberg, P., Kvamstron, M., Nyland, G.J-E. (1980) Occurrence and importance of ectomycorrhizas and nitrogen-fixing nodules in the miombo woodlands of Tanzania. Preliminary report, Department of Forest Mycology and Pathology, Division of Forestry, Uppsala, Sweden.

Howard, J.A.C. (1959) The classification of woodland in western

Tanganyika for type-mapping from aerial photographs. *Commonwealth Forestry Review*, **38**, 348–64.
Huntley, B.J. (1982) Southern African savannas, in *Ecology of Tropical Savannas*, (eds B.J. Huntley and B.H. Walker), Springer-Verlag, Berlin, pp. 101–19.
Jones, J.A. (1989) Environmental influences on soil chemistry in central semi-arid Tanzania. *Soil Science of America Journal*, **53**, 1748–58.
Jones, J.A. (1990) Termites, soil fertility and carbon cycling in African dry woodlands: a hypothesis. *Journal of Tropical Ecology*, **6**, 291–305.
Juo, A.S.R. and Fox, R.L. (1977) Phosphate sorption capacity of some benchmark soils in West Africa. *Soil Science*, **124**, 370–6.
Kellman, M. (1979) Soil enrichment by neotropical savanna trees. *Journal of Ecology*, **67**, 565–77.
Keya, S.O., Mureria, K. and Arshad, M.A. (1982) Population dynamics of soil microorganisms in relation to proximity of termite mounds in Kenya. *Journal of Arid Environments*, **5**, 353–9.
Lal, R. (1987) *Tropical Ecology and Physical Edaphology*, John Wiley, New York.
Lamprey, H.F., Halevy, G. and Makacha, S. (1974) Interactions between *Acacia*, bruchid seed beetles and large herbivores. *East African Wildlife Journal*, **12**, 81–5.
Lee, K.E. (1982) The influence of earthworms and termites on soil nitrogen cycling. New trends in soil biology, in *Proceedings of the VII International Colloquium of Soil Zoology*, (eds P. Lebrun, H.M. Andre, A. De Medts and C. Gregoire-Wibo), Dieu-Brichart, Louvain-la-Néuve, Belgium, pp. 35–48.
Lee, K.E. and Butler, J.H.A. (1977) Termites, soil organic matter decomposition and nutrient cycling. *Ecological Bulletin*, **25**, 544–8.
Lepage, M.G. (1978) *The Impact of Foraging Populations of the Termite* Macrotermes subhyalinus *on a Semi-arid Ecosystem (Kajiado, Kenya)*, International Centre for Insect Physiology and Ecology (ICIPE), Nairobi, Kenya.
Lepage, M.G. (1979) La récolte en strate herbacée de *Macrotermes* aff. *subhyalinus* (Isoptera: Macrotermitinae) dans un ecosysteme semi-aride (Kajiado, Kenya). *Comptes Rendus of the International Union for the Study of Social Insects (Lausanne)*, **1979**, 145–51.
Lind, E.M. and Morrison, M.E.S. (1974) *East African Vegetation*, Longman, Harlow.
Malaisse, F.P. (1978) The miombo ecosystem, in *Tropical Forest Ecosystems: A State of Knowledge Report*, (ed. anon.), Unesco, Paris, pp. 17–152.
Malaisse, F., Freson, R., Goffinet, G. and Malaisse-Mousset, M. (1975) Litterfall and litter breakdown in miombo, in *Tropical Ecological Systems*, (eds F. Golley and E. Medina) Springer-Verlag, New York, pp. 137–52.

Mehra, D.P. and Jackson, M.L. (1960) Iron oxide removal from soil and clays by a dithionite–citrate system buffered with sodium bicarbonate. *Seventh National Conference on Clays and Clay Minerals*, **7**, 317–27.

Meiklejohn, J. (1965) Microbiological studies on large termite mounds. *Rhodesia Zambia Malawi Agriculture Journal of Research*, **3**, 67–79.

Morison, C.G.T., Hoyle, A.C. and Hope-Simpson, J.F. (1948) Tropical soil vegetation catenas and mosaics – A study of the southwestern part of the Anglo-Egyptian Sudan. *Journal of Ecology*, **36**, 1–84.

Morris, B. (1970) The nature and origin of *Brachystegia* woodland. *Commonwealth Forestry Review*, **49**, 155–8.

Nelson, W.L., Mehlich, A. and Winters, E. (1953) The development, evaluation, and use of tests for P availability. *Agronomy*, **4**, 153–88.

Neter, J., Wasserman, W. and Kutner, M.H. (1989) *Applied Linear Regression Models*, 2nd edn, Richard Erwin, Boston, Massachusetts, USA.

Noirot, C. and Noirot-Timothée, C. (1969) The digestive system, in *Biology of Termites*. Vol. 1, (eds K. Krishna and F.M. Weesner), Academic Press, London, pp. 49–88.

Olsen, S.R. and Khasawneh, F.E. (1980) The use and limitations of physical-chemical criteria for assessing the status of phosphorus in soils, in *The Role of Phosphorus in Agriculture*, (eds F.E. Khasawneh, E.C. Sample and E.J. Kamprath), American Society of Agronomy, Madison, Wisconsin, USA.

Olsen, S.R. and Sommers, L.E. (1982) Phosphorus, in *Methods of Soil Analysis. Part 2. Chemical and microbiological methods*, (eds A.L. Page, R.H. Miller and D.R. Keeney), American Society of Agronomy, Madison, Wisconsin, USA, pp. 403–30.

Rapp, A., Murray-Rust, D.H., Christiansson, C. and Berry, L. (1972) Soil erosion and sedimentation in four catchments near Dodoma, Tanzania. *Geografiska Annaler*, **54A**, 255–318.

Rodhe, H. and Virji, H. (1976) Trends and periodicities in East African rainfall data. *Monthly Weather Review*, **104**, 307–15.

Scheiner, D. (1976) Determination of ammonia and Kjeldahl nitrogen by indophenol method. *Water Resources*, **10**, 31–6.

Setaro, F.V. and Jones, J.A. (1990) *Methods of Nutrient Analysis for Ecological Studies of Soils and Vegetation*, Department of Geography, University of California, Santa Barbara, California, USA.

Singh, U.R. and Singh, S.J. (1981) Temperature and humidity relations of termites. *Pedobiologia*, **21**, 211–16.

Soil Survey Staff (1975) *Soil Taxonomy: A Basic System of Soil Classification for Making and Interpreting Soil Surveys*, USDA SCS Agriculture Handbook No. 436, US Government Printing Office, Washington, DC, USA.

Stevenson, F.J. (1986) *Cycles of Soil: Carbon, Nitrogen, Phosphorus, Sulfur, and Micronutrients*, John Wiley, New York.

Strang, R.M. (1974) Some man-made changes in successional trends on the Rhodesian highveld. *Journal of Applied Ecology*, **11**, 207–14.

Stromgaard, P. (1986) Early secondary succession on abandoned shifting cultivators' plots in the miombo of south central Africa. *Biotropica*, **18**, 97–106.

Takhtajan, A. (1986) *Floristic Regions of the World*, University of California Press, Berkeley, California, USA.

Thomas, G.W. (1982) Exchangeable cations, in *Methods of Soil Analysis. Part 2. Chemical and Microbiological Methods*, (eds A.L. Page, R.H. Miller and D.R. Keeney), American Society of Agronomy, Madison, Wisconsin, USA, pp. 159–65.

Trapnell, C.G. (1959) Ecological results of woodland burning experiments in northern Rhodesia. *Journal of Ecology*, **47**, 129–68.

Trapnell, C.G., Friend, M.T., Chamberlain, G.T. and Birch, H.F. (1976) The effects of fire and termites on a Zambian woodland soil. *Journal of Ecology*, **64**, 577–88.

Uehara, G. and Gillman, G.P. (1981) *The Mineralogy, Chemistry, and Physics of Tropical Soils with Variable Charge Clays*, Westview Press, Boulder, Colorado, USA.

Wade, F.B. and Oates, F. (1938) *An Explanation of Degree Sheet No. 52 (Dodoma)*, Tanganyika Territory Department of Lands and Mines, Geological Division, Dar es Salaam, Tanzania.

Walkley, A. (1947) A critical examination of a rapid method for determining organic carbon in soils: Effects of variation in digest conditions and/or inorganic soil constituents. *Soil Science*, **63**, 251–63.

White, F. (1983) *The Vegetation of Africa: A Descriptive Memoir to Accompany the Unesco/AETFAT/UNSO Vegetation Map of Africa*, Unesco, Paris.

Whittig, L.D. and Allardice, W.R. (1986) X-ray diffraction techniques, in *Methods of Soil Analysis. Part 1. Physical and Mineralogical Methods*, (ed. A. Klute), American Society of Agronomy, Madison, Wisconsin, USA, pp. 331–62.

Wild, H. (1952) The vegetation of S. Rhodesian termitaria. *Rhodesian Agriculture Journal*, **49**, 280–92.

Wilkinson, L. (1989) *SYSTAT: The System for Statistics*, Systat, Evanston, Illinois, USA.

PART THREE

Environmental Change and Vegetation History

11

Landscape dynamics and surface deposits arising from Late Quaternary fluctuations in the forest–savanna boundary

MICHAEL F. THOMAS and MARTIN B. THORP

SUMMARY

1. The complex pattern of present-day tropical climates and forest–savanna transitions, the difficulties of extrapolating knowledge of contemporary landscape dynamics, and our sketchy understanding of past climates are problems facing this analysis.
2. Temperature depression at the Last Glacial Maximum at lowland sites was probably (3)4–6(7) °C according to location, while rainfall declined by 30–50%, possibly more in sub-Saharan Africa.
3. Climatic swings equivalent to 8° Lat. brought near-desert conditions to outer zones of the forested tropics in some areas, *c.* 18 000 BP, followed by a rapid transition to wetter conditions after 12 500 BP.
4. Widespread elimination of rainforest due to 'glacial' cooling and drying of tropical climates, followed by rapid increases in rainfall in the early Holocene, led to changes in the nature and intensity surface processes, with major consequences for landforms and soils.
5. Studies of alluvial sediments and sequences indicate major changes in stream discharge and sedimentation, matching known climate shifts.
6. Semi-arid conditions are indicated by colluvial materials, including mudflows; a hiatus in stream sedimentation from *c.* 20 000–12 500 BP (corresponding with low lake levels), and by hillslope gullies, screes, footslope pediments and even dunes.
7. Widespread *stone-lines* in soil profiles are frequently indicators of changes in transport energy on slopes and some may be 'lag' gravels dating from periods of open vegetation cover.

8. So-called *white sands* form extensive alluvial and coastal formations and are regarded as 'giant' podzols requiring high rainfall and/or a hydromorphic regime. But their sandy parent materials may be products of rapid erosion and deposition during periods of increasing rainfall over open vegetation.
9. Forest recolonization appears to have lagged 2000–3000 years behind post-glacial rainfall increases, and this period was undoubtedly one of great landscape instability and change.
10. The cumulative effects of repeated climatic oscillations during the Quaternary were profound and are seen today in patterns of saprolite stripping and preservation; transported and disturbed soil parent materials and in fluvial erosion and sedimentation.
11. The application of geomorphological analysis to studies of soils and land quality is seen as an essential, but often neglected, input to studies of the forest and savanna environments and their utilization.

INTRODUCTION

Difficult problems confront any attempt to determine the questions implied by the title. First, the concept of a continuous forest–savanna boundary, defining large and uninterrupted tracts of country may be misleading. Present-day patterns in transitional zones frequently present mosaics of forest and non-forest communities (Bigarella and Andrade-Lima, 1982; Absy, 1982), though some of these are induced by burning and farming practices. Even if maps of Late Pleistocene forest refugia are accepted as approximations of past reality (Livingstone, 1982; Whitmore and Prance, 1987; Maley, 1987), then these show the occurrence of 'islands' of forest within the savannas. Presumably, therefore, the shrinkage of forest area, resulting from climatic deterioration during the last glacial maximum, was a process of fragmentation, leading to galleries and islands marking favourable local climates, moisture or edaphic conditions. Such vegetation mosaics may have been relatively stable over 10^2–10^3 years and would have contributed to differences of soil and landforming processes within regional landscapes. Although doubts have been expressed concerning the fragmentation of the forest area in south-east Asia (Walker, 1982), Heaney (1991) has recently hypothesized a zone of seasonality and low rainfall associated with open vegetation communities, stretching from southern Thailand, through peninsular Malaysia, western and southern Borneo and Sulawesi, to connect with the seasonal climates of northern Australia. Much of this zone would have been exposed as dry land (to -120 m below present sea level) on to which forest is unlikely to have become established (Walker, 1982).

Present-day anomalies in rainforest–savanna boundaries have also been noted and attributed in the central Ivory Coast to both a legacy of dry paleoclimates and to local edaphic patterns (Peltre, 1977).

Further complications arise from the widely varying climatic regimes associated with different types of evergreen or semi-evergreen broad-leaved forests today. This is not simply a question of differences between climates of <1500 mm and >5000 mm of rainfall; it is also a distinction between equatorial regimes and monsoonal climates with important and severe dry-season conditions. Although many of the latter climates experience high rainfalls (>2300 mm yr^{-1} over much of Sierra Leone, for example), they are in reality marginal forest climates because of the seasonal drought. They also experience high rainfall erosivity. The *erosivity index* of Wischmeier and Smith (1958), based on rainfall intensity during the first 30 minutes of a storm (I_{30}), has been developed by Delwaulle (1973), Lal (1977) and adapted by Roose (1977, Fig. 5.1), who was able to establish a relationship between I_{30}, and daily rainfalls for the whole of West Africa and to compute an annual climatic erosion index $R(EI_{30})$ that approximates half of the mean annual rainfall. The value of R near the forest–savanna boundary in Sierra Leone becomes 1250, and in Ghana and Nigeria, 650–750. This implies that the erosivity of rainfall at the forest-savanna boundary in the monsoonal climate of central Sierra Leone is approaching 80% higher than in southern Ghana or Nigeria. If this differential was preserved during the periods of lowered total rainfalls that led to contractions in the forested area, then more erosive rains would have fallen on the former forest soils of Sierra Leone than in the other areas. Such differences in rainfall pattern, when combined with variations in geology and relief of areas not far from the present-day limits of the rainforest, will have led to differences in vegetal cover and to complex patterns of surface deposits and soils that are the legacies of Quaternary climatic changes.

Second, our knowledge of landscape dynamics associated with tropical rainfall under forests and savanna woodlands remains sketchy. Despite many individual studies of water balance and of runoff coefficients under forested conditions, no clear consensus emerges. Differences in lithology and soil structure, of slope and actual vegetation cover all lead to widely varying measurements. Roche (1981) found that runoff coefficients of 15–27% characterized experimental, forested, watersheds in French Guyana (precipitation, *c.* 3500 mm yr^{-1}) and such figures are not untypical. However, others range from <10% to >35%. Douglas and Spencer (1985) collated a large amount of information and numerical data on rates of denudation and of sediment yield in the forested tropics, but their figures offer few consistent conclusions that allow unambiguous comparison with savanna environments (for which there are fewer data).

Roose (1973) studied rainforest plots in Ivory Coast and obtained figures of 0.03, 0.2 and 1.0 $t\,ha^{-1}\,yr^{-1}$ on slopes of 4°, 12.5° and 24° respectively. On the other hand he obtained a figure of 0.5 $t\,ha^{-1}\,yr^{-1}$ for a 5° slope under semi-deciduous forest. However, although Kesel (1977) obtained figures of 3.3–7.7 $t\,ha^{-1}\,yr^{-1}$ in the Rupununi savannas of Guyana, he also obtained figures of 7.3–8.5 $t\,ha^{-1}\,yr^{-1}$ under rainforest with slopes of 10–15°! What can be stated with certainty, however, is that when vegetation cover is altered at a site, usually by clearance of the forest, runoff and sediment yields increase alarmingly. Lal (1981) quotes Brünig (1975) that, while under virgin forest sediment yield may be as low as 0.2 $t\,ha^{-1}\,yr^{-1}$, this can rise under cultivation to 600 or even 1200 $t\,ha^{-1}\,yr^{-1}$.

The question that must be asked is whether similar increases in soil loss (or sediment yield in catchment experiments) through several orders of magnitude could have arisen in the past from climate-induced changes in vegetation cover. This is not an easy question to answer. First, because climate-induced changes will have been less abrupt and probably less severe than those brought about by deliberate clearance of forest. And, second, because, as Roose (1977) has shown, the aggressivity of tropical rainfall is directly related to annual rainfall totals ($R = \frac{1}{2}H$) and declines from the wetter forested areas into the deciduous woodlands and savannas. His erosivity index (R) is thus 2000 at Monrovia, 800 at Lagos, but only 450 at Ouagadougou. The difficulty with this measure is that it is a result of storm size, intensity and frequency, and it is frequency that declines most obviously with lower rainfalls. In the drier parts of the tropics intense storms still occur and will transport surface sediment, but their infrequency means that sediment transfer into rivers is much less, on an annual basis of measurement, than that measured by wetter stations.

Attempts to relate the incidence of landsliding to deforestation have been only partially successful, and exceptional storms are capable of triggering mainly debris flow/mudflow slides under forested slopes (Temple and Rapp, 1972; Jones, 1973; De Ploey and Cruz, 1979). It is, however, clear that more frequent landsliding will occur on steep, deforested slopes (Lumb, 1975), in response to events of lesser magnitude. However, the anthropogenic clearance of forest does not simulate a natural change of vegetal cover over hundreds of years, and we should be cautious in drawing parallels between these processes. It is also important to recognize that the pluvial conditions of the Pleistocene/Holocene transition were almost certainly much wetter than present-day climates, and that these wetter conditions would probably have increased the incidence of landsliding, whether under closed forest or more open vegetation communities. Evidence from the forested Freetown Peninsula of

Sierra Leone indicates a recent (Holocene) period of deep-seated sliding (mostly rotational slumps) superimposed on older ferricreted debris, probably from debris slides or talus and fan gravels, dating to a drier climate (Thomas, 1983).

Third, the nature of past climates is imperfectly known and this makes deductive reasoning concerning their effects very difficult. The extent, rate and pattern of climate change are all rather poorly understood for the humid tropics, and it is still dispiriting to read literature on the Quaternary of the tropics that depends almost entirely on arid zone (Littmann, 1989) or mountain climates (Flenley, 1979), but the reality remains that relatively few studies of lowland sites in the humid tropics have been attempted. This question is explored further below.

Fourth, the rapid warming of climate that marked the end of the glacial period led to rapid changes in rainfall distribution and amount. These changes were probably more closely linked to major changes in the balance of processes than were either the drier or wetter climates before or after. The increasing erosivity of rainfall with annual totals observed by Roose (1977) must lend force to arguments requiring enhanced sediment yield from non-forested catchments as rainfall amounts increased between 12 500 and 9000 BP, when the final recovery of the forest area was accomplished. Field evidence from lake sedimentation (Talbot *et al.*, 1980, 1984; Soubiès *et al.*, 1989; Absy *et al.*, 1989), the deposition of channel gravels (Thomas and Thorp, 1980; Hall *et al.*, 1985), and off-shore sedimentation (Damuth and Fairbridge, 1970; Pastouret *et al.*, 1978) all support this view.

Modern analogues of this critical phase of climate warming and increased rainfall that commenced world-wide within the intertropical zone *c.* 12 500 BP are not available. As climate dries out, degraded forms of former vegetation cover must persist, perhaps throughout transitional climates over thousands of years but after several millennia of dry conditions, adaptive and open vegetation communities will prevail and a subsequent, rapid increase in rainfall will at first fall on soil surfaces lacking a protective canopy, and perhaps with >40% bare area exposed to raindrop impact. In geologic terms such rainfall changes are thought to have been rapid, following the effective end of the 'glacial' climates *c.* 12 500 BP (Maley, 1987; Street-Perrott *et al.*, 1985; Lézine and Casanova, 1989). Yet in southern Ghana, for example, rainforest conditions were not established until *c.* 9000 BP, representing a lag of 2000–3000 years, during which time rates of erosion may have been very high. Such interpretations may still partly reflect the nature of our slender database.

Consideration of these factors must induce some caution concerning generalizations about tropical climates and climate change; about the spatial variability of landscape and landscape processes (dynamics), and about the vegetational response to environmental crises on different spatial scales and time-scales. The interpretation of palaeo-environmental data in the absence of this caution inevitably leads to apparent contradiction and to error. In the argument that follows, field evidence cannot always be fully evaluated from published details and the authors may have been less cautious on some issues than is strictly prudent. Our objective, however, is to demonstrate, from geomorphological evidence, the direction and magnitude of changes in landforming processes during the Late Quaternary, and to match this evidence to available chronologies for environmental change during this period, such as that compiled for West Africa (Fig. 11.1).

NATURE OF THE LATE QUATERNARY CLIMATE CHANGES

Any discussion of climate change must address several different aspects of the problem, including: (a) the direction of change, whether towards aridification or pluviation; cooler or warmer conditions; (b) the magnitude of the change, or the intensity of aridification or pluviation, involving both the annual means of climate variables and the character of individual weather events; and (c) the spatial pattern and geographical extent of particular climate changes. It is convenient, however, to consider these manifestations of climatic oscillation within a chronological framework which, in this study, has been limited mainly to the last 40 000 years for which we have radiometric dating.

Ice-age cooling and aridity

Lowland temperatures at the last glacial maximum

Menabe and Hahn (1977) produced a map of temperature reduction at the glacial maximum for lowland tropical areas that shows a range from 0 to 4 °C. But this is derived from reconstructions of sea-surface temperatures by the CLIMAP Project Members (1976, 1981) and most estimates of temperature changes on land indicate a greater depression (Broecker and Denton, 1990). Syntheses by Flenley (1979) and Hamilton (1982), based largely on palynological studies of montane sites and the depression of vegetation zones, give much higher estimates of 5–9 °C. In a recent study of equatorial mountain sites in Burundi, Bonnefille and Riollet (1988) estimated a temperature fall of 4–6 °C, between 30 000 and 13 000 BP, but at the lowland site of Lake Bosumtwi in southern Ghana cooling was estimated by Maley and Livingstone (1983) at only 2–3 °C under an increased cloud cover at *c.* 15 000 BP. Verstappen (1980) argued that a

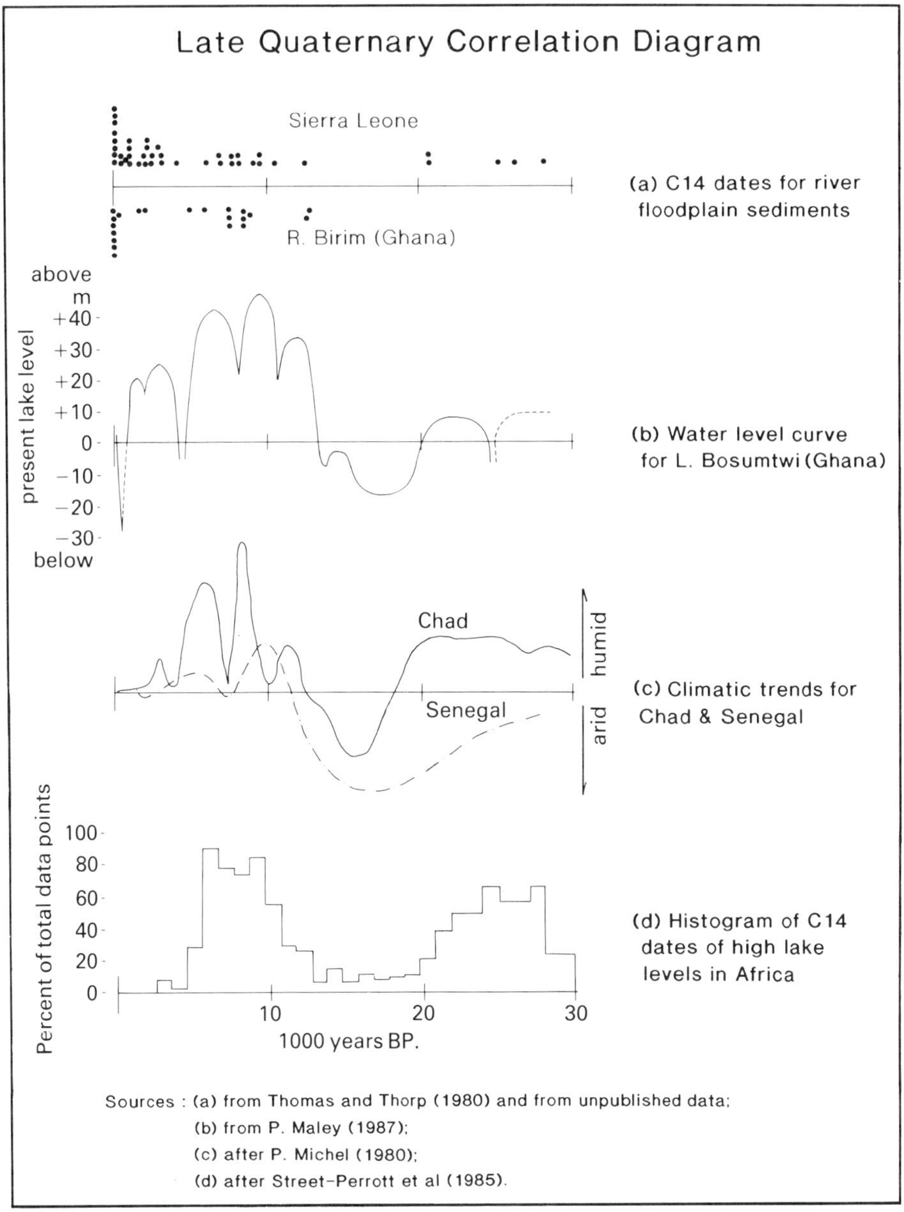

Fig. 11.1 Late Quaternary correlation diagram applicable to intertropical Africa. The graphs illustrate environmental and geomorphological change at continental, regional and local scales. Note that graph (a) shows dates of sediments indicative of periods when river discharges were sufficient for floodplain erosion and deposition within forested areas of Ghana and Sierra Leone. Graph (d) is based on C^{14} dates that record lake levels in Africa, and shows the percentage of dates indicating high lake levels.

depression of temperature of at least 3–5 °C was necessary to explain vegetation changes in SE Asia and quoted Kraus (1973) in support of an increased lapse rate during glacial conditions. However, Stuijts *et al.* (1988) have suggested a lower figure of 2–3 °C, based on the palynology of upland sites in Sumatra and Java. Their pollen studies lacked evidence for drier climates, but the occurrence of coarser sediments after 18 000 BP indicated a change in landscape dynamics, and possibly drier conditions. Figures in the higher range were supported by recent reviews of the evidence from Central and South America (Markgraf, 1989; Van der Hammen, 1991) and from Africa (Hamilton and Taylor, 1991; Maley, 1991). Such reductions on land would have followed reduced global sea-surface temperatures during the glacials, possibly magnified regionally by altered patterns of ocean current upwelling (Littmann, 1989). Oceanic evaporation losses may have been reduced by 18% globally and by much more in the tropics, and the weakening of both the Intertropical Convergence Zone (ITCZ) and the monsoonal circulation would have led to widespread rainfall reductions during the last glacial maximum (22 000–14 000 BP). Van Campo (1986) has evidenced the aridity of SW India from 22 000 to 18 000 BP, associated with the cooling of the Indian Ocean and weakening of the monsoon which gradually extended its influence after 16 500 BP, culminating in very humid conditions *c.* 11 000 BP. Loss of radiative warming due to increases in albedo in drier areas would also have been a major factor leading to cooling of continental climates.

Available evidence thus suggests that global cooling before and during the Last Glacial Maximum (LGM) affected all tropical sites so far analysed in terms of their pollen and phytolith records. The magnitude of that cooling may have been regionally variable, and some sites may have only local significance, but an increasing number of records indicate significant cooling of 4–6 °C. In some records, the timing of the onset of cooler conditions is significantly earlier than the LGM and there is increasing evidence for cooler climates before 30 000 BP in both South America (Bush *et al.*, 1990) and Africa (Bonnefille and Riollet, 1988). From the pollen record alone, some doubts have been voiced concerning concomitant drying out of the cooler climates, but these doubts have mainly been about hilly areas such as Sumatra (Stuijts *et al.*, 1988) and the Andean foothills of Ecuador (Bush *et al.*, 1990) that may be unrepresentative of the wider continental landscapes. There are, however, good theoretical reasons for associating cooling with lowered rainfalls at lowland sites (Williams, 1985).

The severity of Ice-age aridity

If the zone of maximum influx of windblown sand into the Atlantic off West Africa shifted 8° southwards (from 23° to 15°N) during the last glacial maximum as suggested by Kolla *et al.* (1979), then greatly reduced rainfalls will have penetrated well into the forested areas which must have deteriorated to open savanna woodlands or scrub at that time. It does not seem too extreme today to suggest that presently forested areas in southern Ghana at Lat. 6°N then experienced a climate corresponding with Lat. 14°N, and that in central Sierra Leone between Lat. 8°–9°N, a climate similar to 16°–17°N, where today the vegetation is of a Sudano–Sahelian transitional type, may have prevailed for at least 2000–3000 years. Sarnthein (1978) calculated that global deserts covered 50% of the land between 30°N and 30°S at 18 000 BP, compared with only 10% today.

The suggested 8° Lat. southward shift of the rainbelts in West Africa converges with the views of Rossignol-Strick and Duzer (1979) that the ITCZ may not have penetrated far beyond 8°N for 4000–7000 years around the LGM. If true, this would have desiccated areas of Africa north of Lat. 7°30′N, and left many areas to the south with rainfalls diminished by more than 50% (it should be noted that the present-day S–N rainfall gradient in this part of West Africa is around 2.5 mm km^{-1}). Rainfall reductions in SE Asia of 30–50% (Verstappen, 1980) have been widely proposed, but this parameter of climate change will have been regionally and locally variable; some areas experiencing much greater reductions (an 8° latitude shift in eastern Sierra Leone could have reduced the rainfall from more than 2000 mm today to well below 1000 mm). Evidence indicates a hiatus in stream activity between *c.* 20 000 and 13 000 BP, within smaller catchments in Sierra Leone which today receive >2000 mm of rainfall (Thomas and Thorp, 1980), and this must reflect a major loss of runoff to rivers, even in the wet season. If runoff coefficients rise with rainfall decreases sufficient to degrade forest into open woodland or savanna, then such evidence might imply a much greater reduction in rainfall than would be needed to bring this transition about. Bigarella and Andrade (1965) and Bigarella and Becker (1975) deduced from slope deposits, stone-lines and stream sedimentation that semi-arid conditions had affected eastern Brazil from 3–36°S, possibly during the LGM, in areas which today receive 1400–3000 mm precipitation (summarized by Ab'Saber, 1982; Bigarella and Andrade-Lima, 1982). This work is supported by more recent interpretations of hillslope deposits in SE Brazil (De Meis and Monteiro, 1979; De Meis and De Moura, 1984), while the dating of sediments from enclosed depressions in the Serra dos Carajás (720 m), close to the forest–savanna boundary in Pará State (Absy *et al.*, 1989; Soubiès *et al.*, 1989) provides clear evidence for alternating periods of forest and grassland or savanna, one of which is bracketed by the crucial dates 20 000–13 000 BP. A review of evidence from northern South America by Schubert (1988), concluded that severe late glacial aridity affected the Venezuelan and Colombian llanos, where there are large fossil dune fields under a present-day savanna, and that Lake Valencia in the forested Venezuelan Coastal Range was dry before 13 000 BP and surrounded by a dry thorn scrub or savanna (Bradbury *et al.*, 1981). All these indicators provide evidence for major shifts in rainfall totals during the Late Quaternary. In Indonesia (Verstappen, 1980; Caratini and Tissot, 1988; Thorp *et al.*, 1990) and Singapore (Gupta *et al.*, 1987) there are indications of both changed river regimes and rates of sedimentation during the Pleistocene and also clear evidence for much lower sea levels. The probable exposure of the Sunda Shelf during the LGM would not only have affected the regional climates (Heaney, 1991), but may also have produced an equatorial lowland devoid of closed forest (Walker, 1982). This in turn would have influenced stream deposition. The pattern of climate change in Malesia is less certain (Walker, 1982).

To illustrate the sequence and impact of the climatic changes during the last 40 000 years, Fig. 11.1 is presented to summarize knowledge for much of tropical Africa. The world-wide nature of this pattern of change is now acknowledged and mechanisms to explain synchronous climatic oscillations between polar and equatorial areas are now emerging (Broecker and Denton, 1990).

The nature of Early Holocene climates

By contrast, the early Holocene brought an equally significant reverse swing of the climatic zones involving a major extension of humid climates in West Africa. Lézine and Casanova (1989) concluded (p. 49) that the 'general palaeogeographic pattern between 6°30′ and 21°22′N corresponds to a 400–500 km northward extension of the modern latitudinal vegetation zonation'. This figure appears very large (potentially an 8° Lat. 'arid' shift plus 4° Lat. 'humid' shift), but precipitation changes cannot be predicted on a simple latitudinal shift of climatic belts, and fluctuations in other continents may have responded to different stimuli. Such changes may not have been evident everywhere in the humid tropics and particular doubt has been voiced concerning the extent of climate change in SE Asia (Walker, 1982) for example.

Although Lézine and Casanova (1989) have recently argued for evenly spread light rains from *c.* 12 000 to 7700 BP across a wide zone of West Africa, different conclusions arise from studies of deltaic and marine sedimentation along the central/West African coast (Pastouret *et al.*, 1978; Giresse, 1978; Giresse *et al.*, 1982). Field evidence for deep channel cutting in West African rivers at this time (Thomas and Thorp, 1980; Hall *et al.*, 1985) also suggests very high peak discharges, and that prolonged heavy rains were a feature of this period. In addition, there is evidence of very high floods in the Nile valley around 12 500 BP (Rossignol-Strick *et al.*, 1982; Paulissen, 1989). Talbot *et al.* (1984) found evidence at Lake Bosumtwi in Ghana for a period of highly seasonal climates, leading to laminated sediments prior to 9000 BP. By this date most tropical lakes had also attained their highest water levels (Street and Grove, 1979; Street-Perrott *et al.*, 1985).

Evidence from Amazonia and northern parts of South America is, in general, consistent with these views, although there remain many uncertainties (Tricart, 1985; Schubert, 1988). More precise timing of events is possible for the Holocene and more sites are available for analysis. As a result, the picture becomes more complex and Markgraf (1989) records for Central and South America that lowered temperatures and/or increased aridity corresponding to the glacial interval (Younger Dryas) between 11 000 and 10 000 BP is not consistently evidenced from pollen records. Street-Perrott *et al.* (1985) found that some tropical lakes receded rapidly after 10 800 BP but this trend is not evident when all data points are combined. Servant *et al.* (1989) found that forest climates did not return to SE Brazil until *c.* 8000 BP, with evidence for 'a dry climate, characterized by heavy and concentrated rainfalls between 9000 and 8000 years BP'. Markgraf (1989) also noted the occurrence of a mid-Holocene (*c.* 6000 BP) period of aridity lasting up to 1000 years in tropical America. In West Africa, lake levels and stream activity also declined, but this occurred from 4500 to 4000 BP at Lake Bosumtwi (Talbot and Delibrias, 1980; Thomas and Thorp, 1980) and may reflect such a dry interval. Lézine and Casanova (1989) emphasized the occurrence of two positive hydrological phases in the Holocene of West Africa which they date to 8500 BP and 4000–2500 BP respectively. In fact, the sequence of Holocene variations is difficult to generalize and we remain with the impression of regional and even local patterns of response to climatic fluctuations of decreasing magnitude and effect on the general landscape.

LANDFORMS AND DEPOSITS ASSOCIATED WITH ENVIRONMENTAL CHANGE

The addition of geomorphological evidence to the pollen and lake-level records clarifies some of these issues, but adds complexity to the overall picture. In this section, an attempt will be made to identify landforms and deposits that signify the impacts of past environmental change on land and soil-forming processes.

The nature of the field evidence

Three principal field areas will be used to illustrate the response of geomorphic systems to environmental changes. *Koidu* is in the forest zone of Sierra Leone (8°–9°N, monsoonal, 2000–3000 mm yr^{-1}), but within 50–100 km of the present forest–savanna boundary. The area has been widely deforested this century as a consequence of population influx to the diamond mines. Stratigraphic data from stream sediment sections, soil pits and drill logs have been examined for this area, and radiocarbon dating of wood and analysis of sediments from the headwaters of the Bafi/Sewa River system have established beyond reasonable doubt major fluctuations in stream activity during the last 36 000 years, while analysis of surface deposits has shown a complexity of morphogenesis that must parallel events in stream valleys (Thomas and Thorp, 1980, 1985; Thomas *et al.*, 1985; Teeuw, 1989). A similar record has been obtained for the *Birim River* (Hall *et al.*, 1985; Thomas and Thorp, 1991) in southern Ghana (Lat. 6°N, near equatorial, 1750–2000 mm yr^{-1}).

Such results are of limited interest to palynologists and palaeoecologists, but they are highly significant to the interpretation of palaeohydrology and landscape dynamics. The need for independent geomorphological, pedological and sedimentological evidence for landscape change is reinforced by the quite serious problems encountered when the records from several different lakes in a region are compared (see Baumhauer, 1989, on the South Central Sahara). The contrast in approach between Markgraf (1989), using pollen and lake-level records only, and Schubert (1988), who has assembled a wide range of geomorphological, pedological and sedimentological data, regarding Central and South America, is revealing in this context. The compilation of data for tropical Africa by Kadomura (1986) is no less significant in this context.

A more recent study has been made of coastal plain, alluvial sediments in *Kalimantan Barat*, Indonesia (Lat. 0°, equatorial, 3200–5000 mm yr^{-1}). Here a number of small catchments drain from hilly terrain under 900 m altitude. The wet, equatorial climate is probably optimal for evergreen rainforest *sensu stricto*, and a savanna–forest boundary is difficult to conceive. However, the lowered sea-levels of the LGM would have greatly extended the land area, across the Sunda Shelf between Borneo, Sumatra and peninsula Malaysia, and a drier, seasonal climate has been proposed by Verstappen (1980) and Heaney (1991). An extensive 'fringe' of mainly Late Pleistocene sediments has been mapped in this area, and these are almost entirely fluvial in origin and form palaeochannel deposits appearing as terraces.

A problem with many landforms is their indeterminate age, and in the present context this becomes a serious difficulty. It is also important to attempt a distinction between morphogenetically relict features and dynamic forms. But, while some processes such as active dune formation (see Goudie, 1983, Table 9.1) are confined to a particular range of climates, others take place with varying intensity or frequency across a wide range of conditions (landslides, formation of stone-lines). In many cases the 'stratigraphic position' or landscape context can offer relative ages for landforms and deposits, and complex weathering and pedological sequences can be determined from stone-lines, datable artefacts and soil micromorphology (Bigarella *et al.*, 1965; Kadomura and Hori, 1990).

Widespread fans and *glacis*, or 'slope pediments' (Rohdenburg, 1969) are present as major landscape elements in the humid tropics. In eastern Brazil, Bigarella and Andrade (1965) recorded three pediments, each associated with coarse detrital material, and these were attributed to more than one phase of Pleistocene aridity. However, Klammer (1981, 1982) argued that fossil pediments and talus fans in both the Matto Grosso and in southern Brazil dated from the pre-glacial Neogene and recorded a single major arid period which was succeeded by fluctuating humidity during the Pleistocene. Klammer's views on the Pleistocene are difficult to accept, but the notion of prolonged pre-glacial aridity is not. Such phenomena are widespread in West Africa, and Michel (1973) has also described three glacis from the basins of the Gambia and Senegal rivers, and these can be found in the present forest zone of Sierra Leone (Thomas, 1980). The *haut glacis* is present as fragments carrying weathered duricrust blocks on low interfluves and is probably mid-Tertiary; the *moyen glacis* is strongly duricrusted with ferricrete, forming piedmont benches around many residual hill masses (mostly of mafic rocks) and considered Pliocene or early Pleistocene. According to Michel (1973) the *bas glacis* is currently undergoing ferrallitic pedogenesis and is developed within present-day river valleys. 'Slope pediments' described by Rohdenburg (1969) from Nigeria and from West Malaysia by De Dapper (1989) belong to this group of forms. De Dapper and Debaveye (1986) established a date of *c.* 30 000 BP from volcanic tephra for one such pediment or glacis, correlated with an alluvial terrace.

Ferricrete from the middle glacis around the Kasewe Hills in Sierra Leone is at least as old as the last magnetic reversal (700 000 BP), which is recorded in the remnant magnetism of constituent haematite (unpublished research with Dr R. Thompson, Edinburgh), and similar features are present around the mountains of the Freetown Peninsula (Fig. 11.2). These older features of prolonged aridity are widely found and possibly correspond with the greatest extensions of the Kalahari sand sheet (De Dapper, 1988), though these and other dune occurrences such as those of the llanos of Venezuela (Schubert, 1988) appear to be, at least in part, Late Pleistocene. Older forms have been reactivated or modified

Fig. 11.2 Ferricreted *Glacis* (ramps, rasa, pediments) forming the footslopes within the Orugu River valley, Freetown Peninsula, Sierra Leone. These features are widespread in Sierra Leone and appear to mark prolonged aridity which could be early Pleistocene or the cumulative effect of multiple arid/humid cycles during the Quaternary.

by later morphogenesis and there remain many unsolved questions concerning the past rhythms of morphogenetic change.

In SE Brazil De Meis and Monteiro (1979) have described *rampas* as concave slope forms developed by slope retreat in dry climates, and also documented the complex stratigraphies of associated slope deposits in Late Quaternary '*rampa* complexes' (De Meis and De Moura, 1984). *Rampas* appear to relate to areas of fluvial dissection and encroach into deeply weathered, convex interfluve compartments (*meias laranjas* or *demi orange* topography). They are a type of slope pediment, but not identical to the wide, planar forms associated with this term.

Many erosional slopes in granitoid rocks are also widely stripped of saprolite, despite the extensive occurrence of deep weathering throughout the humid tropical zone and the stability of regoliths on present-day slopes of >20° in forested areas such as Malaysia. In Koidu, the *glacis* are commonly surmounted by core boulders and *tors*, and hillslopes also exhibit many boulder outcrops. All of this evidence from the erosional landscape strongly suggests that parts of the humid tropics may have been semi-arid for a long period in the later Tertiary, an aridity probably associated with the growth of the Antarctic ice cap. Clearly such landforms and deposits would have been reactivated and modified repeatedly during the succeeding glacial and interglacial epochs.

River regimes and climate change

There are ferruginized terraces, possibly corresponding with the middle glacis, bordering the larger rivers in Ghana and Sierra Leone, but they remain undated. Radiocarbon dating provides a framework for later sedimentation. In Koidu, black swampy clays provided five dates between 36 000 and 20 500 BP, indicating periods of backswamp wetlands around rivers of quite modest catchment size and corresponding with a low terrace. These must have formed under conditions of sufficient humidity to maintain swampy conditions even in quite small floodplains, though evidence suggests marked oscillations in discharge and sediment loads, maintaining wider floodplains than the present regime. An erosional event separates two depositional phases dated to *c.* 36 000 and 25 000 BP and may correspond with climatic deterioration after 30 000 BP (Thomas and Thorp, 1980). No datable material with radiocarbon ages between 20 500 BP and 12 500 BP was collected from either Ghana or Sierra Leone (from a total of 61 radiocarbon-dated samples). However, the renewed fluvial activity commencing around 12 500 years ago corresponds with timbers jammed under large boulders, exposed in bedrock channels and excavated 3–4 m below previous levels and not reached by present streams (Fig. 11.3). In larger river channels, pot-holes more than 5 m deep were formed, and both these and the deep channels are filled with coarse gravels and clays (Thomas and Thorp, 1980, 1992; Fig. 11.4). These events appear to correspond with high lake levels, extraordinary Nile flood levels, and spectacular increases in the sediment load of rivers such as the Niger (Pastouret *et al.*, 1978; Street, 1981; Rossignol-Strick *et al.*, 1982; Paulissen, 1989). The bulk of this evidence appears to contradict the view that Late Pleistocene climates were typified by low-intensity rainfall (Lézine and Casanova, 1989).

The large accumulations of Pleistocene alluvial sediments deposited in the coastal plains of Kalimanatan fall into at least two groups: an older terrace containing abundant timbers dated >40 000 BP at 6–12 m depth (two dates 51 000 and 54 000 BP, plus six more >40 000), and younger floodplain sediments for which we have two near basal dates of 10 500 BP. The older suite is intriguing and may relate to climate change at the termination of Isotopic Zone C (marine oxygen-isotope record Stage 3) from *c.* 58 000 BP onwards (Jouzel *et al.*, 1987), and to a major rise in

Fig. 11.3 Floodplain sediments in the Upper Moinde valley, Koidu, Sierra Leone. Gravels containing timber were deposited between granite boulders during high discharges *c.* 12 500 BP; typical floodplain facies have formed since that time, though older terrace fragments can be found.

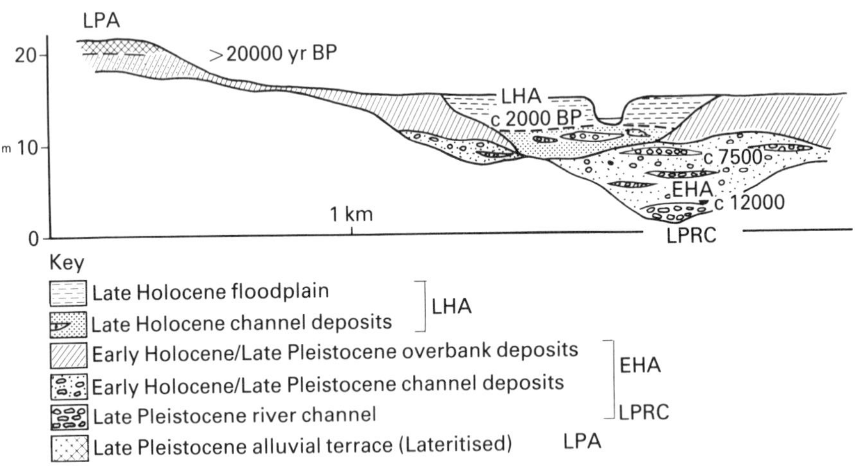

Fig. 11.4 Typical sequence of alluvial deposits found below many floodplains in the humid tropics of West Africa. The deep, Late Pleistocene rock channel is filled with coarse gravels that pass upwards into finer deposits, the present channel being inset within the sequence, marking a reduction in stream power since *c.* 12 500 BP.

ice-rafted debris in the North Atlantic (Lat. 47°N) after 60 000 BP (Heinrich, 1988). It is sufficient here to point out that these large accumulations of sediment have occurred as a result of rapid erosion in small catchments in an equatorial forest environment. The sediments approximate to fans or shallow, braided valley fills and contain little-worn gold and angular quartz shards. They appear today as podzolized 'white sands' forming terrace-like features that fall more steeply towards the coastline (4 m km^{-1}) than the gently inclined floodplains. Sudden flood and rapid deposition appear to have been the conditions of sedimentation, and, whilst neotectonic movements in this area cannot be ruled out, they would not adequately account for the sedimentology of the coastal plain deposits (Thorp *et al.*, 1990). Schubert (1988), and Schubert and Valastro (1980) recorded arid conditions of sedimentation for coarse terrace deposits in the northern Venezuelan Andes during a comparable period (*c.* 53 000 BP). In assessing the significance of such observations, Sutherland's (1985) observation that there have been 23 major glacial–interglacial cycles during the last 2 300 000 years, each of which must have caused dramatic environmental changes in the tropics, is clearly relevant.

This evidence supports the argument that virtually all parts of the landscape within many areas of the humid tropics have experienced major shifts in the balance of landforming (and soil-forming) processes during the Quaternary. More specific evidence for Late Pleistocene dry climates in the forested tropics is discussed below.

Semi-arid processes and deposits

Alluvial and colluvial sediments

Undated (but undoubtedly Late Pleistocene) deposits recording semi-arid conditions in the present-day forest zone of West Africa include mudflow deposits in shallow tributary valleys of the Birim River in Ghana. The Apiansu (near Akwatia) has been mined for diamonds, revealing several intriguing features. The diamonds were found distributed through a clayey, matrix-supported gravel 1–2 m thick, quite unlike conventional stream gravels in the main river and forming a tongue extending along the valley floor. The weathered phyllite bedrock below exhibits polygonal cracks more than 2 m deep, infilled with pale, coarse quartz sand and fine gravel (Fig. 11.5). These features suggest a period of strong desiccation followed by mudflow events, that may have marked the early stages of climate warming after 14 000–13 000 BP (at Lat. 6°N). They were first recognized by Junner in 1943, who found these features rare near the centres of the valley floors but common around the margins, where they 'suggest dry conditions and very little vegetation at the time they were formed'. The deposits do not appear to contain organic matter for dating.

Fig. 11.5 Polygonal block of weathered phyllite, coated with coarse bleached sand, excavated from beneath possible mudflow sediments, in a Birim Valley tributary, central Ghana (*c.* 50 km from Lake Bosumtwi).

In the present-day dry zone of South East Ghana which today experiences a low rainfall and savanna conditions (*c.* 800 mm), Brückner in 1955 described ventifacts in the ubiquitous stone-lines and claimed arid conditions in the Quaternary. Such claims now need to be taken very seriously. On the Jos Plateau of Nigeria, which although 12°N, has a present-day rainfall in excess of 1200 mm and was formerly forested, accumulations of probable mudflows form a major part of the landscape (Hill and Rackham, 1978), and R. Zeese (personal communication) has dated material from the base of one such sequence to 18 000 BP.

In Koidu, Sierra Leone, many headwater valleys appear to have shrunk in effective length and become infilled with colluvial 'head' deposits that bury earlier gravels. In one valley, two phases of colluvial activity are recorded in a situation of minimal ground-surface slope above the valley head (Thomas and Thorp, 1985; Fig. 11.6a), and similar infills of colluvium are widespread both in this area and around many groups of hills including the mountains of the Freetown Peninsula. On more level interfluves, surface gravels indicate strong surface flows of water and removal

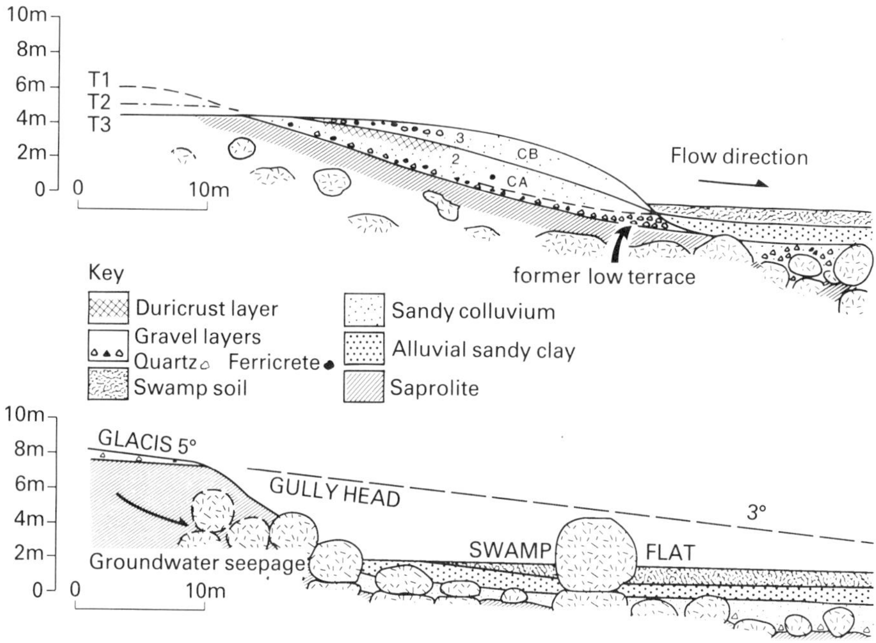

Fig. 11.6 (a) Generalized section through valleyhead sediments, indicating two periods of colluvial activity (CA, CB) and implying a shrinkage of the stream network during a semi-arid climate. (b) Extensions of gullyheads into (dry phase) *Glacis* or slope pediments by sapping and erosion marks a reversal of (a) in more humid conditions. Both diagrams based on field studies in Koidu, Sierra Leone.

of finer sediment. Not all sediment was moved by diffuse sheet flow during the dry climates of the Late Pleistocene; there is evidence also of gullying in the landscapes of both Koidu and the Birim Valley, where the older terrace bluffs are serrated by short, steep 'ravines' up to 30 m deep.

Many headwater valleys in areas of low relief are channelless swamps, known as *bolis* in Sierra Leone, and they are similar to the *dambos* of Central Africa, where clay and sand infills into headwater valleys are common (Thomas, 1986; Thomas and Goudie, 1985). Similar forms have been noted in South East Brazil (Coltrinari and Nogueira, 1989). These alluviated, flat-bottomed valleys often conceal former linear drainage channels and are therefore erosional in origin. However, there is also strong evidence for etching and chemical denudation into the bedrock beneath these valleys (Thomas and Thorp, 1985; McFarlane, in Boast, 1990), and recent extensions of the valley network also appear to have taken place by chemical sapping processes (Fig. 11.6b). In Malawi, Meadows (1985) found evidence for channel cutting prior to 12 000 BP with a fill phase starting shortly after 12 000 years ago and continuing until shortly after 5000 BP. In Inyanga, Zimbabwe the onset of this sedimentation was *c.* 11 700 BP (Tomlinson, 1974). These dates are almost identical with those obtained from Koidu and Birim for basal sediments in some larger streams. However, these areas have different climates and locations and may have responded in different ways to a similar chronology of events (Watson *et al.*, 1983).

It is difficult to place the period(s) of active slope wash and colluviation accurately into a time-frame. In wetter areas, the gradual onset of dry climates would have led to a slow increase in sediment transfer from 20 000 BP onwards, and slope wash would have continued throughout the dry glacial period. The rates of sediment transfer would, however, have increased rapidly as climates became wetter after 13 000 BP, when higher rainfalls possibly also led to widespread gullying and to channel cutting with deposition of coarse sediments in the river channels (Knox, 1972).

The amount of water available for overland flow and stream activity in the Koidu and Birim areas appears to have been insufficient to supply significant discharge to the trunk streams in their upper reaches between 20 500 and 12 500 BP, and wood and other organic material either were not entrained, or were not deposited and preserved, during this period. The survival of older timbers, however, suggests that sediment reworking was minimal during this period. It is possible that the gallery forests largely disappeared.

Hillslope deposits

The Freetown Peninsula and several inland mountain groups in Sierra Leone also reveal dramatic changes in the balance of surface processes (Thomas, 1983; Fig. 11.7). Apart from the widespread duricrust benches that may be Plio-Pleistocene, there are ferruginized hillslope deposits of 15°+ that are becoming fragmented and redistributed under forested conditions. Deep hillslope gullies are also infilled with coarse slope deposits of a strong red (5YR) hue and containing unsorted stones up to boulder size. These features coexist with a pattern of arcuate landslide scars that occupy more than 20% of parts of the main hill range of the Peninsula (reaching 900 m a.s.l.). There are similar sequences around inland hills that can be tentatively related to the main glacis surfaces, and to periods of incision. A duricrusted, ferricrete pseudokarst (Bowden, 1980) has also developed beneath the middle glacis, and the subterranean drainage here excavates a coarse landslide or mudflow deposit beneath the surface.

Close to the coast of southern Sierra Leone in the region of the Sewa River, deep soil pits have revealed what appear to be dry phase slope deposits beneath 2–3 m of recent sediment, probably derived by soil creep and bioturbation (mostly termite mixing). These thin layers contain

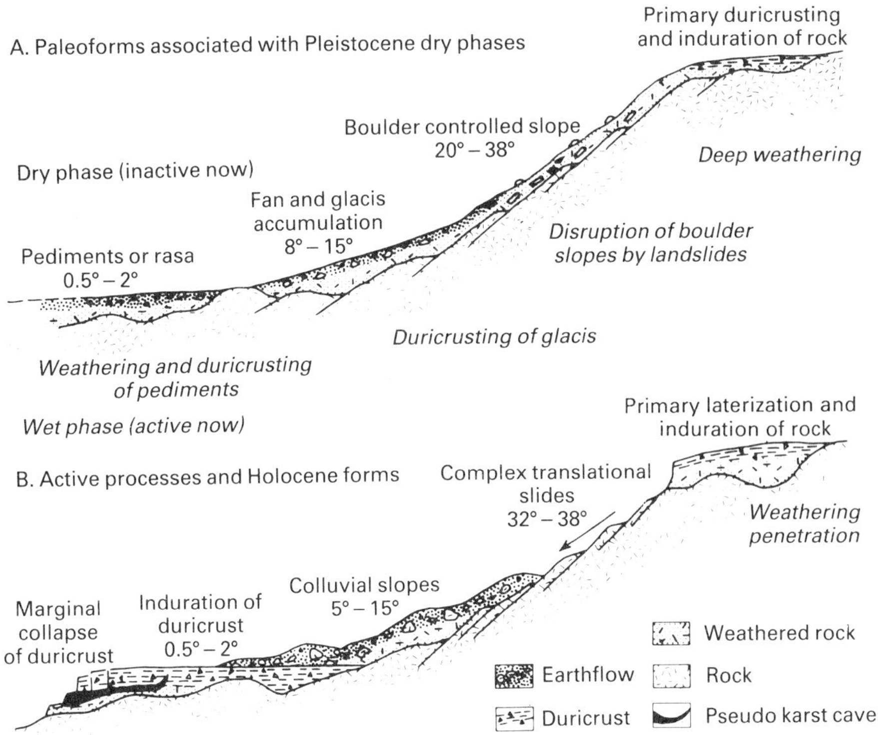

Fig. 11.7 Hillslope deposits found in the Freetown Peninsula, Sierra Leone, marking changes in slope dynamics from dry phases (a) to humid periods (b).

angular partly feldspathic breccia of red-purple hue, and also silcrete flags that denote strong evaporation of soil solutions rich in silica. These features appear to relate to semi-arid processes and, although undated, probably relate to Late Pleistocene aridification.

Interfluve gravels

Large areas of South America and Africa, and significant portions of other tropical landmasses, are underlain by gently sloping plains or dissected multi-convex (demi orange) topography. On these land surfaces, complex mosaics of surface deposits are found, and stone-lines are almost ubiquitous beneath the valley sides (Ab'Saber, 1982; Kadomura and Hori, 1990). Another variety of residual material occurring in such areas is the so-called white sand, discussed by Fairbridge and Finkl (1984). However, complex patterns of surface materials occur on tropical cratons, large areas being occupied by complex mixtures of coarse quartz clasts, iron nodules and duricrust fragments, rolled pebbles and resistant heavy minerals (Teeuw, 1989).

Deposits with polygenetic or disputed origins

Stone-lines

Taken literally, 'stone-lines' are thin stringers of little weathered clasts, found within many soil profiles at a depth of 0.5–2 m below the ground surface (Fig. 11.8). But the reality is often more complex and gravel layers may be more than 1 m thick in places and merge upslope into residual, interfluve gravels. Two or more stone-lines may be superposed within sections 5 m or more in depth. Some stone-lines are mainly unrolled quartz or quartzite clasts, but others contain rolled pebbles, clasts of other resistant minerals or rocks, pieces of ferricrete and iron oxide (haematite/goethite) nodules.

Fig. 11.8 Thin stone-line within the forest zone of southern Nigeria. Note that the stone-line dips at a steeper angle compared with the present landsurface and is covered by colluvium. A second, surface gravel may mark increased sheetwash erosion from anthropogenic causes.

Most theories for the origin of stone-lines require the removal of finer material downslope, allowing the stony material to accumulate as a 'lag' deposit which is later re-covered by sandy clays under a lower energy regime or by bioturbation. To explain such a sequence environmental change is usually invoked. Ab'Saber (1982) concluded from observations in Brazilian Amazonia that a majority of occurrences, 'represent a single widespread period of climatic aridity' (p. 44), while Bigarella and Andrade (1965) described a 'palaeopavement' carrying coarse stony detritus over pediment landforms in eastern Brazil. Bigarella *et al.* (1965) attributed the palaeopavement and associated stone-line complexes to mass movement and surface wash processes in an arid climate, and it can be noted that Fairbridge and Finkl (1984) considered stone-lines to be 'winnowed' mudflow deposits. Carefully dated stone-line stratigraphies with palaeolithic implements have been documented by Kadomura and Hori (1990) from southern Cameroon and western Kenya, and they concluded that such deposits record intense 'glacial' aridity in the humid tropics (20 000–12 000 BP).

Shallow trenching of interfluve surfaces in Koidu, Sierra Leone, revealed a complex lag gravel within 0.5 m of the ground surface, in places resting directly on a bedrock surface with a strong iron patina, and exhibiting clusters of heavy clasts in rill-like depressions. The complex lithology of these gravels (Teeuw, 1989) indicates their derivation from a variety of local rock types, while their packing and variations in thickness suggest transport by localized water flows under an open vegetation canopy.

However, residual accumulation of vein quartz and other resistant clasts can occur as weathering penetrates the subjacent rocks and the clays and solutes produced are removed by surface and sub-surface water flow, while simultaneously bioturbation renews the topsoil. No environmental change is required to explain this process which is a function of the *dynamic etching* (Thomas and Thorp, 1985; Thomas, M.F. 1989b) of tropical land surfaces, whereby rock weathering, chemical denudation and the gradual removal of fine-grained materials (mostly clays) effect the downwasting of the landscape under conditions of low energy surface erosion, under forest. These same conditions lead to bioturbation and the renewal of fine topsoil. This appears to be comparable with the interpretation of stone-lines in Malawi and Zimbabwe by McFarlane and Pollard (1989), who considered them to have formed at a dissolution front separating the saprolite from the superficial cover. Stone-lines can be traced within soil profiles of connecting valley-side slopes which may be regarded as conveyors for coarse clasts moving downslope, while other stone-lines clearly have alluvial origins, and are the degraded remnants of river terraces subjected to colluviation. Stone-lines may occur in complex sequences as in the valley-head colluvium observed in Koidu (Fig. 11.6a), or in gully fills, and often contain no datable material. Thus while the palaeoclimatic significance of stone-lines is not straightforward, many probably relate to the aridity of the LGM, while others record shorter periods of (possibly localized) landscape instability.

It is probable that many stone-line complexes evolve through several stages including: (a) residual accumulation resulting from dissolution and removal of weatherable and finer material under humid conditions; (b) redistribution and concentration of the gravel by high energy surface water flows and related colluvial activity, and (c) modification and burial by bioturbation, low energy surface wash, tree throw and possibly by anthropogenic activity. A model for their development was developed by De Dapper (1989), building on earlier work by Rohdenburg (1969) and Fölster (1969), and postulating the probability of landscape instability due to aridification, leading to the retreat of low scarps in the saprolite covered slopes. The fine materials are washed into the valley floors while the vein quartz, duricrust fragments and other clasts are left as a lag deposit which may later become buried by fine soil material. A similar model integrating events on interfluves, valley sides and small stream channels was developed for West Africa (Thomas, 1989b; Fig. 11.9).

Podzolized white sands

Similar problems and pitfalls beset the interpretation of the so-called white sands that appear to represent podzolized sandy sediments within both forested and savanna areas of today. According to Fairbridge and Finkl (1984, p. 64) 'large areas of the tropics became covered by sands that had been progressively leached of their soluble components during repeated wet phases of the Pleistocene, but distributed largely by wind action during the arid phases'. The reactivation of ancient sand sheets such as the Kalahari Sands during each arid phase and the subsequent leaching during pluvial periods must have occurred many times, the latest movements of the Kalahari Sand probably being Late Pleistocene (De Ploey, 1965).

Ab'Saber (1982) noted that different modes of genesis attach to contrasting types of white sand formation in the Brazilian Amazon. Some result from pedogenesis on intraformational sandy lenses in (Tertiary) bedrock, a feature seen also in Kalimantan. But the most important occurrences are fluvial sands which Ab'Saber considered'were deposited during a savanna environment' (p. 49). These include old channel fills containing fossil dunes, and near Rio Caeté sands extending beneath present-day riverine plains dominated by clays and organic sediments, a feature similar to coastal Kalimantan (Thorp *et al.*, 1990). Ab'Saber

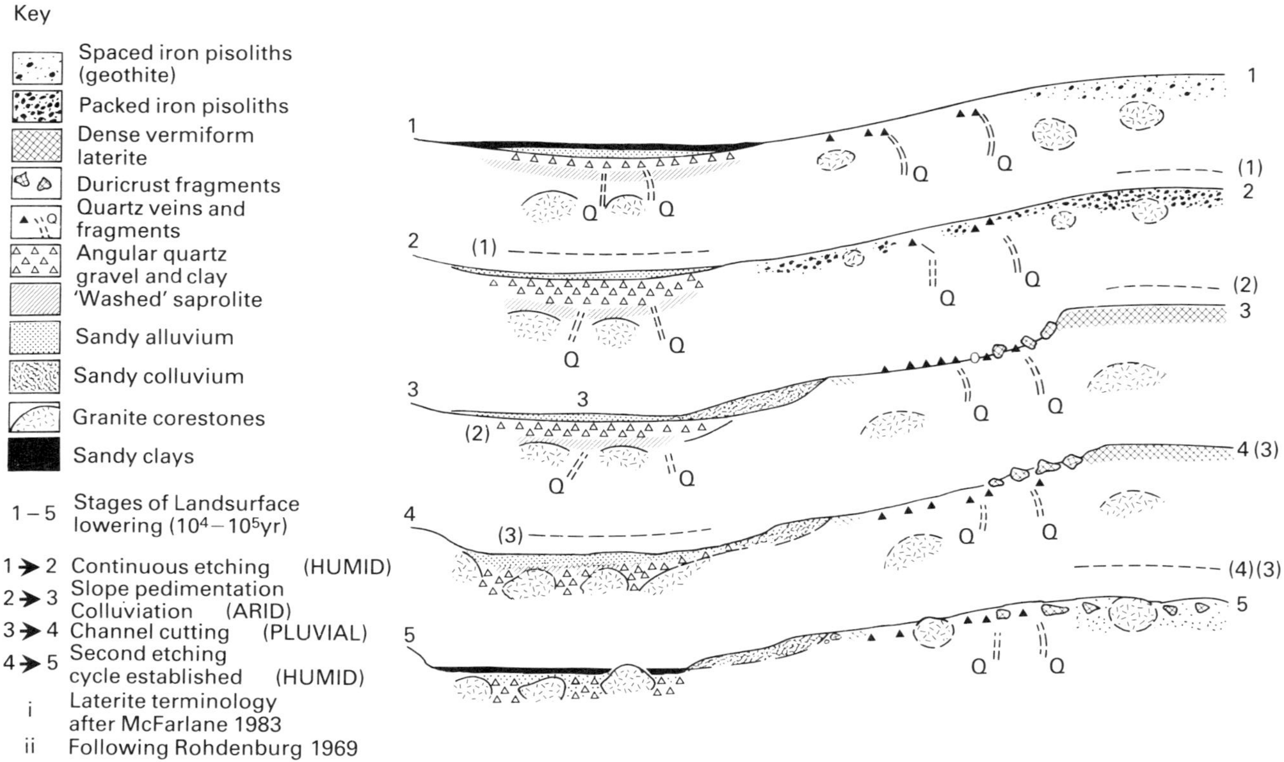

Fig. 11.9 Schematic representation of slope forms and deposits resulting from one Quaternary humid–arid–humid climatic cycle, and based on studies of low relief areas in Koidu, Sierra Leone. Lowering of interfluves and valley floors under humid tropical weathering and low energy erosion is referred to as 'continuous etching' (1–2) and (here) leads to accumulation of Fe_2O_3. A switch to dry conditions initiates a phase of slope pedimentation (3–4). Colluvial and stoneline complexes result from repeated climatic oscillation.

attributed different occurrences to dates ranging from Early Quaternary to Holocene for sands beneath small tributary valleys.

In other areas 'white sands' typify coastal plain alluvium, as in the Guyanas (Heyligers, 1963) and in Kalimantan (Sieffermann, 1988; Thorp *et al.*, 1990). Here aeolian transfer may be ruled out, but rapid deposition of already leached quartz sands derived from hillslope saprolites by sudden floods in open savanna environments probably contributed the bulk of the sediment which has subsequently become podzolized under humid tropical conditions. In this connection, the work of Sieffermann (1988) emphasized the important role of the peat swamps in leading to loss of iron and aluminium in a process described as 'deferrification'. Large areas of Amazonia and the Guyana coastal plains, together with those of Kalimantan, thus exhibit sediments containing sequences of pure sands intercalated with kaolinitic and gibbsitic clays. Their particular features appear to reflect the influence of the humid tropical climate within a hydromorphic pedogenic regime (Bravard and Righi, 1990). Their palaeoenvironmental significance must therefore lie in the nature of the sediments, their morphology and sedimentology, rather than in their striking colour and their recognition as 'giant' podzols.

In coastal Kalimantan the *Kerangas* forests reflect the occurrence of 10–30 m of these podzolized quartz sands, deposited in braided river plains and as low-angled fans during the Pleistocene, and quite probably during periods of rapid climate change towards wetter conditions (Thorp *et al.*, 1990). Additional interest centres on the low sea levels of the (Late) Pleistocene in this area, where the Sunda and Sahul shelf areas were exposed during low sea levels of −100 m (*c.* 18 000 BP). Walker (1982) has pointed out that the exposure of the sea floor would not have resulted in rainforest colonization. He stated (p. 565) that 'with mean sea level often oscillating vertically at rates between 1.9 m and 1.5 m per century, large parts would frequently be swept by extensive mangrove and other coastal formations and, where exposed, have hyper-saline soils.' The effects of the much increased land area during the last glacial maximum, combined with possible falls of temperature of at least 3–5 °C, may have led to increased seasonality of rainfall, (Verstappen, 1980; Thorp *et al.*, 1990; Heaney, 1991), although Walker (1982) has voiced doubts on this question.

LANDSCAPE DYNAMICS AND CHANGING CLIMATES OF THE LAST GLACIAL CYCLE

The evidence of landforms and deposits from large areas of the humid tropics suggests major shifts in the balance of processes during the

Quaternary, with clear evidence of aridity during the LGM, possibly lasting 4000–7000 years, according to locality. Most pollen records show that forest was re-established between 10 000 and 7000 years ago and widespread by 9000 years ago. Severe river flooding in both small and larger catchments is evident from 12 500 BP with a hiatus between 11 000 and 10 000 BP. Peak rainfalls may have occurred around 9000–7500 years ago, but the results were less spectacular than the earlier floods, presumably because the forest canopy was established by this time.

Consequences of glacial cooling

The cooling of lowland tropical climates by (3)4–6(7) °C, after 24 000–22 000 BP, accompanied by a sharp decline in rainfall between 22 000 and 13 000 BP must have led to the loss of forest trees and much of the herbaceous/shrub layer, resulting in an open woodland or scrub, periodically swept by fire. On local elevations and permeable substrates, the fall of water table would have led to quite severe pedological aridity. Some sites, including Lake Botsumturi (Maley, 1991), seem to indicate a greater influence of cooling with a change towards montane communities after 28 000 BP, and there is new evidence of earlier cooling in tropical and subtropical forest sites (Bush *et al.*, 1990; Heusser, 1990; Taylor, 1990), and we have shown evidence of older periods of rapid sedimentation, that correspond with signals from certain ice-cap and marine isotope data. The nature, extent and stability of pre-22 000 BP forest cover in the humid tropics must therefore remain in question.

As climate became drier after 20 000 BP, it is likely that a slow increase in runoff and sediment yield from slopes took place, stream discharges were reduced and shrinkage of the drainage network with colluviation of lower slopes and valleys ensued. At the LGM, conditions were severe enough in some areas for localized dune activity, and elsewhere mudflows occurred, affecting many small valleys. In the wetter areas, soil and groundwater reserves were sufficiently depleted in catchment headwaters for many valleys to become dry, experiencing only ephemeral or brief seasonal flows of water. The steeper hillslopes would have lost soil and saprolite cover almost continuously once the forest cover had been lost. Debris slides and mudflows would have increased in frequency, but it is not clear whether deep-seated slides were accelerated.

Induration of iron segregations and the formation of nodules would have been characteristic of many soils, while older duricrust cappings which may have become fragmented under a previous forest cover would have been exposed as benches and breakaways around valley sides and interfluves. The loss of topsoil would have been widespread on slopes of only 1°–2°, as rapid runoff from occasional storms produced sheet floods and concentration of colluvium in depressions, valley heads and foot-slopes. Stone pavements were formed and sandy 'pedisediment' would have accumulated on distal slopes and in valley floors.

It is necessary to visualize a landscape at around 13 000 years ago that had experienced such conditions for several millennia. This landscape would have been one of bare rocky hillslopes and duricrust cappings; boulder-strewn slopes with localized debris fans and talus accumulation; extensive glacis cut across truncated soil and weathering profiles and exhibiting boulders and coarse gravelly sediments upslope and sandy fills downslope, merging with an atrophied valley network. In some areas, surface sands will have been blown into dunes. It is not possible to chart the exact distribution of this landscape, and it would have been gradational from its drier margins at or beyond the present forest–savanna boundary towards any forest refugia. A review of evidence for former extensions of desert conditions can be found in Thomas (1989a).

Effects of post-glacial warming

The onset of rapid climate warming must have led to early increases in rainfall, and possibly to very unstable climatic conditions. The notion of gentle rains (Lézine and Casanova, 1989) at this time remains speculative, but there is a possibility of prolonged 'monsoonal' downpours in place of individual convectional storms (Maley and Livingstone, 1983). Initially, the increasing rainfall may have shifted more sediment from slopes into valleys, leading to sedimentation, but, as the duration or frequency of precipitation events increased, groundwater recharge would have led to base flows in many formerly dry river beds, and to high flood peaks resulting from up to 60% runoff from sparsely vegetated slopes. As vegetation responded to reduced drought stress and greater seasonal water availability, so sediment yield may have declined relative to runoff, leading to erosional trenching of valley floors and of colluvium on slopes and glacis. Increasing density of tree cover would have been influenced by the properties of site and substrate. Floods in rivers both small and large may have been quite exceptional. Evidence from the Nile suggests that deposition of silts took place well above older terrace levels (Paulissen, 1989), while in the Birim River of Ghana and the Sewa head-waters in Sierra Leone, older sediments were swept out and river channels incised up to 4 m into bedrock, exposing massive core stones which were later concealed by accumulations of channel sediments. Large timbers with radiocarbon dates of 12 500–11 500 BP were jammed under these boulders. These timbers were presumably derived mainly from bank erosion and therefore from gallery forests. In Kalimanatan, the braided river plains and fans became incised by rivers with lower gradients.

Floods of this magnitude and effect appear not to have recurred, probably because of recovery of the vegetation. The wetter conditions after 11 500 BP must have started a process of forest extension which gained momentum after 10 000 BP and was virtually completed over the following 1000–2000 years. Sediment yields from slopes would have been steadily reduced and immediate runoff replaced increasingly by throughflow. Rivers formed stable meandering channels, the smaller and seasonal streams often flowing on earlier sedimentary fills.

On interfluves and on the coastal plain terraces, sandy sediments became subject to rapid leaching and podzolization, while iron mottling affected the clayey mudflows and other colluvial deposits. Bioturbation, and particularly the work of termites, will have brought fine sand and silt to the surface, burying the stone pavements.

On steeper hillslopes, where saprolite remained, shallow translational landslides continued and slumping will have affected deeply weathered areas. In some cases landslide debris came to rest on ferricrete benches, exposed during the glacial dry period around many residual hill masses in monsoonal West Africa. Landslide activity will have been influenced by the occurrence of prolonged heavy rains (typically 200–250 mm day^{-1} over five days) and it is not possible at present to offer dates for periods when these were more common than at the present time.

Cumulative results of repeated climate change

With such major shifts in the dynamics of the ground surface taking place repeatedly during the Quaternary, it is remarkable that such widespread and deep saprolite is found beneath much of the present-day forest. It has been suggested (Lichte, 1989) that much of this saprolite was formed during the Holocene, but, where depths exceed 50 m as in South East Brazil, there has been insufficient time to produce observed depths. It is more likely that the saprolite mantle was dissected and selectively stripped during the Quaternary, with weathering penetration continuing into a progressively compartmented landscape. In southern Nigeria, however, the acid gneiss, migmatites and many granites are widely stripped of saprolite (Thomas, 1965), and throughout West Africa, rocky inselbergs, tors and boulder-strewn slopes abound both in the forest and the savanna. It is tempting to suggest that such contrasts in the preservation of weathering profiles may reflect varying duration and intensities of arid conditions during the Quaternary. In this case, Atlantic Brazil may have been dry for shorter periods (and perhaps less dry) than the monsoonal coast of West Africa, even where present-day rainfall totals are comparable (1500–2000 mm).

Under favourable rainfall regimes forest can and does colonize such boulder-strewn slopes, and the loss of older saprolite and soil covers may not have prevented the forest from claiming back the savannas. However, the different views of the rate of forest spread may be in part a result of the differential speed of colonization onto slopes with different soil hydrological regimes. In areas of marginal rainforest climate, as in South Western Nigeria (*c*. 1400 mm) there do appear to be sudden changes in saprolite and soil development close to the present-day forest boundary, and it may be that 9000 years of soil development has continued within the forest, while the savannas beyond have remained open woodlands for more than 20 000 years.

Thus soil patterns have been suggested by Haffer (1987) as contributing in a major way to the persistence of forest refuges in tropical America. In the same way the 'reclamation' of the savannas by forest has undoubtedly been influenced by similar patterns. But between those areas that remained under open woodland or scrub throughout the later Quaternary at least, and those which kept their forest canopy more or less intact, there were much larger areas that experienced the fluctuations of process similar to those described here.

In debates about future policies for rainforest areas, an understanding of the landforms, sediments and soils that are legacies of past climates should inform project planning and land evaluation. Humid tropical landscapes are particularly fragile and the inherited Quaternary deposits have already been destabilized in many areas by activities that have ignored these widespread features.

REFERENCES

Ab'Saber, A.N. (1982) The paleoclimate and paleoecology of Brazilian Amazonia, in *Biological Diversification in the Tropics*, (ed. G.T. Prance), Columbia, New York, pp. 41–59.

Absy, M.L. (1982) Quaternary palynological studies in the Amazon Basin, in *Biological Diversification in the Tropics*, (ed. G.T. Prance), Columbia, New York, pp. 67–73.

Absy, M.L., Van Der Hammen, T., Soubiès, F. *et al.* (1989) Data on the history of vegetation and climate in Carajás, eastern Amazonia, in *International Symposium on Global Changes in South America During the Quaternary, São Paulo, 1989*, Special Publication, Vol. 1, pp. 129–31.

Baumhauer, R. (1989) Paleolakes of the south central Sahara – problems of paleoclimatic interpretation. *Abstracts of Papers. Second International*

Conference on Geomorphology, Frankfurt/Main, 1989, Vol. 1, Geöko-plus, p. 216.

Bigarella, J.J. and Andrade, G.O. de (1965) Contribution to the study of the Brazilian Quaternary. *Geological Society of America, Special Paper*, **84**, 435–51.

Bigarella, J.J., Mousinho, M.G. and Da Silva, J.X. (1965) Processes and environments of the Brazilian Quaternary, *Symposium on Cold Climate Processes and Environments, Fairbanks, Alaska, VII INQUA Congr., 1965*, Univ. Paraná, Curitiba.

Bigarella, J.J. and Becker, R.D. (eds) (1975) International Symposium on the Quaternary (Southern Brazil, July 15–31, 1975), *Boletim Paranaense de Geociências*, **33**, 370.

Bigarella, J.J. and Andrade-Lima, D. de (1982) Palaeoenvironmental changes in Brazil, in (ed. G.T. Prance), *Biological Diversification in the Tropics*, Columbia, New York, pp. 27–39.

Boast, R. (1990) Dambos: a review. *Progress in Physical Geography*, **14**, 153–77.

Bonnefille, R. and Riollet, G. (1988) The Kashiru pollen sequence (Burundi) palaeoclimatic implications for the last 40 000 yr BP in Tropical Africa. *Quaternary Research*, **30**, 19–35.

Bowden, D.J. (1980) Sub-laterite cave systems and other pseudokarst phenomena in the humid tropics; the example of the Kasewe Hills, Sierra Leone. *Zeitschrift für Geomorphologie*, **24**, 77–90.

Bradbury, J.P., Leyden, B., Salgado-Labouriau, M.L. *et al.* (1981) Late Quaternary environmental history of Lake Valencia, Venezuela. *Science*, **214**, 1199–205.

Bravard, S. and Righi, D. (1990) Podzols in Amazonia. *Catena*, **17**, 461–75.

Broecker, W.S. and Denton, G.H. (1990) The role of ocean–atmosphere reorganisations in glacial cycles. *Quaternary Science Reviews*, **9**, 305–41.

Brückner, W.D. (1955) The mantle rock (laterite) of the Gold Coast and its origin. *Geologische Rundschau*, **43**, 307 27.

Brünig, E.F. (1975) Tropical ecosystems: state and targets of research into the ecology of humid tropical ecosystems. *Plant Research and Development*, **1**, 22–38.

Bush, M.B., Colinvaux, P., Weimann, M.C. *et al.* (1990) Late Pleistocene temperature depression and vegetation change in Ecuadorian Amazonia. *Quaternary Research*, **34**, 330–45.

Caratini, C. and Tissot, C. (1988) Paleogeographical evolution of the Mahakam Delta in Kalimantan, Indonesia, during the Quaternary and late Pliocene. *Review of Paleobotany and Palynology*, **55**, 217–28.

CLIMAP Project Members (1976) The surface of the ice-age earth. *Science*, **191**, 1131–8.

CLIMAP Project Members (1981) *Seasonal Reconstruction of the Earth's*

Surface During the Last Glacial Maximum. Geological Society of America, Map and Chart Series, No. 36.

Coltrinari, L. and Nogueira, F. (1989) Dambo-like landforms in southeastern Brazil. *Abstracts of Papers. Second International Conference on Geomorphology, Frankfurt/Main, 1989,* Vol. 1, Geöko-plus, p.58.

Damuth, J.E. and Fairbridge, R.W. (1970) Equatorial Atlantic deep-sea arkosic sands and ice-age aridity in tropical South America. *Bulletin of the Geological Society of America,* **81**, 189–206.

De Dapper, M. and Debaveye, J. (1986) Geomorphology and soils of the Padang Terap District, Kedah, Peninsular Malaysia. *Bulletin of the Geological Society of Malaysia,* **20**, 765–90.

De Dapper, M. (1988) Geomorphology of the sand-covered plateaux in southern Shaba, Zaire, in (eds G.F. Dardis and B.P. Moon), *Geomorphological Studies in Southern Africa,* Balkema, Rotterdam, pp. 115–35.

De Dapper, M. (1989) Pedisediments and stone-line complexes in Peninsular Malaysia, in (eds J. Alexandre and J.J. Symoens), *Stone-Lines. Journée d'Étude, Bruxelles, 24 mars 1989, Académie Royale des Sciences D'Outre-Mer*, Geo-Eco-Trop, **11**, 37–59.

Delwaulle, J.C. (1973) Résultats de six années d'observations sur l'érosion au Niger. *Bois et Forêts des Tropiques,* **150**, 15–37.

De Meis, M.R.M. and De Moura, J.R. da Silva (1984) Upper Quaternary sedimentation and hillslope evolution: southeastern Brazilian plateau. *American Journal of Science,* **284**, 241–54.

De Meis, M.R.M. and Monteiro, A.M.F. (1979) Upper Quaternary 'rampas': Doce river valley, southeastern Brazilian plateau.*Zeitschrift für Geomorphologie NF,* **23**, 132–51.

De Ploey, J. (1965) Position géomorphologique, génèse et chronologie de certains dépots superficiels au Congo occidentale. *Quaternaria,* **7**, 131–54.

De Ploey, J. and Cruz, O. (1979) Landslides in the Serra Do Mar, Brazil. *Catena,* **6**, 111–22.

Diester-Haas, L. (1976) Late Quaternary climatic variations in NW Africa deduced from east Atlantic sediment cores. *Quaternary Research,* **6**, 299–314.

Douglas, I. and Spencer, T. (1985) Present-day processes as a key to the effects of environmental change, in (eds I. Douglas and T. Spencer), *Environmental Change and Tropical Geomorphology,* Allen & Unwin, London, pp. 39–73.

Fairbridge, R.W. and Finkl, C.W. Jr (1984) Tropical stone lines and podzolised sand plains as palaeoclimatic indicators for weathered cratons. *Quaternary Science Reviews,* **3**, 41–72.

Flenley, J.R. (1979) *The Equatorial Rainforest: a Geological History*, Butterworth, London.

Fölster, H. (1969) Slope development in SW Nigeria during the late Pleistocene and Holocene. *Göttinger Bodenkundliche Berichte*, **10**, 3–56.

Giresse, P. (1978) Le control climatique de la sedimentation marine et continentale en Afrique centrale Atlantique à la fin du Quaternaire. *Palaeogeography, Palaeoclimatology and Palaeoecology*, **23**, 57–77.

Giresse, P., Bongo-Passi, G., Delibrias, G. and Du Plessy, J.C. (1982) La lithostratigraphie des sediments hemiplagiques du delta profond du fleuve Congo et ses indications sur les paléoclimats de la fin du Quaternaire.*Bulletin Société Géologie française*, Série, **24**, 803–15.

Goudie, A.S. (1983) The arid earth, in (eds R. Gardner and H. Scoging), *Mega-geomorphology*, Clarendon, Oxford, pp. 152–71.

Gupta, A., Rahman, A., Poh Poh, W. and Pitts, J. (1987) The Old Alluvium of Singapore and the extinct drainage system to the South China Sea. *Earth Surface Processes and Landforms*, **12**, 259–75.

Haffer, J. (1987) Quaternary history of tropical America, in (eds T.C. Whitmore and G.T. Prance), *Biogeography and Quaternary History in Tropical America*, Clarendon, Oxford, pp. 1–18.

Hall, A.M., Thomas, M.F. and Thorp, M.B. (1985) Late Quaternary alluvial placer development in the humid tropics: the case of the Birim Diamond Placer, Ghana. *Journal of The Geological Society*, **142**, 777–87.

Hamilton, A.C. (1982) *Environmental History of East Africa: A Study of the Quaternary*, Academic Press, New York.

Hamilton, A.C. and Taylor, D. (1991) History of climate and forests in tropical Africa during the last 8 million years. *Climate Change*, **19**, 65–78 (Special Issue: Tropical Forests and Climate).

Heaney, L.R. (1991) A synopsis of climatic and vegetation change in southeast Asia. *Climate Change*, **19**, 53–61 (Special Issue: Tropical Forests and Climate).

Heinrich, H. (1988) Origin and consequences of cyclic ice rafting in the northeast Atlantic Ocean during the past 130 000 years. *Quaternary Research*, **29**, 142–52.

Heusser, C.J. (1990) Ice age vegetation and climate of subtropical Chile. *Palaeogeography, Palaeoclimatology and Palaeoecology*, **80**, 107–27.

Heyligers, P.C. (1963) *Vegetation and Soil of a White-Sand Savanna in Suriname*. Verhandelingen der koninkluke Nederlandse, Akademie van Wetenschappen, AFD Natuurkunde, NV Noord-Hollandshce Uitgevers Maatschappij, Amsterdam.

Hill, I.D. and Rackham, I.J. (1978) Indications of mass movement on the Jos Plateau, Nigeria. *Zeitschrift für Geomorphologie*, **NF 22**, 258–74.

Jones, F. (1973) *Landslides of Rio de Janeiro and the Serra das Araras escarpment, Brazil*, US Geological Survey Professional Paper, No. 697.

Jouzel, J., Lorius, C., Petit, J.R. *et al.* (1987) Vostok ice core: a continuous isotope temperature record over the last climatic cycle (160 000 years). *Nature*, **329**, 403–8.

Junner, N.R. (1943) The diamond deposits of the Gold Coast. *Gold Coast Geological Survey, Bulletin*, **12**, 55.

Kadomura, H. (1986) Late glacial–early Holocene environmental changes in tropical Africa: a comparative analysis with deglaciation history. *Geographical Reports, Tokyo Metropolitan University*, **21**, 1–21.

Kadomura, H. and Hori, N. (1990) Environmental implications of slope deposits in humid tropical Africa: evidence from southern Cameroon and western Kenya. *Geographical Reports, Tokyo Metropolitan University*, **25**, 213–36.

Kesel, R.H. (1977) Slope runoff and denudation in the Rupununi savanna, Guyana. *Journal of Tropical Geography*, **44**, 33–42.

Klammer, G. (1981) Landforms, cyclic erosion and deposition, and Late Cenozoic changes in climate in southern Brazil. *Zeitschrift für Geomorphologie*, **NF 25**, 146–65.

Klammer, G. (1982) Die Päleowüste des Pantanal von Mato Grosso und die pleistozäne Klimageschichte der brasilianischen Randtropen. *Zeitschrift für Geomorphologie*, **NF 26**, 393–416.

Knox, J.C. (1972) Valley alluviation in southwestern Wisconsin. *Annals of the Association of American Geographers*, **62**, 401–10.

Kolla, B., Biscaye, P.E. and Hanley, A.F. (1979) Distribution of quartz in Late Quaternary Atlantic sediments in relation to climate. *Quaternary Research*, **11**, 261–77.

Kraus, E.B. (1973) Comparison between ice age and present general circulation. *Nature*, **245**, 129–33.

Lal, R. (1977) Analysis of factors affecting rainfall erosivity and soil erodibility, in (eds D.J. Greenland and R. Lal), *Soil Conservation and Management in the Humid Tropics*, John Wiley, Chichester, pp. 49–56.

Lal, R. (1981) Deforestation of tropical rainforest and hydrological problems, in (eds R. Lal and E.W. Russell), *Tropical Agricultural Hydrology*, John Wiley, Chichester, pp. 131–40.

Lézine, A.-M. and Casanova, J. (1989) Pollen and hydrological evidence for the interpretation of past climates in tropical West Africa during the Holocene. *Quaternary Science Reviews*, **8**, 45–55.

Lichte, M. (1989) Arid processes in the southeast Brazilian relief evolution during the last ice-age. *International Symposium on Global Changes in South America During the Quaternary, São Paulo, 1989, Special Publication*, **1**, 60–4.

Littmann, T. (1989) Climatic change in Africa during the Last Glacial: facts and problems. *Palaeoecology of Africa*, **21**, 163–79.

Livingstone, D.A. (1982) Quaternary geography of Africa and the Refuge Theory, in (ed. G.T. Prance), *Biological Diversification in the Tropics*, Columbia, New York, pp. 523–36.

Lumb, P. (1975) Slope failures in Hong Kong. *Quarterly Journal of Engineering Geology*, **8**, 31–65.

Maley, J. (1987) Fragmentation de la forêt dense humide Africaine et extension des biotopes montagnards au Quaternaire récent: nouvelles données polliniques et chronologiques. Implications paleoclimatiques et biogéographiques. *Palaeoecology of Africa*, **18**, 307–34.

Maley, J. (1991) The African rainforest vegetation and palaeoenvironments during Late Quaternary. *Climate Change*, **19**, 79–98 (Special Issue: Tropical Forests and Climate).

Maley, J. and Livingstone, D.A. (1983) Extension d'un élément montagnard dans le sud du Ghana (Afrique de l'Ouest) au Pléistocène supérieur et à l'Holocene inférieur: premières données polliniques. *Comptes Rendus, Académie des Sciences, Paris*, **296, Serie II**, 1287–92.

Markgraf, V. (1989) Palaeoclimates in central and south America since 18 000 BP based on pollen and lake-level records. *Quaternary Science Reviews*, **18**, 1–24.

McFarlane, M.J. and Pollard, S. (1989) Some aspects of stone-lines and dissolution fronts associated with regolith and dambo profiles in parts of Malawi and Zimbabwe, in (eds J. Alexandre and J.J. Symoens), *Stone-Lines. Journée D'Étude, Bruxelles, 24 mars 1987*, Académie Royale des Sciences D'Outre-Mer, Geo-Eco-Trop **11**, pp. 23–35.

Meadows, M.E. (1985) Dambos and environmental change in Malawi, Central Africa, in (eds M.F. Thomas and A.S. Goudie), *Dambos: Small Channelless Valleys in the Tropics, Zeitschrift für Geomorphologie Supplementband*, **52**, 147–69.

Menabe, S. and Hahn, D.G. (1977) Simulation of the tropical climate of an ice age. *Journal of Geophysical Research*, **82**, 3889–911.

Michel, P. (1973) Les Bassins des fleuves Sénégal et Gambie – étude géomorphologique. *ORSTOM, Memoir*, Vol. 63.

Michel, P. (1980) The southwestern Sahara margin: sediments and climatic changes during the recent Quaternary. *Palaeoecology of Africa*, **12**, 297–314.

Pastouret, L., Charnley, H., Delibrias, G. *et al.* (1978) Late Quaternary climatic changes in western tropical Africa deduced from deep-sea sedimentation off the Niger Delta. *Oceanologica Acta*, **1**, 217–32.

Paullissen, E. (1989) The inundations of the Saharan Nile around 12 500 BP: unique catastrophic floods of an exotic river. *Abstracts of Papers, Second International Conference on Geomorphology, Frankfurt/Main, 1989*, Vol. 1, Geöko-plus, p. 216.

Peltre, P. (1977) Le 'V Baoulée' (Côte – d'Ivoire centrale) – Héritage géomorphologique et paléoclimatique dans le tracé du contact forêt-savane. *Travaux et Documents de L'ORSTOM*, Vol. 80, Paris, p. 198.

Roche, M.A. (1981) Watershed investigations for development of forest resources of the Amazon Region in French Guyana, in (eds R. Lal and E.W. Russell), *Tropical Agricultural Hydrology*, John Wiley, Chichester, pp. 75–92.

Rohdenburg, H. (1969) Hangpedimentation und Klimawechsel als wichstigste Faktoren der Flächen und Stufenbildung in den wechselfeuchten Tropen an Beispielen aus Westafrika, besonders aus dem Schichtstufenland Sudost-Nigerias. *Göttinger bodenkundliche Berichte*, **10**, 58–78.

Roose, E.J. (1973) *Dix-sept années de mesures éxperimentales de l'érosion et du russellement sur un sol ferrallitique sableux de basse Côte d'Ivoire*. Thèse Docteur Ingénieur, Faculté des Sciences, Université d'Abidjan, ORSTOM, Abidjan.

Roose, E.J. (1977) Application of the universal soil loss equation of Wischmeier and Smith in West Africa, in (eds D.J. Greenland and R. Lal), *Soil Conservation and Management in the Humid Tropics*, John Wiley, Chichester, pp. 177–87.

Rossignol-Strick, M. and Duzer, D. (1979) A late Quaternary continuous climatic record from palynology of three marine cores off Senegal. *Palaeoecology of Africa*, **11**, 185–8.

Rossignol-Strick, M., Nesteroff, W., Olive, P. and Vergnaud-Grazzini, C. (1982) After the deluge: Mediterranean stagnation and sapropel formation. *Nature*, **295**, 105–10.

Sarnthein, M. (1978) Sand deserts during glacial maximum and climatic optimum. *Nature*, **272**, 43–6.

Schubert, C. (1988) Climatic changes during the last glacial maximum in northern South America and the Caribbean: a review. *Interciência*, **13**, 128–37.

Schubert, C. and Valastro, S. Jr (1980) Quaternary Esnujaque Formation, Venezuelan Andes: preliminary alluvial chronology in a tropical mountain range. *Zeitschrift für Deutsche Geologie, Gesselschaft*, **131**, 927–47.

Servant, M., Soubiès, F., Suguio, K. *et al.* (1989) Alluvial fans in southeastern Brazil as an evidence for early Holocene dry climatic period. *International Symposium on Global Changes in South America During the Quaternary, São Paulo, 1989*, Special Publication, Vol. 1, pp. 75–7.

Sieffermann, R.G. (1988) Le système des grandes tourbières equatoriales. *Annales de Géographie*, **544**, 642–66.

Street F.A. (1981) Tropical palaeoenvironments. *Progress in Physical Geography*, **5**, 157–85.

Street, F.A. and Grove, A.T. (1979) Global maps of lake level fluctuations since 30 000 BP. *Quaternary Research*, **12**, 83–118.

Street-Perrott, F.A., Roberts, N. and Metcalfe, S. (1985) Geomorphic implications of Late Quaternary hydrological and climatic changes in the Northern Hemisphere tropics, in (eds I. Douglas and T. Spencer), *Environmental Change and Tropical Geomorphology*, Allen and Unwin, London, pp. 164–83.

Stuijts, I., Newsome, J.C. and Flenley J.R. (1988) Evidence for Late Quaternary vegetational change in the Sumatran and Javan highlands. *Review of Palaeobotany and Palynology*, **55**, 207–16.

Soubiès, F., Suguio, K., Martin, L. *et al.* (1989) The Quaternary lacustrine deposits of the Serra dos Carajás (State of Pará, Brazil): ages and other preliminary results, in *International Symposium on Global Changes in South America During the Quaternary, São Paulo, 1989*, Special Publication, Vol. 1, pp. 125–8.

Sutherland, D.G. (1985) Geomorphological controls on the distribution of placer deposits. *Journal of The Geological Society*, **142**, 727–37.

Talbot, M.R. and Delibrias, F. (1980) A new late Pleistocene – Holocene water-level curve for lake Bosumtwi, Ghana. *Earth and Planetary Science Letters*, **47**, 336–44.

Talbot, M.R., Livingstone, D.A., Palmer, P.G. *et al.* (1984) Preliminary results from sediment cores from Lake Bosumtwi, Ghana. *Palaeoecology of Africa*, **16**, 173–92.

Taylor, D.M. (1990) Late Quaternary pollen records from two Ugandan mires: evidence for environmental change in the Rukiga Highlands of southwest Uganda. *Palaeogeography, Palaeoclimatology and Palaeoecology*, **80**, 283–300.

Teeuw, R.W. (1989) Variations in the composition of gravel layers across the landscape. Examples from Sierra Leone, in (eds J. Alexandre, and J.J. Symoens), *Stone-Lines. Journée D'Étude, Bruxelles, 24 mars 1987*, Académie Royale des Sciences D'Outre-Mer, Geo-Eco-Trop **11**, pp. 151–69.

Temple, P.H. and Rapp, A. (1972) Landslides in the Mgeta area, western Uluguru Mountains, Tanzania. *Geografiska Annaler*, **54 A**, 157–93.

Thomas, D.S.G. (1989a) Reconstructing ancient arid environments in (ed. D.S.G. Thomas), *Arid Zone Geomorphology*, Belhaven, London, pp. 311–34.

Thomas, M.F. (1965) An approach to some problems of landform analysis in tropical climates, in (eds J.B. Whittow and P.D. Wood), *Essays in Geography for Austin Miller*, Univ. Reading, pp. 118–44.

Thomas, M.F. (1980) Timescales of landform development on tropical shields – a study from Sierra Leone, in (eds R.A. Cullingford, D.A. Davidson and J. Lewin), *Timescales in Geomorphology*, John Wiley,

Chichester, pp. 333–54.
Thomas, M.F. (1983) Contemporary denudation systems and the effects of climatic change in the humid tropics – some problems from Sierra Leone, in (eds D.J. Briggs and R.S. Waters), *Studies in Quaternary Geomorphology*, Geo Books, Norwich, pp. 195–214.
Thomas, M.F. (1986) Savanna, in (eds P.G. Fookes and P.R. Vaughan), *Handbook of Engineering Geomorphology*, Surrey Univ. Press, Guildford, pp. 125–36.
Thomas, M.F. (1989b) The role of etch processes in landform development I. Etching concepts and their applications. *Zeitschrift für Geomorphologie*, **NF 33**, 129–42.
Thomas, M.F. and Goudie, A.S. (eds) (1985) Dambos: small channelless valleys in the tropics. *Zeitschrift für Geomorphologie, Supplementband*, **52**, 222.
Thomas, M.F. and Thorp, M.B. (1980) Some aspects of the geomorphological interpretation of Quaternary alluvial sediments in Sierra Leone. *Zeitschrift für Geomorphologie, Supplementband*, **36**, 140–61.
Thomas, M.F. and Thorp, M.B. (1985) Environmental change and episodic etchplanation in the humid tropics: the Koidu etchplain, in (eds I. Douglas and T. Spencer), *Environmental Change and Tropical Geomorphology*, Allen and Unwin, London, pp. 239–67.
Thomas, M.F. and Thorp, M.B. (1992) The geomorphology of some Quaternary placer deposits. *Zeitschrift für Geomorphologie Supplementband*, (in press).
Thomas, M.F., Thorp, M.B. and Teeuw, R.M. (1985) Palaeogeomorphology and the occurrence of diamondiferous placer deposits in Koidu, Sierra Leone. *Journal of The Geological Society*, **142**, 789–802.
Thorp, M.B., Thomas, M.F., Martin, T. and Whalley, W.B. (1990) Late Pleistocene sedimentation and landform development in western Kalimanatan (Indonesian Borneo). *Geologie en Mijnbouw*, **69**, 133–50.
Tomlinson, R.W. (1974) Preliminary biogeographical studies on the Inyanga Mountains, Rhodesia. *South African Geographical Journal*, **56**, 15–26.
Tricart, J. (1985) Evidence of Upper Pleistocene dry climates in northern South America, in (eds. I. Douglas and T. Spencer), *Environmental Change and Tropical Geomorphology*, Allen and Unwin, London, pp. 197–217.
Van Campo, E. (1986) Monsoon fluctuations in two 20 000 Yr BP oxygen-isotope pollen records off southwest India. *Quaternary Research*, **26**, 376–88.
Van der Hammen, T. (1991) Palaeoecological background: neotropics. *Climate Change*, **19**, 37–47 (Special Issue: Tropical Forests and Climate).

Verstappen, H.Th. (1980) Quaternary climatic changes and natural environment in SE Asia. *Geo Journal*, **4**, 45–54.

Walker, D. (1982) Speculations on the origin and evolution of Sunda-Sahul rain forests, in (ed. G.T. Prance), *Biological Diversification in the Tropics*, Columbia, New York, pp. 554–75.

Watson, A., Price-Williams, D. and Goudie, A.S. (1983) Palaeoenvironmental interpretation of colluvial sediments and palaeosols of the Late Pleistocene hypothermal in Southern Africa. *Palaeogeography, Palaeoclimatology and Palaeoecology*, **45**, 225–50.

Whitmore, T.C. and Prance G.T. (eds) (1987) *Biogeography and Quaternary History in Tropical America*, Clarendon, Oxford.

Williams M.A.J. (1985) Pleistocene aridity in tropical Africa, Australia and Asia, in (eds. I. Douglas and T. Spencer), *Environmental Change and Tropical Geomorphology*, Allen and Unwin, London, pp. 219–33.

Wischmeier, W.H. and Smith, D.D. (1958) Rainfall energy and its relationship to soil loss. *Transactions of The American Geophysical Union*, **39**, 285–91.

12

The development of rainforest–savanna boundaries in tropical Australia

A.P. KERSHAW

SUMMARY

1. In northern Australia, a variety of rainforest types exist as small, isolated patches surrounded by open sclerophyll or savanna vegetation.
2. Pollen diagrams from close to a major rainforest–savanna boundary on the Atherton Tableland provide a substantial record of vegetation changes within the later part of the Cainozoic period (i.e. the last few million years).
3. Through the late Cainozoic there have been increases in savanna relative to rainforest and changes in the nature of the savanna vegetation due mainly to reduced precipitation, and increased climatic variability.
4. Fire has probably always been a component of savanna environments but it is considered that an increase in burning, probably associated with the activities of Aboriginal people, brought about the destruction of drier rainforest communities and a substantial expansion of savanna.
5. The present nature of rainforest–savanna boundaries within the area most likely dates from the early Aboriginal colonization although there have been subsequent marked changes in boundary position due to continuing climatic change.
6. Although the Atherton Tableland landscapes and climate are somewhat atypical of those of much of northern Australia, the scenario advanced for this area goes some way to explaining patterns of distribution in the region generally.

INTRODUCTION

Little attention has been paid to rainforest–savanna boundaries in Australia. This is perhaps because of the present limited extent of rainforest on the continent and the unique nature of more open vegetation, dominated by species *Eucalyptus*, which disguises underlying similarities with savanna vegetation on other continents. In addition, Australia is unusual in having a continuum of rainforest vegetation along the east coast of the continent, albeit as small isolated patches (Fig. 12.1). This has tended to focus research on coastal/inland variation in vegetation rather than on that occurring specifically within tropical or temperate latitudes.

Despite a lack of background research, scientists in Australia are well placed to examine the relative importance of various factors involved in the development of rainforest–savanna boundaries over a long period of

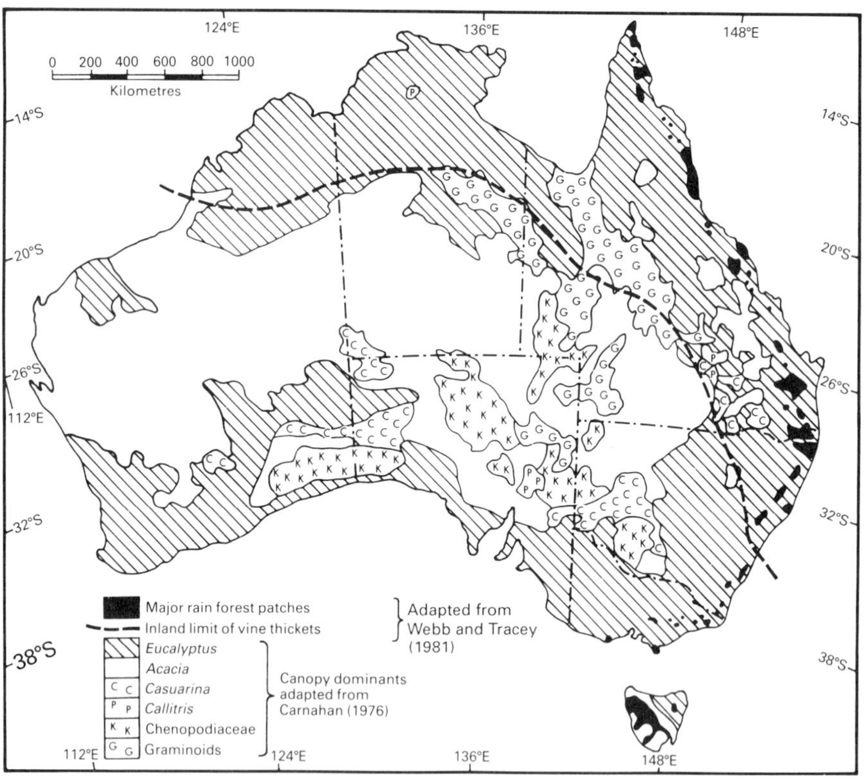

Fig. 12.1 The general distribution of vegetation types in Australia based on canopy dominants.

time from critically located palynological sites. Australia also has a good archaeological record and there is some consensus as to the time of arrival of people and their possible impact on the environment.

This chapter will initially provide a picture of vegetation changes on the Atherton Tableland of north-eastern Queensland from a number of pollen diagrams and then examine the regional application of the results to the development of present-day rainforest–savanna patterns within northern Australia in general.

NORTHERN AUSTRALIAN VEGETATION AND ENVIRONMENTS

The tropical region of Australia is characterized by a highly seasonal, summer rainfall climate, with mean annual rainfall ranging from about 300 mm towards the centre of the continent to greater than 1000 mm along the north and north-eastern coastlines (Nix, 1982). Values in excess of 2000 mm are recorded on the mountains along the eastern seaboard and on adjacent coastal plains. The bulk of rainfall is derived from north-west monsoon winds that pick up their moisture from the warm waters of the shallow continental shelves to the north of the continent. This is supplemented along the east coast by precipitation from the persistent south-easterly trade winds. Some additional rain from this source, combined with mist and fog on the higher mountains, produces islands of less seasonal moisture conditions. An additional significant, though erratic, source of summer precipitation through the region is derived from cyclones. Mean annual air temperature is consistently above 24 °C except on the mountains, while mean minimum temperatures of the coldest month range from about 8 °C in the interior to about 22 °C along the coast (Nix, 1983).

The landscape is generally old and stable, and this, combined with the high temperatures and strong seasonality, has produced deep lateritic soils of low nutrient status (Jones and Bowler, 1980). Exceptions include the more fertile, heavy alluvial soils of the Artesian Basin and more recent soils of the eastern highlands, particularly those derived from basalt outpourings.

In general terms, major vegetation types reflect the variation in physical environments. Savanna woodlands dominate, but give way to *Acacia* open shrublands with a hummock grassland understorey in the arid centre, tussock grasslands on the heavy clay soils and rainforest in the moister parts of the eastern highland region (Fig. 12.1). This broad picture, however, masks a great deal of local variation, particularly in savanna and rainforest distributions. Rainforest, as defined for Australia, covers a whole range of floristically related, closed canopied communities (Webb and Tracey, 1981). These include not only the floristically diverse and structurally complex communities requiring high rainfall, but also semi-evergreen and deciduous monsoon forests and thickets that form a continuum for coastal to inland regions. These have very restricted distributions that extend through much of the savanna-dominated area to about the 600 mm isohyet (Fig. 12.1). Generally, they are found in locally moist and fire-proof patches, but can occur under apparently identical environments to surrounding savanna particularly on basaltic soils. Similarly, patches of savanna vegetation occur within the east coast major rainforest patches, generally, but not exclusively, where rainforest development is inhibited by poor or excessive drainage and topographically fire-prone situations.

Other distinctive communities are found scattered through the savanna region. These are dominated by *Acacia*, *Callitris* or Casuarinaceae and like *Eucalyptus*, the major canopy component of savanna, are regarded as sclerophylls, but grasses tend to be less common in the understorey.

Although savanna and rainforest are considered to be floristically, as well as structurally, distinct, some rainforest-related taxa do exist in savanna. Some of these may be recent invaders, or more likely survivors in areas taken over by savanna, while others have had a longer association with savanna vegetation. These include species of *Bauhinia*, *Bombax*, *Brachychiton*, *Flindersia*, *Gyrocarpus* and *Terminalia*, many of which demonstrate relationships with other dry tropical parts of the world (Webb and Tracey, 1981).

Boundaries between rainforest and savanna woodland are typically sharp, especially where they coincide with a topographic or edaphic change. The western slopes of major forested areas on lower nutrient status soils can have ecotones up to 500 m wide due to exposure to frequent fires and relatively open rainforest canopy. However, environmental conditions in the north of Australia seldom lead to the intense fire responsible for the development and maintenance of extensive wet sclerophyll or tall open forests found in the southern part of the continent.

THE HISTORY OF THE RAINFOREST–SAVANNA BOUNDARY ON THE ATHERTON TABLELAND

Pollen analysis of sediments resulting from volcanic activity on the Atherton Tableland in north-eastern Queensland provides a unique opportunity to examine the changing nature and position of a major rainforest–savanna boundary through a substantial period of time. The boundary, on the basaltic Tableland, presently marks the western edge of

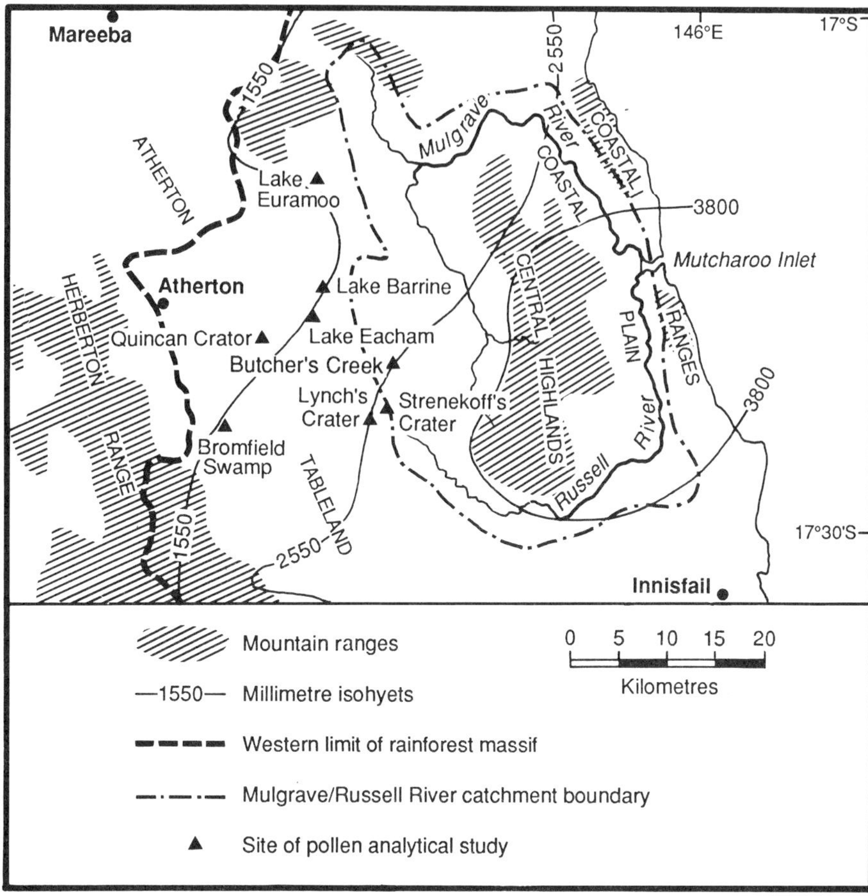

Fig. 12.2 The location of pollen analytical sites on the Atherton Tableland, north-eastern Queensland.

part of the largest rainforest area in Australia and approximates to the 1300 mm isohyet (Fig. 12.2). It crosses a steep climatic gradient so that any effects of changing precipitation levels should be clearly reflected in movement of the boundary. Along the gradient, true tropical rainforest (complex mesophyll vine forest) grades into a drier facies (complex notophyll vine forest) before being replaced by medium woodland dominated by a variety of eucalypt species (Tracey, 1982). As the Tableland is composed almost entirely of one parent material, soil differences should not be an important influence. In characterizing past vegetation distributions, some complications are likely to arise from pollen derived from surrounding vegetation on soils derived from granite and metamorphics. Here the boundary tends to occur under somewhat higher rainfall and is not as abrupt, being characterized by the presence of an ecotonal sclerophyll forest between the rainforest and open savanna woodland. In addition, the highly variable topography contrasts with the flat to undulating Tableland, situated at an altitude of about 700 m.

A number of sites have been examined: all are volcanic crater lakes or swamps whose accumulated sediments have revealed pollen records extending back from present to between 7500 and *c.* 200 000 BP except for an older lake and swamp deposit of Late Pliocene or Early Pleistocene age. It is considered most likely that this older deposit formed as a result of drainage disruption due to basalt extrusion. All sites are presently located within rainforest but not further than 30 km from its boundary (Fig. 12.2).

The Plio-Pleistocene spectra from Butcher's Creek are dominated by rainforest pollen reflecting very different communities from those now occurring within the region (Kershaw and Sluiter, 1982) (Fig. 12.3). Parent species of the dominant taxon, *Nothofagus*, with '*brassii*' type pollen, are now confined to highland New Guinea and New Caledonia, while the conifers *Phyllocladus* and *Dacrydium* have also become extinct on the Australian mainland. Casuarinaceae is now almost entirely represented in northern Australia by the sclerophyll genus *Allocasuarina*, which occurs within many open vegetation communities as well as dominating some vegetation types with restricted distribution. Its high representation in association with rainforest pollen suggests that parent species belonged mainly to another genus of the family, *Gymnostoma*, which is now largely restricted to rainforests of the Indomalesian region. It has been determined from palaeobotanical evidence that the bulk of Tertiary Casuarinaceae representatives in Australia were *Gymnostoma* (Christophel, 1982) and one species still survives in a few isolated patches in rainforest about 50 km north of the Atherton Tableland today.

Evidence for the presence of savanna is equivocal. Grasses were certainly present and Poaceae pollen occurs at levels consistent with a rainforest distribution not significantly greater than that of today, considering the likely high pollen contribution from major existing rainforest taxa. However, the pollen may also have derived from local swamp grasses. If open vegetation was present it would have been different from that of today. There is little evidence of eucalyptus; almost all the Myrtaceae pollen can be assigned to rainforest types. Casuarinaceae species may have formed a woodland canopy in addition to being components of the rainforest vegetation. Regardless of the nature of any open vegetation, it may not have been regarded as tropical savanna due to the nature of the prevailing climate. From estimates of temperature ranges for

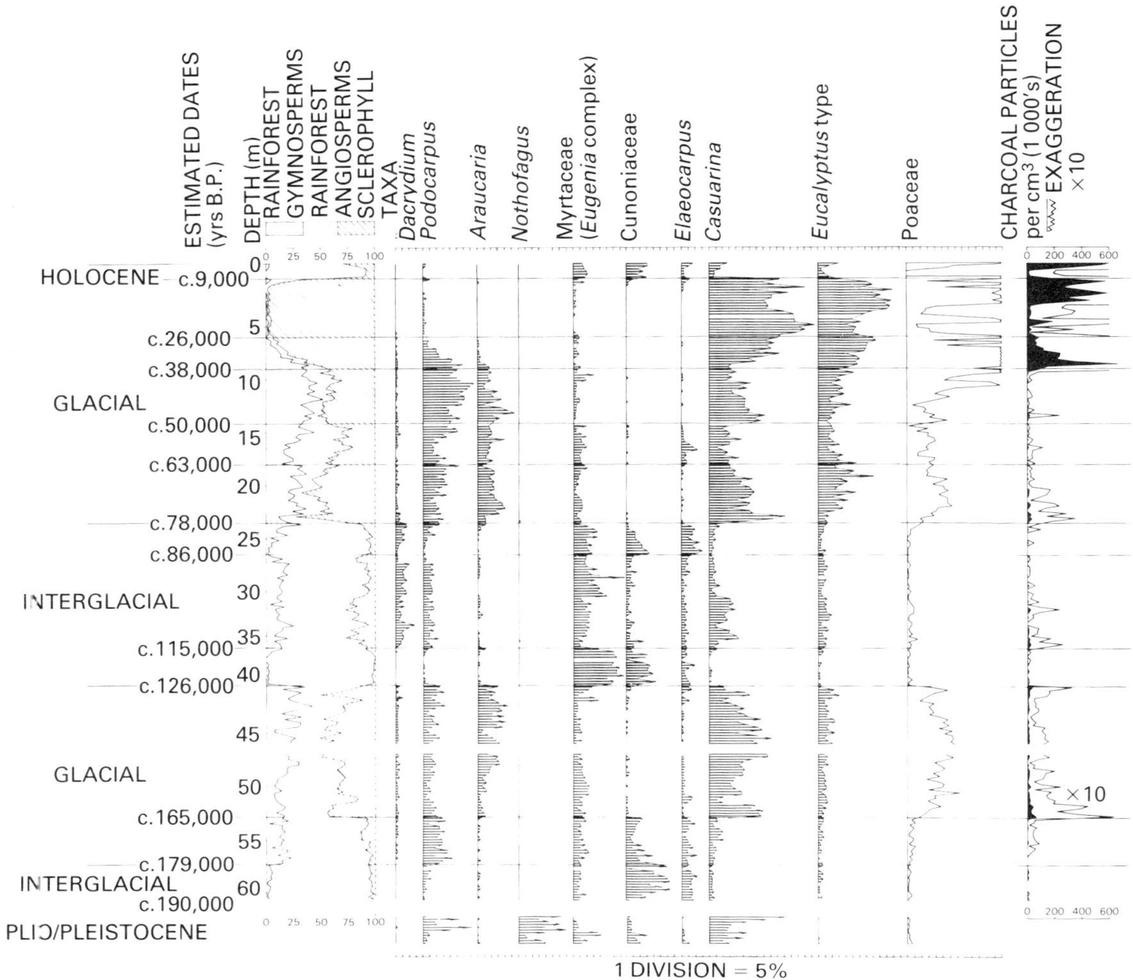

Fig. 12.3 Selected results of pollen analyses from the Pliocene or Early Pleistocene of Butcher's Creek and of pollen and charcoal analyses from the Mid Pleistocene to Holocene of Lynch's Crater on the Atherton Tableland. Values of individual taxa are expressed as percentages of the dry land woody plant pollen sum, for the appropriate spectrum.

component taxa (Ash, 1982; Kershaw and Nix, 1988) it is considered that annual mean temperatures were 3–4 °C lower than the 20–21 °C experienced on the Tableland today, i.e. 'warm temperature' climate.

The record from Lynch's Crater (Fig. 12.3) indicates substantial fluctuations in the distribution and extent of rainforest types and savanna vegetation that can be related broadly to Pleistocene climatic oscillations (Kershaw, 1986). In general terms, complex rainforest was dominant around the sites during the interglacials when rainfall was as high or higher than today. The regional vegetation during glacial times, for the most part, was dominated by drier rainforest types, which included microphyll vine forest with araucarian emergents. Savanna vegetation had restricted presentation during the interglacial periods but was prominent during the glacials. This could indicate a simple boundary across the Tableland in response to prevailing moisture levels. However, the situation is likely to have been more complex as there is little representation of drier rainforest types in the vegetation sequence today.

One explanation is that these types were squeezed out during interglacials and savanna abutted directly against complex rainforest. The high degree of similarity between Holocene and previous interglacial spectra suggests that this is a feasible hypothesis. However, pollen characteristic of drier rainforest types has limited dispersal and significant representation would not be expected away from parent vegetation. In any case, savanna pollen may have been derived from vegetation on surrounding lower nutrient status soils rather than on basalt.

A second possibility is that drier forests have been largely eliminated and the rainforest sequence truncated. Support for this is provided by vegetation changes during the latter part of the last glacial period. Here, savanna gradually replaces rainforest around Lynch's Crater between about 38 000 and 26 000 BP and it might be inferred that rainforest of all kinds was very restricted in distribution for the remainder of the glacial period. Although the more complex rainforest types expanded again with the increase in precipitation in the Holocene, drier forests remain as relict pockets lacking certain taxa. As no such replacement or extinction occurred during the previous glacial period, it appears that this was a unique event demanding a cause other than climatic change. The substantial increase in charcoal particles within the core suggests a major increase in fire which can most economically be explained by the activities of people who were known to have been present over much of the continent by this time (Allen, 1989).

A complicating factor in this interpretation is a successional change around 38 000 years ago from open water to swamp in the basin of deposition. The marked increase and subsequent extreme variability in Poaceae values illustrates the establishment of grasses on the swamp and masks the regional dry land pollen input. It is likely also that the increase in burning is exaggerated by the representation of charcoal from local fires occurring over the swamp surface. However, this is not considered to detract from the interpretation of dry land vegetation changes as recent evidence from nearby Strenekoff's Crater, not affected by significant successional changes at this time, shows a similar pattern of vegetation and burning (Kershaw *et al.*, 1991). Although the condensed nature of the Strenekoff's record and problems associated with dating the whole of it prohibit detailed correlation with the Lynch's Crater sequence, it generally confirms the regional significance and explanation of the replacement of drier rainforest by savanna vegetation. Further support is provided from charcoal preserved in a soil profile on the Windsor Tableland, about 100 km north of the Atherton Tableland. The charcoal, much of which has been identified as *Eucalyptus*, is dated between 12 000 and 27 000 BP and demonstrates that sclerophyll vegetation then covered the area now occupied by rainforest (Hopkins *et al.*, 1990).

A number of sites provide a better indication of spatial patterns of vegetation changes through the Holocene period. Selected features of these sites are shown on Fig. 12.4 with the sites arranged approximately along the present-day precipitation gradient from lowest at Lake Euramoo to highest at Lynch's Crater. The beginning of organic sedimentation or increase in the sedimentation rate at most of these sites marks an increase in precipitation, about or slightly before 10 000 BP. A later date for Quincan Crater can be attributed to the porous nature of the surrounding scoria, inhibiting early development of a permanent lake or swamp. Low levels for pollen attributable to rainforest and high values for both Poaceae and sclerophyll taxa (mainly *Casuarina* and *Eucalyptus*) indicate that savanna vegetation covered the Tableland during the early Holocene. Very high Poaceae values at Lynch's Crater and Quincan Crater suggest the presence of grasses in the local swamp vegetation. Within the sclerophyll phase at both Lake Euramoo and Bromfield Swamp, there is a decline in Poaceae percentages that could indicate increased density of the sclerophyll canopy. A similar explanation is forwarded for the Lake Barrine record by Walker and Chen (1987), although Poaceae values are not provided by them. At this site, it is considered, from time series analysis of charcoal influx values, that fires occurred on average once every ten years before 9300 BP within grassy savanna and once every 220–240 years in the denser sclerophyll forest, after this time.

All available evidence points to the fact that precipitation increased from the beginning of the Holocene to about 6000 years ago, but that rainforest replaced sclerophyll vegetation around the sites at different times.

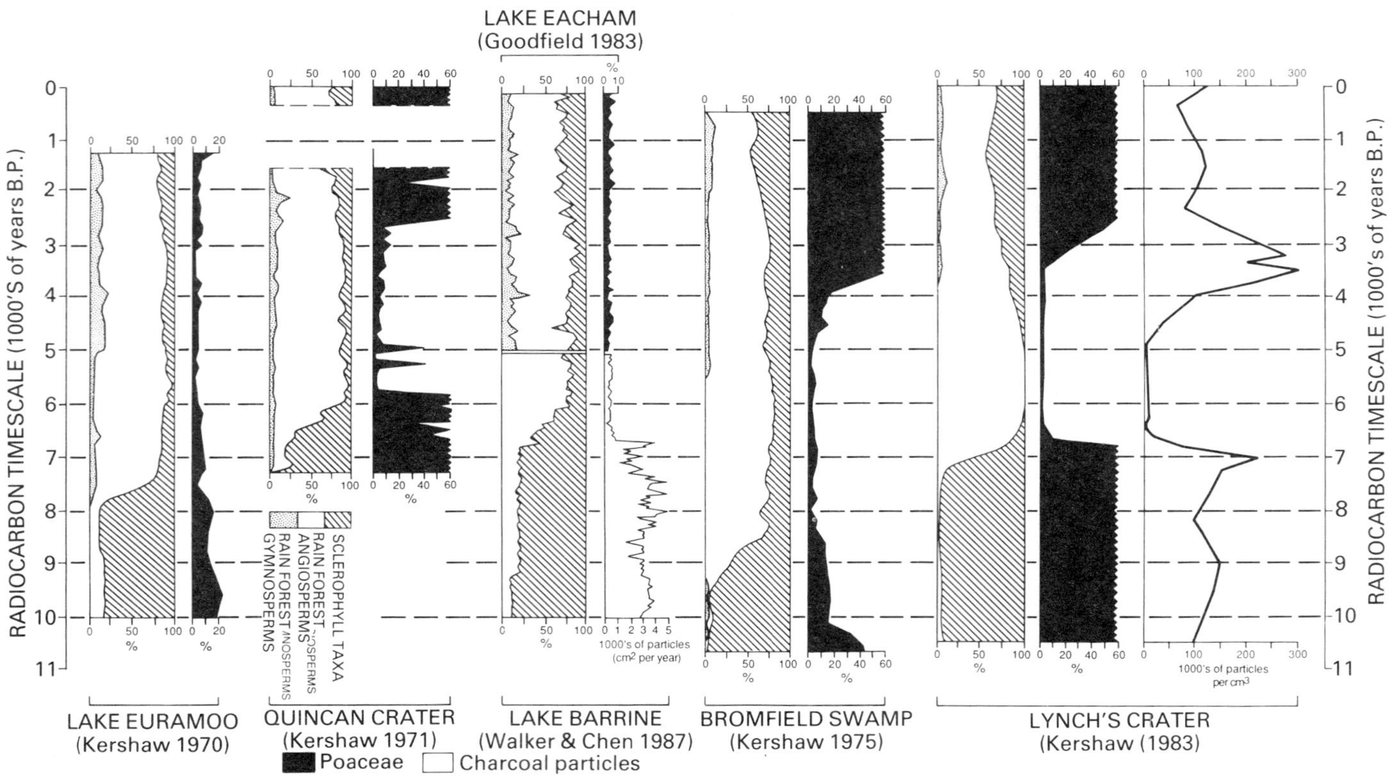

Fig. 12.4 Ratios of dry land woody plant pollen components, Poaceae percentages of dry land woody pollen, and charcoal particle concentrations in Holocene pollen diagrams from the Atherton Tableland.

As the time of arrival of rainforest does not correspond with the inferred climatic gradient across the Tableland, it is considered that factors other than precipitation increase were involved in the transition. One explanation is that it took rainforest some time to migrate from retreats occupied during the glacial period, and that the actual time of arrival around any one site depended on the distance from the retreat area (Ash, 1988). Migration rates would have depended on intrinsic characteristics of the vegetation and on burning of rainforest margins. It is interesting to note that, at both Quincan Crater and Lynch's Crater, swamp grasses virtually disappear around the time that rainforest becomes dominant within the catchments, despite the continued presence of swamp environments. At Lynch's Crater, rainforest covered the site, effectively shading out all herbaceous elements, but at Quincan Crater, ferns and sedges became prominent. This could indicate either the establishment of more consistent water levels or the elimination of fire access to the swamp vegetation.

Rainforest occupied its maximum Holocene extent between at least 6000 and about 3000 years ago under effectively higher precipitation levels than those of today. A slight reduction in percentages of rainforest pollen is recorded about 5000 years ago at a time when both rainfall and temperature are estimated to have reached their highest values (Kershaw and Nix, 1988). This could be explained climatically by either a decrease in effective precipitation or increased seasonality. Alternatively it may represent higher disturbance levels within the rainforest. There are increases in proportions of pollen from early successional forest trees, the most notable being Urticaceae/Moraceae at Quincan Crater and *Blepharocarya involucrigera* at Lake Euramoo. The dynamics of this latter taxon are unusual for a rainforest tree in that it is capable of forming and maintaining monospecific stands. Presently, it is spreading beneath eucalyptic vegetation in some areas. There was also an increase in *Agathis* around Lake Euramoo, and this coniferous emergent requires a relatively open rainforest canopy to allow regeneration. Evidence of probable human activity within rainforest at this time is provided from Lynch's Crater where the disappearance of swamp forest is associated with a massive influx of charcoal particles. As the site was surrounded by non-flammable rainforest, and precipitation would have been in excess of 2500 mm per annum, the conclusion drawn is that people must have burnt the forest for some reason – perhaps to encourage the growth of edible plants or even to cultivate them! It may also be no coincidence that the earliest date for the presence of humans within the rainforests of north Queensland is *c.* 5000 BP (Horsfall, 1987).

Swamp grasses re-establish at three sites in the late Holocene and, although this re-establishment is not entirely synchronous between sites, attainment of high values occurs around the beginning of climatic deterioration, around 3000 years ago. The fact that Poaceae does not increase at Lynch's Crater until some 2000 years after the destruction of swamp forest on site, together with the maintenance of rainforest vegetation around all sites, suggests that the presence of grasses is more likely to be the result of low and possibly fluctuating water levels than fire. However, disturbances resulting from the activities of people cannot be dismissed.

Most of the evidence for vegetation change comes from the palynologically diverse rainforest but, from the few indicators present, there is some indication of variation within the savannas. In the Butcher's Creek sequence, *Eucalyptus* was barely recorded, so that any open dry land vegetation was likely to have been pure grassland or Casuarinaceae forest or woodland with some grassy understorey. By the time of the beginning of the Lynch's Crater record, *Eucalyptus* was present with the grasses and similar patterns of fluctuation in these extra taxa through the sequence suggest that they formed components of the same communities. Fair correspondence with Casuarinaceae variations indicates also that these pollen grains were derived from open rather than closed vegetation. All three taxa have their highest representation in the later part of the last glacial period, consistent with the proximity of savanna vegetation to the site. However, in comparison with Casuarinaceae, *Eucalyptus* values may indicate the development of a more continuous canopy. Considering that the eucalypts are superbly adapted for fire promotion (Gill, 1975), and that their proliferation has been linked with increased fire activity (Singh *et al.*, 1981; Singh and Geissler, 1985) it is likely that fire frequency and/or intensity has increased within savanna vegetation through the Quaternary period. The activities of Aboriginal people can then be viewed as causing an acceleration of an existing trend.

THE BROAD REGIONAL DEVELOPMENT OF RAINFOREST–SAVANNA BOUNDARIES

In the absence of substantial palaeobotanical data for most of the tropical region, a consideration of the development of present vegetation patterns within the tropics as a whole depends heavily on an examination of the composition, distribution and relationships of extant communities as well as on inferences that can be drawn from the Atherton Tableland records. In consideration of the Atherton data, it is clear from Fig. 12.5 that, climatically, this area is not typical of the tropical region of Australia and it cannot be assumed that the derived climatic history has broad regional application. However, the Quaternary records from Atherton can be

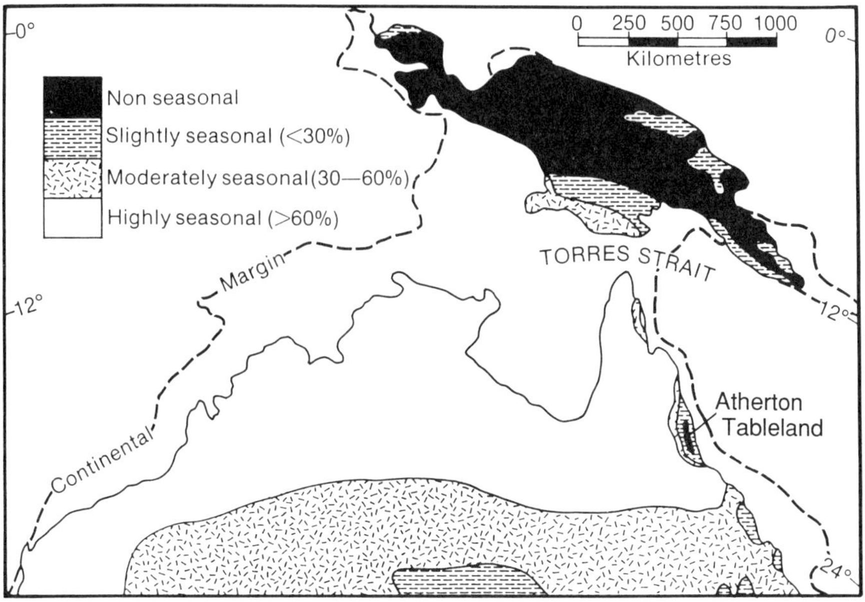

Fig. 12.5 Seasonality of water regime (coefficient of variation of weekly mean moisture index) for northern Australia (from Nix, 1982).

explained broadly by global changes in temperature and sea level and there is good correspondence with simulated climatic regimes constructed for the whole of northern Australia (Nix and Kalma, 1972; Webster and Streten, 1972). Furthermore, palaeoclimatic estimates from an examination of fossil dunes and relict shorelines in the Northern Territory and Western Australia demonstrate that hydrological changes have been of a similar magnitude to those on the Atherton Tableland (Jones and Bowler, 1980). One unresolved problem here is the apparent evidence for very much more extensive lakes at some stage during the last glacial period when generally lower effective precipitation levels prevailed in the Atherton area. However, dating of the inland lake features is fraught with difficulties and it is possible that highest lake levels were associated with periods during the last interglacial complex. An additional limitation to the use of the Atherton data is that this area provides information on the dynamics of only a small proportion of the rainforest-related types occurring in the region.

The more extensive distribution of rainforest in the Butcher's Creek samples is consistent with the general Australian pattern that rainforest was widespread during the Tertiary period (Kemp, 1978; Kershaw, 1988). Unfortunately there is little direct information available to demonstrate the actual extent in the tropical region. In fact, the lack of suitable deposits could indicate that the continuously wet conditions suitable for pollen preservation were lacking. The intense weathering and laterization indicative of seasonal climatic conditions over a long period of geological time adds support to this view. Considering the large number of genera shared with the drier forests of Africa and particularly India it might be reasonable to assume a long history for these forests, dating back to Late Cretaceous continental contact. There is no doubt, though, from their present fragmented distribution, that they were more extensive in the past. Fragmentation might have been initiated by increasingly more seasonal conditions associated with the general drying trend throughout Australia in the latter part of Cainozoic. The effects of fire would have been progressively more important with the spread of grasses and subsequently eucalypts. Some of the less fire-sensitive canopy species of the drier forests would have survived to form the savanna woodland canopy, although these may now being out-competed by the fire-promoting eucalypts.

The influence of the Aboriginal people, through their use of fire, is likely to have been greatest during the early stages of colonization, when many fire-sensitive communities, particularly rainforests, would have already been under stress because of the relatively arid conditions existing during the last glacial period. There is evidence also for a substantial reduction in the size of some forest patches during the mid–late Holocene from the dating of charcoal within abandoned mounds constructed by rainforest-dwelling scrub fowl, now surrounded by eucalypt woodland (Stocker, 1971; Russell-Smith, 1985). This could be the result of increased firing with 'intensification' of occupation. Conversely, there is some indication of conservation of vine thicket patches by controlled backburning around them to prevent fire entry (Jones, 1975). This may explain the survival of those patches not protected in fire proof topographic situations. Although there is little direct evidence of the impact that Aboriginal burning has had on the extent of regional forest vegetation generally, there is little doubt that the present nature of savanna vegetation and the actual locations of the rainforest–savanna boundaries have been determined by Aboriginal burning practices. Aboriginal fire regimes and patterns differ significantly from fires initiated by lightning strikes in the seasonality of burning, the frequency and intensity of fires and areas which are burnt (Haynes, 1985). In many areas, disruption of the Aboriginal burning system is resulting in an expansion of rainforest vegetation and blurring of boundaries (Russell-Smith and Dunlop, 1987).

The simplistic picture of gradual forest destruction and savanna expansion in response to deteriorating climatic conditions and increase in fire, masks the real situation which would have been far more complex.

There has inevitably been substantial variation in the composition and relative importance of the diversity of communities now represented in the region. The status of sclerophyll communities dominated by *Allocasuarina, Callitris* and *Acacia* has not been addressed but it is likely that they were more important within the region before fire became very prevalent and that they, rather than savanna, were the initial successful competitors of the drier vine forests. Some vine thicket species are opportunistic and rather than being relict, have managed to form communities in coastal plain environments that can only date from the mid-Holocene (Russel-Smith and Dunlop, 1988), while significant latitudinal changes have been detected, presumably in response to past temperature variations, from floristic analyses on large rainforest data sets (Webb and Tracey, 1981; Webb *et al.*, 1984).

ACKNOWLEDGEMENTS

I thank Geoff Tracey for his interest and very valuable comments on a draft of this paper and Gary Swinton for drawing the figures.

REFERENCES

Allen, J. (1989) When did humans first colonize Australia? *Search*, **20**, 149–54.

Ash, J. (1982) The *Nothofagus* Blume (Fagaceae) of New Guinea, in *Biogeography and Ecology in New Guinea*, (ed. J.L. Gressitt), Junk, The Hague, pp. 355–80.

Ash, J. (1988) The location and stability of rainforest boundaries in north-eastern Queensland, Australia. *Journal of Biogeography*, **15**, 619–30.

Carnahan, J.A. (1976) Natural vegetation, in *Atlas of Australian Resources*, Second Series, Canberra.

Christophel, D.C. (1980) Occurrence of *Casuarina* megafossils in the Tertiary of southeastern Australia. *Australian Journal of Botany*, **28**, 249–59.

Gill, A.M. (1975) Fire and Australian flora: a review. *Australian Forestry*, **38**, 4–25.

Goodfield, M.L. (1983) *A Holocene Pollen Analytical Study of Lake Eacham, Atherton Tableland, North-east Queensland*, Unpublished B.Sc. (Hons) thesis, Monash University, Melbourne.

Haynes, C.D. (1985) The pattern and ecology of *munwag*: traditional Aboriginal fire regimes in north-central Arnhemland. *Proceedings, Ecological Society of Australia*, **13**, 203–14.

Hopkins, M.S., Graham, A.W., Hewett, R. *et al.* (1990) Evidence of late Pleistocene fires and eucalypt forest from a north Queensland humid tropical rainforest site. *Australian Journal of Ecology*, **15**, 345–7.

Horsfall, N. (1987) *Living in Rainforest: the Prehistoric Occupation of North Queendsland's Humid Tropics*, Unpublished Ph.D. thesis, James Cook University, Townsville.

Jones, R. (1975) The neolithic, palaeolithic and the hunting gardeners: man and land in the antipodes, in *Quaternary Studies*, (eds R.P. Suggate and M.M. Cresswell), Royal Society of New Zealand, Wellington, pp. 21–34.

Jones, R. and Bowler, J.M. (1980) Struggle for the savanna: northern Australia in ecological and prehistoric perspective, in *Northern Australia: Options and Implications* (ed. R. Jones), Research School of Pacific Studies, Australian National University, Canberra, pp. 3–31.

Kemp, E.M. (1978) Tertiary climatic evolution and vegetation history in the southeast Indian Ocean. *Palaeogeography, Palaeoclimatology, Palaeoecology*, **24**, 169–208.

Kershaw, A.P. (1970) A pollen diagram from Lake Euramoo, northeast Queensland, Australia. *New Phytologist*, **69**, 785–805.

Kershaw, A.P. (1971) A pollen diagram from Quincan Crater, northeast Queensland, Australia. *New Phytologist*, **70**, 669–81.

Kershaw, A.P. (1975) Stratigraphy and pollen analysis of Bromfield Swamp, northeast Queensland, Australia. *New Phytologist*, **75**, 173–91.

Kershaw, A.P. (1983) A Holocene pollen diagram from Lynch's Crater, north-eastern Queensland, Australia. *New Phytologist*, **94**, 669–82.

Kershaw, A.P. (1986) The last two glacial–interglacial cycles from northeastern Australia: implications for climatic change and Aboriginal burning. *Nature*, **332**, 47–9.

Kershaw, A.P. (1988) Australasia, in *Vegetation History*, (eds B. Huntley and T. Webb), 111, Kluwer, Dordrecht, pp. 237–306.

Kershaw, A.P. and Nix, H.A. (1988) Quantitative palaeoclimatic estimates from pollen taxa using bioclinic profiles of extant taxa. *Journal of Biogeography*, **15**, 589–602.

Kershaw, A.P. and Sluiter, I.R. (1982) Late Cenozoic pollen spectra from the Atherton Tableland, northeastern Australia. *Australian Journal of Botany*, **30**, 279–95.

Kershaw, A.P., Baird, J.G., D'Costa, D.M. *et al.* (1991) A comparison of long Quaternary pollen records from the Atherton Tableland and Western Plains volcanic provinces, in *The Cainozoic of the Australian Region: a Re-appraisal of the Evidence*, (eds M.A.J. Wiliams, P. De Deckker and A.P. Kershaw), Geological Society of Australia, Sydney, pp. 288–301.

Nix, H.A. (1982) Environmental determinants of biogeography and evolution in Terra Australis, in *Evolution of the Flora and Fauna of Arid Australia*, (eds W.R. Barker and P.J.M. Greenslade), Peacock, Adelaide, pp. 47–66.

Nix, H.A. (1983) Climate of tropical savannas, in *Tropical Savannas*, (ed. F. Bourlière), Elsevier, Amsterdam, pp. 37–62.

Nix, H.A. and Kalma, J.D. (1972) Climate as a dominant control in the biogeography of northern Australia and New Guinea, in *Bridge and Barrier: the Natural and Cultural History of Torres Strait*, (ed. D. Walker), Research School of Pacific Studies, Published BG3, Australian National University, Canberra, pp. 61–92.

Russell-Smith, J. (1985) A record of change: studies of Holocene vegetation history in the South Alligator River region, Northern Territory. *Proceedings, Ecological Society of Australia*, **13**, 191–202.

Russell-Smith, J. and Dunlop, C. (1987) The status of monsoon vine forests in the Northern Territory: a perspective, in *The Rainforest Legacy: Australian National Rainforests Study*, Vol. 1, (eds G.L. Werren and A.P Kershaw), Special Australian Heritage Published 7(1), Australian Government Publication Service, Canberra, pp. 227–88.

Singh, G. and Geissler, E.A. (1985) Late Cainozoic history of vegetation, fire, lake levels and climate at Lake George, New South Wales, Australia. *Philosophical Transactions, Royal Society of London*, **311**, 379–447.

Singh, G., Kershaw, A.P. and Clark, R. (1981) Quaternary vegetation and fire history in Australia, in *Fire and Australian Biota*, (eds A.M. Gill, R.A. Groves and I.R. Noble), Australian Academy of Science, Canberra, pp. 23–54.

Stocker, G.C. (1971) The age of charcoal from old jungle fowl nests and vegetation change on Melville Island. *Search*, **2**, 28–30.

Tracey, J.G. (1982) *The Vegetation of the Humid Tropical Region of North Queensland*, CSIRO, Melbourne.

Walker, D. and Chen, Y. (1987) Palynological light on tropical rainforest dynamics. *Quaternary Science Reviews*, **6**, 77–92.

Webb, L.J. and Tracey, J.G. (1981) Australian rainforests, patterns and change, in *Ecological Biogeography in Australia*. (ed. A. Keast), Junk, The Hague, pp. 607–94.

Webb, L.J., Tracey, J.G. and Williams, W.T. (1984) A floristic framework of Australian rainforest. *Australian Journal of Ecology*, **9**, 169–98.

Webster, P.J. and Streten, N.A. (1972) Aspects of late Quaternary climate in tropical Australasia, in *Bridge and Barrier: the Natural and Cultural History of Torres Strait*, (ed. D. Walker), Research School of Pacific Studies, Published BG3, Australian National University, Canberra, pp. 39–60.

13

Ultraviolet-B insolation and the altitudinal forest limit

J.R. FLENLEY

SUMMARY

1. The chapter examines the hypothesis that vegetation zonation over the upper parts of high tropical mountains is related to insolation by ultraviolet light.
2. Evidence concerning the UV-B waveband is discussed through experiments on plant growth and related to Pleistocene altitudinal zonations.
3. It is concluded that the hypothesis deserves more detailed examination and that UV-B or temperature, or a combination of both, might account for current and past forest limits.

INTRODUCTION

On tropical mountains there is a vegetation type at high altitude called the alpine or tropic-alpine grassland (Troll, 1959; Smith, 1974). Although it is usually dominated by grasses or sedges or both, it is not a savanna in the usual sense. I suggest, however, that the junction between the alpine grassland and forest bears comparison with forest–savanna boundaries. In this chapter, New Guinea (especially Mount Wilhelm) will be used as an example, but the principles are thought to be the same on other tropical mountains.

The alpine zone is simply one in a series of altitudinal zones (Fig. 13.1). The lowland rainforest gives way at about 1000 m to the lower montane rainforest. This continues up to *c.* 2800 m, although the altitude, like those of all these zone boundaries, varies markedly, being especially lower on isolated mountains near the sea (the Massenerhebung effect, Fig. 13.2). Above this is the upper montane rainforest or mossy forest.

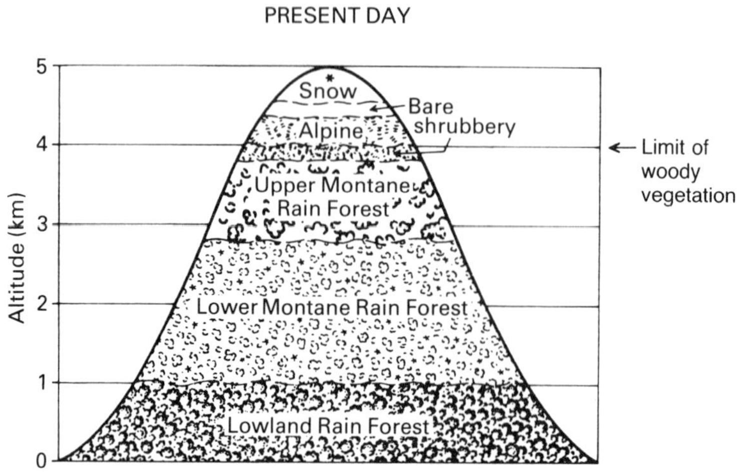

Fig. 13.1 The generalized altitudinal zonation on mountains in New Guinea (after various authors).

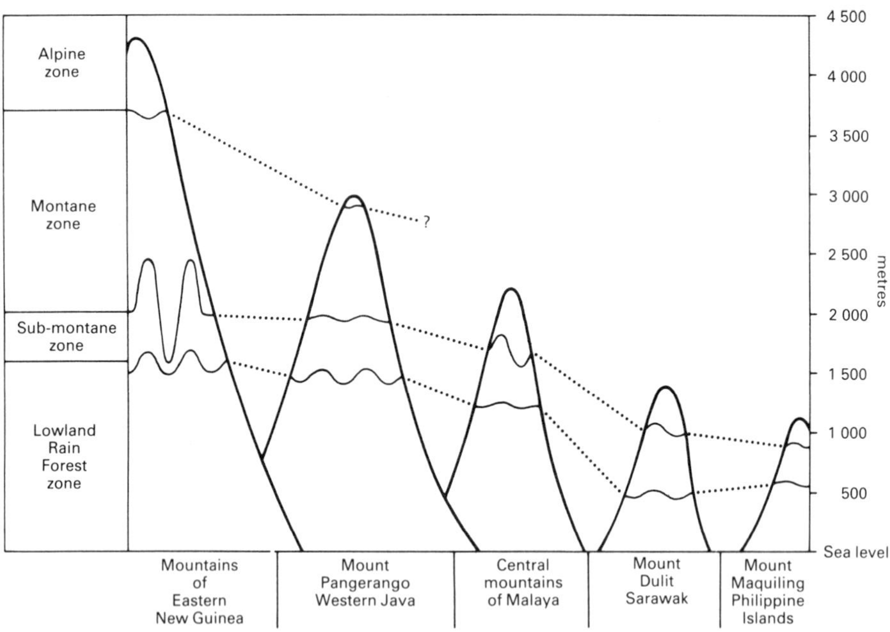

Fig. 13.2 The Massenerhebung Effect in South-east Asia and New Guinea (modified after Eyre, 1968).

This is characterized not only by abundance of epiphytic bryophytes, but also by morphological peculiarities such as stunted growth and small, thick leaves with a hypodermis (Grubb, 1977). Between this and the alpine grassland is a subalpine zone. Sometimes this includes a subalpine forest, sometimes a shrub zone; in places it is replaced by subalpine tussock grassland (with abundant tree ferns) which has been shown sometimes to result from burning of woody vegetation (Corlett, 1984). The zone may extend over about 200 m vertically, before the alpine grassland begins at *c.* 3800 m (the altitudinal forest limit).

What are the controlling factors of this zonation? It is often assumed that temperature must exert overall control. Mean annual values decline at about 0.6 °C per 100 m, with increasing altitude, and the alpine grassland begins at about 6 °C mean annual temperature (Walker and Flenley, 1979). Diurnal variation is extreme, so that the grassland experiences nightly frost and high temperatures in the day. Even the Massenerhebung effect may partly be explicable by temperature, for temperature lapse rates can be steeper on isolated mountains near the sea than on large ranges (Hastenrath, 1968; Flenley and Richards, 1982). How temperature operates other than by frost action is not well understood. Various workers (e.g. Brass, 1964; Grubb and Whitmore, 1966) have suggested that the upper montane rainforest is associated with cloudiness. It has also been suggested that temperature could operate via the soil (Grubb, 1977), lower temperatures leading to greater accumulation of organic matter and to changes in nutrient supply.

Most of the explanations so far offered are based mainly on association. Experimental evidence of causation is usually scanty, or rather inconclusive. Alternative possible causative factors, such as the decline in atmospheric pressure with altitude, or the increase in ultraviolet light with altitude, have rarely been considered.

HYPOTHESIS

This chapter investigates the hypothesis that the zonation of vegetation on the upper parts of tropical high mountains (above *c.* 2000 m) is related to insolation by ultraviolet light. This hypothesis was advanced by the late Francis Merton (personal communication) in 1973, but there was no serious consideration given to it at the time because of lack of evidence. Recently, however, new discoveries have justified a revival of the hypothesis.

Specifically, the hypothesis relates to the UV-B waveband, i.e. 290–315 nm. The longer (>315 nm) wavelength light, UV-A, is not very significant biologically. UV-C, with very short waves (<290 nm), although

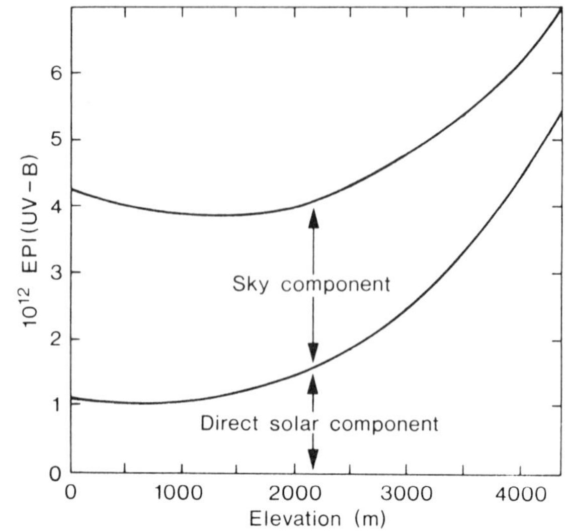

Fig. 13.3 Solar UV-B irradiance as a function of elevation in the USA for air mass of 2.00. Values calculated for cloudless conditions with an atmospheric ozone concentration of 0.301 atm.cm and an atmospheric turbidity coefficient of 0.04 (after Caldwell, 1971).

highly destructive to organisms, does not reach the surface of the Earth in sunlight to any significant extent. In between is UV-B, which is biologically damaging, and is present in sunlight to a significant, and varying, extent.

EVIDENCE

The relevant evidence falls into three categories: data about the relationship between UV-B and altitude; data about the experimental effects of UV-B on plant growth; and data about the altitudinal zonations in the Pleistocene.

UV-B and altitude

It is well known to photographers that it is desirable to use a UV filter at high altitude. This can be explained theoretically by the lesser amount of atmosphere which has to be traversed by the incoming rays. For UV-B, Caldwell (1971) calculated the relationship between insolation and altitude (Fig. 13.3). This was for the USA, and therefore used an air mass of 2.0, i.e. the angle of solar elevation was such that the rays traversed twice the amount of atmosphere compared with the vertical. For the tropics this would be appropriate early and late in the day. Clearly the total UV-B is significantly higher above *c.* 2000 m, and by 4000 m is roughly 50% higher than at 2000 m. This is for cloudless conditions, but the insolation can be increased further by reflection from clouds (Caldwell, 1971).

UV-B and plant growth

Controlled experiments which simulate natural conditions are difficult with UV-B, for artificial light sources do not correctly reproduce the solar spectrum, and species differ widely in their response to individual wavelengths. Nevertheless, it has been possible to experiment on a range of crop plants (Teramura, 1983; Murali and Teramura, 1986). In general the plants became stunted and developed small, thick leaves with a hypodermis: precisely the characteristics of the upper montane and subalpine forests. They also developed extra flavonoid pigments.

Pleistocene zonations

Palynologically-based palaeoecology has given strong evidence in recent years about the altitudinal zonations in the period *c.* 20 000–15 000 years ago, the coldest period of the last glaciation. The evidence from the New Guinea Highlands consists of pollen diagrams from a variety of altitudes between 1900 and >4000 m. These are consistent with a lowering of the altitudinal forest limit to *c.* 2000 m in the Late Pleistocene (Fig. 13.4). This could be explained by a lowering of temperatures.

There is, however, an anomaly which cannot be explained by temperature change alone: the fact that in the Late Pleistocene the upper montane rainforest apparently disappeared almost completely (Walker and Flenley, 1979). Its constituent taxa must have survived somewhere, presumably as rare individuals near the altitudinal forest limit. In its place, the alpine grassland was greatly expanded, and was apparently rich in tree ferns.

If this absence of the upper montane rainforest had been an isolated instance, one could perhaps have ignored it. A similar phenomenon has, however, now been reported from the Colombian Andes (Salomons, 1986). In this case it was the subpáramo, the subalpine shrubbery of the Andes, which was discontinuous in the Pleistocene. Both cases amount, however, to a great reduction in the Late Pleistocene of the upper woody formations which are characterized by stunted growth, small thick leaves and a hypodermis.

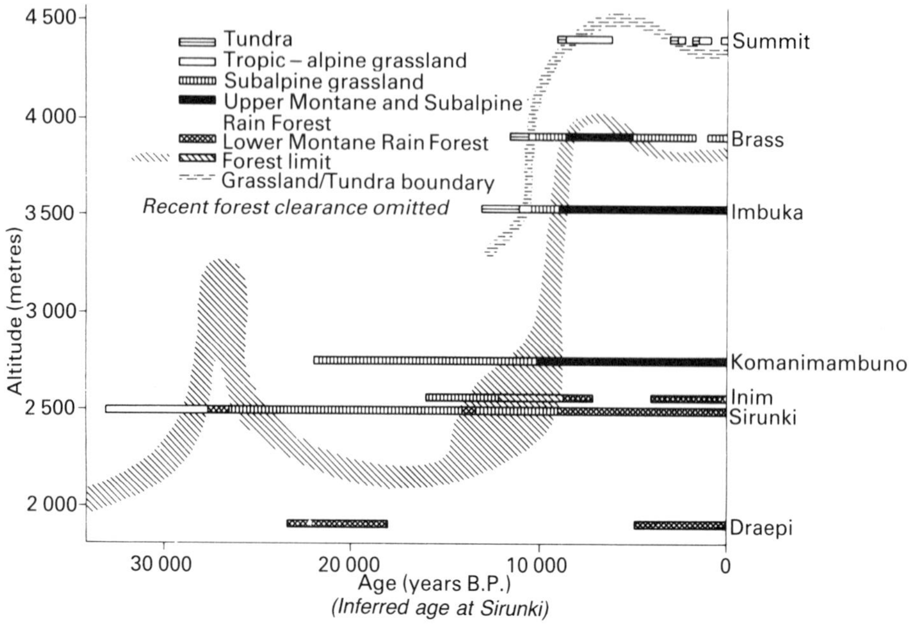

Fig. 13.4 The Late Quaternary changes in vegetation on New Guinea mountains, as evidenced by palynology (after Flenley, 1979).

DISCUSSION

The correspondence between the features induced in crop plants, and the features present in the upper montane and subalpine forests, is rather striking. It must be remembered, however, that the former is phenotypic and the latter (presumably) genotypic. This need not be an insuperable difficulty. Probably genetic fixation of an initially induced feature would happen by natural selection. Is it possible, therefore, that the upper woody vegetation of tropical mountains is genetically adapted for resistance to UV-B? If this were so, its disappearance during the Late Pleistocene could be explained in the following way (Fig. 13.5). At lower altitudes, below *c.* 3000 m, these genetically stunted trees would be at a selective disadvantage compared with the larger trees of the lower montane rainforest. Above *c.* 3000 m they would be at an advantage, because of greater UV-B insolation there. In the Late Pleistocene, lower temperatures brought the forest limit below 3000 m, therefore the stunted species became rare. In the Holocene, warmer temperatures allowed forests to expand uphill, but only those species genetically adapted to the high UV-B insolation could take advantage of this. The present forest limit

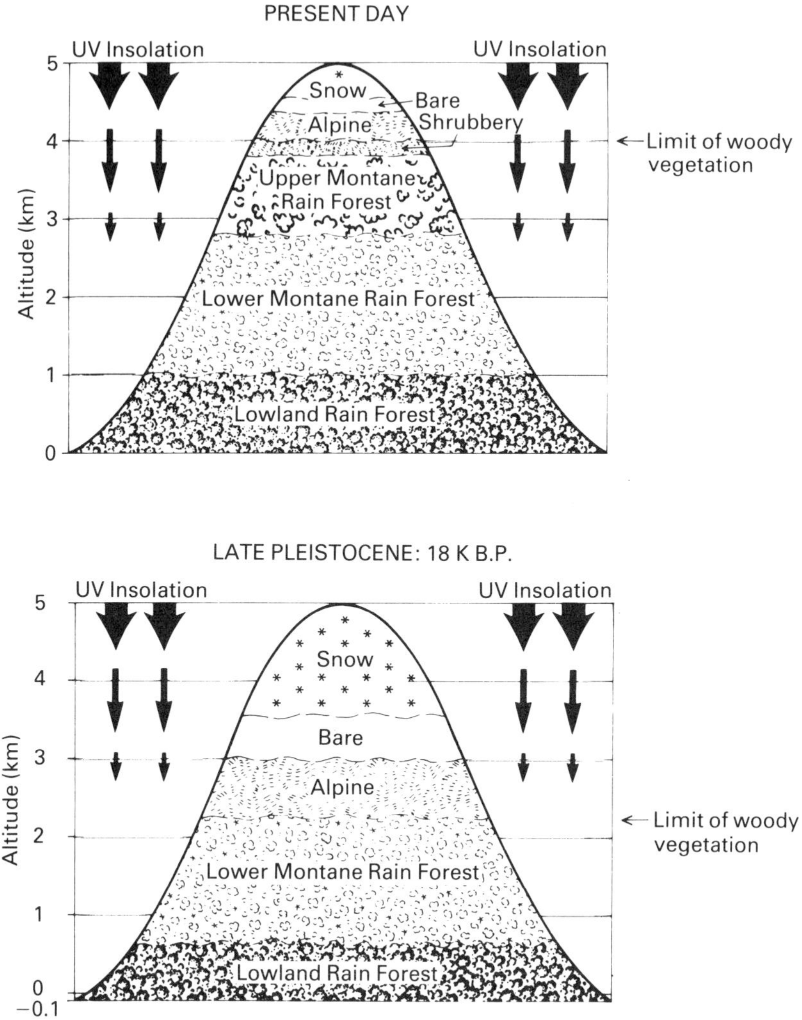

Fig. 13.5 The hypothesis that UV-B insolation controls the upper woody vegetation formations on tropical mountains.

thus could be controlled by temperature or UV-B, or a combination of both.

The meteorological conditions experienced by many tropical mountains could be particularly favourable to UV-B damage. Clouds form in the valleys during the night (Fig. 13.6) and the early morning insolation is increased by reflection from them. Only later in the day do the clouds

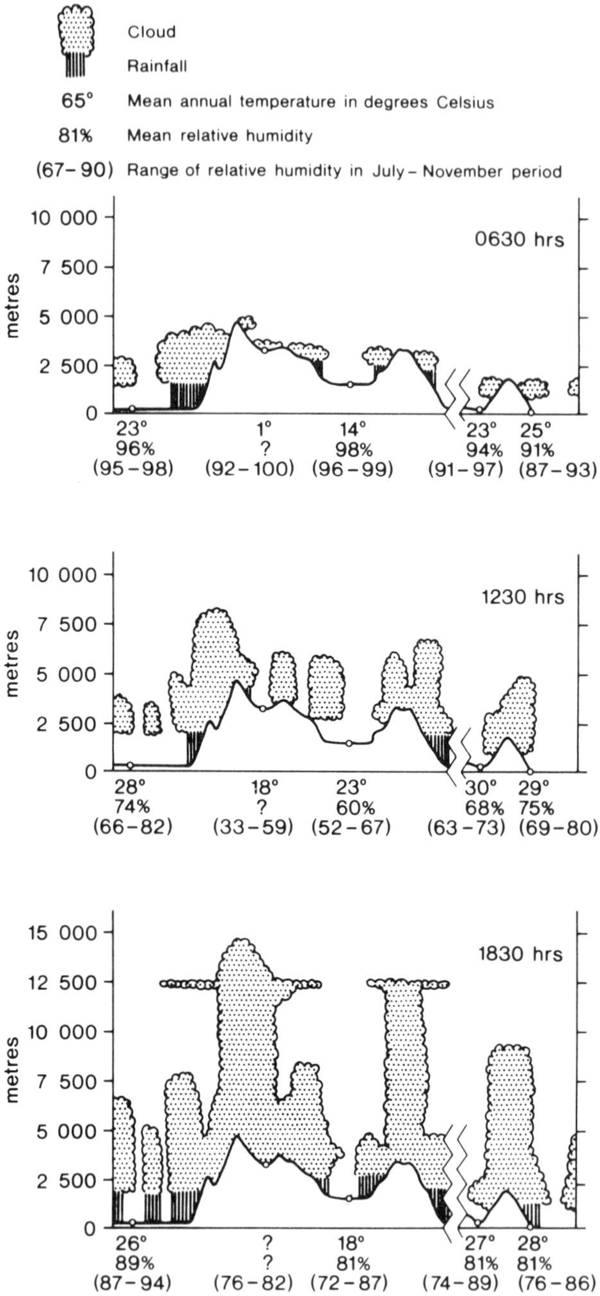

Fig. 13.6 The daily weather regime in the New Guinea Highlands (modified after Brookfield, 1964).

move uphill and envelop the upper forests, reducing them almost to darkness (Hope, 1986). This later reduction of insolation may serve to exacerbate the effects of any UV-B damage caused earlier in the day. This is because of photo-reactivation, a process by which plants repair themselves from UV-damage. Photo-reactivation is strongly dependent on visible light insolation (Caldwell, 1971). The daily tropical regime of a heavy dose of UV-B in the morning, followed by semi-darkness, may therefore be particularly harmful to plants that are not adapted to it.

CONCLUSION

It is concluded that the hypothesis is worthy of further consideration. This should include measurements of UV-B insolation over extended periods on tropical mountains at various altitudes, along with morphological assessment of the vegetation. It could also include experiments to exclude UV-B from vegetation at high altitude. This is conveniently done by glass enclosures. Measurement of the concentration of flavonoid pigments would also be relevant.

REFERENCES

Brass, L.J. (1964) Results of the Archbold Expeditions No. 86. Summary of the Sixth Archbold Expedition to New Guinea (1959). *Bulletin of the American Museum of Natural History*, **127**, 145–216.

Brookfield, H.C. (1964) The ecology of highland settlement: some suggestions. *American Anthropologist*, **66**, 20–38.

Caldwell, M.M. (1971) Solar U.V. irradiation and the growth and development of higher plants, in *Photophysiology*, Vol. VI. (ed. A.C. Giese), Academic Press, New York, pp. 131–268.

Corlett, R.T. (1984) Human impact on the subalpine vegetation of Mt. Wilhelm, Papua New Guinea. *Journal of Ecology*, **72**, 841–54.

Eyre, S.R. (1968) *Vegetation and Soils: a World Picture*, 2nd edn, Arnold, London.

Flenley, J.R. (1979) *The Equatorial Rain Forest: a Geological History*, Butterworths, London.

Flenley, J.R. and Richards, K. (eds) (1982). *The Krakatoa Centenary Expedition. Final Report*, University of Hull, Department of Geography, Miscellaneous Series No. 25.

Grubb, P.J. (1971) Interpretation of the Massenerhebung effect on tropical mountains. *Nature*, **229**, 44–5.

Grubb, P.J. (1977) Control of forest growth and distribution on wet tropical mountains. *Annual Review of Ecology and Systematics*, **8**, 83–107.

Grubb, P.J. and Whitmore, T.C. (1966) A comparison of montane and lowland rainforest in Ecuador. 2 The climate and its effects on the distribution and physiognomy of the forests. *Journal of Ecology*, **54**, 303–33.

Hastenrath, S. (1968) Certain aspects of the three-dimensional distribution of climate and vegetation belts in the mountains of Central America and southern Mexico. *Colloquium Geographicum*, **9**, 122–30.

Hope, G.S. (1986) Development of present day biotic distribution in the New Guinea mountains, in *Flora and Fauna of Alpine Australasia: Ages and Origins*, (ed. B.A. Barlow), CSIRO, Melbourne, pp. 129–45.

Murali, N.S. and Teramura, A.H. (1986) Intraspecific differences in *Cucumis sativus* sensitivity to ultraviolet-B radiation. *Physiologia Plantarum*, **68**, 673–7.

Salomons, J.B. (1986) Paleoecology of volcanic soils in the Colombian Central Cordillera (Parque Nacional Natural de los Nevados). *The Quaternary of Colombia*, **13**, 1–212.

Smith, J.M.B. (1974) *Origins and Ecology of the Non-forest Flora of Mt Wilhelm, New Guinea*, Unpublished Ph.D. thesis, Australian National University.

Teramura, H. (1983) Effects of ultraviolet-B radiation on the growth and yield of crop plants. *Physiologia Plantarum*, **58**, 415–27.

Troll, C. (1959) *Die tropischen Gebirge. Ihre dreidimensionale klimatische und pflanzengeographische Zonierung*, Bonn, Dummiers.

Walker, D. and Flenley, J.R. (1979) Late Quaternary vegetational history of the Enga District of upland Papua New Guinea. *Philosophical Transactions of the Royal Society of London, Series B, Biological Sciences*, **286**, 265–344.

14

How old are South Africa's grasslands?

W.N. ELLERY and M.T. MENTIS

SUMMARY

1. The age of the grassland biome in South Africa is examined in the light of studies over the last 30–40 years in a wide variety of fields including palaeobotany, archaeology, ecology and biogeography.
2. The classical view is that many of South Africa's grasslands have arisen as a result of the extensive removal of forest trees by Bantu and European farmers during the last 600 years. The view in this paper is that the grassland biome as it is presently mapped has been in existence for at least 1000 years, but probably for much longer.
3. This makes it necessary to re-examine the determinants of the grassland biome as a whole, and a number of suggestions are made, including climatic and edaphic conditions, and naturally occurring and anthropogenic fires initiated by Stone Age people before the subcontinent was settled by Iron Age people.

INTRODUCTION

This article examines Acocks's (1953) supposed recent anthropogenic origin of his 'false' grasslands, and proposes a substantially greater antiquity of this vegetation type.

Acocks (1953) identified two broad types of grassland, namely the 'pure' and 'false' grasslands (Fig. 14.1). His classification was implicitly based on the reigning ecological paradigm of succession (Weaver and Clements, 1938). According to this theory, the climax was 'the highest expression of the vegetation in a climatic region' (Acocks, 1964), and prevailed except where disturbance had occurred, for example by grazing of domestic livestock and by fire. Accordingly, Acocks (1953) recognized

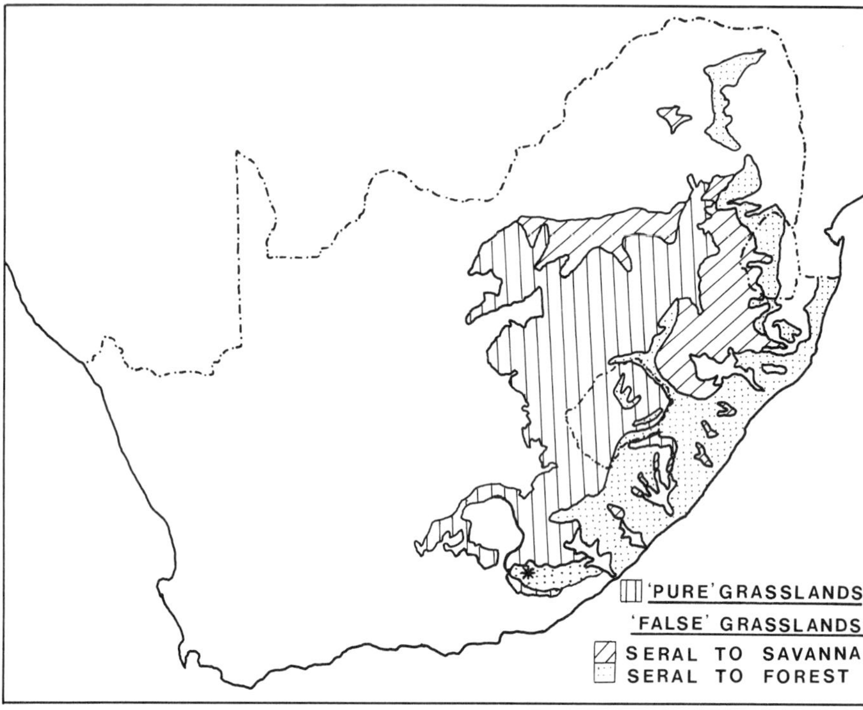

Fig. 14.1 The major vegetational divisions of the grassland biome (after Mentis and Huntley, 1982), showing the approximate location of the Winterberg (*), where palaeobotanical studies have been used to describe the vegetation history.

the grassy vegetation mostly to the west of the continental divide (his veld types 48–60) to be 'pure' in the sense that grassland there was the highest expression of the vegetation that the climate permitted. To the east of the continental divide, Acocks (1953) also observed grasslands, despite in his view there being climate suitable (warmer in winter and wetter than to the west; Rutherford and Westfall, 1986) for the development of 'higher' forms of vegetation, namely forest and scrub-forest. He explained the anomaly by invoking anthropogenic effects, implying that the former climax forests and scrub-forests of as little as 600 years ago had been reverted to disclimax or 'false' grasslands. Within this group he included his veld types 61–68.

At the time of Acocks's (1953) initial proposal, Bantu-speaking farmers were considered to have been in the area for many hundreds of years when they were encountered by Europeans immigrating into the interior from the southern Cape in the early 19th century (Fig. 14.2a). This was

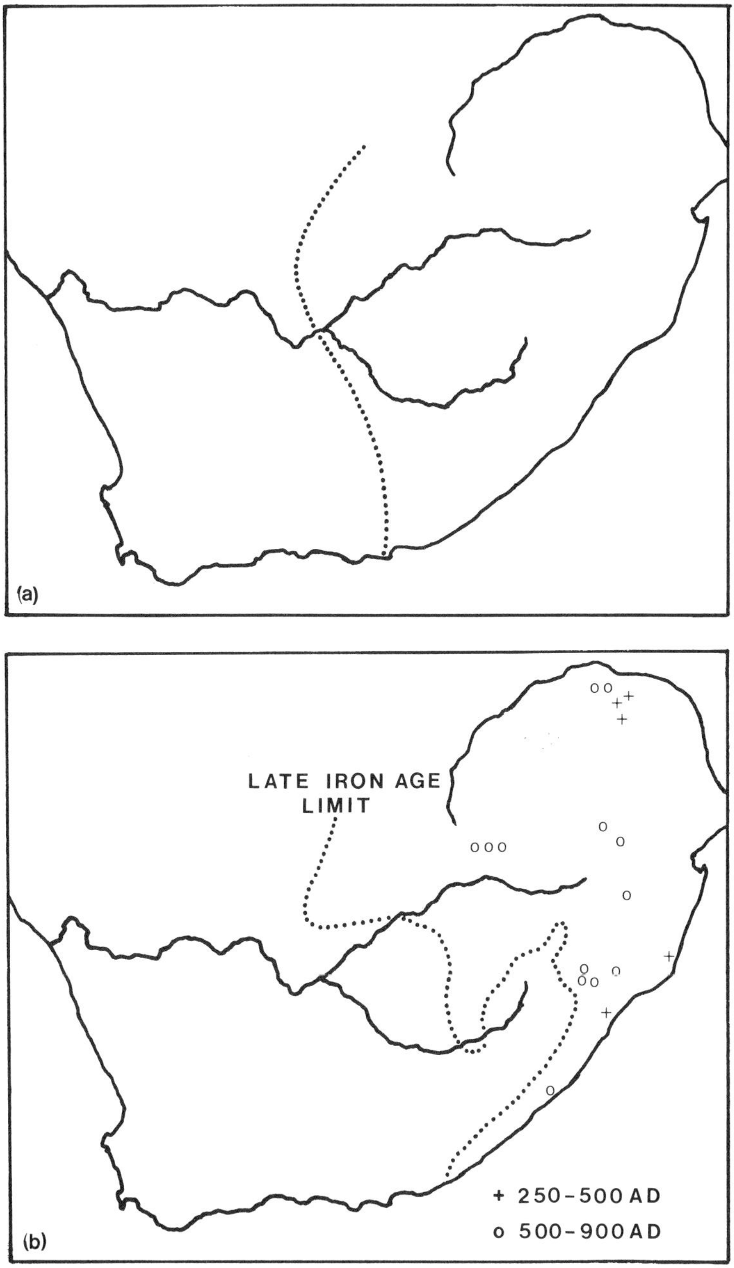

Fig. 14.2 The distribution of Iron Age Bantu-speaking farmers prior to 1870: (a) based on the work of Stow (1905); and (b) at various intervals during the Iron Age; (b) based on Maggs (1984).

based largely on the work of Stow (1905), and was the prevalent view for many decades thereafter (Soga, 1930; Krige, 1936). Since the 1950s, studies in a variety of disciplines including palaeobotany, archaeology, biogeography and ecology have improved our understanding of the age of the grassland biome as a whole. This information suggests that the grassland biome as presently mapped (including the 'false' grasslands) is considerably older than the 600 years suggested by Acocks (1953).

PALAEOBOTANICAL EVIDENCE

Palaeobotanical studies would provide the most useful test of Acock's (1953) suggestion that during the last 600 years there has been widespread destruction of forest east of the continental divide, yet little work has been done in this regard. The vegetation history of the Winterberg mountains in the eastern Cape (Fig. 14.1), in an area considered to be seral to forest (Acocks, 1953), has been described for the last +12 000 years by means of pollen analysis from organic sediments (Meadows and Meadows, 1988). Throughout the period covered by cores taken at two sites, pollen from the Poaceae and Cyperaceae have been dominant, suggesting that vegetation similar to that in existence today has been predominant throughout the period of organic accumulation (Meadows and Meadows, 1988). During the period 12 000–8000 BP the grassland flora contained more xeric elements than at present. Montane forest elements appeared for the first time approximately 8000 BP, at which time the presence of xeric elements declined. Throughout the remaining period the forest elements appear to have been present in small patches, similar to the mosaic that exists in the region at present, although sediments from the last few hundred years show signs of more intensive agriculture and disturbance (Meadows and Meadows, 1988).

On the basis of this evidence, Meadows and Meadows (1988) suggest that montane forest communities of the Winterberg have not expanded much beyond their present-day patches during the last 12 500 years. In addition, archaeological studies indicate that the Winterberg area has not been exploited by agropastoralists until occupation by European settlers during the last 200 years. Prior to that it may have been inhabited by hunter-gatherer communities (Hall, 1986) whose impact on the vegetation was probably not as marked.

A further study on the grasslands of the Nyika plateau in Malawi, also considered seral to forest (Chapman and White, 1970; White, 1983), reveals the grasslands there to have been in existence for at least 12 500 years (Meadows, 1983), which once again is considerably longer than they have been occupied by man. Meadows (1983) therefore suggests that this is also a stable vegetation type, and recognizes that its distribution is poorly understood.

ARCHAEOLOGICAL EVIDENCE

The assumed destruction of woody vegetation by Bantu-speaking farmers during the last 600 years in the areas considered to be seral to forest and savanna (Acocks, 1953) is based on the assumption that these areas have been inhabited for at least that length of time. The people that Acocks considered to have been responsible would have been Iron Age Bantu-speaking people who relied on the use of wood for many of their activities, including the construction of semi-permanent villages and the production of pottery and metal products (Maggs, 1984; Hall, 1987). These people also grew crops and reared domestic livestock.

Historiographic and archaeological studies since the 1950s indicate that the immigration of these people into the subcontinent was a much more recent phenomenon than was considered the case before that time, and an understanding of their patterns of immigration provides strong circumstantial evidence that the grassland biome as it is presently distributed has been in existence for much longer than it has been inhabited by Iron Age people.

Features characteristic of Iron Age people were introduced into the subcontinent in about AD 250. At that time Iron Age settlement sites were predominantly east of the great escarpment. Sites show a clear relationship with environmental conditions. Prior to AD 300 they were all located within relatively close proximity of the coast, on dunes which would appear to have supported forest (Maggs, 1984). Many sites in the Early Iron Age were distributed along the coast as far south as the eastern Cape Province, but they appear to have been confined to well wooded valleys on sites with colluvial soils which would have supported agriculture. Within these areas there is little evidence of any activity in the surrounding grasslands (Feely, 1987). Subsequent immigration into the Transvaal from the north and north-west between AD 400 and 500 extended as far south as the present northern limit of the grassland biome. In the Transvaal, sites were situated beside permanent rivers and on colluvial soils which would once again have supported agriculture. In Natal there was some movement to the interior at much the same time, from settlements further east, into the hot, dry, incised valleys of larger rivers in the areas presently mapped as valley bushveld by Acocks (1953). These sites were probably associated with a mixed bushveld vegetation with an understorey of sweet grasses suitable for grazing (Maggs, 1984) and would therefore be analogous to the present vegetation at these sites.

At about AD 1000–1200 the distribution of Iron Age settlement sites was confined to areas north and east of the grassland biome as indicated on Rutherford and Westfall's (1986) map showing the distribution of South Africa's biomes. The movement of Late Iron Age people after this period into areas currently mapped as the grassland biome was accompanied by a number of adaptations indicative of a tree-less environment. Stone walling replaced the traditional use of wood for construction purposes, bone tools were used instead of iron tools, and dung was used as a major source of fuel (Hall and Maggs, 1979; Maggs, 1976). Furthermore, there are very few iron smelting sites in the area presently mapped as the grassland biome, presumably due to a shortage of wood. The increase in trade with people further north would have enabled the acquisition of iron tools by people in the grasslands. There is evidence of limited wood use for domestic purposes, however, and it would appear that within the grassland biome it would have been wooded sites that were selected (Hall, 1987). This may have resulted in the reduction of woody elements in what would appear to have been an essentially grassland habitat, but it seems unlikely that forest and savanna were converted to grassland by the activities of these people.

Further archaeological evidence is obtained in an examination of faunal skeleton remains obtained from Stone Age deposits in the grasslands in Natal which Acocks (1953) considered seral to forest. Of the ungulate mammals identified from these deposits, many of which are greater than 5000 years old, the majority are considered to be associated with grasslands (Cable *et al.*, 1980; Maggs and Ward, 1980; Klein, 1984).

BIOGEOGRAPHICAL AND ECOLOGICAL EVIDENCE

A number of typical grassland animals such as the redwing francolin (*Francolinus levaillantii*) have their distribution centred in the humid grasslands, and others such as the bald ibis (*Geronticus calvus*) are endemic (Tainton and Mentis, 1984; Siegfried, 1971; Manry, 1982) to the same area. It seems incongruous that these animals have their distribution centred in an area which was considered by Acocks (1953) to have been forest or scrub forest just 600 years ago. The abundance and species richness of wildfowl, small mammals and antelope in the Natal Drakensberg are related to the time elapsed since the last fire (Mentis and Rowe-Rowe, 1979). Faunal abundance and richness is greatest up to three years after fire and in areas never or rarely burned. Under an intermediate fire frequency, faunal abundance and richness are low. This bimodal distribution of faunal abundance and richness is unlikely if the biota of the fire-accessible areas had not been exposed to fire for a great deal longer than the age that Acocks (1953) suggested as the age of these grasslands

(Tainton and Mentis, 1984). Rather it suggests a long evolutionary history.

The flora of the grasslands in the areas suggested by Acocks (1953) to be seral to forest, scrub-forest and savanna is also adapted to frequent defoliation by fire. Many grasses produce rhizomes or stolons at or close to the soil surface, and the tiller initials of these structures are stimulated by fire so that the plant is able to recover rapidly (Tainton and Mentis, 1984). Tufted perennial grasses appear better adapted to defoliation by fire than the sod-forming species (Tainton and Mentis, 1984) and they occur over much of the biome as a whole, suggesting that frequent fire has long been a part of the grassland habit. The vernal aspect geophytes of the eastern grasslands depend on regular defoliation for their survival, in the absence of which they do not reach reproductive maturity and eventually die (Bayer, 1955). The widespread distribution of *Themeda triandra*, a grass which has a very poor dispersal capability, in the grasslands of Natal is not easily explained if the dominant vegetation in this area was forest or scrub-forest just 600 years ago (Mentis and Huntley, 1982).

Some soils on the interfluves of the coastal hinterland of Natal have deep (over 1 m) humic A-horizons (Mentis and Huntley, 1982). In well drained upland sites, such depth of humified organic matter is thought to develop only under grass. Based on likely rates of pedogenesis, the development of this horizon in a soil profile would take much longer than the few centuries proposed as the age of the 'false' grasslands.

CONCLUSIONS

The evidence indicates that the grassland biome as presently mapped has been in existence for millennia, and that the savannas and forest have been subject to much greater disturbance than the grasslands themselves (Feely, 1980). It may be that the boundaries of savannas, forests and grasslands have been modified from what they were prior to settlement by Iron Age people, but to invoke the biome-wide alteration of forest and savanna to grassland, within the areas currently mapped as 'false' grasslands, over the last five or six centuries appears unwarranted.

This raises the question of the determinants of the distribution of the grassland biome, as many grasslands do not appear to have been the result of human activity in recent times as has often been suggested. Are the 'false' grasslands the result of human activity over a period much longer than the subcontinent has been inhabited by Iron Age people? Certainly it appears that fire has been a servant of humans in Africa for at least 100 000 years (Dimbleby, 1977) and may have been used for 1–1.5

million years (Brain and Sillen, 1988). The subcontinent has been inhabited by man since our early evolution, and it is not unlikely that we have modified its vegetation by using fire (Hall, 1984). In areas with a prolonged dry season it appears almost impossible to protect vegetation against man-made fires (Dimbleby, 1977). Grasses have, however, been an important component of the vegetation for considerably longer than 50 million years (Stebbins, 1974), and it is possible that natural fires have been responsible for the development of fire-maintained grasslands prior to the time when humans could have initiated fires at will.

Within the grasslands considered seral to forest (Acocks, 1953) hot, dry *berg* winds are common during the dry winter season. This appears to account for the confinement of forest patches of berg wind 'shadow' areas (van Daalen and Geldenhuys, 1988). These authors suggest that natural fire patterns in relation to wind and topography have had a major influence on the distribution of forest patches in much of Southern Africa.

Alternatively, it has been argued that climate is of overriding importance in determining the distribution of Southern Africa's grasslands (Ellery *et al.*, 1991). The existence of highly seasonal summer rainfall, and therefore of a prolonged dry season in the grasslands considered seral to forest or scrub-forest by Acocks (1953) and White (1983), results in seasonal desiccation of the vegetation. The frequency of burning would therefore appear to be at least partially related to the mean duration or intensity of the dry season. Of the Southern African biomes, it is evident that mean annual lightning ground strike densities are highest in the grassland biome (Manry and Knight, 1986), and may account for the frequent occurrence of natural fires.

An alternative view is that the entire grassland biome may be edaphically determined (Feely, 1987) rather than being maintained by either fire or climate.

It seems that in order to resolve the issue of the determinants of Southern Africa's grasslands, generalized models are needed that provide some biologically meaningful insights into the historical and presently observed existence of grasslands in the regions in which they are presently mapped in Southern Africa. Experimental and modelling approaches are required to elucidate the interactions between soil, climate and natural and man-made fire, if the determinants of Southern Africa's grasslands are to be satisfactorily determined.

ACKNOWLEDGEMENTS

The authors acknowledge financial assistance provided by the Council for Scientific and Industrial Research. Prof. T.N. Huffman and Mr E.R. Robinson provided valuable comments on a draft of the manuscript.

REFERENCES

Acocks, J.P.H. (1953) Veld types of South Africa. *Memoirs of the Botanical Survey of South Africa*, No. 28.

Acocks, J.P.H. (1964) Karoo vegetation in relation to the development of deserts, in *Ecological Studies in Southern Africa*, (ed. D.H.S. Davis), *Monographiae Biologicae*, Vol. 14, pp. 100–112.

Bayer, A.W. (1955) The ecology of grasslands in *The Grasses and Pastures of South Africa*, (ed. D. Meredith), CNA Ltd, Parow, pp. 539–50.

Brain, C.K. and Sillen, A. (1988) Evidence from the Swartkrans cave for the earliest use of fire. *Nature*, **336**, 464–6.

Cable, J.C.H., Scott, K. and Carter, P.L. (1980) Excavations at Good Hope shelter, Underberg district, Natal. *Annals of the Natal Museum*, **24**, 1–34.

Chapman, J.D. and White, F. (1970) *The Evergreen Forests of Malawi*, Commonwealth Forestry Institute, Oxford.

Dimbleby, G.W. (1977) *Ecology and Archaeology*, Studies in Biology No. 77, Edward Arnold, London.

Ellery, W.N., Scholes, R.J. and Mentis, M.T. (1991). An initial approach to predicting the sensitivity of the grassland biome of South Africa based on climatic indices. *South African Journal of Science*, **87**, pp. 499–503.

Feeley, J.M. (1980) Did Iron Age man have a role in the history of Zululand's wilderness landscapes? *South African Journal of Science*, **76**, 150–2.

Feely, J.M. (1987) The early farmers of Transkei, Southern Africa. *Cambridge Monographs in African Archaeology* No. 24.

Hall, M. (1984) Man's historical and traditional use of fire in southern Africa, in *Ecological Effects of Fire in South African Ecosystems*, (eds P. de V. Booysen and N.M. Tainton), Springer-Verlag, Berlin, pp. 39–52.

Hall, M. (1987) *The Changing Past: Farmers, Kings and Traders in Southern Africa*, pp. 200–1860, David Phillip, Cape Town.

Hall, M. and Maggs, T. (1979) Nqabeni, a Late Iron Age site in Zululand. *South African Archaeological Society Goodwin Series*, 3, 159–75.

Hall, S.L. (1986) Pastoral adaptations and forager reactions in the eastern Cape. *South African Archaeological Society Goodwin Series*, **5**, 42–9.

Klein, R.G. (1984) The large mammals of southern Africa: Late Pliocene to recent, in *Southern African Prehistory and Palaeoenvironments*, (ed. R.G. Klein), Balkema, Rotterdam, pp. 107–46.

Krige, E.J. (1936) *The Social System of the Zulus*, Shuter & Shooter, Pietermaritzburg.

Maggs, T. (1976) Iron Age communities of the southern Highveld. *Natal Museum Occasional Paper* No. 2.

Maggs, T. (1984) The Iron Age south of the Zambezi, in *Southern African*

Prehistory and Palaeoenvironments, (ed. R.G. Klein), Balkema, Rotterdam pp. 329–60.

Maggs, T. and Ward, V. (1980) Driel Shelter: rescue at a late Stone Age site on the Tugela River. *Annals of the Natal Museum*, **24**, 35–70.

Manry, D.E. (1982) Habitat use by foraging bald ibises *Geronticus calvus* in western Natal. *South African Journal of Wildlife Research*, **12**, 85–93.

Manry, D.E. and Knight, R.S. (1986) Lightning density and burning frequency in South African vegetation. *Vegetatio*, **66**, 67–76.

Meadows, M.E. (1983) Past and present environments of the Nyika Plateau, Malawi. *Palaeoecology of Africa*, **16**, 353–90.

Meadows, M.E. and Meadows, K.F. (1988) Late Quaternary vegetation history of the Winterberg mountains, eastern Cape, South Africa. *South African Journal of Science*, **84**, 253–9.

Mentis, M.T. and Huntley, B.J. (1982) A description of the Grassland Biome Project. *South African National Scientific Programmes Report* No. 62.

Mentis, M.T. and Rowe-Rowe, D.T. (1979) Fire and faunal abundance and diversity in the Natal Drakensberg. *Proceedings of the Grassland Society of South Africa*, **14**, 75–7.

Rutherford, M.C. and Westfall, R.H. (1986) Biomes of southern Africa – an objective categorization. *Memoirs of the Botanical Survey of South Africa* No. 54.

Siegfried, W.R. (1971) The status of the bald ibis of southern Africa. *Biological Conservation*, **3**, 88–91.

Soga, J.M. (1930) The South-eastern Bantu. *Bantu Studies* No. 4, Witwatersrand University Press, Johannesburg.

Stebbins, G.L. (1974) *Flowering Plants: Evolution above the Species Level*, Edward Arnold, London.

Stow, G.W. (1905) *The Native Races of South Africa*, Swan Sonnenschein, London.

Tainton, N.M. and Mentis, M.T. (1984) Fire in grassland, in *Ecological Effects of Fire in South African Ecosystems*, (eds P. de V. Booysen and N.M. Tainton), Springer-Verlag, Berlin, pp. 115–47.

van Daalen, J.C. and Geldenhuys, C.J. (1988) Southern African forests, in *Long-term Data Series Relating to Southern Africa's Renewable Natural Resources*, (eds I.A.W. Macdonald and R.J.M. Crawford), South African National Scientific Programmes Report No. 157.

Weaver, J.E. and Clements, F.E. (1938) *Plant Ecology*, McGraw-Hill, New York.

White, F. (1983) *The Vegetation of Africa. A Descriptive Memoir to Accompany the Unesco/AETFAT/UNSO Vegetation Map of Africa*, Unesco, Paris.

PART FOUR

Vegetation Structure and Dynamics at the Boundary

15

The phytogeography of savanna species of neotropical Chrysobalanaceae

GHILLEAN T. PRANCE

SUMMARY

1. Neotropical Chrysobalanaceae contain 388 species of which 40 are savanna or savanna-margin species.
2. The phytogeography and the relationships of the savanna species were analysed.
3. The closest relatives of savanna species come from five habitats, the most important of which are riverine forest and other savannas.
4. Savanna species are generally in the evolutionary more advanced part of the genera to which they belong and can usually be traced back to affinities with forest species.
5. The species of savanna margins tend to be far more wide-ranging than those that are true savanna species and they often also occur in gallery forest.
6. Species of the Brazilian *cerrado* do not occur in Amazonian savannas, indicating a degree of phytogeographic difference between these savannas.

INTRODUCTION

During the course of monographic work in the Chrysobalanaceae (Prance, 1972, 1989), Lecythidaceae (Prance and Mori, 1979; Mori and Prance, 1990) and other woody families, the distributions of many neotropical species have been mapped. In this chapter the savanna species of the Chrysobalanaceae have been selected to consider their distribution and taxonomic relationships in order to learn more about their evolution and affinities.

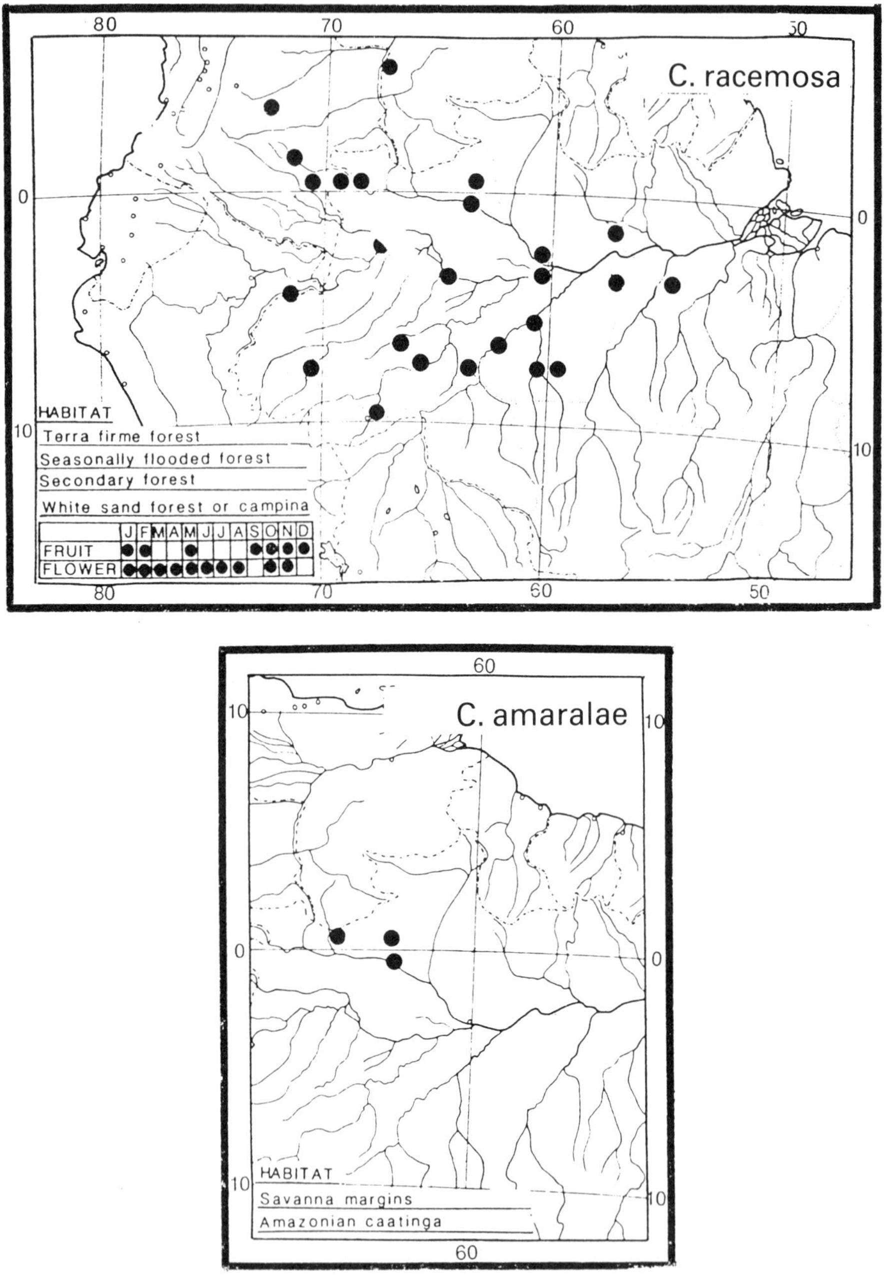

Fig. 15.1 Distribution of *Couepia racemosa* and the closely related *C. amaralae*, both of which are white sand specialists but the former species also occurs in a much wider range of habitats.

Table 15.1 The distribution, habitat and closest related species of the 40 savanna and savanna-margin species of Chrysobalanaceae

Species	*Savanna habitat*	*Closest relative(s)*	*Habitat*	*Fig.*
1. *Couepia amaralae* Prance	Upper Rio Negro sandy caatinga	*C. racemosa* Benth. ex Hook. f.	Various, especially white sand campina	15.1
		2. *C. cognata* Steud.	Guiana/Venezuela savannas	15.2
3. *C. multiflora* Benth.	Guiana savannas	*C. ovalifolia* (Schott) Benth.	East Brazil coastal restinga forest	15.3
4. *C. parillo* DC.	Guiana and Amazon savanna	*C. canomensis* (Mart.) Benth.	Forest, Amazonia	15.4
5. *C. grandiflora* (Mart. & Zucc.) Benth.	Cerrado	*C. paraensis* (Mart. & Zucc.) Benth.	Riverine Amazonian forest	15.5
6. *C. uiti* (Mart. & Zucc.) Benth.	Savanna gallery forest	*C. cataractae* Ducke	Rheophyte	15.6
7. *Exellodendron cordatum* (Hook. f.) Prance	Cerrado	*E. barbatum* (Ducke) Prance	Forest, Amazonia	15.7,
		E. coriaceum (Benth.) Prance	Riverine forest and savanna	15.8
8. *E. gardneri* (Hook. f.) Prance	Cerrado	*E. gracile* (Kuhlm.) Prance	East Brazil forest	15.7
9. *H. adderleyi* Prance	Savanna	10. *H. ulei* Pilg. and	Savanna, campina	15.9
		11. *H. glabrata* Pilg.	Savanna, campina	
12. *H. bullata* Benth.	Savanna-margins	*H. americana* L.	Dry forest around Caribbean	15.10
13. *Hirtella burchellii* Britton	Savanna gallery forest	*H. lancifolia* Ducke *H. mucronata* Prance *H. sprucei* Benth.	Forest	15.11
14. *H. ciliata* Benth.	Cerrado, E. Amazonian savanna			15.12

Table 15.1 *(continued)*

Species	*Savanna habitat*	*Closest relative(s)*	*Habitat*	*Fig.*
15. *H. cordifolia* Prance	Savanna	*H. deflexa* Maguire and	Gallery forest	15.13
		16. *H. paniculata*	Savanna	
17. *H. corymbosa* Cham. & Schlecht.	Cerrado	18. *H. scabra* Benth.	Savanna	15.14
		19. *H. punctillata* Ducke	Savanna	
20. *H. glandulosa* Spreng.	Cerrado and Amazon savanna-margins	21. *H. hoehni* Pilg.	Cerrado gallery	15.15
22. *H. gracilipes*	Cerrado	*H. brachystachya* [Spruce ex] Benth.	Riverine forest	15.16
		H. caduca Fansh. & Maguire	Riverine forest	
23. *H. pimichina*	Savanna-margins Venezuela	*H. subscandens* [Spruce ex] Hook. f.	Amazonian caatinga	15.17
24. *Licania dealbata* Hook. f.	Cerrado	*L. pallida* [Spruce ex] Sagot	Amazonian forest	15.18
25. *L. foldatsii* Prance	Savanna	26. *L. steyermarkii* Maguire	Savanna	15.19
		L. subrotundata Maguire	Montane forest	
27. *L. gardneri* Hook. f.	Cerrado gallery	*L. apetala* (E. Meyer) Fritsch	Amazonian riverine and terra firme forest	15.20
28. *L. cordata* Prance	Savanna	*L. hebantha* [Mart. ex] Hook. f.	Amazonian caatinga	15.21
		L. spicata Hook. f.	East Brazil forest	

29. *L. incana* Aubl.	Savanna	*L. leptostachya* Benth.	Riverine forest	15.22
30. *L. lanceolata* Prance	Savanna	*L. kunthiana* Hook. f. and *L. spicata* Hook. f.	Wide-ranging incl. forest and savanna; East Brazil forest	15.23
31. *L. nitida* Hook. f.	Cerrado	*L. pruinosa* R. Ben.	Riverine forest	15.24
32. *L. humilis* Cham. & Schlecht.	Cerrado	*L. octandra* (Hoffmgg. ex R. & S.) Kuntze	Riverine and terra firme forest	15.25
33. *L. orbicularis* [Spruce ex] Hook. f.	Savanna-margins	*L. coriacea* Benth.	Terra firme forest	15.26
34. *L. savannarum* Prance	Savanna	*L. stewardii* Prance	Caatinga and campina	15.27
36. *L. sclerophylla* (Mart. ex Hook. f.) Fritsch	Cerrado	*L. longistyla* (Hook. f.) Fritsch	Riverine forest	15.28
37. *L. silvatica* [Glaziou ex] Prance	Cerrado	*L. araneosa* Taub.	Gallery forest	15.29
38. *Parinari maguirei* Prance	Savanna			15.30
39. *P. obtusifolia* Hook. f.	Cerrado			15.30
40. *P. campestris* Aubl.	Savanna and river margins	*P. sprucei* Hook. f.	Riverine forest	15.31

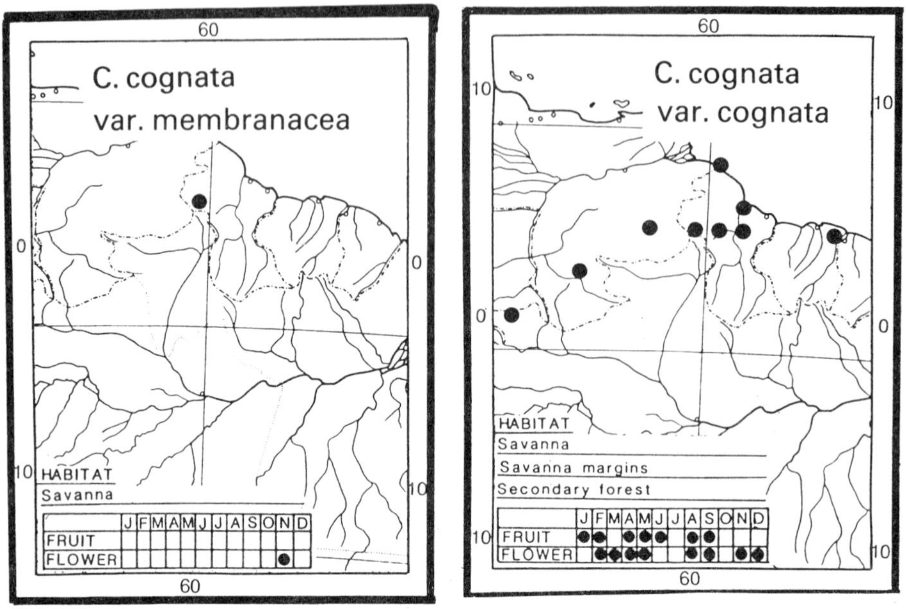

Fig. 15.2 Distribution of *Couepia cognata*, a savanna species that is close to *C. racemosa* (Fig. 15.1).

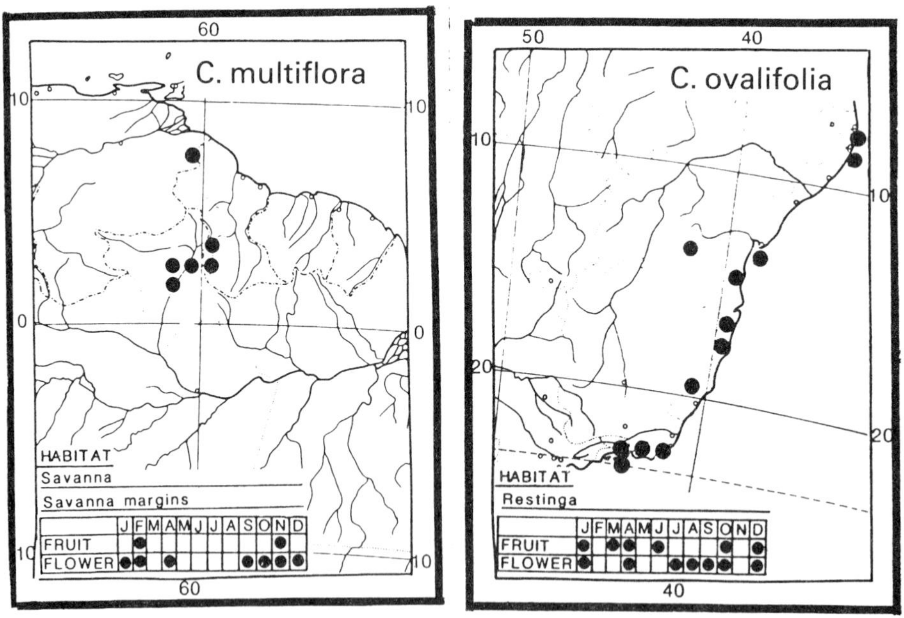

Fig. 15.3 Distribution of *Couepia multiflora* of sandy savannas of the Guianas and its closest, but widely disjunct, relative, *C. ovalifolia*, of similar habitats in the coastal restinga of eastern Brazil.

The neotropical Chrysobalanaceae contain 388 recognized species in eight genera. The majority of the species are forest trees but the family has radiated into various non-forest habitats, especially savanna, white sand formations and lower-montane habitats. Forty species, or 10% of the total, can be considered as savanna or savanna-margin species (Table 15.1). In addition, several more wide-ranging ecological transgressor species occur in savanna and forest habitats. The purpose of this paper is to consider the phytogeography and relationships of the 40 savanna species which are summarized in Table 15.1 and mapped in Fig. 15.1–15.31, and to provide a series of distribution maps of savanna species.

ORIGIN OF SAVANNA SPECIES

An analysis of the closest relatives of the savanna species shows that they are from five different habitats (Table 15.2), two of which, riverine forest and other savanna areas, are much more important than the other three.

However, the savanna and the savanna-margin species and groups of species in the Chrysobalanaceae were ultimately derived from forest species. Savanna species have more advanced adaptations that are rare in the family as a whole and these have probably evolved independently in individual savanna species. Such characters include more coriaceous leaves, thick, corky, fire-resistant bark (for example in *Couepia grandiflora*) or suffrutex habit (for example in *Licania dealbata*). There have been little or no floral or fruit adaptations for the savanna habitat. The general form of the fruit of *Hirtella* or *Parinari* is the same whether in a savanna of forest habitat, and in both habitats there are pollinators and dispersal agents adapted to the basic floral and fruit structure of the different genera. For example, in the cerrado rheas disperse the fruit of *Parinari obtusifolia*, whereas a similar fruit in *P. excelsa* is distributed in the rainforest by primates, and that of the riverine *P. campestris* is dispersed by fish.

Table 15.2 Analysis of habitats of closest relatives of savanna species of Chrysobalanaceae

Habitat	*No. of species*
Riverine forest (igapó or várzea)	*10*
Other savanna habitats	*9*
Rainforest on terra firme	*6*
Atlantic coastal restinga forest	*1*
Semi-deciduous dry forest	*1*

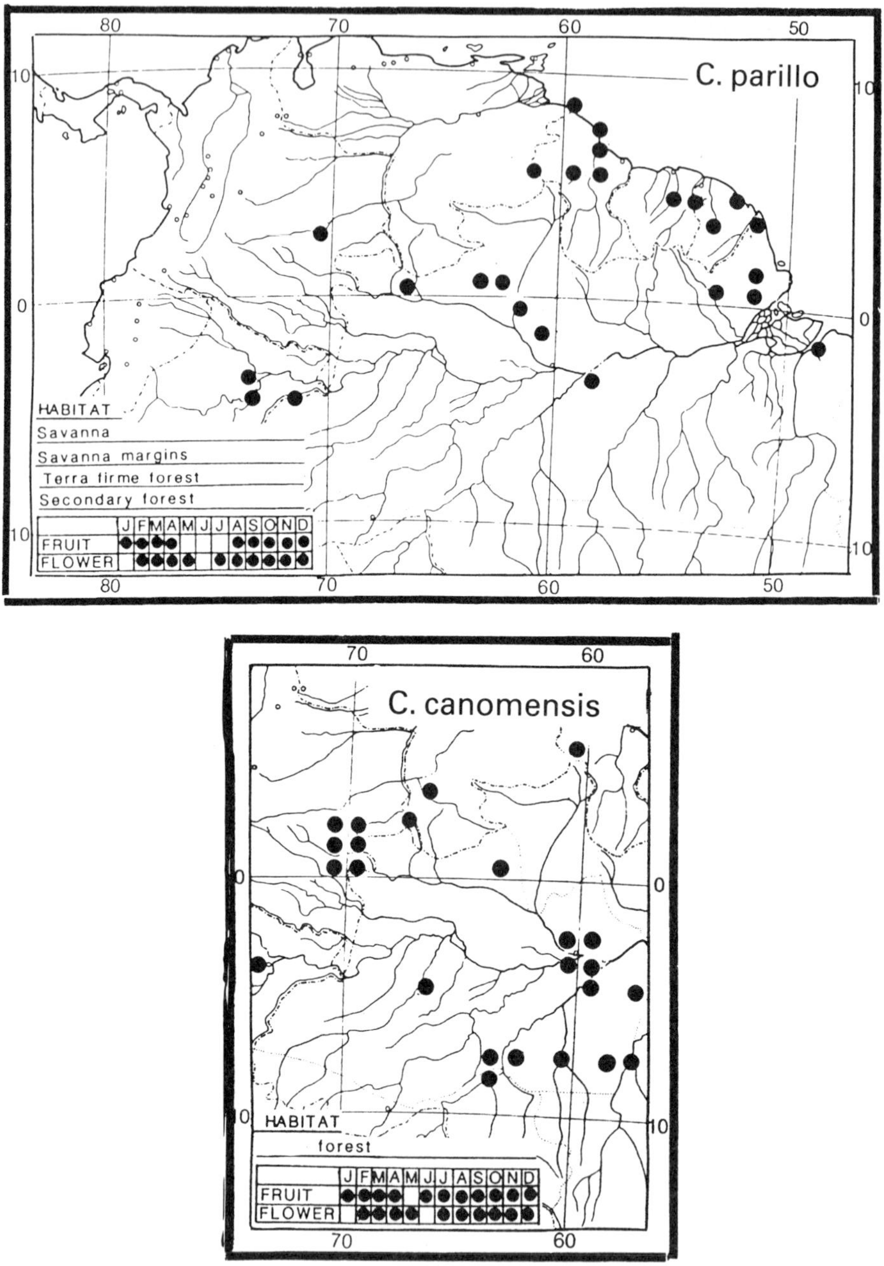

Fig. 15.4 Distribution of *Couepia parillo*, a savanna species, and its forest relative *C. canomensis*.

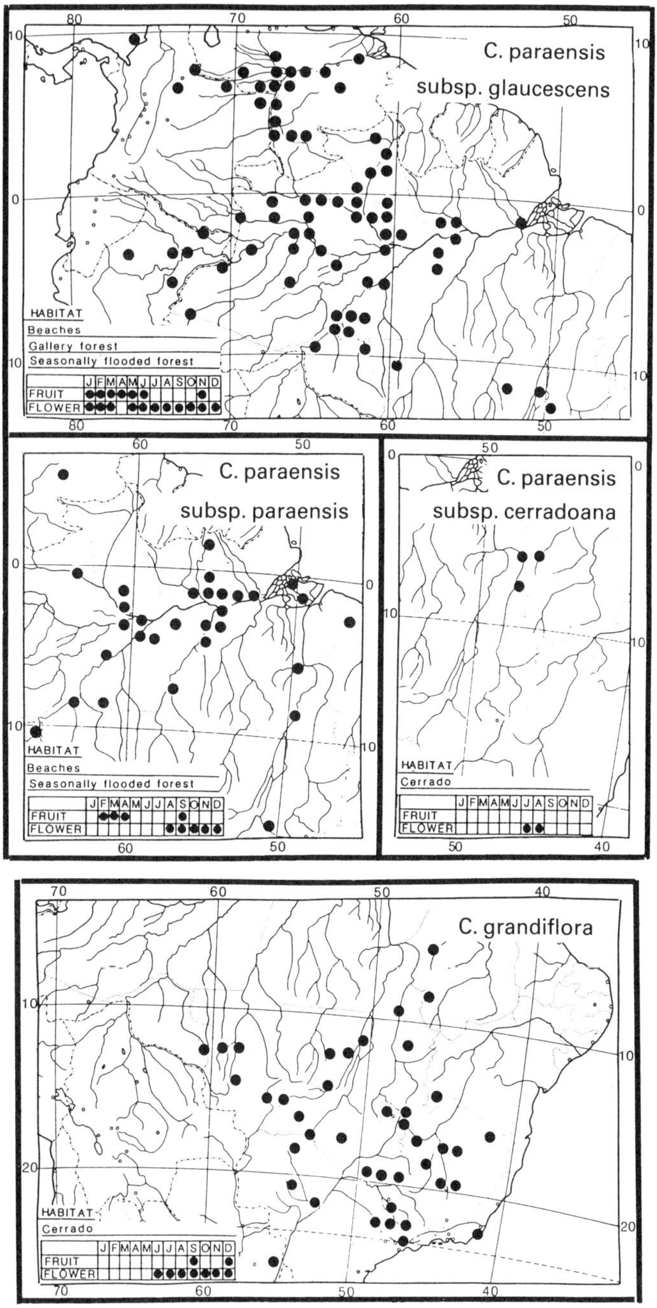

Fig. 15.5 Distribution of the riverine forest species *Couepia paraensis* and the closely related *C. grandiflora*, a cerrado species with a thick corky bark.

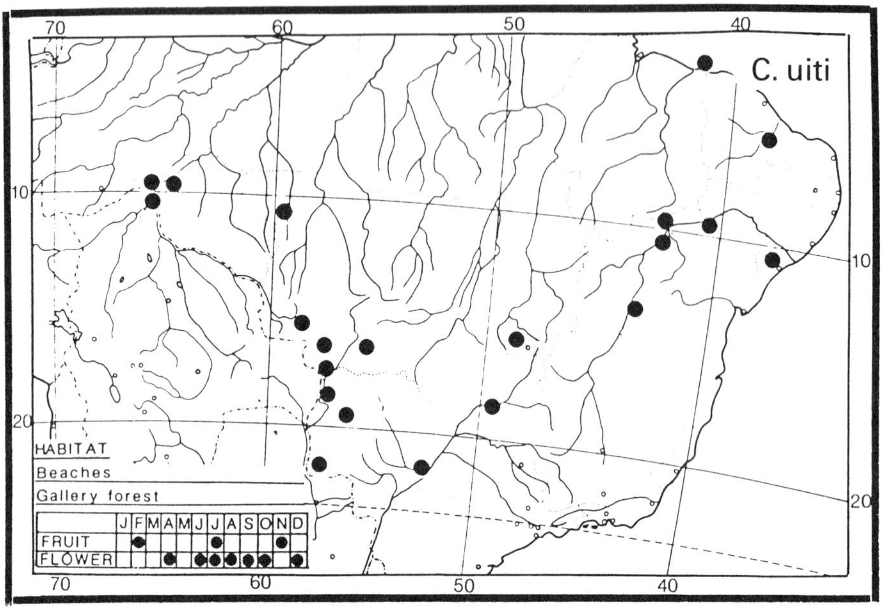

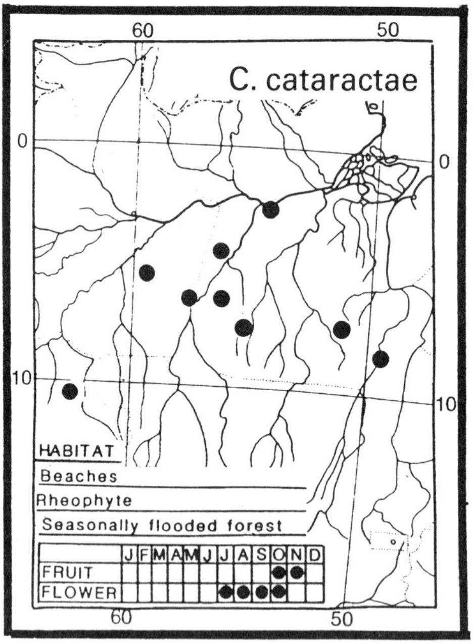

Fig. 15.6 Distribution of *Couepia uiti* and *C. cataractae*, closely related gallery forest species of central Brazil.

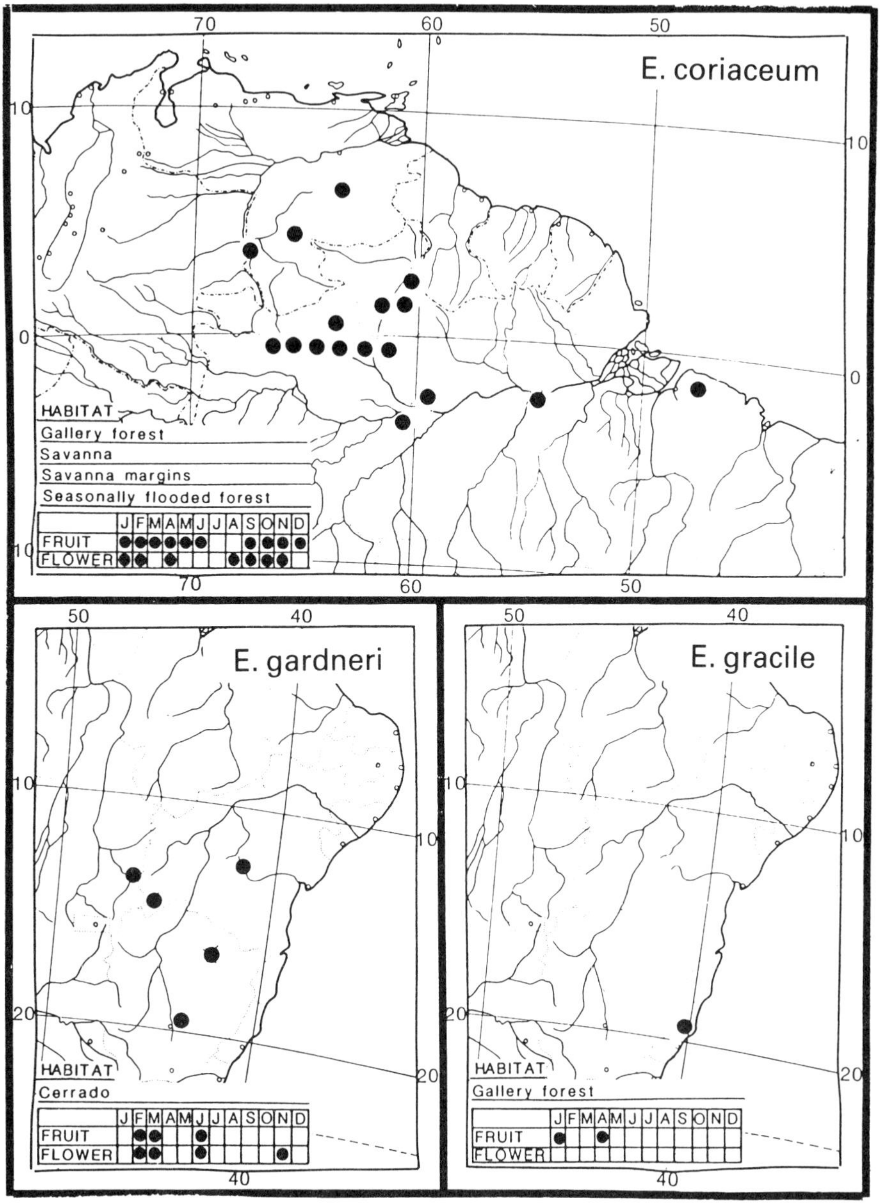

Fig. 15.7 Distribution of *Exellodendron gardneri* from cerrado, *E. gracile* from gallery forest and *E. coriaceum* from various habitats.

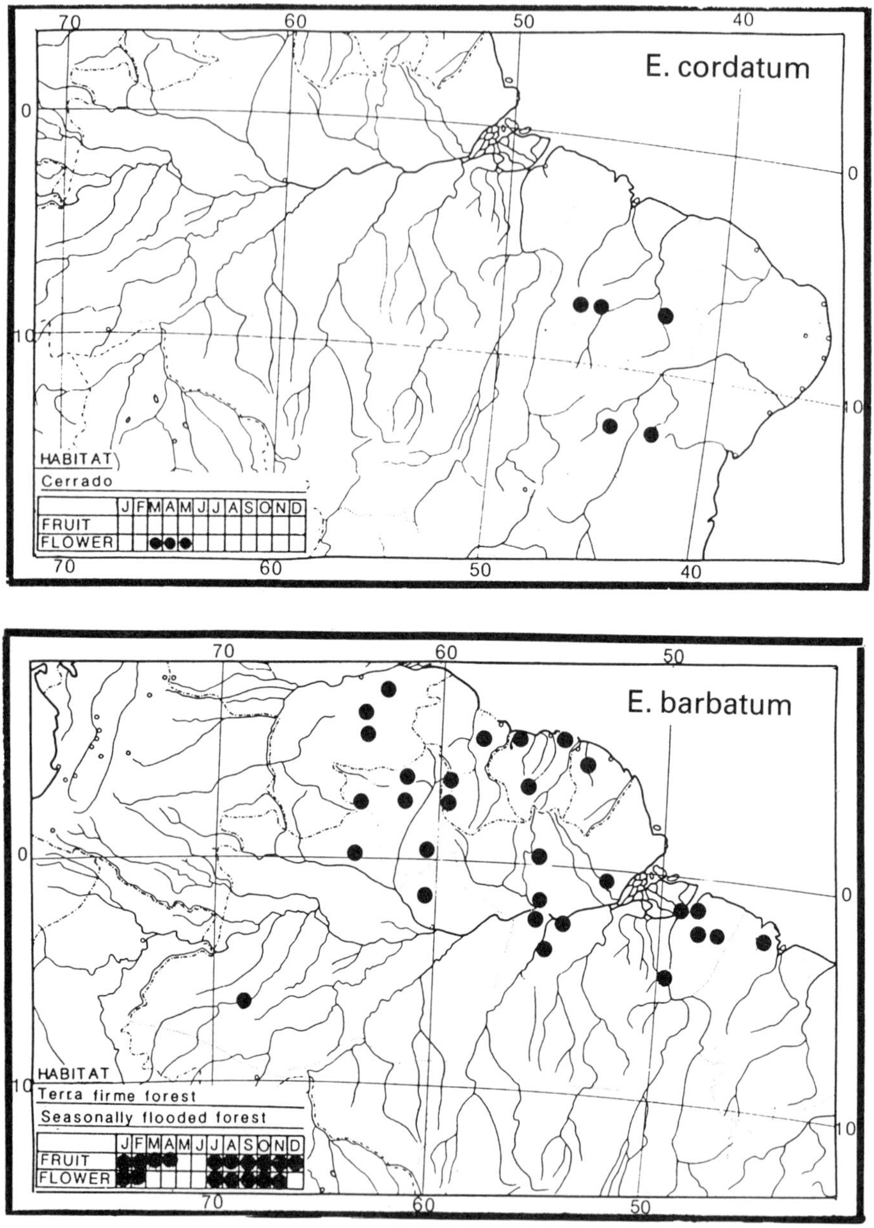

Fig. 15.8 Distribution of *Exellodendron cordatum* of cerrado and the related *E. barbatum*, an Amazonian forest species. The five species of this genus form a closely related species complex with three rainforest and two cerrado species.

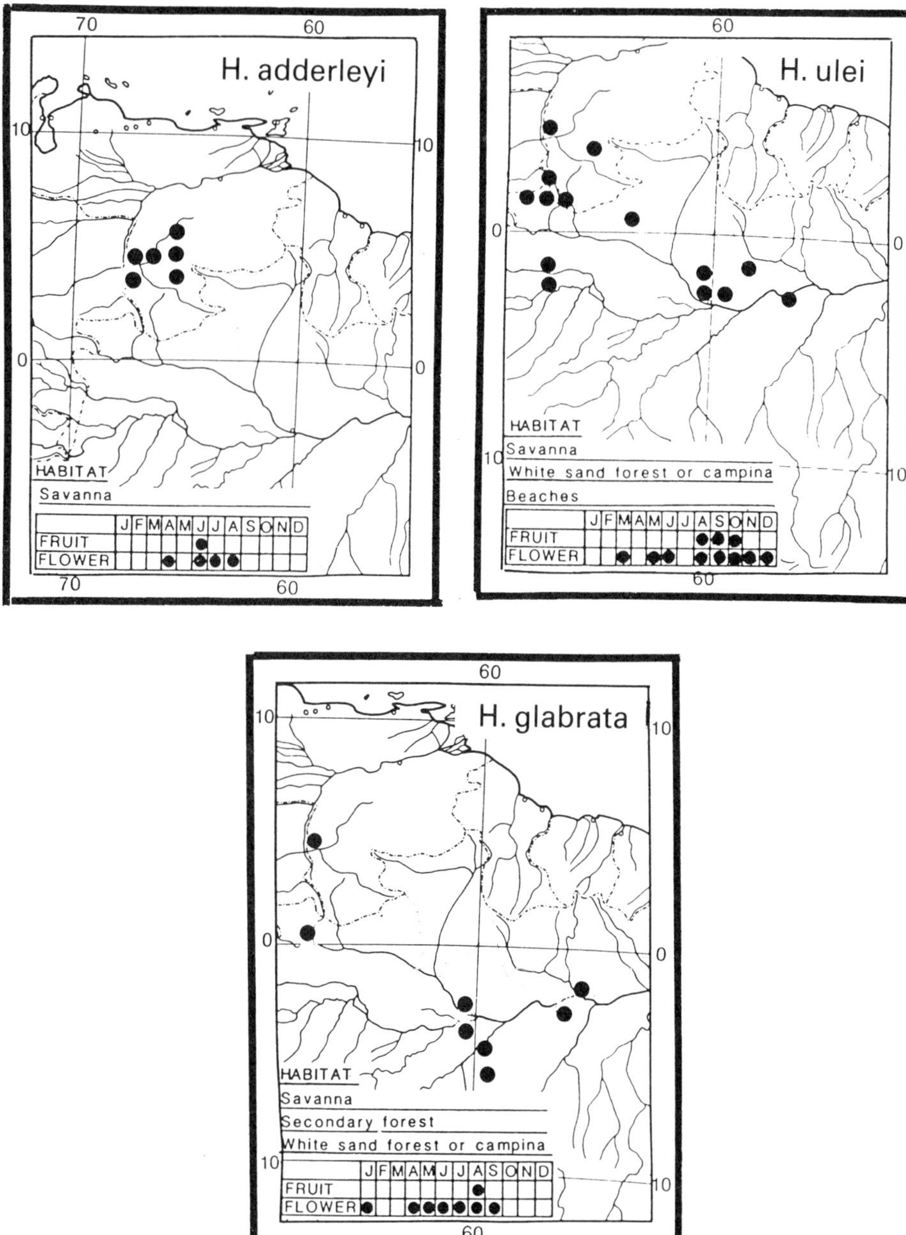

Fig. 15.9 Distribution of a complex of three closely related savanna species of *Hirtella* of the Amazonian region.

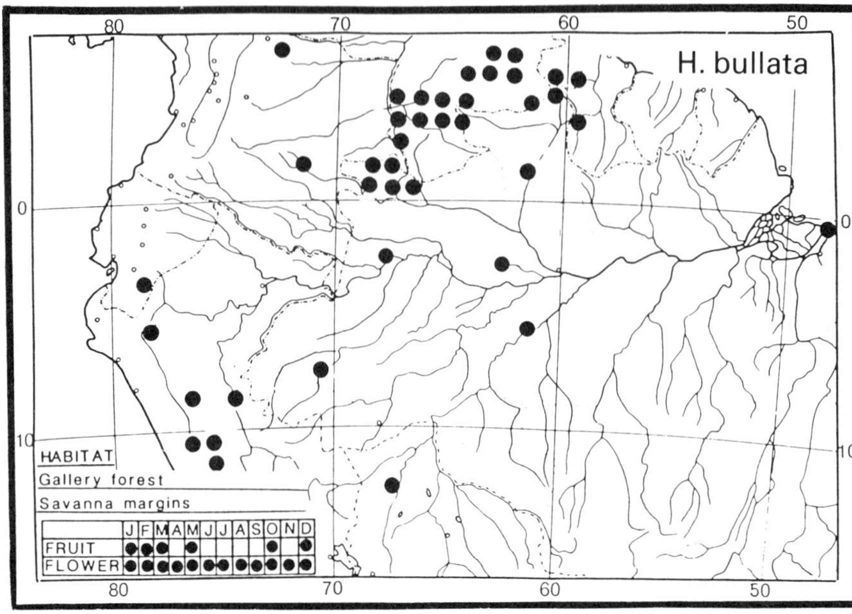

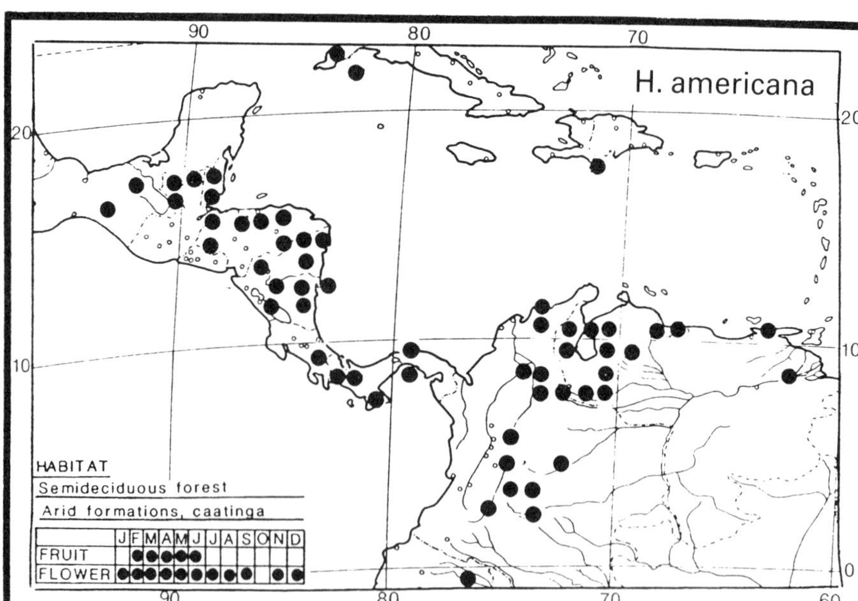

Fig. 15.10 Distribution of *Hirtella bullata*, a savanna-margin species, and *H. americana*, a deciduous forest species, which are closely related.

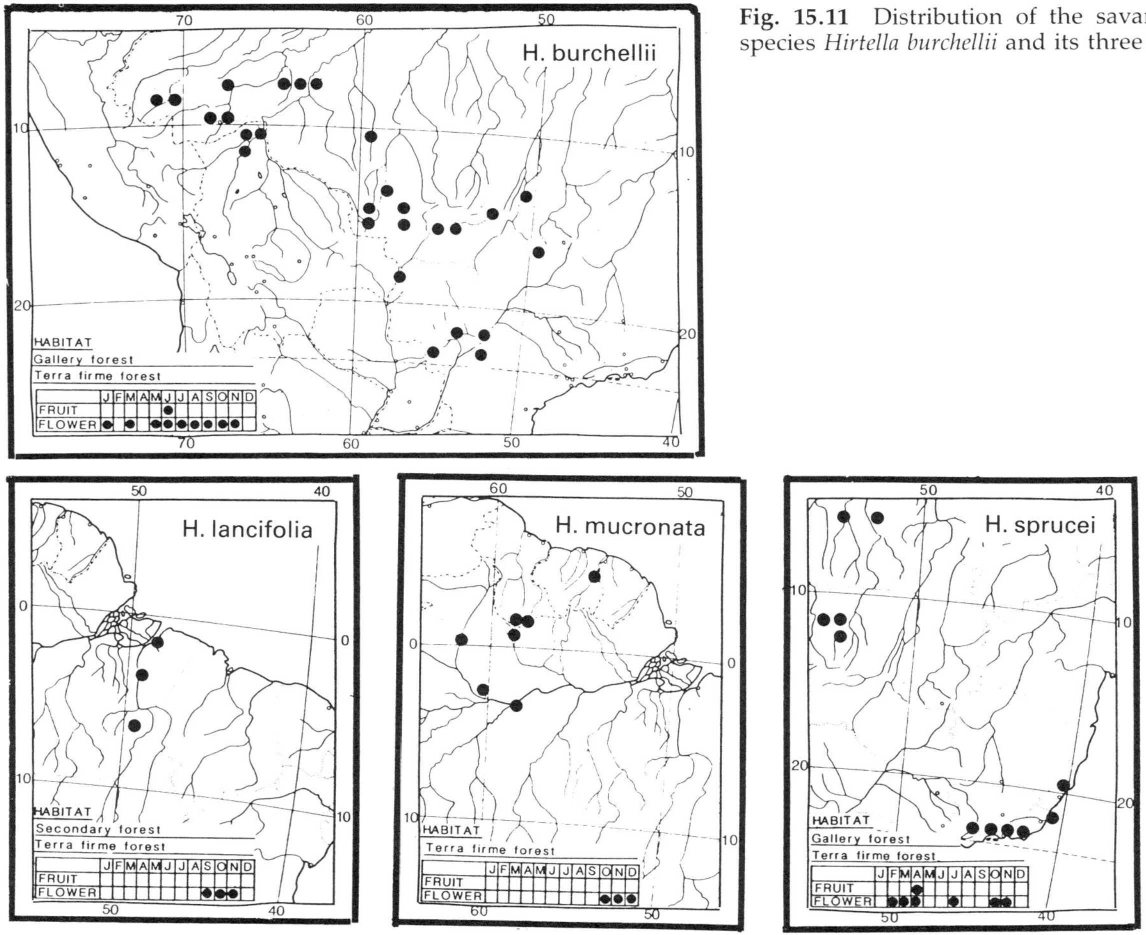

Fig. 15.11 Distribution of the savanna gallery forest species *Hirtella burchellii* and its three closest relatives.

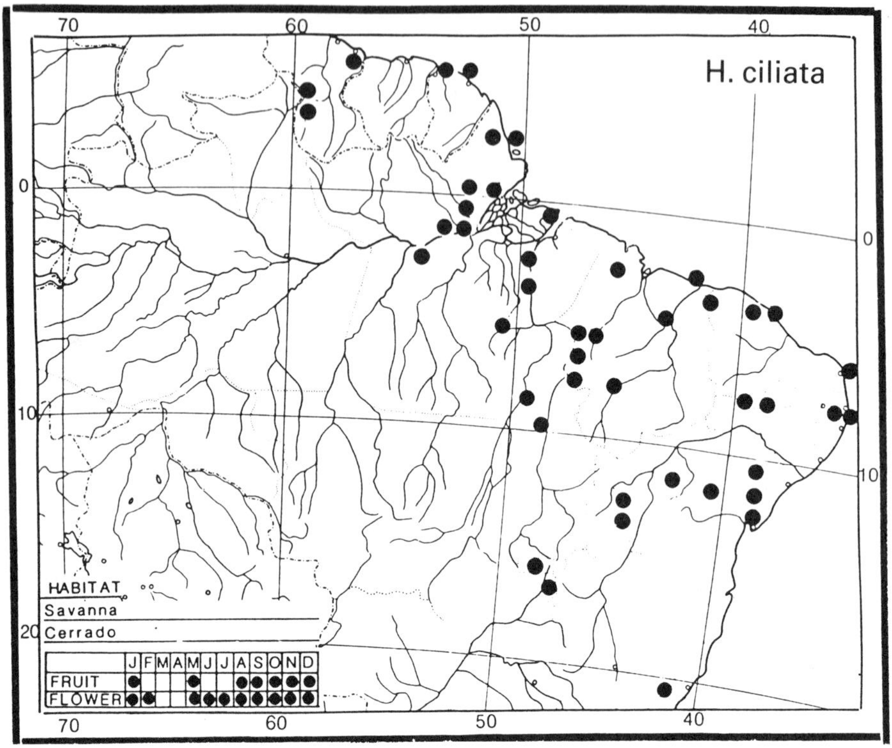

Fig. 15.12 Distribution of *Hirtella ciliata*, a widespread savanna species with a thick corky bark adapted to fire.

SAVANNA-MARGIN AND RIVERINE SPECIES

One of the most characteristic and abundant savanna-margin species is *Hirtella glandulosa* (Fig. 15.15) whose distribution represents almost a complete vegetation map of the cerrado and Amazonian savannas. Such widespread distribution also occurs in the true savanna species *H. ciliata* (Fig. 15.12). *H. glandulosa* has been successful in dispersal to even some of the most remote savannas such as the Humaitá and Puciari savannas of the Rio Purus basin and the Savana de Amélia near Manaus. Its closest relative is another savanna-margin and gallery forest species, *H. hoehnei* of the cerrado region of central Brazil, which grows in wetter places than *H. glandulosa*. The two other close relatives of these species are *H. bullata*, a savanna-margin and gallery forest species, and *H. americana*, a species of the semi-deciduous forest of Central and Northern South America

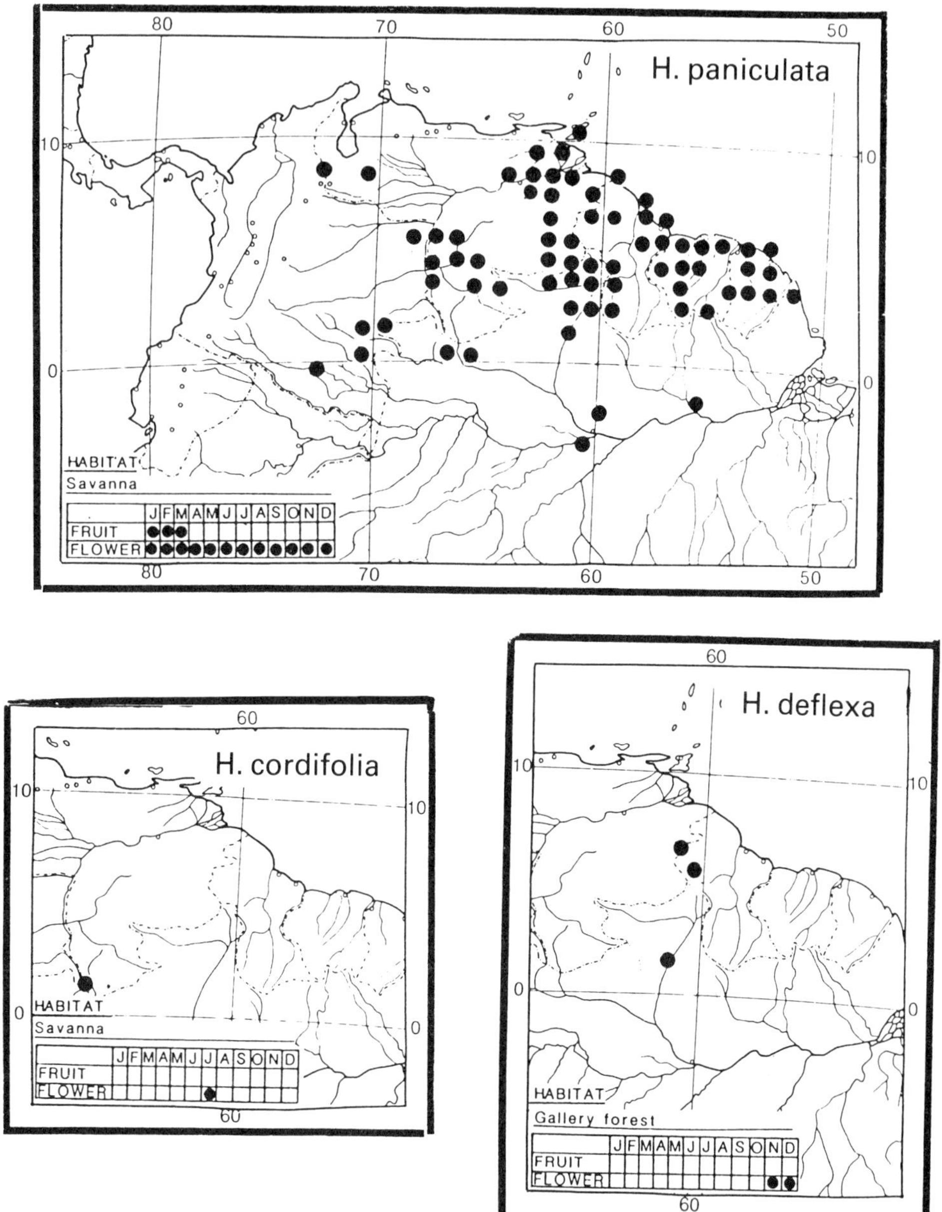

Fig. 15.13 Distribution of the widespread savanna species *Hirtella paniculata* and two derived local endemic species, *H. cordifolia* of savanna and *H. deflexa* of gallery forest.

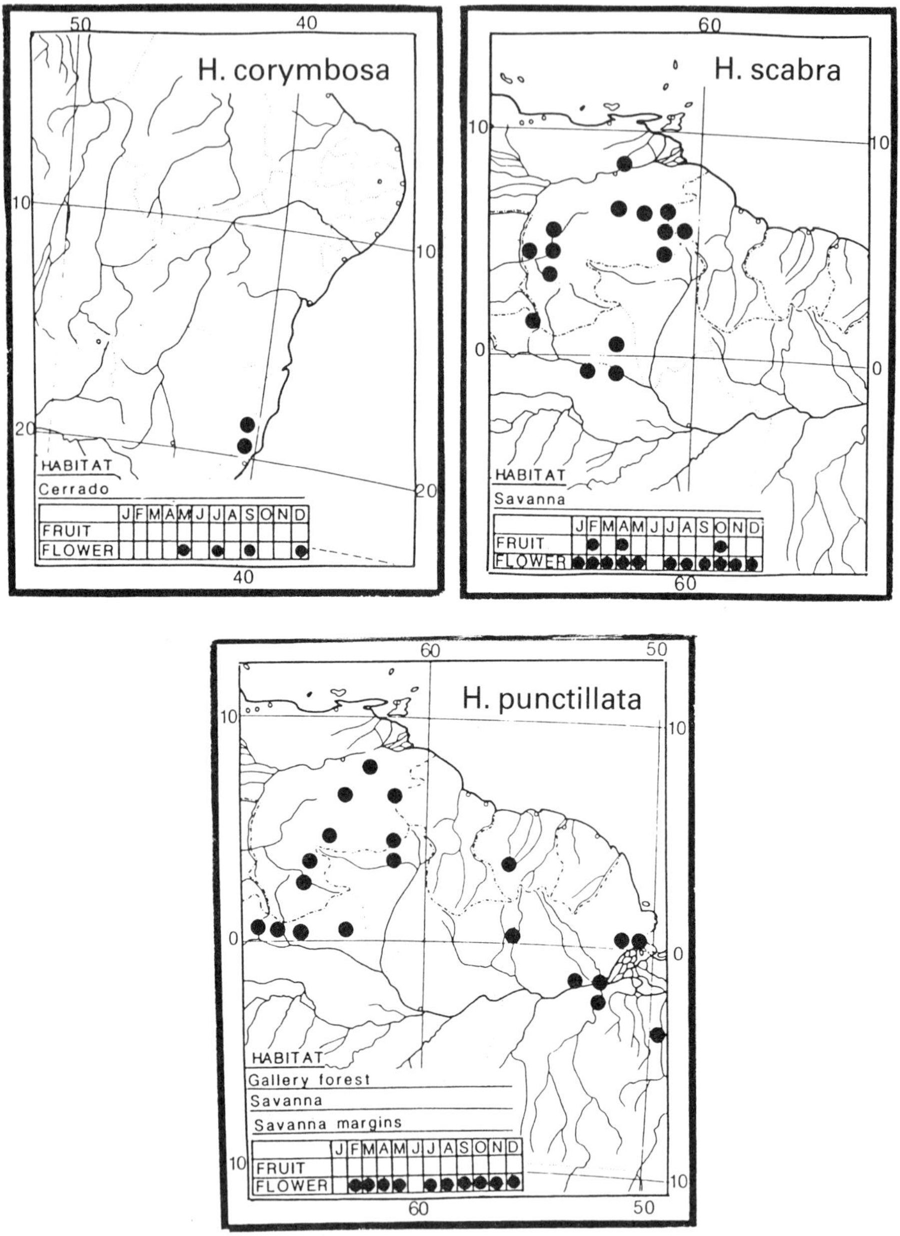

Fig. 15.14 Distribution of a complex of savanna species, *Hirtella corymbosa*, *H. scabra*, and *H. punctillata*, and their close forest relative, *H. bicornis*.

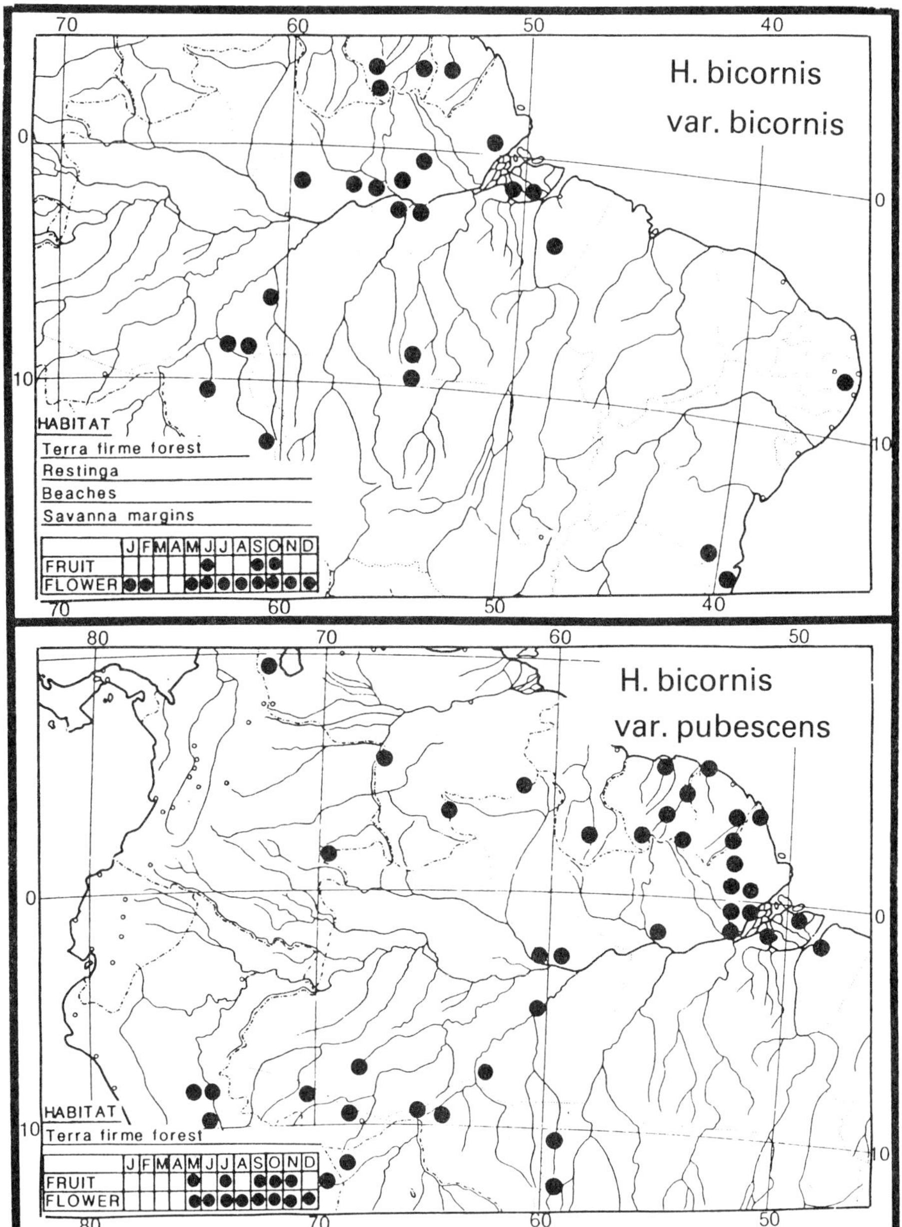

Fig. 15.14 *Continued*

(Fig. 15.10). It is notable that several savanna-margin species also occur in gallery and riverine forests within the rainforest areas. Other examples of this dual habitat preference are *Parinari campestris*, whose closest relative is a várzea forest species, *P. sprucei* (Fig. 15.31), and *Licania gardneri*, a typical savanna-margin and gallery forest species of central Brazil, whose closest relative is the widespread riverine species *L. apetala* (Fig. 15.20). *L. orbicularis* (Fig. 15.26) also occurs in savanna margins and gallery forest and is closest to a terra firme rainforest species, *L. coriacea*.

Hirtella caduca (Fig. 15.16) of gallery forest and savanna margins in Guyana is closely related to *H. brachystachya* of gallery forests of the upper Rio Negro region and *H. gracilipes* of the cerrado of central Brazil. These three species form an extremely closely related species group. Other savanna/gallery forest species pairs are:

1. *Licania humilis*, cerrado, and *L. octandra*, Amazonian riverine (Fig. 15.25);
2. *L. sclerophylla*, Amazon savanna and cerrado margins, and *L. longistyla*, Amazonian riverine (Fig. 15.28);
3. *L. incana*, Guianan savanna, and *L. leptostachya*, Guianan riverine (Fig. 15.22);
4. *L. nitida*, cerrado, and *L. pruinosa*, French Guianan and Amapá várzea (Fig. 15.24);
5. *Couepia grandiflora*, cerrado, and *C. paraensis*, Amazonian riverine (Fig. 15.5).

These examples clearly show that the adaptive radiation involving savanna and gallery forest species is a common occurrence and this is supported by our data from Lecythidaceae, Caryocaraceae and other woody plant families.

SAVANNA SPECIES COMPLEXES

Many of the savanna species of Chrysobalanaceae are more closely related to other savanna species which in turn are derived from riverine or terra firme forest species. The existence of several savanna species groups in Chrysobalanaceae indicates that considerable diversification has taken place between the different neotropical savannas.

The Venezuelan Amazonas savanna species *Hirtella adderleyi* (Fig. 15.9) is most closely related to *H. ulei* of savanna, beach and white sand areas of the Rio Negro and the sympatric *H. glabrata* which occurs in the same habitats as *H. ulei*, but is more common in white sand areas.

Hirtella paniculata (Fig. 15.13) is one of the most common savanna species of Venezuela, the Guianas and northern Amazonian Brazil. It is closest to

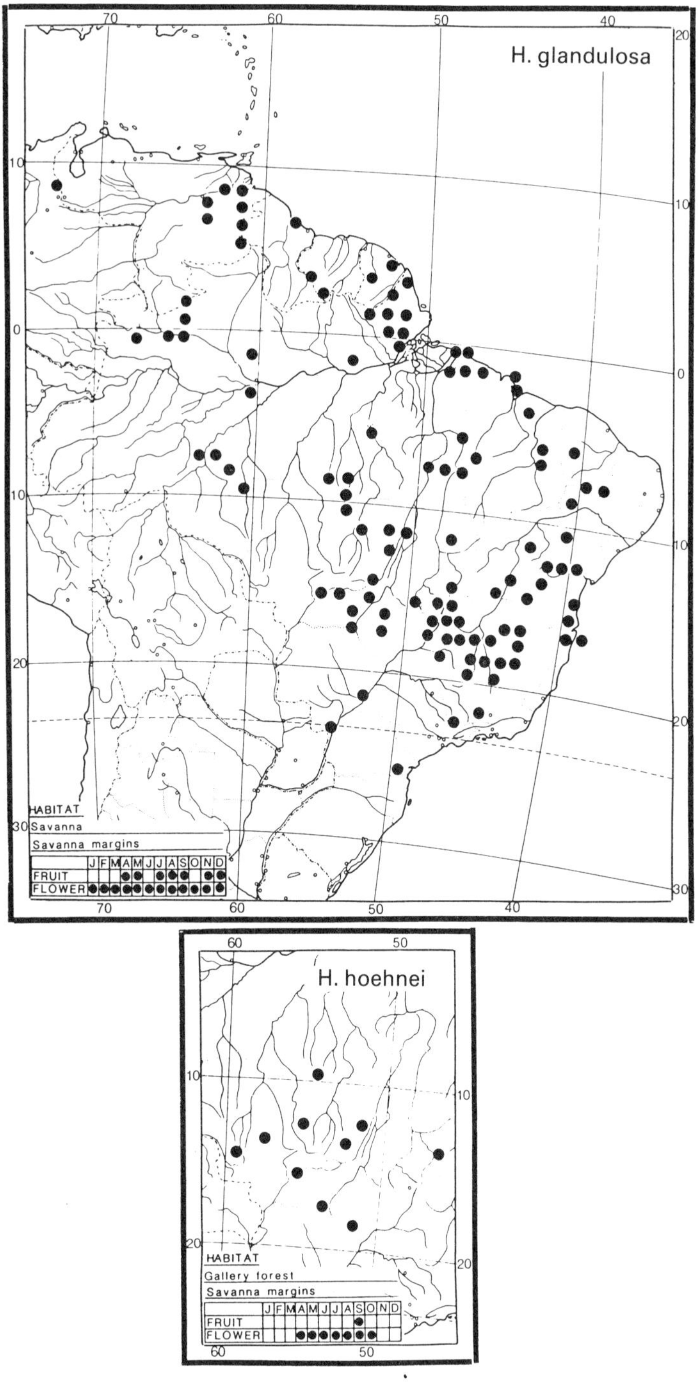

Fig. 15.15 Distribution of the widespread savanna-margin species *Hirtella glandulosa* and the gallery forest and humid savanna species *H. hoehnei*, its closest relative.

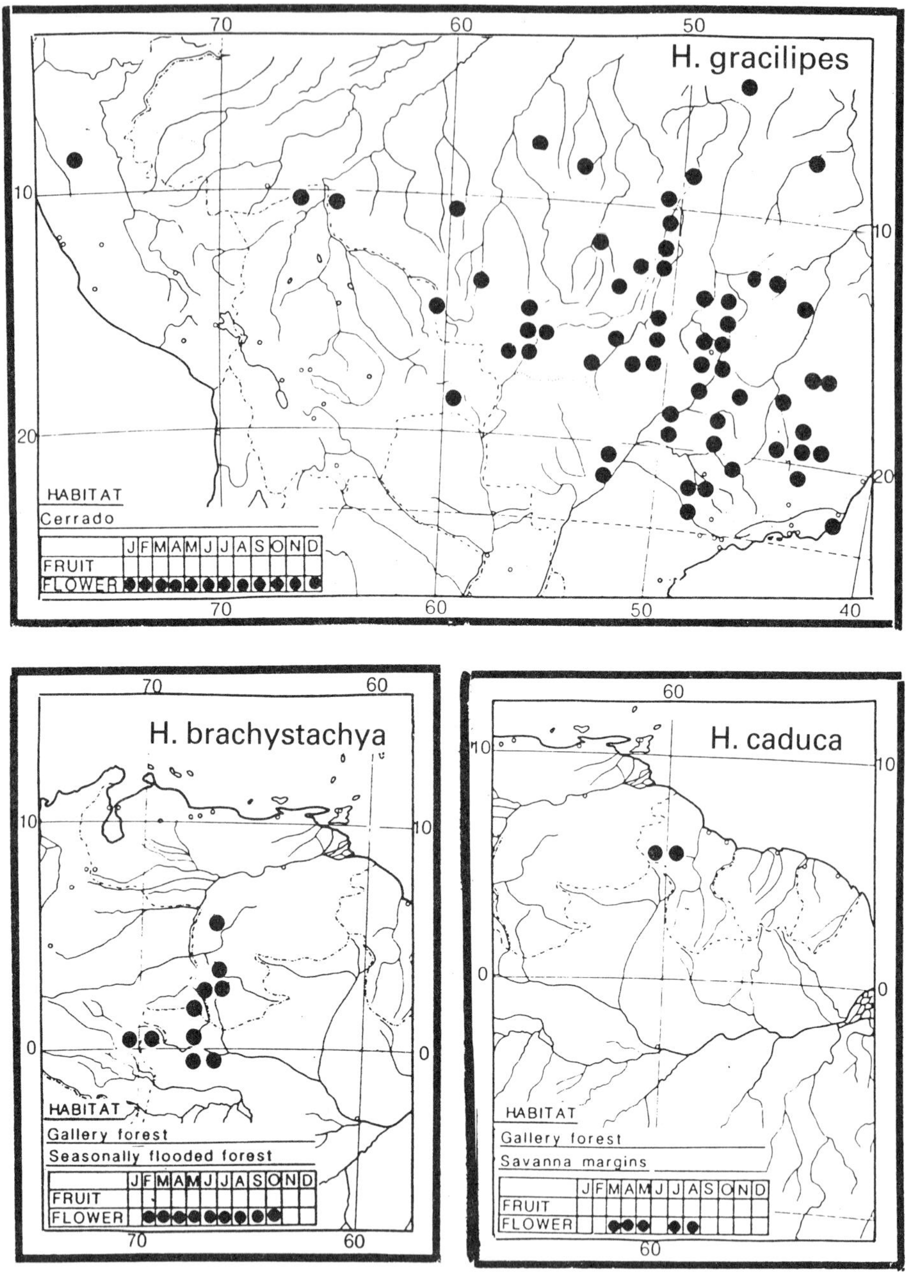

Fig. 15.16 Distribution of three closely related species of Hirtella, *H. gracilipes* of the cerrado, *H. brachystachya* of the Venezuelan savannas, and *H. caduca* of savanna-margins and gallery forest in Guyana.

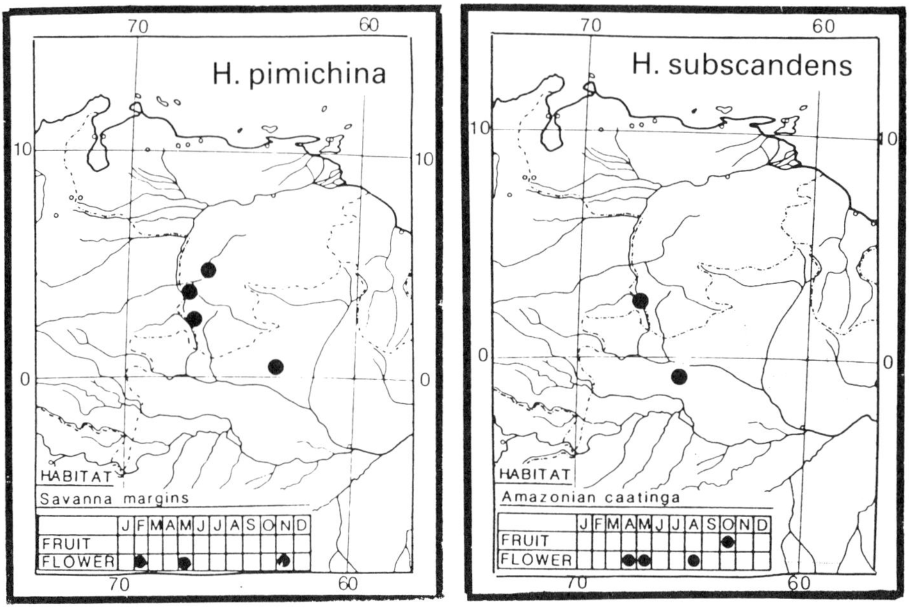

Fig. 15.17 Distribution of the closely related savanna species pair *Hirtella pimichina* and *H. subscandens*.

H. cordifolia of restricted range on the upper Rio Negro and *H. deflexa* of Guyana and Roraima. The two related species have almost certainly diversified from the wider ranging *H. paniculata*.

The *Hirtella bicornis* species group (Fig. 15.14) includes the widespread Amazonian and Guianan *H. bicornis* var. *pubescens*, one of the largest forest trees of the genus, and *H. bicornis* var. *bicornis* of beaches and savanna-margins of Amazonia and coastal restinga of eastern Brazil. Closely related to *H. bicornis* are three savanna species, *H. corymbosa* of Atlantic coastal savannas of Brazil, *H. scabra* of Venezuelan savannas and *H. punctillata* of the Guianas, Venezuela and Pará, Brazil.

Couepia racemosa (Fig. 15.1) is a widespread white sand, gallery forest and secondary forest species of central and western Amazonia and the Rio Negro region. It is replaced in the Guianas by its closest relative, *C. cognata* (Fig. 15.2), a typical savanna species. *C. amaralae* is a related local endemic of the caatinga of the upper Rio Negro in Brazil.

The distributions of *Licania foldatsii*/*L. steyermarkii* (Fig. 15.19) and of *L. hebantha*/*L. cordata* (Fig. 15.21) show typical allopatric species pairs which have obviously differentiated through isolation in different savannas.

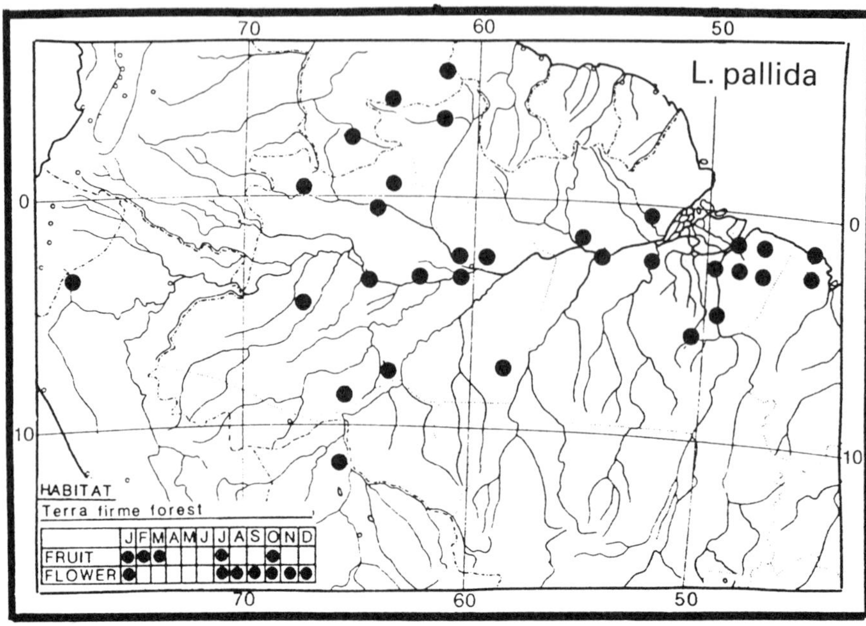

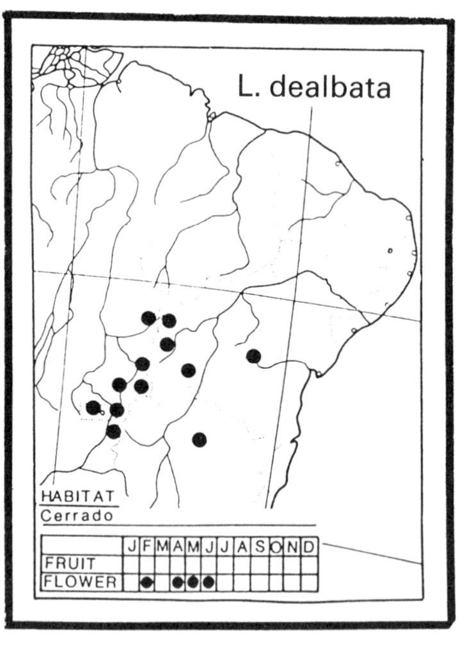

Fig. 15.18 Distribution of the cerrado suffrutex species *Licania dealbata* and its forest relative *L. pallida*.

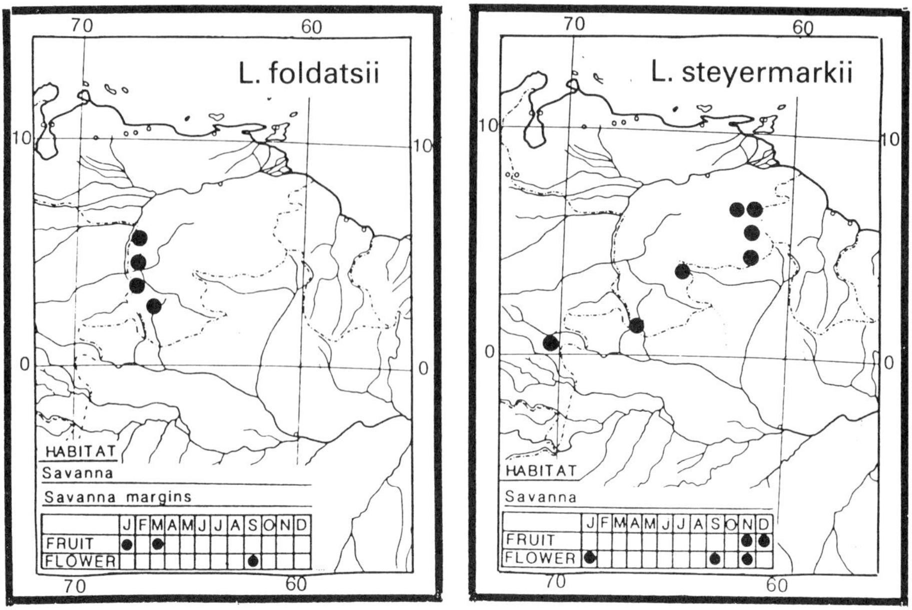

Fig. 15.19 Distribution of the closely related savanna species pair *Licania foldatsii* and *L. steyermarkii*.

DISTRIBUTION OF SAVANNA SPECIES AND THEIR RELATIVES

So far habitat differences between savanna species and their relatives have been dealt with. Now, focus is placed on distributions.

Licania dealbata (Fig. 15.18) is a suffrutex of the cerrado of central Brazil which is closely related to *L. pallida*, a tall tree of Amazonian terra firme forest. This type of diversification has occurred in many other groups where a forest tree is closely related to a savanna suffrutex (for example in *Parinari obtusifolia*, Fig. 15.30, and in other families, e.g. *Anacardium*).

The small genus *Exellodendron* consists of five very closely related species, which are either geographically or habitat-isolated from each other. *Exellodendron coriaceum* (Fig. 15.7) and *E. barbatum* (Fig. 15.8) are sympatric Amazonian forest trees of savanna/gallery forest and terra firme forest respectively. *Exellodendron gardneri* (Fig. 15.7) and *E. cordatum* are allopatric cerrado species, while the fifth species, *E. gracile*, is a restricted endemic of gallery forests in the Atlantic coastal rainforest of Espírito Santo. Clearly species differentiation in this group has taken place both by habitat and geographic isolation.

Couepia multiflora of the savannas of Guyana is closely related to

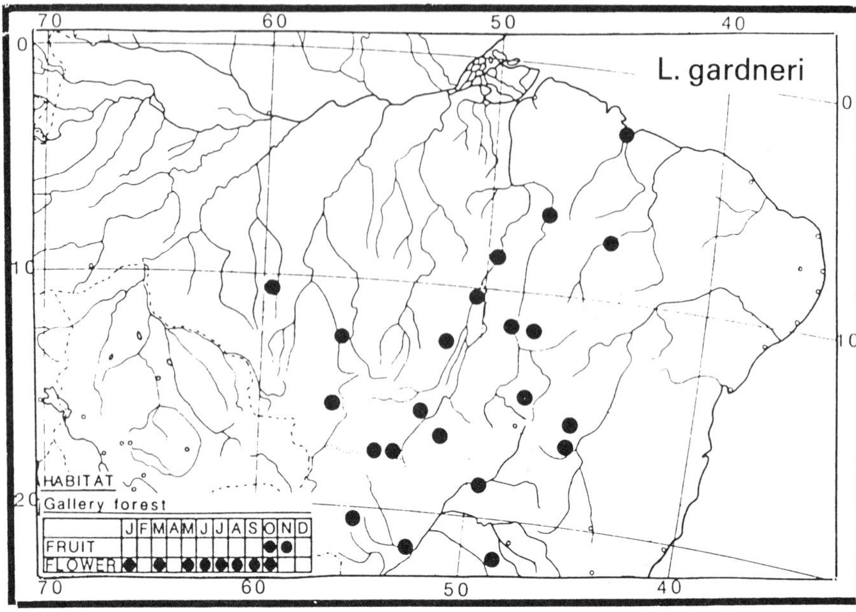

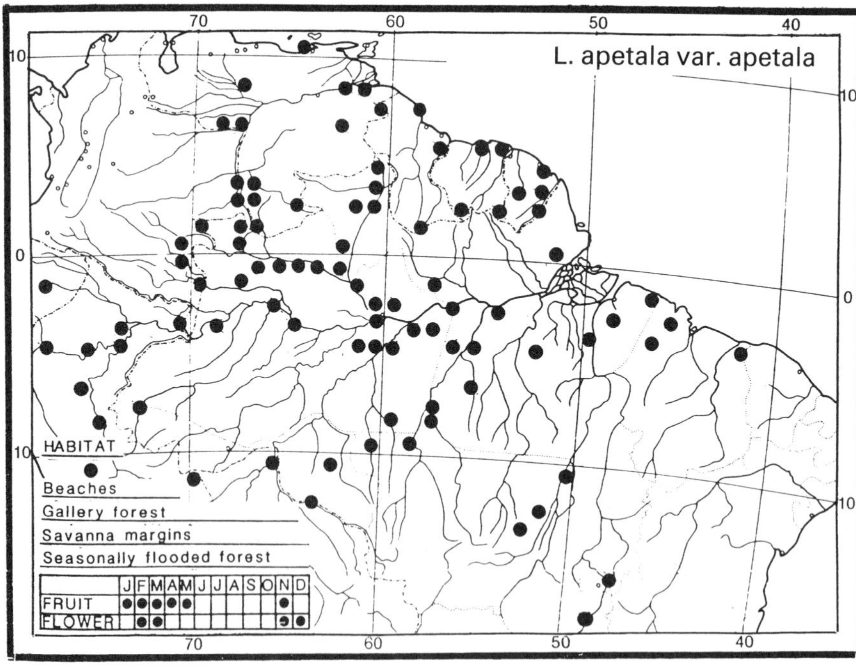

Fig. 15.20 Distribution of the cerrado gallery forest species *Licania gardneri* and its closest relative, *L. apetala*.

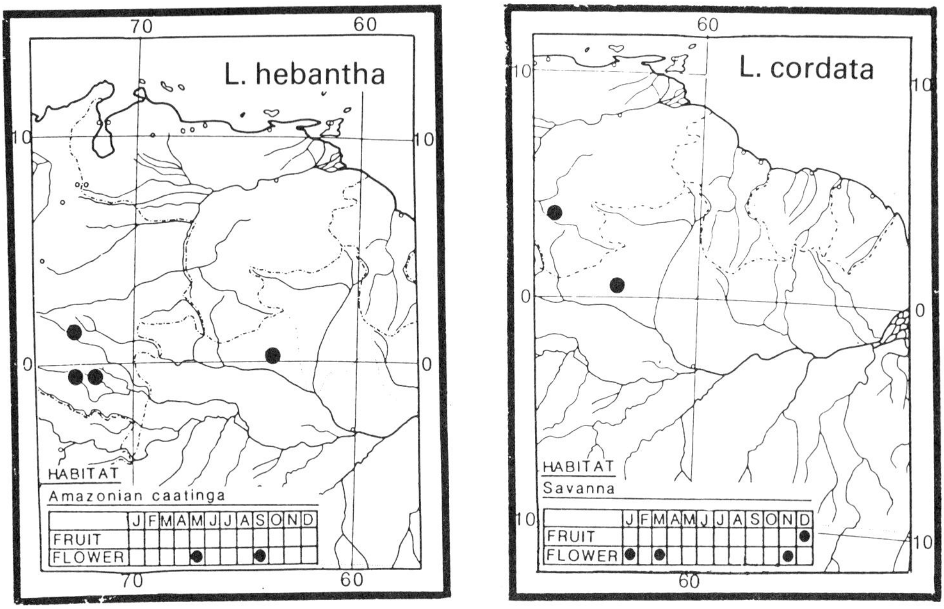

Fig. 15.21 Distribution of the savanna species pair *Licania hebantha* and *L. cordata*.

C. ovalifolia (Fig. 15.3) of the coastal restingas of eastern Brazil. This disjunction between two closely related species is quite common and indicative of past climate changes and former more continuous distributions.

CONCLUSIONS

This review of the savanna species of Chrysobalanaceae shows that the abundance of the family in the neotropics is partially due to its ability to adapt to a large range of habitats, including savanna. The savanna species are generally in the evolutionary more advanced part of the genera, and their affinities can usually be traced back to forest species, amongst which the maximum diversity occurs. The family has been in savannas for long enough for savanna species complexes to have evolved in isolated savannas. Transition to savanna has most frequently been via gallery forest into the drier savanna areas.

The species that occupy savanna margins tend to be far more wide-ranging than true savanna species and many also occur in gallery forests. They seem to be adapted to these more open habitats, rather than to the closed forest. Most of the cerrado species in the families I have studied

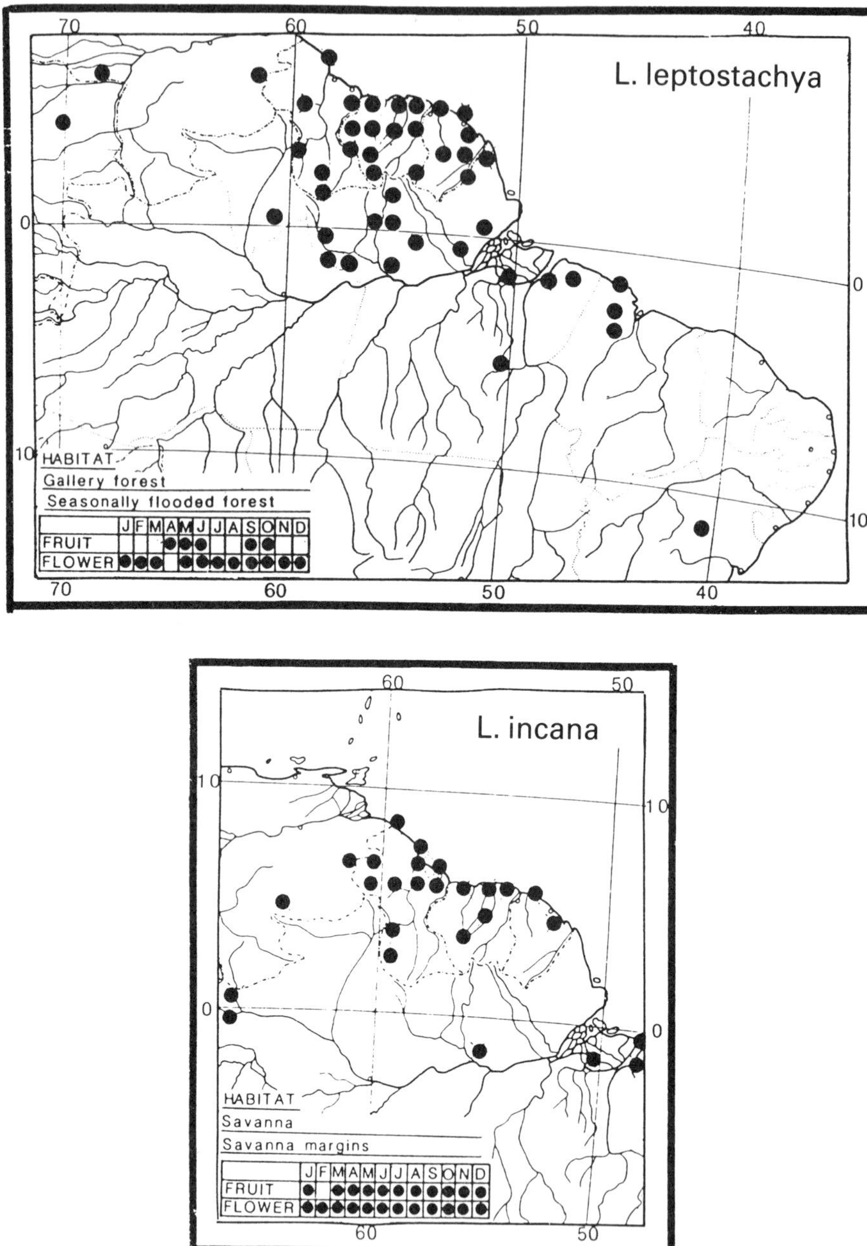

Fig. 15.22 Distribution of the savanna species *Licania incana*, and its closest relative *L. leptostachya*, which is sympatric but in gallery forest.

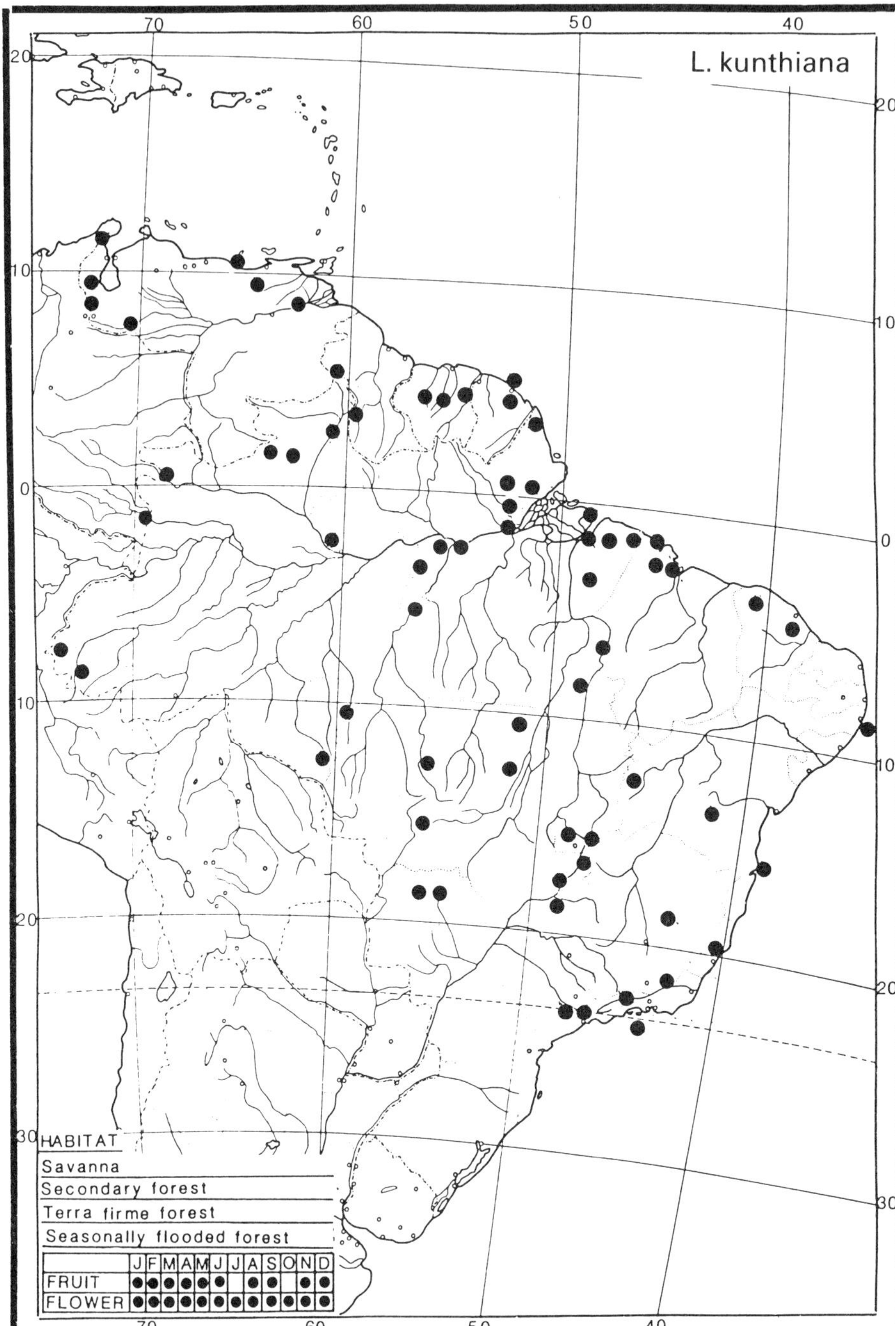

Fig. 15.23 Distribution of the savanna species *Licania lanceolata* and its closest relatives *L. kunthiana* and *L. spicata*.

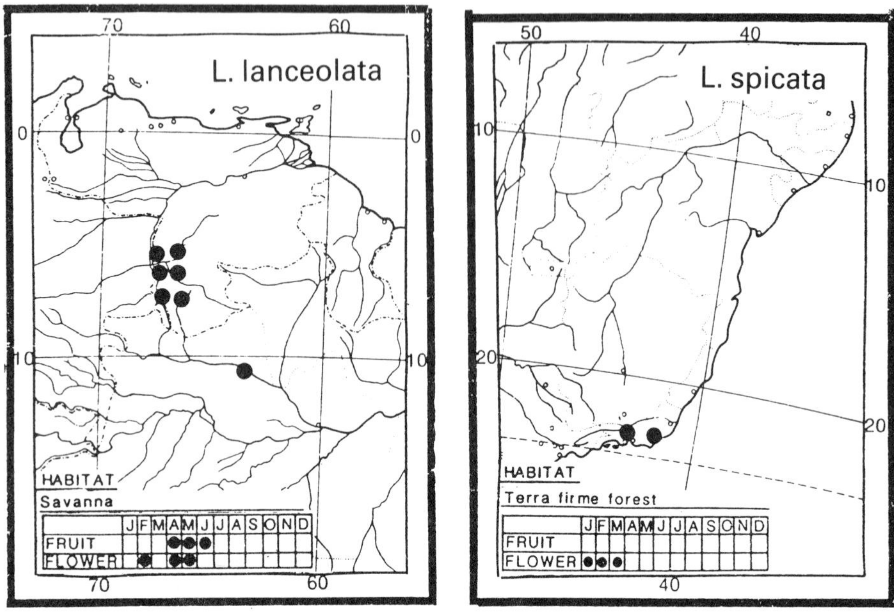

Fig. 15.23 *Continued*

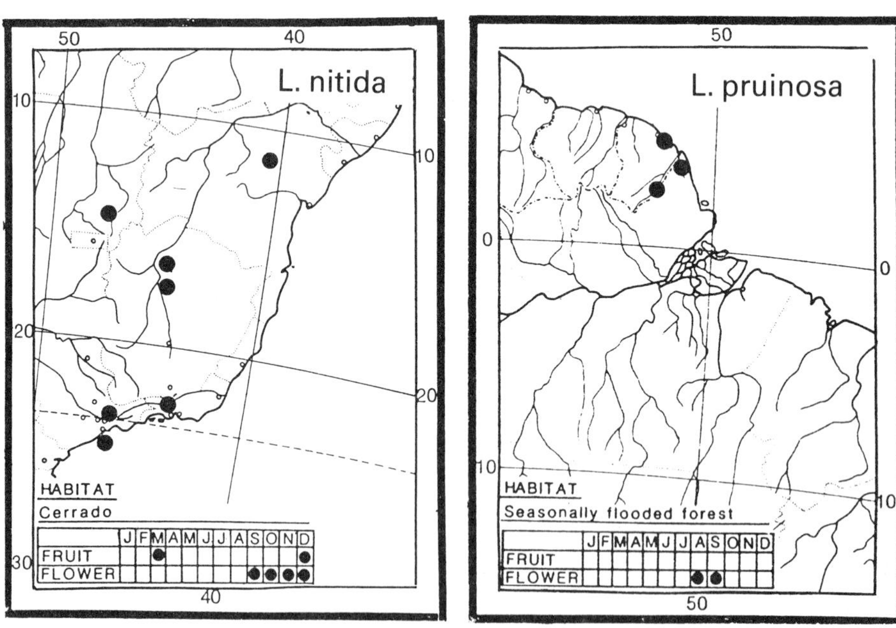

Fig. 15.24 Distribution of *Licania nitida* of cerrado and the extremely closely related but disjunct flooded forest species, *L. pruinosa*.

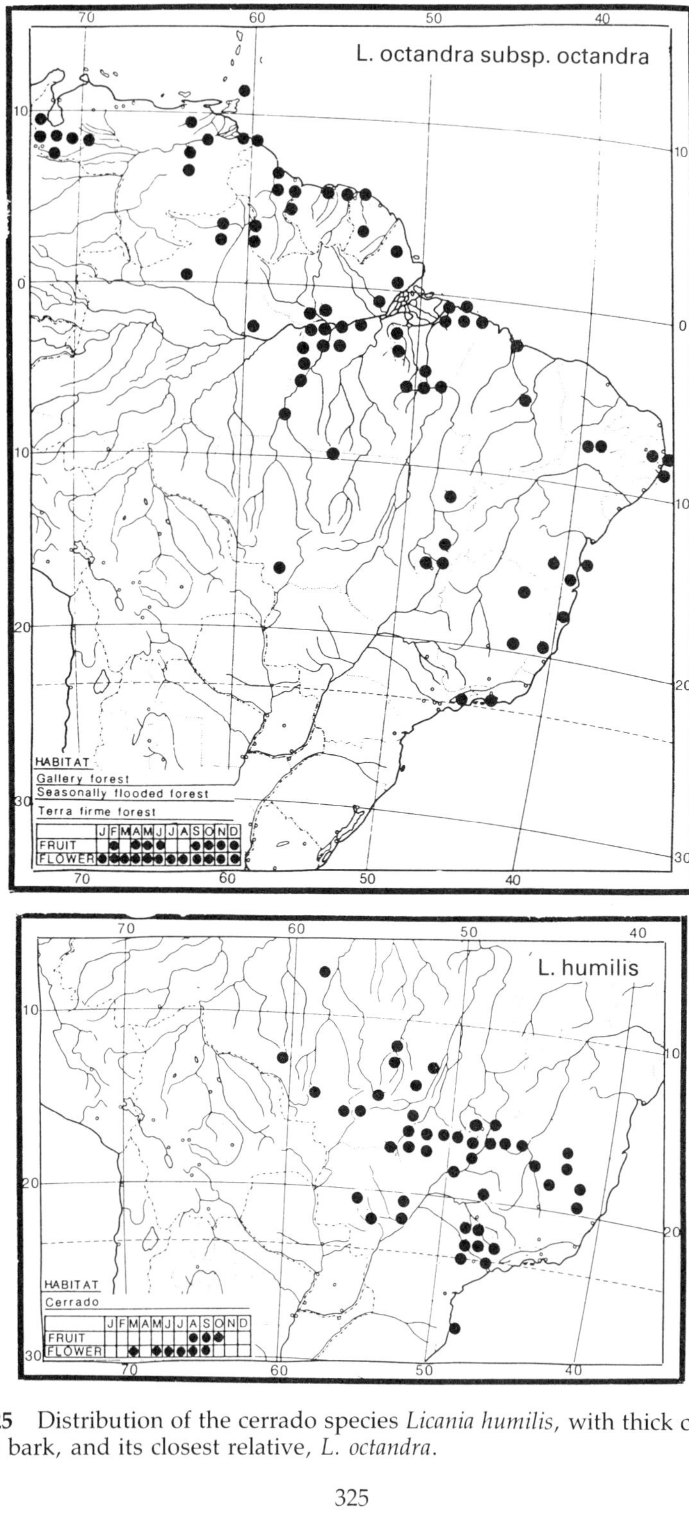

Fig. 15.25 Distribution of the cerrado species *Licania humilis*, with thick corky fire resistant bark, and its closest relative, *L. octandra*.

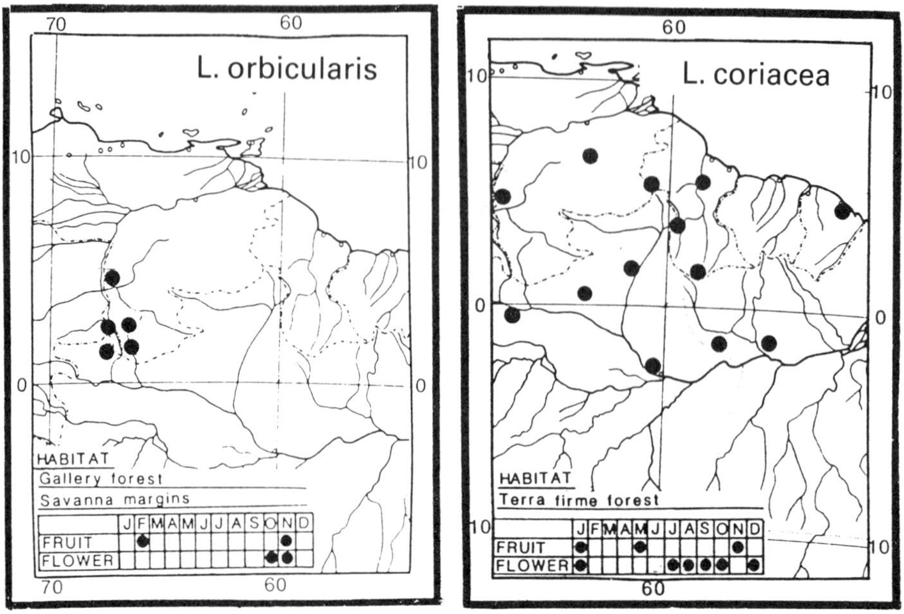

Fig. 15.26 Distribution of the savanna-margin species *Licania orbicularis* and the closely related terra firme forest species *L. coriacea*.

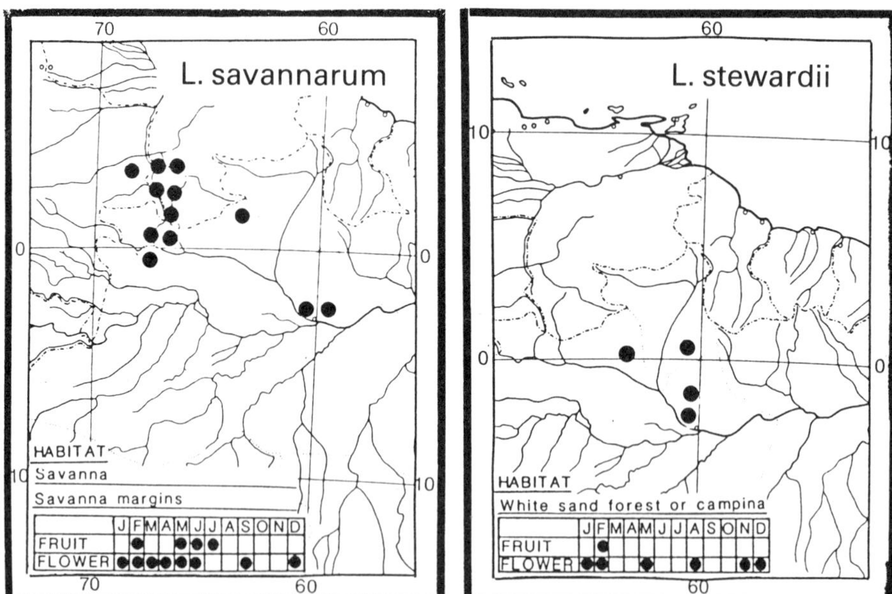

Fig. 15.27 Distribution of the closely related savanna species pair, *Licania savannarum* and *L. stewardii*.

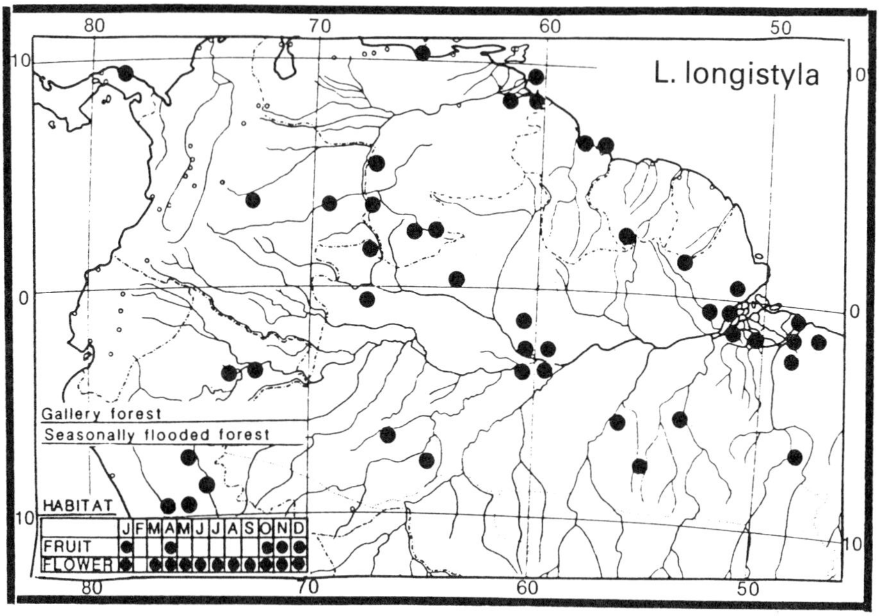

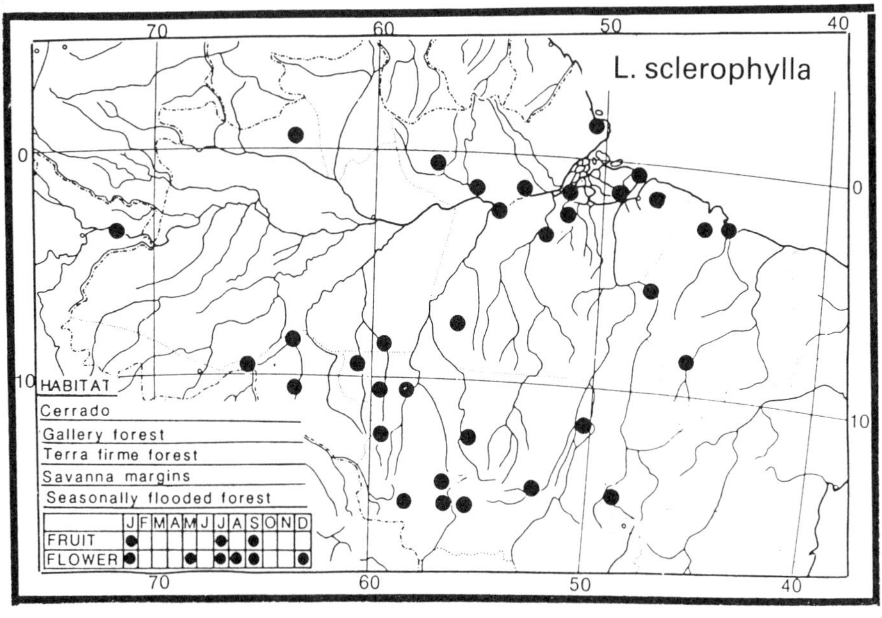

Fig. 15.28 Distribution of the savanna-margin and gallery forest species, *Licania sclerophylla* and the closely related flooded forest species, *L. longistyla*.

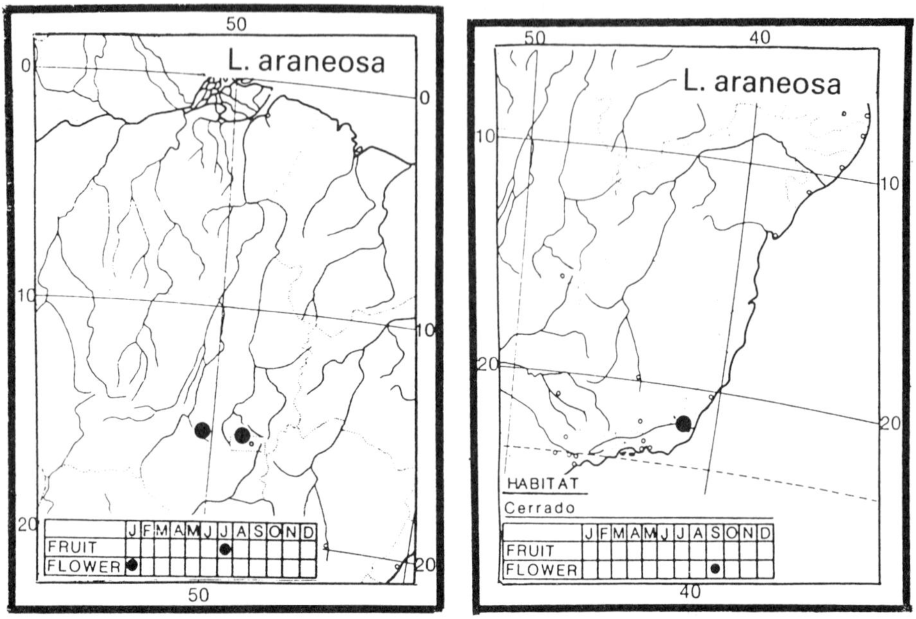

Fig. 15.29 Distribution of the savanna species *Licania silvatica* and the closely related forest species *L. araneosa*.

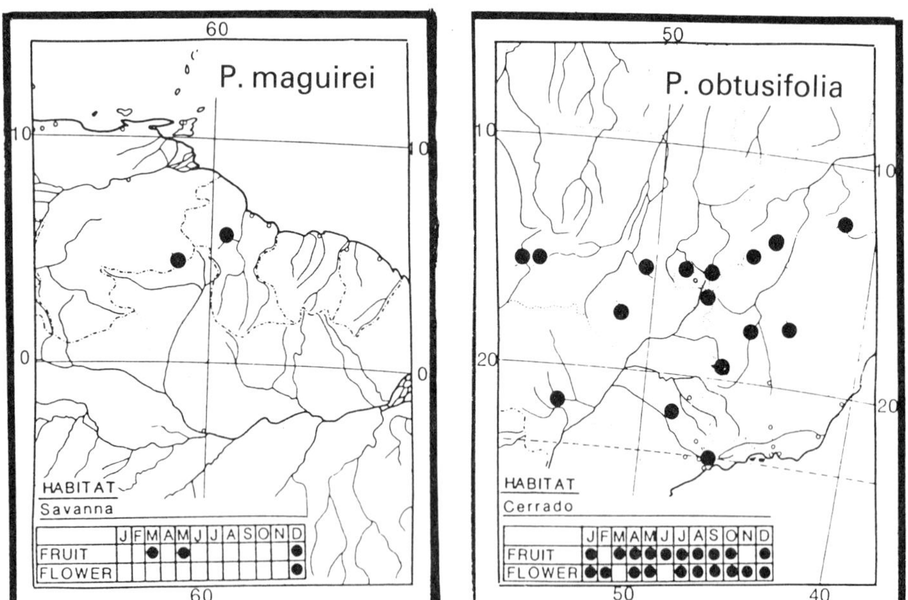

Fig. 15.30 Distribution of *Parinari maguirei* of the Venezulea–Guyana savanna and of *P. obtusifolia*, a suffrutex of the cerrado region.

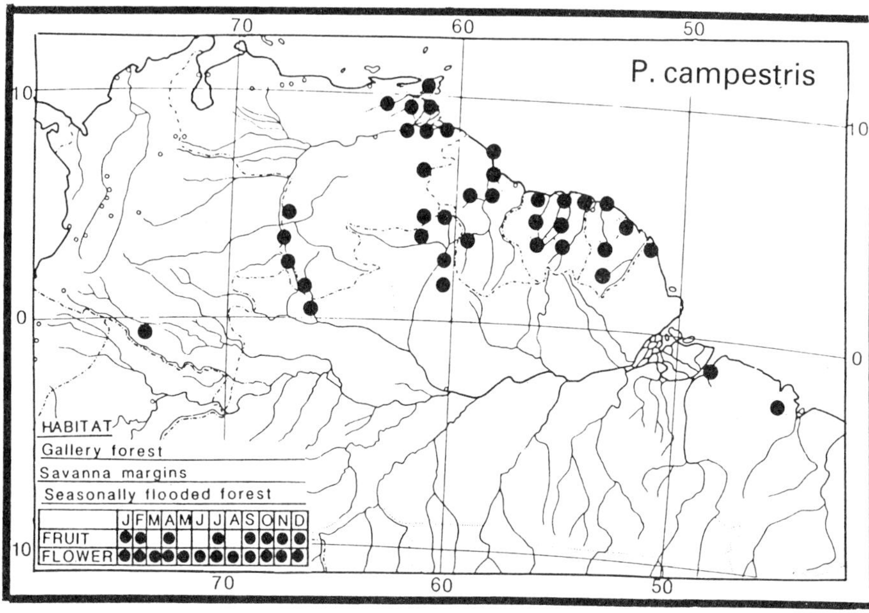

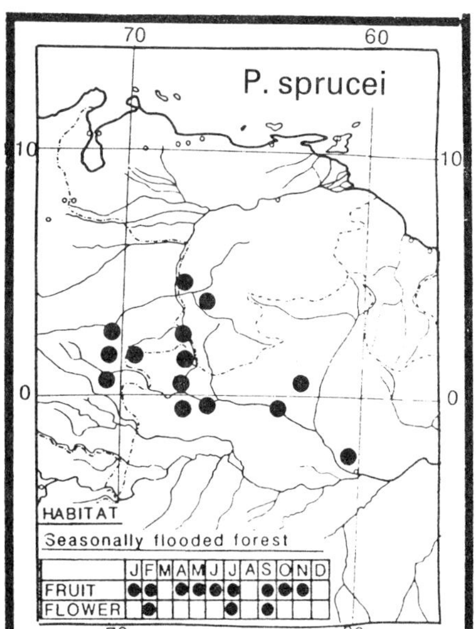

Fig. 15.31 Distribution of *Parinari campestris* of the savanna-margin and gallery forest, and the closely related *P. sprucei* of the Rio Negro gallery forest.

are confined to the cerrado area and the Guianas. This would indicate a certain degree of phytogeographic difference between the cerrado region of the Planalto of central Brazil and the northern savannas.

The savannas are being destroyed at an alarming rate and the current attention being given to the destruction of the rainforest tends to deflect interest from them. This chapter has shown that savannas contain many endemic species derived from rainforest ancestors. Perhaps they represent the more adaptable end of the genetic spectrum of rainforest groups and are thus particularly significant. It is extremely important not to neglect conservation of the pockets of endemism represented by many of the isolated savannas scattered throughout the neotropics.

REFERENCES

Mori, S.A. and Prance, G.T. (1990) Lecythidaceae Part 2: The zygomorphic-flowered New World Lecythidaceae. *Flora Neotropica Monograph*, No. 52, The New York Botanical Garden, Bronx, New York.

Prance, G.T. (1972) Chrysobalanaceae. *Flora Neotropica Monograph*, No. 9, Hafner, New York.

Prance, G.T. (1989) Chrysobalanaceae. *Flora Neotropica Monograph*, No. 9S, The New York Botanical Garden, Bronx, New York.

Prance, G.T. and Mori, S.A. (1979) Lecythidaceae Part 1: The actinomorphic-flowered New World Lecythidaceae. *Flora Neotropica Monograph*, No. 21, The New York Botanical Garden, Bronx, New York.

16

The nature and the dynamics of the forest–savanna boundary in south-western Nigeria

J.O. ADEJUWON and F.A. ADESINA

SUMMARY

1. This chapter examines the nature and dynamics of the forest–savanna boundary in Nigeria in terms of its vegetation characteristics, environmental factors moderating it, and its relative stability.
2. Physiognomically, the vegetation of the boundary is largely made up of forest and savanna. Floristically, some of the units of the vegetation are mixed, comprising varying combinations of forest and savanna species. The farmland and tree crop mosaics are also important components of the vegetation at the boundary.
3. The dynamics of the vegetation at the boundary are determined by anthropic, edaphic and climatic factors. Climate sets the regional boundary between the two vegetation types. The boundary detail is, however, largely influenced by anthropic and edaphic factors. Which vegetation type occupies which site depends on the intensity of cultivation and the characteristics of the soil.
4. The boundary in general is stable. It has remained relatively in equilibrium with the environmental factors that have produced it, within historical times.

INTRODUCTION

Bourlière and Hadley (1983) defined savanna as a community having a continuous grass layer scattered with trees; Hopkins (1974) described forest as consisting predominantly of woody plants and from which

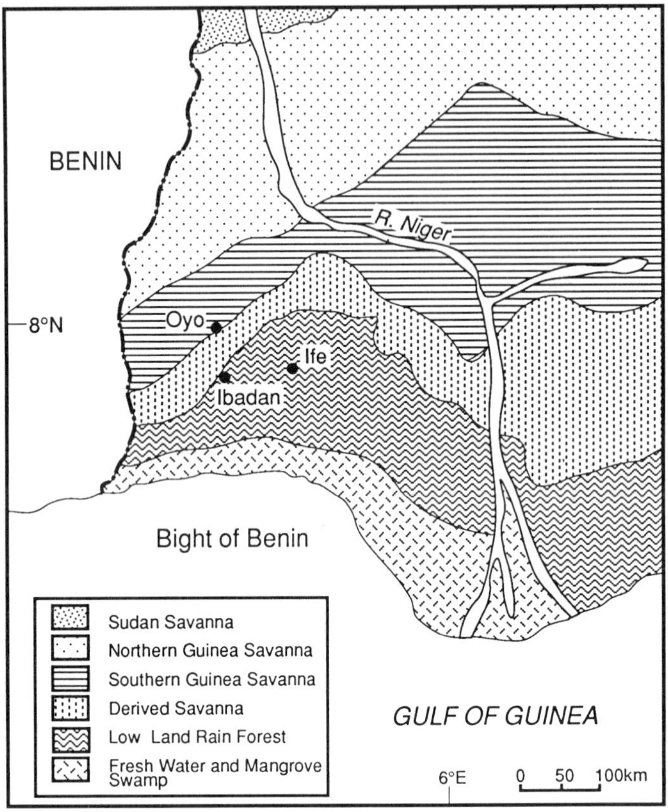

Fig. 16.1 Vegetation map of south-western Nigeria (after Keay, 1959a).

grasses are virtually absent. Figure 16. 1 shows the distribution of forest and savanna in south-western Nigeria.

A basic principle of community ecology is that 'the community has boundaries in space' (Daubenmire, 1968). This principle is very useful to the geographer who deals with spatial data and phenomena that have areal expression. An examination of the boundaries of a geographical phenomenon will help to identify the limiting factor(s) on which it depends, and reveal the pattern of its areal dynamics (i.e. its tendency to expand or contract).

Vegetation boundaries are usually mapped as lines, but in reality are rarely linear. Linear boundaries are found in rare situations where they are caused by abrupt edaphic discontinuities, such as geological variations (e.g. the boundary between limestone and shale) which produce contrasting soil types. In general, however, one vegetation type gradually changes to another across a zone of transition, the ecotone. Three types of ecotones can be recognized. First, the zone may bear its own unique vegetation, different from the larger communities it separates. Such zones become problematic in vegetation classification and mapping since the decision as to whether to recognize them as separate vegetation zones or as ecotones depends on the scale of mapping. A separate plant community at the regional scale may become an ecotone at the continental scale.

A more common type of ecotone is one where the vegetation is transitional, in the sense that it combines the characteristics of the two communities it separates. Across such a zone, the main features of one vegetation gradually give way to those of the other, and species found in both communities mix.

The third type of ecotone, which has not been well documented, is one in which two contrasting communities, which are unable to produce intermediates, compete for space over areas that are marginal for the existence of both. In such a zone, units of vegetation with characteristics similar to those of the two major plant communities are mixed in various proportions. Which community is found where depends on subtle habitat differences which may be difficult to discern or may, in fact, be fortuitous. The forest–savanna boundary in Nigeria appears to belong mainly to this third ecotone category.

The objectives of this chapter are:

1. to analyse and describe the vegetation characteristics and patterns of the forest–savanna boundary in Nigeria;
2. to demonstrate the effect of human activities on the dynamics of the vegetation of the boundary;
3. to evaluate the relation between vegetation and climate as it affects the stability of the boundary; and
4. to show the nature and relevance of soil–vegetation relations along the boundary.

THE FOREST–SAVANNA BOUNDARY IN SOUTH-WESTERN NIGERIA

As observed elsewhere (Adejuwon, 1968), the forest–savanna boundary is better described as an ecotone consisting of a vegetation mosaic in which units of forest and savanna are mixed. In certain places, the ecotone in Nigeria may be a few kilometres across while in others it may be as wide as 50 km (Fig. 16.2). In general, the mosaic of forest and savanna units separates an area of continuous forest on the wetter side,

Fig. 16.2 The forest–savanna ecotone in south-western Nigeria.

from another zone of continuous savanna on the drier side. This view is shared with Morgan and Moss (1965) who stated that 'the boundary between forest and savanna is a mosaic of communities representative of each vegetation type'. They noted that the zone of change is made up of 'islands, salients and enclaves of forest and savanna sharply separated from one another'. It is the local boundaries between these smaller units of the vegetation belonging to the two contrasting communities that they describe as 'remarkable for its clarity, whether seen from the ground or the air', rather than the ecotone which is a broad zone of change. It is hence necessary to recognize both a regional boundary, which is a zone of change, and the numerous local boundaries, which are sharp separations.

Figure 16.3 has been produced from aerial photographs to show savanna patches in the forest section of the forest–savanna ecotone in south-western Nigeria. The patches vary from about 80 km^2 and tend to occur in groups.

The above concept of an ecotonal mosaic disagrees with Keay (1959a, p. 8) who noted that 'an important local type of vegetation is intermediate in character between forest and savanna and is known as *Transitional woodland*'. Hopkins (1974) also recognized a 'boundary vegetation consisting of a mixture of both fire tolerant (savanna) and fire tender (forest) species with *Andropogon tectorum* and *Aframomum latifolium* characterising the ground layer'.

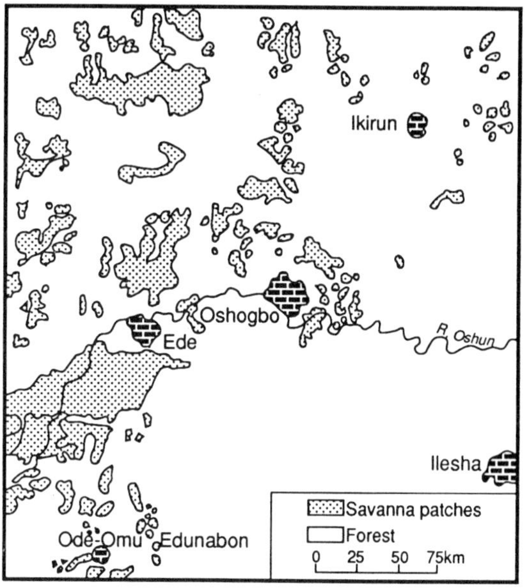

Fig. 16.3 Distribution of savanna patches in Oshogbo-Ilesha area.

VEGETATION CHARACTERISTICS OF THE ECOTONE

Ecotonal forests

The forest units associated with savanna within the ecotone are not in any important way different from those characterizing the main forest areas. These have been earlier described as the 'anthropic derivatives' of the tropical rainforest (Adejuwon, 1973). Three major forest types can be recognized in the ecotone:

1. secondary forest mosaic;
2. farmland mosaic; and
3. tree crop mosaic.

Secondary forest mosaic

The forest units included in this mosaic have characteristics that are intermediate between forest and farmland. The vegetation is dominated by trees which are 12–20 m high. The understory is sparse, except on the ground where there are occasional shrubs and seedlings of bigger tree species. The flora in the understorey is highly varied and is most often used for subdividing the secondary forest. The most common tree is the oil palm, *Elaeis guineensis*. It seems that any of the high forest species can be present, but the following are very frequent: *Albizzia ferruginea, A. glaberrima, Alstonia congensis, Cola cordifolia, Ficus exasperata, Funtumia elastica, Lecaniodiscus cupanioides, Musanga cecropioides, Pycanthus kombo, Physocarpus coccineus* and climbers such as *Hippocratea* spp. and *Paullinia pinnata*.

Farmland mosaic

The vegetation of the areas for food crops is recognized by the dominance of herbaceous species. Also, saplings growing from the stumps are very common, and other remnants of trees and shrubs which had to be felled to make way for food crops. The farmland mosaic can be called an oil palm bush, given the dominance of *E. guineensis* in it. The tree sometimes forms an upper tree stratum under which cropland and early successional species are found. Apart from the various food crops such as yam, cassava and maize, the dominant elements of the farmland mosaic include fallow regrowth and thickets.

Fallow regrowths include vegetation of less than five years old. A similar vegetation may, however, be older on poorer soils. During the first year after cessation of cropping in the rotational-bush fallow system, the vegetation consists of a thick growth of foliage up to 2.4 m high. On most of such fallows, the dominant herb is *Chromolaena odorata*. Also common in the first-year fallow vegetation are *Ageratun conyzoides, Aspilia africana, A. laterifolia, Asystasia* spp., *Dissotis rotundifolia, Melanthera escandens, Monechma depauperatum, Stachytarpheta cajannensis, Triumfetta cordifolia* and *T. rhomboidea*.

In general, thickets represent the normal limit reached by secondary succession before clearing and recultivation and they develop into secondary forests where recultivation is delayed. They consist of mainly woody vegetation less than 8 m tall which usually has a continuous foliage canopy between 2–6 m above the ground. Below this, there may or may not be much plant growth. Above the continuous foliage layer are usually found scattered trees of the same species as in nearby forests.

Tree crop mosaic

Many tree crops are grown in the area but only cocoa, kola and citrus are important in the tree crop mosaic. In floristic composition, the vegetation is not limited to the individual tree crops but contains weeds including: *Aneilema beninense, Aspilia* spp., *Brillantaisia lamium, Cyathula pedicellata, Dissotis rotundifolia, Fleurya* spp., *Crassocephalux biafrae, Melanthera scandens, Monechma depauperatun, Phaulopsis falcisepala, Pteris togoensis* and *Synedrella nodiflora*.

Ecotonal savannas

Floristic composition

The woody elements in the savanna units found within the ecotone include both typical forest and typical savanna species. The relative importance of each group and the constituent species vary from one unit to the other. The most frequent forest tree in the savanna vegetation is the oil palm. A majority of the others are small trees normally found in the lower strata of the rainforest. In the examples examined in south-western Nigeria, the commonest are *Albizzia* spp., *Blighia sapida, B. unijugata, Deinbollia pinnata, Fagara rubescens, Ficus exasperata, Phyllanthus discoideus* and *Rauvolfia vomitoria*.

Among the typical savanna elements, *Ficus capensis* and *Vitex doniana* are the most common. They are to be found in almost all the savanna units. The other species which are fairly widespread include *Anogeissus leiocarpus, Bridelia* spp., *Milletia thonningii* and *Parkia clappertoniana*. Almost all the savanna trees found in the boundary are among the most widely distributed in the savanna zones of West Africa. Four out of the 38 species characteristic of the ecotonal savannas are found in all the main savanna zones including the Sahel. These are *Adansonia digitata, Anogeissus leiocarpus, Borassus aethiopium* and *Terminalia laxiflora*. Fifteen other species are found in all the savanna zones with the exception of the Sahel, and 15 more throughout the sub-Sudan areas of West Africa.

With a few exceptions, the common grass species in the ecotonal savanna belong to the same genera as those of the main savanna zone. The most common is *Andropogon tectorum*, which dominates the grass layer of most of the savanna enclaves during the rainy season. Next in importance is *Imperata cylindrica*. Because of its smaller size, *Imperata* may not be as obvious as *Andropogon*, especially during the rainy season, but it is universally present. It is always the first grass to sprout after fires, so that surveys immediately after the first rains give the impression that it is the dominant ground layer species. *Andropogon gayanus* and *A. pseudapricus*, which are very important in the grass layer of the Guinea savanna zones to the north of the forest, are not as prominent in the ecotonal savanna. Invariably they tend to be overshadowed by the more robust *Andropogon tectorum*. In addition to *A. tectorum* and *I. cylindrica*, certain species occur as local dominants. These include *Brachiaria deflexa, Cyperus distans, Digitaria horizontalis, Loudetia* spp., *Hyparrhenia subplumosa, Panicum burmanii* and *Sorghum arundinaceum*. There are many grass species which are common but only rarely dominant; among them are *Hyparrhenia chrysagera, Panicum maximum, Pennisetum purpureum, P. polystachyum* and *Setaria barbata*. Sedges including *Bulbostylis, filamentosa* and *Fimbristylis* spp. are nearly always present but are never important physiognomically because of their relatively small sizes.

On the basis of their floristic composition, one can classify the units of savanna into three main types:

1. savanna in which both fire-tolerant and fire-tender species are prominent;
2. savanna dominated by fire-tolerant species;
3. savanna dominated by fire-tender forest species.

Physiognomy

The most notable physiognomic feature of the vegetation is the fairly tall ground layer relative to the height attained by the trees and shrubs. Because of this, the grass foliage is, in most of the units, not separated from that of the trees. *Andropogon tectorum* sometimes grows to a height of 3 m, enough to submerge all shrubs and the smaller trees. *Sorghum arundinaceum* grows as tall as 4 m, while tufts of *Panicum maximum* and *Pennisetum purpureum* in the wetter locations can raise the general upper grass layer to over 3 m.

FLORISTIC ANALYSIS AND THE ORDINATION OF THE VEGETATION ALONG THE BOUNDARY

In this section, the results of a study of a part of the forest–savanna boundary in a part of south-western Nigeria are considered. The objective was to examine quantitatively the vegetation characteristics and patterns of the boundary in terms of its floristics. Fifty sites in the boundary area were selected randomly. These sites were recognized in the field as either forest or savanna following the definitions given earlier. Lists of all the vascular plant species in each of the sites were made. From these, a presence/absence data table of the 65 species which occurred in at least two of the sites was prepared. The data were subjected to a two-way indicator species analysis (TWINSPAN) and detrended correspondence analysis (DECORANA) with default runs of the relevant programs prepared by Hill (1979a,b).

Table 16.1 shows the results of the TWINSPAN while Table 16.2 gives the full names of the species abbreviated in Table 16.1. The TWINSPAN produces three main groupings of the species in relation to the sites studied.

1. Species distributed mainly among forest sites (Table 16.1(a)). These are typically found in forest formations.
2. Species which appear not to discriminate forest or savanna sites (Table 16.1 (b)). Such species include, among others, *Daniellia oliveri* which is a savanna species, and *Elaeis guineensis* which is a forest species.

3. Species distributed among savanna sites only (Table 16.1(c)). These, exemplified by *Burkea africana* and *Sarcocephalus esculentus*, are known to be fire-tolerant and characteristic of the savanna environment.

Figure 16.4 shows the ordination of the 50 sites based on the results of the DECORANA. The savanna cluster is designated 'A' while the forest is 'B'. The ordination indicates that the sites are broadly separable into forest and savanna. However, some savanna sites located in the forest clusters and vice versa contain mixtures of forest and savanna species.

Table 16.1 A Twinspan classification of plants along a section of forest–savanna ecotone in Western nigeria. Abbreviated species names (in full in Table 16.2) are in the left-hand column. 1 indicates presence, – absence. The three group a, b and c are separated with solid lines

Species		*Sites, relative numbers*
		1 12 12 111 221243313422212343333331245444444
(a)		*93501475561242813246301594667978017892368030489275*
Pil.	thon.	-----1-------1---1--------------------------------
Ery.	mil.	--------------------1-----------------------------
Pyc.	kom.	-----1-1-----1------------------------------------
Imp.	cyli.	-------1----1-------------------------------------
Dan.	oli.	------------1-------------------------------------
Chr.	afri.	------1---
Cel.	pen.	-------1--
Alb.	lei.	-------1-----1------------------------------------
Law.	unif.	---1---111-11111----1-----------------------------
Gym.	sen.	--1----1-1--11------------------------------------
Bli.	sapi.	----------1--11-----------------------------------
Fun.	ela.	--1----1--11----111--1------------------------1---
Cel.	zenk.	1--1-1111---1---1-11-1-1--------------------------
Asp.	lat.	-1----11--11--1-1----1--------------------------1-
Anch.	dif.	1-1---11--1-1--11------1--------------------------
Mil.	thon.	-1-----11---
Fic.	asp.	11-1-1111----11-------------------------------1---
Eryt.	gui.	-1----11--
Comb.	mic.	1-----11--
Afz.	afri.	--11-1-1------------1-----------------------------
Ann.	sen.	--11-1-11111--------------------------------------
Fag.	spp.	1111-----11-1-------------------------------------
New.	lea.	111--------1--------------------------------------
Gly.	lat.	1---
Entr.	spp.	--1---
Cro.	febr.	-1--
Chy.	alb.	11-1--
Bri.	mic.	----1---

Table 16.1 (continued)

```
Species            Sites, relative numbers
               1 12   12 111    221243313422212343333331245444444
          93501475561242813246301594667978017892368030489275

(b)
Spo.  mom.  ||||-||||||||-|||-|-||-----------|--|||-----------
Lec.  cup.  -----|---------|-----|--|-------------------------
Ent.  afri. -----|-|||-----|-|||----|-----|-------------------
Ela.  gui.  ||||||||||||||||||||||-|||-----||-|||------------|
Anc.  mon.  ||||||||||-|||||||||||---------|-|||--------------
Sta.  sti.  |-||------------------|---------------------------
Pte.  spp.  |-||---|----------------|-------------------------
And.  tec.  --|||--|-----------------------|------------------
Chr.  odo.  ------|-|------|---------|------------------|-----
Chl.  exc.  ------||-----------------|------------------------
Bli.  unj.  ---------------|--|-------------|-----------------
Ric.  afr.  -------|-----|-------------------------------|--|-
Kha.  gra.  ----------------|--------------------------------|
Pho.  rec.  --|--------------|--------|-----------------------
Pau.  pin.  --|---|---|----------|||-------------------|------
Lon.  cya.  --------------------|||--|------------------------
Col.  nit.  ---||---|------------|-|-|------------------------
Pte.  osu.  ---|--------------------------------------------|-
Pen.  pur.  |||-|||-|||||||||||---|||||||||||||||---|--|-|----
Par.  fil.  --------|----------||---||-||-||--||----|||------|
Fic.  cap.  -|----|--|||-|-|---|-||||||------|||||||||-|----||

(c)

Cas.  spp.  -----------------------|--------------------------
Als.  bon.  ----------------------|---------------------------
Sch.  spp.  ---------------------|--||----------|---|-----|--|
Sac.  esc.  ----------------------|-||--------||||||-|---||-||
Pen.  pur.  ----|----------------|||||||-|----|-----||||-||-|-
Sid.  ver.  ----------------------------------|--------------|
Hym.  aci   ------------------|----|||||||-|||||||||||||||||||
Hom.  dol.  ----------------------|-|--||||||||||||||||||||---
Con.  pen.  -------------------|-|-----|||||||---|--||-||-|||-
Bur.  afri. -----------------------------------|--------------
Asp.  afr.  --|-----------------------------|-||---|-----||---
Bra.  spp.  -----------------------------------|-------||||||||
Ant.  afr.  ------------------------|---------|-|||-||||||||||
Ana.  occ.  -----------------------|--------------------||----
Pen.  ram.  ---||-------------------|----|----|--||-|-||-|----
Ano.  leo.  --|----------------------||--|-|-----------|------
```

Table 16.2 Full names of species (abbreviated in Table 16.1) and their affinities: f = forest tree species; fh = forest herb species; s = savanna tree species; sh = savanna herb species

Acanthus montanus (fh)	*Erythrina mildbraedii* (s)
Afzelia africana	*Erythrophleum guineese* (f)
Albizzia ferruguinea (f)	*Fagara* spp. (f)
Alstonia boonei (f)	*Ficus asperifolia* (f)
Anacardium occidentale (s)	*Ficus capensis* (s)
Anchomanes difformis (fh)	*Funtumia elastica* (f)
Andropogon tectorum (sh)	*Glyphaea laterifolia* (f)
Annona senegalensis (s)	*Gymnosporia senegalensis* (f)
Anogeissus leiocarpa (s)	*Homalium dolichophyllum* (f)
Antiaris africana (f)	*Hymenocardia acida* (s)
Aspilia africana (fh)	*Imperata cylindrica* (sh)
Aspilia latifolia (fh)	*Khaya grandifolia* (f)
Blighia sapida (f)	*Lawsonia* sp. (f)
Blighia unijugata (f)	*Lecaniodiscus cupanioides* (f)
Brachystegia spp. (f)	*Lonchocarpus cyanescens* (s)
Bridelia micrantha (s)	*Millettia thonningii* (s)
Burkea africana (s)	*Newbouldea leavis* (f)
Cassia spp. (sh)	*Parkia filicoidea* (s)
Ceiba pentandra (f)	*Paullinia pinnata* (f)
Celtis zenkeri (f)	*Pennisetum purpureum* (sh)
Chlorophora excelsa (f)	*Pennisetum ramosum* (sh)
Christiana africana (f)	*Phoenix reclinata* (s)
Chromolaena odorata (fh)	*Piliostigma thonningii* (s)
Chrysophyllum albidum	*Pterocarpus osun* (f)
Cola nitida (f)	*Pterygota* spp. (f)
Combretum micranthum (f)	*Pycnanthus kombo* (f)
Conopharyngia penduliflora (s)	*Ricinodendron africanus* (f)
Crossopteryx febrifuga (s)	*Sacrocephalus esculentus* (s)
Daniellia oliveri (s)	*Schizachyrium* spp. (sh)
Elaeis guineensis (f)	*Sida veronicifolia* (sh)
Entada africana (s)	*Spondias mombin* (f)
Entrandophragma spp. (f)	*Staudtia stipitata* (f)

The foregoing points to the fact that the boundary vegetation complex consists of two sharply contrasting physiognomic types. Floristically, three types are recognized, including one with typical forest species, another with typical savanna species, and a third composed of a mixture of forest and savanna species. Therefore, using the two criteria gives us four vegetation types including:

1. savanna including only fire-tolerant species;
2. savanna including both fire-tolerant and fire-tender species;
3. forest including both fire-tolerant and fire-tender species;
4. forest composed mainly of fire-tender species.

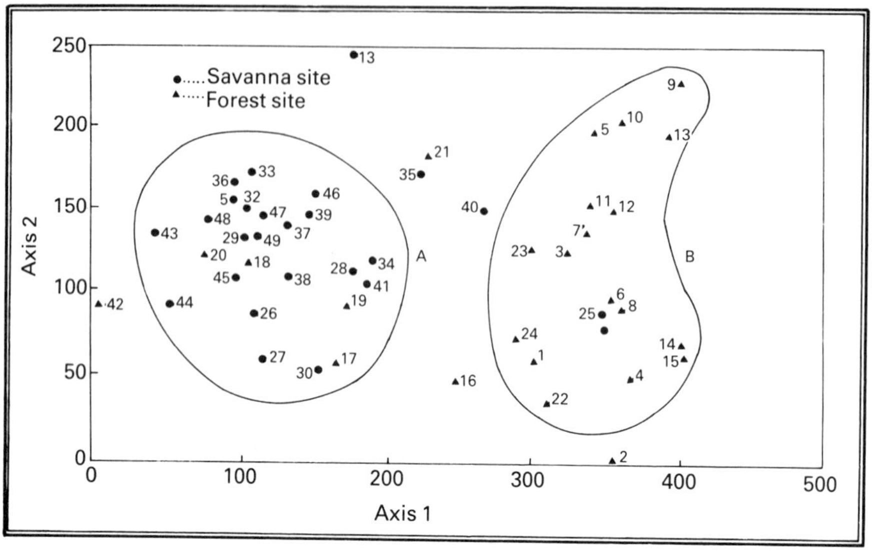

Fig. 16.4 Site ordination along the boundary by Decorana.

ENVIRONMENTAL FACTORS AT THE BOUNDARY

The main anthropic factor within the ecotone is shifting cultivation. Under this system, cultivation commences by the clearance of the vegetation and, after a crop or two, the land is abandoned. The woody plants which were much depleted during the cropping start to grow back as soon as the farmer abandons the plot. The period of fallow varies; the longer the period, the taller and denser is the woody regrowth. In areas with intensive shifting cultivation, the vegetation, compared with the mature rainforest, becomes more herbaceous. Fast-growing tree species become more frequent, and the diversity of species becomes greater.

Two types of vegetation gradients, related to the system of cultivation, could be noticed in the areas of forest regrowth. First, there is the gradient in which biomass increases with the length of fallow. This is the basis of the segregation of the vegetation into fallow regrowth, thickets and secondary forests. It is also the principle on which traditional shifting cultivation is based. In general, the greater the length of fallow, the better restored the soil is for cultivation.

Second, there is the gradient in which sustained and intensified cultivation with short fallows leads to a proportional reduction in the biomass. This agricultural practice produces savanna in forest areas (e.g. Clayton, 1958). In other words, the climax vegetation is high forest which can be re-established with prolonged fallow. The process of savannization could be observed in the fallow, which gradually acquires a field layer dominated by grasses and a tree layer composed of fire-tolerant species. The process can be broken down into three stages.

1. Savanna species begin to appear in the fallow regrowth. The field layer is still made up of forbs such as *Aspilia* spp. and *Chromolaena odorata*, but typical savanna trees like *Daniellia oliveri*, *Ficus capensis* and *Parkia filicoides* begin to appear.
2. The floristic composition becomes more significantly a blend of forest and savanna species. The savanna species may be fewer. The field layer especially may have a mixture of grasses and forbs. Further intensive use of such units produce those described in stage 3.
3. The field layer of the regrowth becomes predominantly grassy. Above this layer, some forest species may survive, but typical savanna species dominate.

Prolonged fallow is expected to result in a reversal of the above process. It appears, however, that a state of dynamic equilibrium has been established between the vegetation and human activities, and that neither savannization nor its reverse process is much in evidence.

The first vegetation gradient is not discernible in areas with savanna vegetation. Because of the effect of annual fires, there is invariably no appreciable increase in the biomass during the period of fallow. However, the second gradient differentiates areas with higher intensities of burning which bear the more open types of savanna from the areas with the lower intensities carrying the woodland types of savanna.

Climate and the forest–savanna boundary

The regional boundary between forest and savanna corresponds in general to the pattern of climate. The linear approximation of the zone as well as its 'drier' and 'wetter' boundaries follow the same general course as the isopleths of climatic elements, especially those controlling the supply of water to plants. In the Ibadan area it has been observed that the forest–savanna ecotone seeks to occupy areas having the same type of climate (Adejuwon, 1971). The southern 'wetter' boundary follows an isopleth of an average number of 100 rainfall days. The northern, 'drier' boundary is closely related to an isopleth of 80 days with rain. The 1250 m mean annual rainfall isohyet and the isopleths representing 25 mm mean annual water deficit and the 1300 mm mean annual potential evapotranspiration all have the same shape as the zone and pass through it (Adejuwon, 1971).

Soil factors and the forest–savanna boundary

It has been demonstrated earlier that the savanna units within the ecotone correspond to the distribution of the poorer soil types. While savanna is absent from clayey soils such as the Araromi, Egbeda, Etiomi, Fagbo and Itagunmodi Series, it has its widest distribution on such sandy series as Apomu Ibadan, and Asejire, derived from nutrient-deficient parent rocks (Adejuwon, 1968). One can infer from this that the topographical boundary, or the 'boundary in detail', might have equivalents in soil discontinuities.

Soil–vegetation complexes within the forest–savanna ecotone

The study of a part of the forest–savanna boundary earlier referred to (Adejuwon 1968, 1971) included soil sampling. Five soil samples were randomly taken to a depth of 20 cm from each of the 50 sites studied. The soil samples from each site were pooled, mixed thoroughly and sufficient quantities taken to the laboratory for analysis. The soil properties measured include particle size distribution by the hydrometer method; pH, in 0.01 M $CaCl_2$ solution with 1:2 soil:$CaCl_2$ solution ratio; available phosphorus by the Bray No. 1 method; organic matter by the Walkley–Black dichromate method; nitrogen in the form of nitrates by adding 2,4,phenoldisulphuric acid to the prepared solution; exchangeable calcium, sodium and potassium by flame photometry. The resulting soil data were paired with species occurrence data for canonical correlation analysis and ordination, to relate the soil and vegetation subsystems and assess the ecological relations among them in terms of their combined soil and vegetation properties. Canonical correlation analysis (SPSSX, 1986), related the soil characteristics to the vegetation gradients identified with DECORANA.

The canonical ordination of the first pair of variates is shown in Fig. 16.5. This separates the plots broadly into forest and savanna on the basis of the relationships in their soil and flora. Savanna plots dominate the negative end of the ordination plane, while forest plots dominate the positive end. The canonical correlation, which is significant at the 0.01% level, is 0.78 with exchangeable sodium making the greatest contribution to the definition of the soil variate. The other significant soil properties with respect to the first soil variate include exchangeable calcium, potassium, organic matter and nitrate-nitrogen.

The first DCA (detrended correspondence analyses) axis, which by definition is the most significant, makes the greater contribution to the contribution to the first vegetation variate. A diagonal distribution of the plots through the origin on a canonical ordination plane is indicative of a

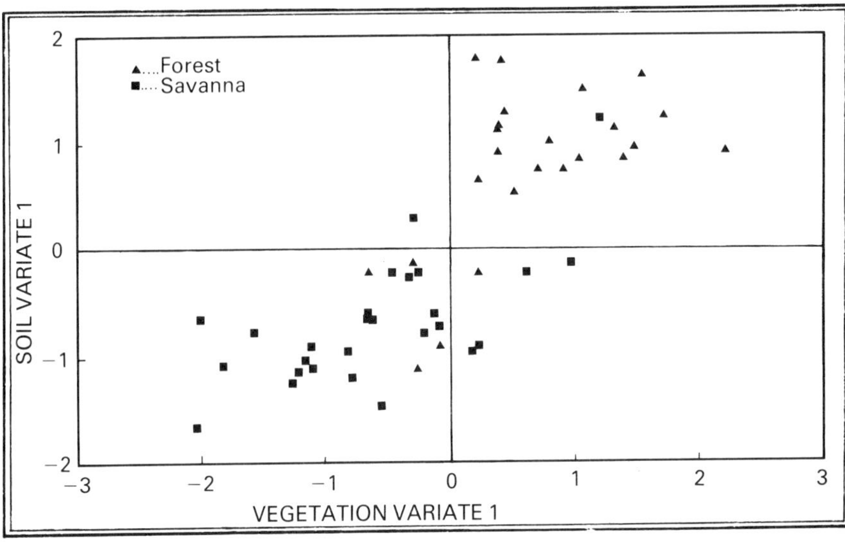

Fig. 16.5 Canonical ordination of soil and vegetation properties along the boundary.

strong relationship between the two sets of data under consideration (Gittings, 1985). Thus the distribution of the plots in relation to the first set of canonical axes shows that soil and vegetation are significantly related. The soil–vegetation complexes at the boundary are also sharply separated into forest and savanna, while a different set of soil characteristics and soil types supports the forest.

The canonical analysis does not show clearly which of soil and vegetation is the cause and which is the effect. However, it is conceivable that the soil pattern, especially as is determined by its parent rock material, existed under the climax rainforest, and that it was this that influenced the types of anthropic derivatives of the forest vegetation that developed on the various sites.

STABILITY OF THE BOUNDARY

According to Odum (1971) 'a plant, or plant community, extends its range of distribution in all directions until some detrimental aspect of the environment prevents a normal existence'. Usually, the particular deficiency in a given direction can be traced to an individual 'limiting factor'. As this factor approaches the limit of tolerance for the plant community, the existence of the latter is threatened. The 'limiting factor' is defined as 'the factor of most immediate importance in the biophysical complex on which the existence or otherwise of the plant or plant community depends' (Daubenmire, 1968).

A geographical analysis should depict a boundary and its limiting factor. If such a limiting factor is subject to change or fluctuation, then the boundary will be in dynamic equilibrium in responding to the 'factor of most immediate importance in the biophysical complex'. If, on the other hand, the limiting factor(s) are such that they do not change readily, then the associated boundary could be considered as in static equilibrium. Since regional climate is a relatively stable element of the environment, boundaries limited by specific values of mean annual rainfall or mean annual temperature should be considered to be in static equilibrium. On the other hand, boundaries limited by the intensities of human activities, which are bound to change with densities of human population, are likely to be in dynamic equilibrium.

Factor interaction and factor compensation

A limiting factor should not be interpreted as the only one providing life support for a biological phenomenon. Its relevance at that specific location is that it is the factor most likely to result in environmental failure while the other factors are still active. At the forest–savanna boundary, the occurrence of the savanna is a function of biotic, climatic and edaphic factors. The original vegetation is forest which has been changed by the process of repeated cultivation. It is hence obvious that the biotic or the human factor is a necessary ingredient of the conditions leading to the existence of the savannas. However, like any other plant community, the savanna subsists on inputs from a constellation of climatic and edaphic variables.

It is conceivable that a given magnitude of human activity will reduce an area of forest to savanna when climatic and edaphic conditions remain constant. Under such a situation, the intensity of human activities would be the limiting factor since savanna would be found where the threshold magnitude was achieved and would be absent where it was not. However, in areas that are differentiated on the basis of climatic and edaphic factors, the status of human activity as the limiting factor may be undermined as a result of factor compensation. The threshold magnitude of human activities is valid only with regard to a specific composition of the ecosystem. Thus within the same general area, where a degree of savannization has been achieved, a unit of forest may not deteriorate to savanna because of better climatic support or relatively more fertile soils. In situations such as these, it can be assumed that climate and soils have compensated for the destructive forces generated by man. More human activity would be required to convert forest to savanna, but meanwhile the boundary is in static, rather than dynamic, equilibrium.

ECOLOGICAL STATUS OF THE BOUNDARIES

The regional boundary

Climate appears to be the factor limiting the forest–savanna boundary. It is the factor that corresponds spatially and meaningfully with the ecotonal boundaries as demonstrated earlier. Thus, as is the case with the boundaries between savanna vegetation zones throughout the world (Adejuwon, 1968), the forest–savanna boundary is basically bioclimatic and specifically limited by the elements of water balance. One can thus observe that the change represented by the forest–savanna ecotone is inherent in the natural ecological system, even though its manifestation has had to await human intervention.

If climate is the limiting factor of the forest–savanna ecotone, then the boundary can be considered as being in static equilibrium. Any substantial migration of the ecotone will be expected to follow similar changes in climate, and since climatic change is measurable in terms of natural rather than human history, the linear approximation of the ecotone could be deemed to be in equilibrium owing to the characteristics of the climate. Unfortunately, few ecological workers have appreciated this. Much effort has been directed at the question of whether the savanna in a given area is anthropic, edaphic or climatic. The preceding paragraphs indicate that the fact that a unit of savanna is anthropic in origin does not necessarily imply that the boundary will be determined by the intensity of human activities.

In Nigeria, the forest–savanna boundary has remained stable in historical times. The records of early explorers (Clapperton, 1829) indicate that, in general, no important changes have taken place in the position of the boundary over the past 200 years. Morgan and Moss (1965), making observations in respect of the same area, arrived at similar conclusions. They noted that the boundary is relatively stable, both in general and in detail. They also contended that biotic factors in the form of cultivation were not at present capable of effecting any rapid degeneration of forest, despite an increase in population. They, however, did not explain why such degeneration was not taking place. Such an explanation would have inevitably led them to the limiting factor, which has been identified here mainly as climatic water supply.

The topographical boundary

The observation by Morgan and Moss (1965) that the forest–savanna boundary is relatively stable both in general and in detail has not been fully substantiated. The 'boundary . . . in general' is the linear approximation of the ecotone and refers to the regional boundary between forest and savanna. The 'boundary . . . in detail' means the patterns within the ecotone and refers to the boundaries between the individual units of savanna and the surrounding forest vegetation. It is the regional boundary or the 'boundary in general' that is limited by climate and judged to be stable. The limiting factors of the topographical or the 'detail' boundaries need to be similarly identified and discussed.

It has been established that the distribution of forest and savanna within the ecotone is such that savanna is located on the poorer soils while forests are found on the relatively richer soils. Also, ordination techniques have given us two sharp soil vegetation complexes based on the forest and savanna dichotomy without an intermediate or transitional complex.

The limiting factor of the topographical boundary between forest and savanna is, without doubt, edaphic. If soil is the limiting factor at the topographical boundary, then the boundary can be judged to be in static equilibrium in as much as the soil is stable with regards to its basic structure and characteristics. Even though some soil characteristics change as a result of human intervention or climate (e.g. waterlogging in American savannas), such changes extend as far as the coverage of the intervention concerned. Any vegetation discontinuity resulting from such changes are more likely to be referred to human activities rather than to consequent soil characteristics.

Thus, as in the case of the regional boundary, the change represented by the topographical boundary between forest and savanna is inherent in the natural ecological system even though its manifestation has had to await human intervention. We have therefore come to agree with Morgan and Moss (1965) that the forest–savanna boundary is stable in 'detail'.

TRANSITIONAL VEGETATION AT A STABLE VEGETATION BOUNDARY

It has been suggested that some of the units of the ecotonal vegetation classified as forest might be regarded as a successional stage between savanna and forest. Such forest units include, in their tree layer, typical savanna species including *Daniellia oliveri* and *Parkia filicoides* along with typical forest species such as *Funtumia elastica* and *Chlorophora excelsa*. The forests are generally similar to the transitional woodlands of Keay

(1959b) and the *Anogeissus* woodlands of Clayton (1958). Even though these forests include typical savanna tree species, like other forests they lack a grassy ground layer.

The idea of a transitional vegetation at a vegetation boundary in static equilibrium appears to be a classical example of a contradiction in terms. If there are vegetation units within the ecotone, that could be described as successional stages between forest and savanna, then the concept of static equilibrium at the boundary must be abandoned. At the best we can speak of a boundary that has a large measure of stability. For example, while the ecotonal savannas on the poorest soils and their boundaries may be permanent, those on the relatively better sites and their boundaries may be subject to obliteration during a pluvial period or as a result of a decrease in the intensity of human activity. Such changes can probably be accommodated in the concept of a 'relatively stable boundary'.

There is, however, another interpretation of the ecological status of the so-called transitional vegetation which may be compatible with a stable vegetation boundary. The forest units which include elements of savanna might have acquired such elements through local seed dispersal. Similarly, areas of savanna with elements of forest vegetation could have benefited from the abundant supply of forest propagules nearby. Such interactions may occur given the way units of forest and savanna are mixed within the ecotone. Moreover, such typical savanna species as *Albizzia ferruginea*, *Anogeissus leiocarpa* and *Daniellia oliveri* may find forest areas that are occasionally opened up through cultivation more conducive for growth and development than a savanna environment that is annually ravaged by grass fires. Thus the dynamics of the forest–savanna boundary may include transformations in the form of exchange of genetic materials, but which do not necessarily lead to a change of forest to savanna or vice versa.

Another important point to note is the fact that savanna and forest are defined on the basis of physiognomy, that is, the presence or absence of a grassy ground layer. The concept of a floristically defined transitional vegetation between the two may hence be illogical and inappropriate especially when the floristic composition considered is that of the tree layer. Thus structural stability may not be incompatible with floristic fluctuations.

CONCLUSIONS

It appears, therefore, that a particular point in the bioclimatic profile from north to south in Nigeria separates (to the south) areas where primary biological productivity is enough to maintain a forest vegetation in the face of intensive food crop cultivation and areas (to the north) where it is not. At the local level, there is an amount of factor compensation based on the soil. Thus south of the linear approximation of the forest–savanna boundary, poor soils permit the occurrence of enclaves of savanna, while to the north, very good soils allow forest to flourish. This appears to be the basis for the creation of an ecotone which consists of 'islands, salients and enclaves of forest and savanna sharply separated from one another'. The main regional boundary, that is the ecotone, which is in static equilibrium with the climate, is relatively stable. Also within the ecotone, the topographical boundaries separating the units of savanna from the units of forest are also relatively stable, as they are limited by elements of the physical environment which are not subject to dramatic changes. However, within the ecotone, normal autogenic and allogenic influences result in species movement between the two main vegetation units which may not necessarily involve marginal gains or losses in areal coverage.

REFERENCES

Adejuwon, J.O. (1968) *Savanna Patches in the Forest Areas of Western Nigeria: The Place in the Dynamics of Nigerian Vegetation*, Ph.D. thesis, University of London.

Adejuwon, J.O. (1971) Savanna patches within forest areas in Western Nigeria: A study of the dynamics of forest/savanna boundary. *Bulletin d'Institut Fondamentale d'Afrique Noir*, **33 Ser. A**, 237–344.

Bourlière, F. and Hadley, M. (1983) Present day savannas: An overview, in *Ecosystems of the World, Tropical Savannas, 13*; (ed. F. Bourlière), Elsevier, Amsterdam, pp. 1–17.

Clapperton, H. (1829) *Journal of a Second Expedition into the Interior of Africa*, London.

Clayton, W.D. (1958) Secondary vegetation and the transition to savanna near Ibadan, Nigeria. *Journal of Ecology*, **46**, 217–38.

Daubenmire, R.F. (1968) *Plant Communities*, Harper & Row, London.

Gittings, R. (1985) *Canonical Analysis*, Springer-Verlag, Berlin.

Hill, M.O. (1979a) *TWINSPAN – a FORTRAN Programme for Arranging Multivariate Data in an Ordered Two-way Classification of Individuals and Attributes*, Ecology and Systematics, Cornell University, Ithaca, USA.

Hill, M.O. (1979b) *DECORANA – a FORTRAN Programme for Detrended Correspondence Analysis and Reciprocal Averaging*, Ecology and Systematics, Cornell University, Ithaca, USA.

Hopkins, B. (1974) *Forest and Savanna*, Heinemann, Ibadan.

Keay, R.W.J. (1959a) *An Outline of Nigerian Vegetation*, Government Printer, Lagos.

Keay, R.W.J. (1959b) Derived savanna: derived from what? *Bulletin d'IFAN*, **Ser. A, 21**, 427–38.
Morgan, W.B. and Moss, R.P. (1965) Savanna and forest boundary in western Nigeria. *Africa*, **35**, 286–94.
Odum, E.P. (1971) *Fundamentals of Ecology*, 3rd edn, W.B. Saunders.
SPSSX (1986) *Statistical Package for Social Sciences*, McGraw-Hill, Wokingham.

17
Effects of trees on understorey vegetation and soils at forest–savanna boundaries

A. JOY BELSKY and ROBERT G. AMUNDSON

SUMMARY

1. Changes in the biotic and abiotic environments that occur at the boundaries between forests (or woodlands) and savannas are caused not only by discontinuities in the physical environment, but also by the trees themselves. If the boundaries result from the destruction of forests by fire or large animals, the trees may be responsible for most of the environmental discontinuities occurring at the boundary.
2. By intercepting rain and solar radiation, trees reduce the amount of water and light reaching the ground layer. Consequently, average soil temperatures and potential evapotranspiration are reduced linearly from grassland to forest along a gradient of increasing tree density.
3. Average herbaceous-layer productivity changes curvilinearly with increasing tree density. At low tree densities, trees increase herbaceous-layer productivity of the savanna by increasing soil fertility and improving the water balance of plants growing in the shade. At higher densities, trees reduce herbaceous-layer productivity by sequestering nutrients in standing woody biomass and by competing with understorey vegetation for light, nutrients and water.
4. As tree densities increase, grazing animals are replaced by browsing animals. Since these two types of herbivores consume different plant species, they may be responsible for many of the changes in the vegetation at forest–savanna boundaries.
5. Isolated trees in savannas facilitate the growth of forest species under their canopies, thereby acting as 'regeneration nuclei' for the expansion of forest vegetation into savannas.

INTRODUCTION

Biotic and abiotic environments change abruptly at the boundary between forests and savannas. These boundaries are often caused by steep gradients or sharp discontinuities in water availability or geological substrate; but they may also result from the destructive effects of fire or animal activity on forest vegetation. In cases where the boundary is created by steep physical gradients, the environmental changes are due both to the physical factors that create the boundary and to the boundary trees themselves. In the case of fire- and animal-created boundaries, the changes are due almost entirely to the effects of the trees. Because the effects of trees on forest–savanna boundaries have received little attention in the literature and because these effects can be confounded with the physical discontinuities forming the boundaries, we wish to draw attention to the many ways that trees alter their environment. We will concentrate on the effects of trees in East African savannas on understorey vegetation, microclimate and soil properties. Since few studies of changes across savanna–forest boundaries have been made, we will mostly extrapolate from what is known about the effects of individual trees on their environment in East Africa and discuss how these effects may change along an increasing gradient of tree density. We will also discuss whether physical and biological variables of the savanna respond linearly with increasing tree density and whether isolated trees in the savannas of East Africa are realistic models for the investigation of communities with higher tree densities.

FORESTS AND SAVANNAS IN EAST AFRICA

East Africa, owing to its aridity, supports few true forests besides those along water courses and in mountains. Instead, most of the vegetation of East Africa can be described as grassland, savanna, woodland, shrub thicket or barren land (Pratt and Gwynne, 1977). Savannas are differentiated from grasslands by having trees, and from woodlands by having discontinuous rather than continuous canopy strata. In areas with rainfall sufficient to support tree growth, communities fluctuate temporally between grasslands, savannas and woodlands (Belsky, 1990). Periodic fires and elephants convert woodlands to savannas and grasslands, which succeed back to woodlands when fire and browsing pressures are reduced (Buechner and Dawkins, 1961; Laws, 1970; Norton-Griffiths, 1979; Pellew, 1983; Belsky, 1984, 1989). A wide variety of plant community types exists within the savanna biome, changing along rainfall, topographic and edaphic gradients. Any point along these gradients, however, may be vegetated by grassland, savanna or woodland, depending on the length of time since the last fire and the history of elephant disturbance. In conditions such as these, the physical environment may be relatively homogeneous except for the changes produced by the trees themselves.

EFFECTS OF TREES ON TREE UNDERSTOREY ENVIRONMENT

Effects of increasing tree density on microclimate

As tree density increases along a gradient from grasslands to woodlands, their canopies intercept increasing amounts of rain and sunlight, reducing the average rainfall and solar radiation reaching the herbaceous layer (Pressland, 1973) (Fig. 17.1). As a result, average soil temperatures and potential evapotranspiration of the herbaceous layer are also reduced. Although the overall decreases in temperature and evapotranspiration are linear along the tree density gradient, the pattern is heterogeneous since trees create discrete light- and rain-shadows on the ground. At midday in Tsavo National Park in Kenya, for example, solar radiation is reduced by 55% and soil temperatures by 8–11 °C in the shade of isolated *Acacia tortilis* trees compared to nearby unshaded areas. In addition, rainfall may be reduced from 0 to 100% under tree canopies depending on the duration and intensity of the event. These light- and rain-shadows are not fixed: light-shadows move west to east daily, and north to south seasonally (Fig. 17.2); and rain-shadows move with changing wind direction. As tree density increases, the probability that a patch of herbaceous vegetation will be shaded or receive reduced throughfall increases.

Effects of trees on understorey productivity

Although herbaceous-layer productivity has been found to decrease with increasing tree density in Australia (Beale, 1973; Walker *et al.*, 1986), average herbaceous-layer productivity in arid regions of Africa is higher in savannas with low densities of trees and shrubs than in treeless grasslands or in savannas with high tree densities (Fig. 17.3) (Kennard and Walker, 1973; Sanford *et al.*, 1982). This higher herbaceous-layer productivity is a result of grassland productivity being significantly higher under the canopies of isolated trees than in nearby grasslands (Christie, 1975; Pratchett, 1978; Georgiadis, 1989; Belsky *et al.*, 1989; Weltzin and Coughenour, 1990). In Tsavo National Park, Kenya, herbaceous-layer

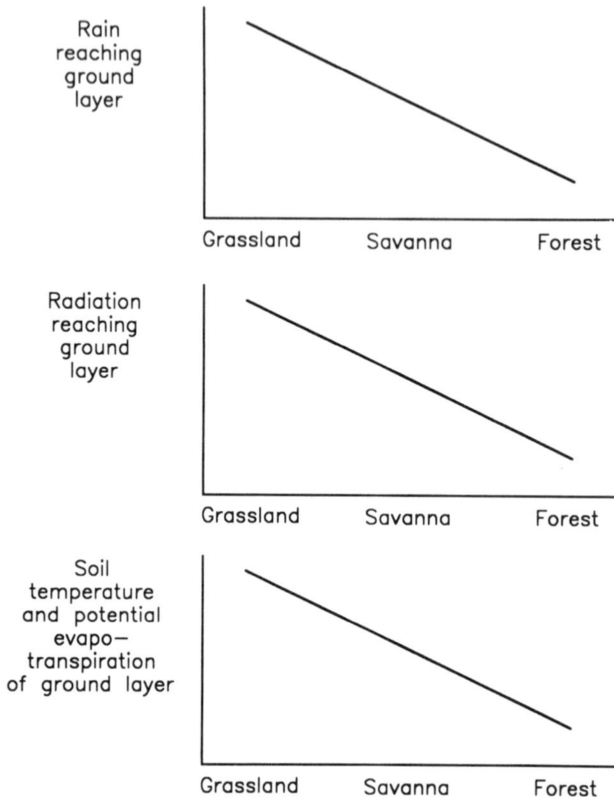

Fig. 17.1 Relationships between climatic variables and increasing tree density.

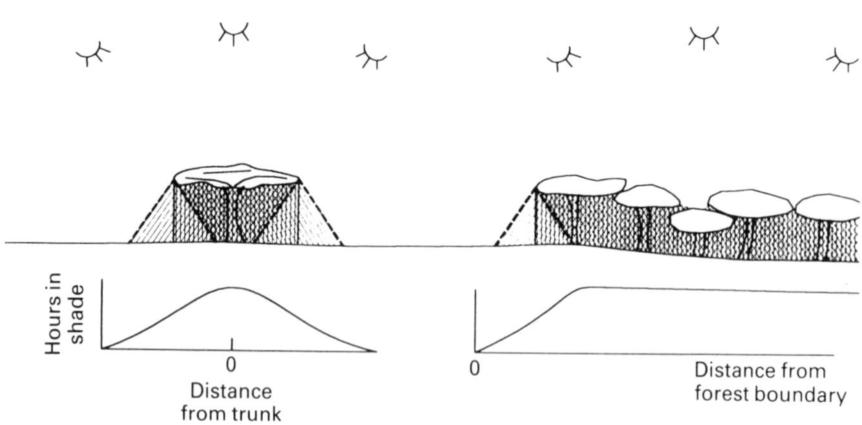

Fig. 17.2 Effects of tree density on ground-level radiation.

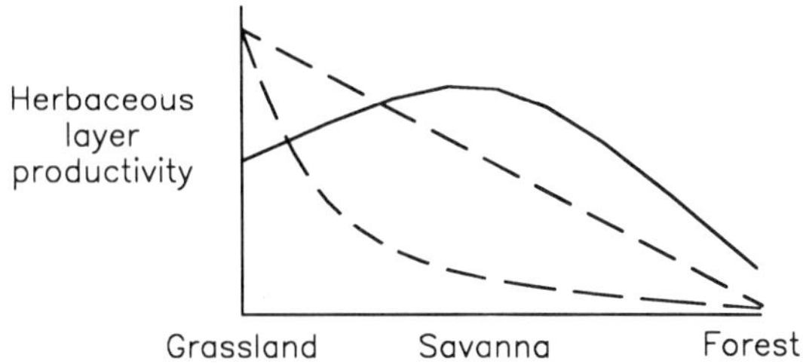

Fig. 17.3 Relationship between herbaceous-layer productivity and increasing tree density in East African (—) and Australian (---) savannas. The Australian curves are from Walker *et al.* (1986).

productivity is nearly twice as high under the canopies of mature *Acacia tortilis* (umbrella thorn acacia) and *Adansonia digitata* (baobab) trees than in the surrounding open grassland in a low-rainfall site (450 mm annual rainfall) and 30% higher under these same tree species in a high-rainfall site (750 mm annual rainfall) (Belsky *et al.*, 1989, 1992). Since increased productivity under tree canopies has not been associated with reduced productivity beyond the edge of the canopies (Belsky *et al.*, 1989), it can be concluded that the herbaceous layer in communities containing trees is more productive than in communities without trees.

The trend for higher understorey productivity with increasing numbers of trees begins to reverse itself at some (unknown) tree density, at which point understorey productivity begins to decline. Consequently, herbaceous-layer productivity of more densely wooded savannas has often been found to be lower than that of grasslands (Fig. 17.3) (Ward and Cleghorn, 1964; Grossman *et al.*, 1980; Dye and Spear, 1982). In the Irima district of Tsavo National Park, the herbaceous-layer productivity in a wooded thicket that had been protected from fire and large mammals for 17 years was less than half that of the grassland outside the exclosure (Table 17.1). Similar reductions in herbaceous-layer productivity in woodlands compared to grasslands have been found in other parts of East Africa (Dye and Spear, 1982).

Causes of changes in understorey productivity along a tree-density gradient

The initial increases in herbaceous-layer productivity along a tree density gradient (Fig. 17.3) can be attributed to a number of factors. First, one of us (A.J.B.) has observed that the species growing under tree canopies are often larger and more deeply rooted than the species growing in the open grassland. In fact, shrubs often dominate this tree understorey habitat.

Table 17.1 Differences in herbaceous-layer productivity and soil properties inside a wooded plot that had been protected from fire and large herbivores for 17 years and in adjacent unprotected grassland plots in Irima District, Tsavo National Park, Kenya (standard error in parentheses)

	Wooded plot	*Grassland plot*	*Probability*
Herbaceous-layer productivity ($g\,m^{-2}$)	280 (88)	648 (147)	<0.05
Infiltration rate ($mm\,sec^{-1}$)	0.65 (0.18)	0.11 (0.04)	<0.05
Organic matter (%)	2.61 (0.17)	2.46 (0.40)	NS
pH (log units)	6.80 (0.15)	7.14 (0.07)	<0.005
Available phosphorus ($mg\,kg^{-1}$)	0.55 (0.16)	0.48 (0.17)	NS
Available potassium ($mg\,kg^{-1}$)	260 (28)	222 (11)	NS
Available calcium ($mg\,kg^{-1}$)	985 (73)	1310 (40)	<0.05
Available magnesium ($mg\,kg^{-1}$)	151 (9)	145 (9)	NS
Available manganese ($mg\,kg^{-1}$)	5.03 (0.88)	2.94 (0.44)	NS

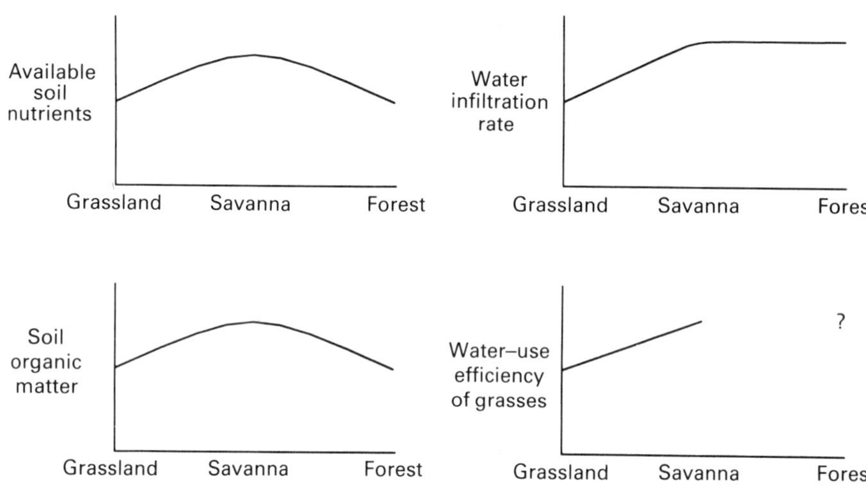

Fig. 17.4 Effects of tree density on soil properties and grass water-use efficiency.

Second, soil properties are usually improved by the presence of trees. Available soil nutrients, including nitrogen, phosphorus potassium and calcium, are nearly always higher under the canopies of isolated trees than in the open grassland (Fig. 17.4) (Kennard and Walker, 1973; Tiedemann and Klemmedson, 1973, 1977; Kellman, 1979; Bernard-Reversat, 1982; Belsky *et al.*, 1989, 1992; Georgiadis, 1989). The sources of the nutrients that enrich the soil under tree canopies have not been thoroughly investigated, but most likely they are transported into the canopy zone by a combination of (a) tree roots that absorb nutrients over a large soil volume and concentrate them into the canopy zone, (b) birds that feed in the grassland but perch in the trees, and (c) large mammals that graze over large areas but rest in the shade of the trees. The nutrients reach the soil in the form of tree litter and animal droppings. In addition, soils under savanna trees also have increased levels of soil organic matter, greater total cation exchange capacities, higher water infiltration rates, higher microbial biomasses and greater water-holding capacities than soils in adjacent herbaceous communities (Fig. 17.4) (Kennard and Walker, 1973; Belsky *et al.*, 1989, 1992; Young, 1989). Third, shade from canopies reduces soil and plant temperatures and evapotranspiration, increasing the moisture of surface soil horizons and possibly improving the water balance of plants in arid environments (Kennard and Walker, 1973; Belsky *et al.*, 1989; Weltzin and Coughenour, 1990). Finally, understorey grass species have been found to be more water-use efficient than species that grow in the open (R.G. Amundson and A.J. Belsky, unpublished).

The reduction of herbaceous-layer productivity that occurs at higher tree densities may also be due to several factors. First, even though isolated trees are associated with increased soil fertility in savannas, concentrations of available soil nutrients in woodlands are often equal to or lower than those found in open grasslands or sparsely wooded savannas (Harrington and Ross, 1974; Trapnell *et al.*, 1976; Brookman-Amissha *et al.*, 1980). In the small protected woodland at Irima in Tsavo National Park, none of the soil nutrients was significantly higher in the woodland than in the open grassland (Table 17.1) and calcium was significantly lower in the woodland. Similar results were also found in long-term burned (grassland) and unburned (wooded) plots in Zambia (Trapnell *et al.*, 1976) and in forest soils compared to grassland soils on Mount Meru in Tanzania (Lundgren and Lundgren, 1972). This reduced fertility in wooded plots may be due to nutrients being tied up in the standing woody biomass of the trees or to reductions in densities of birds and mammals (and their droppings) in forests.

A second reason that herbaceous-layer productivity may decline at higher tree densities is that the intensity of solar radiation is reduced during most of each day, even under trees with sparse canopies. In contrast, plants growing under isolated trees in savannas are in full sunlight during part of each day in some seasons and during the entire day in other seasons (Fig. 17.2) (Obot, 1988). Similarly, the continuous overhead canopy of woodlands may significantly reduce the amount of rainfall reaching the herbaceous stratum.

Finally, it is probable that woody and herbaceous species compete for water and nutrients in woodlands (Walker, 1971; Walker *et al.*, 1981;

Walker and Noy-Meir, 1982). Knoop and Walker (1985) found in tree- and grass-removal experiments in a South African savanna that acacia trees and their understorey grasses reciprocally reduced each other's growth. And in a high rainfall site in Tsavo National Park, Kenya, experimental trenching to sever tree roots significantly increased herbaceous productivity in plots under the canopies of *Acacia tortilis* (A.J. Belsky, unpublished data). Both studies suggest that trees in African savannas compete with herbaceous species. But competition does not always occur. In a low rainfall site in Tsavo National park, trenching around herbaceous plots had no effect on understorey productivity, suggesting that trees in this site did not reduce the resources available to understorey species. Further research revealed that competition between grasses and trees in Tsavo could be predicted by the location of fine roots of the trees. In the low rainfall site, tree roots extended far into the grassland where their fine, absorptive roots were widely dispersed and their effects on herbaceous species were slight. In the high rainfall site where trenching increased herbaceous growth, tree roots were concentrated near the boles of trees where they were dense enough to reduce available resources substantially. These results suggest that competition between trees and their understorey species increases with increasing tree-root density. At higher tree densities, tree roots extend into all available space, including under and around neighbouring trees (Fig. 17.5). This competition may explain the low productivity of the herbaceous layer in sparse woodlands where sunlight appears to be sufficient.

RELATIONSHIP BETWEEN INCREASING TREE DENSITY AND ANIMALS

As grasslands grade into woodlands, browsing mammals, which modify the composition of woody vegetation, replace grazing animals (Lamprey, 1963), which modify the composition of herbaceous vegetation. Browsers

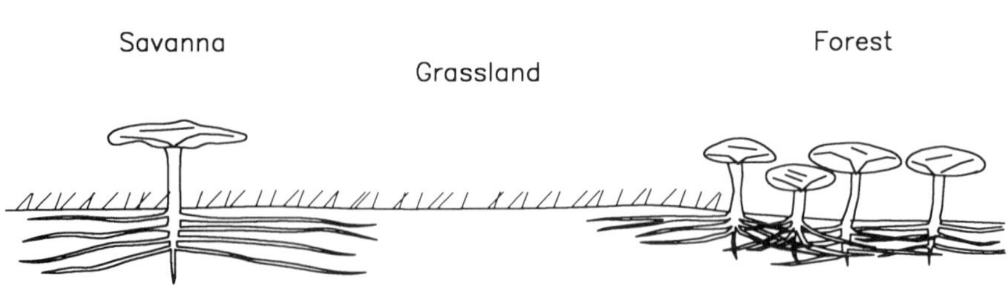

Fig. 17.5 Tree-root distribution in savannas and woodlands.

such as elephants (*Loxodonta africana*), giraffe (*Giraffa camelopardalis*) and eland (*Taurotragus oryx*) push over mature trees, denude and kill small trees, and prevent saplings from growing tall enough to escape the destructive effects of grass fires (Norton-Griffiths, 1979; Pellew, 1983). As a result, browsers alter the composition and density of shrub and tree strata in savannas and woodlands. Grazers such as wildebeest (*Connochaetes taurinus*), zebra (*Equs burchelli*) and hartebeest (*Alcelaphus buselaphus*) alter herbaceous communities by reducing the abundance of palatable species and the dominance of taller species (Belsky, 1986). As a result, palatable species become less abundant and sub-dominant species more abundant in communities with high density of grazers. Grazers also alter the woody community by reducing standing plant biomass, thereby reducing the amount of fuel on the ground and, consequently, the frequency of fires. Many of the compositional changes in the plant community at savanna–forest boundaries therefore may be due to the replacement of grazers by browsers.

Large mammals may also differentially affect the physical properties of soils along tree density gradients. In grasslands and savannas where animals are not restricted in their movements, trampling by large mammals creates little disturbance. In woodlands, however, animals are restricted to the narrow spaces between trees where they create highly disturbed and compacted trails. This compaction and concentrated disturbance may also be partially responsible for the reduced herbaceous-layer productivity in woodlands.

Bird populations also vary along tree density gradients as ground-nesting birds are joined or replaced by tree-nesting species. In many parts of East Africa, trees in savannas attract and provide perches for colonially nesting, graminivorous species. As trees increase in density, the numbers of perches increase but the food supply decreases, and often bird densities decline. In an open savanna in Tsavo National Park, for example, acacia and baobab tree canopies contained 13–16 buffalo weaver (*Bulbalornis niger*) colonies per tree (Belsky *et al.*, 1989), but in a nearby and more heavily wooded savanna, the same tree species supported only 3–5 weaver colonies per tree. Nutrient enrichment of the soil beneath individual tree canopies due to bird droppings can therefore be expected to decline along a gradient from open savanna to woodland.

ISOLATED TREES AS REGENERATION NUCLEI

In the savannas of the New World, isolated trees have been found to create specialized habitats that facilitate the dispersion of woodland and forest species into savannas and subsequently facilitate the formation of forest patches (Kellman, 1979; Guevara *et al.*, 1986; Archer *et al.*, 1988). In these studies Kellman (1979) speculated that the enriched soil fertility found under savanna trees provided a suitable substrate for the invasion and growth of forest species; and Archer *et al.* (1988) and Guevara *et al.* (1986) speculated that isolated savanna trees provided perches for birds that bring seeds of woody species into the savanna. The isolated trees therefore act as 'regeneration nuclei' for the development of small islands of woodlands or forests that later coalesce into larger forest patches. This conversion of isolated savanna trees into woodland clumps has not been reported in East Africa, possibly because fire and browsing animals prevent the seedlings of sub-canopy woody species from growing large enough to enter the canopy stratum. But during periods of low fire frequency and low browser densities, savanna trees may also act as nuclei for the regeneration of woodlands.

CONCLUSION

In this chapter we asked how trees themselves may be creating the environmental discontinuities found at forest–savanna boundaries. Owing to biological processes, herbaceous-layer productivity, densities of shade-intolerant species, soil nutrient concentrations and soil temperatures can be expected to be higher on the savanna side of the boundary; and surface soil moisture and densities of shade-tolerant species can be expected to be higher on the forested side of the boundary. Consequently, before the environmental changes found at forest–savanna boundaries are attributed to physical gradients such as soil type or depth of water table, biological factors must be considered.

We have asked whether the continuum in tree densities from grassland to woodland (or forest) was associated with continua in other biotic and abiotic factors. Although microclimatic factors such as average available throughfall, solar radiation and soil temperatures decline along this continuum (Fig. 17.1), herbaceous-layer productivity (Fig. 17.3), available soil nutrients and water-infiltration rates (Fig. 17.4) reach a maximum at intermediate tree densities. The increase in herbaceous-layer productivity at low tree densities is mostly likely a result of nutrient enrichment, improved water infiltration and the amelioration of the hot, arid climate by the tree shade. The decline in herbaceous-layer productivity and nutrient availability at higher tree densities is most likely due to increased canopy interception of rainfall and sunlight, increased belowground competition from trees, and decreased soil nutrient availability because nutrients are tied up in standing woody biomass.

The curvilinear relationship between tree density and a few of the biological and physical properties of the environment suggests that the investigation of isolated trees in savannas may not allow the quantitative extrapolation of their effects to that of closely spaced trees in woodlands and forests. However, the study of isolated trees provides insights into how trees alter the biological, chemical and physical properties of plant communities and suggests how they may qualitatively be affect the forest–savanna boundary.

ACKNOWLEDGEMENTS

We thank C. Jensen for reviewing an early draft of this chapter. We also thank the Government of Kenya, the Director of the Department of Wildlife Conservation and Management and the Senior Wardens of Tsavo National Park for permission to carry out research in Tsavo National Park. Travel to Glasgow, Scotland, and research for this paper were supported by the US National Science Foundation (BSR-8516982).

REFERENCES

Archer, S., Scifres, C., Bassham, C.R. and Maggio, R. (1988) Autogenic succession in a subtropical savanna: conversion of grassland to thorn woodland. *Ecological Monographs*, **58**, 111–27.

Beale, I.F. (1973) Tree density effects on yields of herbage and tree components in South West Queensland mulga (*Acacia aneura* F. Muell.) scrub. *Tropical Grasslands*, **7**, 135–42.

Belsky, A.J. (1984) The role of small browsing mammals in preventing woodland regeneration in the Serengeti National Park, Tanzania. *African Journal of Ecology*, **22**, 271–9.

Belsky, A.J. (1986) Revegetation of artificial disturbances in grasslands of the Serengeti National Park, Tanzania. II. Five years of successional change. *Journal of Ecology*, **74**, 937–51.

Belsky, A.J. (1989) Landscape patterns in a semi-arid ecosystem in East Africa. *Journal of Arid Environments*, **17**, 265–70.

Belsky, A.J. (1990) Tree/grass ratios in East African savannas: a comparison of existing models. *Journal of Biogeography*, **17**, 483–9.

Belsky, A.J., Amundson, R.G., Duxbury, J.M. *et al.* (1989) The effects of trees on their physical, chemical and biological environments in a semi-arid savanna in Kenya. *Journal of Applied Ecology*, **26**, 1005–24.

Belsky, A.J., Mwonga, S.M., Amundsen, R.G. *et al.* (1992) Comparative effects of isolated trees in their undercanopy environments in high- and low-rainfall savannas. *Journal of Applied Ecology*, in press.

Bernard-Reversat, F. (1982) Biogeochemical cycles of nitrogen in a semi-arid savanna. *Oikos*, **38**, 321–32.

Brookman-Amissah, J., Hall, J.B., Swaine, M.D. and Attakorah, J.Y. (1980) A reassessment of a fire protection experiment in north-eastern Ghana Savanna. *Journal of Applied Ecology*, **17**, 85–99.

Buechner, H.K. and Dawkins, H.C. (1961) Vegetation change induced by elephants and fire in Murchison Falls National Park, Uganda. *Ecology*, **42**, 752–66.

Christie, E.K. (1975) A note on the significance of *Eucalyptus populnea* for buffel grass production in infertile semi-arid rangelands. *Tropical Grasslands*, **9**, 243–246.

Dye, P.J. and Spear, P.T. (1982) The effects of bush clearing and rainfall variability on grass yield and composition in south-west Zimbabwe. *Zimbabwe Journal of Agricultural Research*, **20**, 103–18.

Georgiadis, N.J. (1989) Microhabitat variation in an African savanna: effects of woody cover and herbivores in Kenya. *Journal of Tropical Ecology*, **5**, 93–108.

Grossman, D., Grunow, J.O. and Theron, G.K. (1980) Biomass cycles, accumulation rates and nutritional characteristics of grass layer plants in canopied and uncanopied subhabitats of *Burkea* savanna. *Proceedings of the Grassland Society of South Africa*, **15**, 157–61.

Guevara, S., Purata, S.E. and van der Maarel, E. (1986) The role of remnant forest trees in tropical secondary succession. *Vegetatio*, **66**, 77–84.

Harrington, G.N. and Ross, I.C. (1974) The savanna ecology of Kidepo Valley National Park. I. The effects of burning and browsing on the vegetation. *East African Wildlife Journal*, **12**, 93–105.

Kellman, M. (1979) Soil enrichment by neotropical savannas trees. *Journal of Ecology*, **67**, 565–77.

Kennard, D.G. and Walker, B.H. (1973) Relationships between tree canopy cover and *Panicum maximum* in the vicinity of Fort Victoria. *Rhodesian Journal of Agricultural Research*, **11**, 145–53.

Knoop, W.T. and Walker, B.H. (1985) Interactions of woody and herbaceous vegetation in a Southern African savanna. *Journal of Ecology*, **73**, 235–54.

Lamprey, H.F. (1963) Ecological separation of the large mammal species in the Tarangire Game Reserve, Tanganyika. *East African Wildlife Journal*, **1**, 63–92.

Laws, R.M. (1970) Elephants as agents of habitat and landscape change in East Africa. *Oikos*, **21**, 1–15.

Lundgren, B. and Lundgren, L. (1972) Comparison of some soil properties in one forest and two grassland ecosystems on Mount Meru, Tanzania. *Geografiska Annaler*, **54** Ser. A, 227–40.

Norton-Griffiths, M. (1979) The influence of grazing, browsing, and fire on the vegetation dynamics of the Seregenti, in *Serengeti: Dynamics of an Ecosystem*, (eds A.R.E. Sinclair and M. Norton-Griffiths), University of Chicago Press, Chicago, pp. 310–52.

Obot, E.A. (1988) Estimating the optimum tree density for maximum herbaceous production in the Guinea Savanna of Nigeria. *Journal of Arid Environments*, **4**, 267–73.

Pellew, R.A.P. (1983) The impacts of elephants, giraffe, and fire upon the *Acacia tortilis* woodlands of the Serengeti. *African Journal of Ecology*, **21**, 41–74.

Pratchett, D. (1978) Effects of bush clearing on grasslands in Botswanna, in *Proceedings of the 1st International Rangeland Congress*, (ed. D.N. Hyder), Society for Range Management, Denver, Colorado, pp. 667–70.

Pratt, D.J. and Gwynne, M.D. (1977) *Rangeland Management and Ecology in East Africa*, Hodder & Stoughton, London.

Pressland, A.J. (1973) Rainfall partitioning by an arid woodland (*Acacia aneura* F. Muell.) in south-western Queensland. *Australian Journal of Botany*, **21**, 235–45.

Sanford, W.W., Usman, S., Obot, E.A. *et al.* (1982) Relationship of woody plants to herbaceous production in Nigerian Savanna. *Tropical Agriculture (Trinidad)*, **59**, 315–18.

Tiedemann, A.R. and Klemmedson, J.O. (1973) Effects of mesquite trees on physical and chemical properties of the soil. *Journal of Rangeland Management*, **26**, 27–9.

Tiedemann, A.R. and Klemmedson, J.O. (1973) Effects of mesquite trees on physical and chemical properties of the soil. *Journal of Rangeland Management*, **30**, 361–7.

Trapnell, C.G., Friend, M.T., Chamberlain, G.T. and Birch, H.F. (1976) The effects of fire and termites on a Zambian woodland soil. *Journal of Ecology*, **64**, 577–88.

Walker, B.H. and Noy-Meir, I. (1982) Aspects of stability and resilience of savanna ecosystems, in *Ecology of Tropical Savannas*, (ed. B.J. Huntley and B.H. Walker), Springer-Verlag, Berlin, pp. 577–90.

Walker, B.H., Ludwig, D., Holling, C.S. and Peterman, R.H. (1981) Stability of semi-arid savanna grazing systems. *Journal of Ecology*, **69**, 473–98.

Walker, H. (1971) Natural savannas, in *Ecology of Tropical and Subtropical Vegetation*, Oliver & Boyd, Edinburgh.

Walker, J., Robertson, J.A., Penridge, L.K. and Sharpe, P.J.H. (1986) Herbage response to tree thinning in a *Eucalyptus crebra* woodland. *Australian Journal of Ecology*, **11**, 135–40.

Ward, H.K. and Cleghorn, W.B. (1964) The effects of ring-barking trees in

Brachystegia woodland on the yield of veld grasses. *Rhodesian Agricultural Journal*, **61**, 98–107.

Weltzin, J.F. and Coughenour, M.B. (1990) Savanna tree influence in understorey vetetation and soil nutrients in northwestern Kenya. *Journal of Vegetation Science*, **1**, 325–32.

Young, A. (1989) *Agroforestry for Soil Conservation*, International Council for Research in Agroforestry, Nairobi.

18

The forest–savanna boundary on Maracá Island, Roraima, Brazil: An investigation of two contrasting transects

J. THOMPSON, J. PROCTOR, V. VIANA, J.A. RATTER and

D.A. SCOTT

SUMMARY

1. Two transects were set up across forest–savanna boundaries on Maracá Island. Transect A was 275 m long and crossed savanna in a shallow depression which had forest on the higher ground on each side. Transect B was 140 m long and, by contrast, crossed savanna on raised ground with forest on the lower areas at each side.
2. Seven vegetation zones were identified on transect A: evergreen forest, marginal woodland, low scrub (1), savanna grassland, low scrub (2), dense thicket and semi-evergreen forest. On transect B the vegetation zones were: swamp forest, tree/shrub savanna, marginal woodland and evergreen forest. The swamp forest had a prolonged high water table, the evergreen forest was drier and continuous with that of transect A.
3. Soil pits were dug along each transect in each of the vegetation zones. The soil water levels in the pits were measured for several months over the wet season.
4. Four replicate soil samples were collected at 0–10 cm depth on two sampling occasions from each vegetation zone on transect A and were analysed chemically and for particle size. Analyses were made of samples collected from soil pits on one sampling occasion on transect B and the data for the surface horizons compared with those for the 0–10 cm depth samples from transect A.

5. The savanna along transect A fitted the 'hyperseasonal' model of Sarmiento (Chapter 30) and, of soil factors analysed, the best correlations with vegetation zones were with soil waterlogging. The savanna along transect B fitted the 'seasonal' model of Sarmiento (Chapter 30) and the soils under it were mainly drier and more nutrient-poor than those under the forest.
6. Studies of temporal variation in nutrients on transect A showed that the 'hyperseasonal' savanna may have particularly low supplies of nitrogen in the wet season although soil pH and the supply of other nutrients are higher at that time.

INTRODUCTION

Maracá island (3°22′N, 61°26′W) has an area of about 110 000 ha in the Uraricoera River in Roraima State, Brazil (Fig. 18.1). The east of the island is at the zone of transition between the Amazonian rainforest and the Rio Branco–Rupununi savannas (Sarmiento, 1983) and has several examples of forest–savanna boundaries. The present study was made, in an area where such boundaries are frequent, in the north-east of the island about 1 km south of the river at Santa Rosa (Fig. 18.2).

In one locality (traversed by our transect A) the forest occurs on higher ground surrounding savanna in a shallow (up to 2.8 m below the forest) depression. In a second area about 200 m from transect A, there is a contrasting situation with savanna on raised ground up to 2.2 m higher than the forest. This area was crossed by our transect B. On both transects there was a central zone of more or less open savanna. On transect A this was 'savanna grassland' (*campo limpo*) (Sarmiento, 1983), on transect B the central zone was the more wooded 'tree/shrub savanna' (*campo sujo*) (Sarmiento, 1983).

The vegetation on both transects is likely to have been disturbed (Proctor and Miller, in preparation). There were Indians living nearby in the 19th and early 20th centuries and a farmer had a manioc grating plant between 1960 and 1980 within 200 m of the east of transect A and the north of transect B. There were remains of kitchen detritus and a flowering *Hippeastrum* plant (in 1988) within a few metres of the line of transect B, and it seems likely that there was a dwelling there sometime between 1950 and 1980. Fires must have been frequent on the transcts during the times of human occupation but there is no evidence that the savannas are caused by fire. A small fire occurred close to the transects in June 1986.

Although forest–savanna boundary soils and vegetation have been studied in detail elsewhere in the neotropics, e.g. in Belize (Furley, 1974a,b, 1976; Furley and Ratter, 1989) and Mato Grosso (central Brazil)

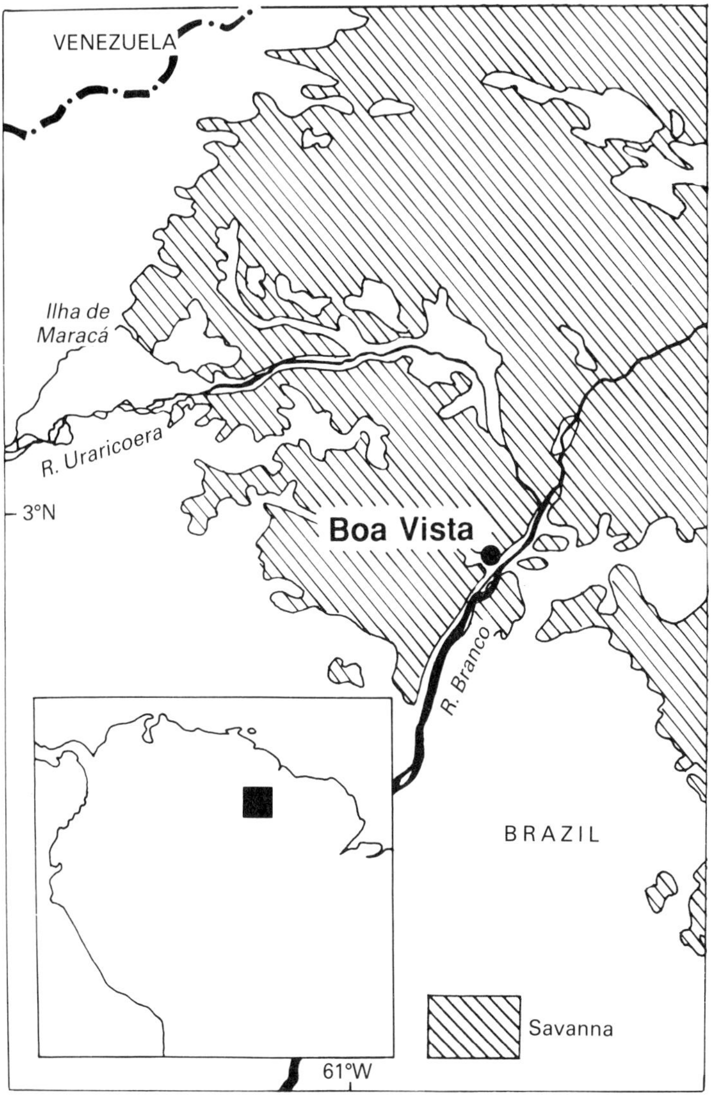

Fig. 18.1 The location of Maracá island and the forest–savanna boundary in its vicinity.

(Ratter *et al.*, 1973, 1978; Furley *et al.*, 1988), little information on vegetation and soils is available for the boundaries of the Rio Branco–Rupunini savannas. Some preliminary work has been done on the Maracá transect A by Furley and Ratter (1990), Marrs *et al.* (1991) and Milliken and Ratter (1989), (who all refer to it as the Santa Rosa transect). Ross *et al.*

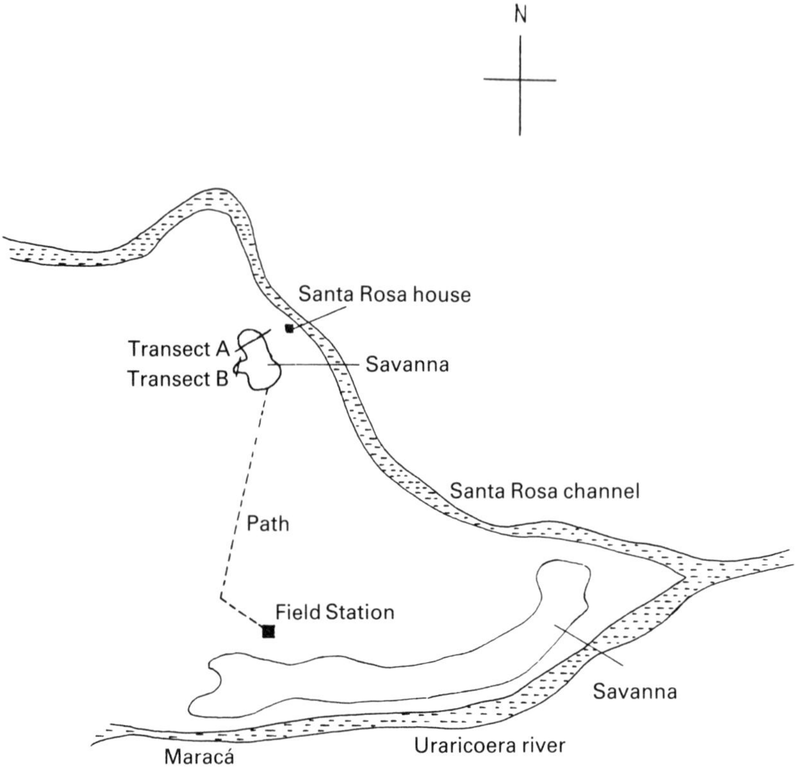

Fig. 18.2 The location of savanna at the eastern end of Maracá island, showing transects A and B.

(Chapter 8) described work on a forest–savanna transect which is about 900 m south of the transects described here. Proctor (1992) has given a brief overview of most of the ecological work on the Maracá savannas.

Our aims in this chapter are first to present the results of detailed studies on transect A and to relate them to the other work. Secondly, we compare transect A with the less intensively studied transect B. Thirdly, we attempt to explain the causes of the forest–savanna boundaries on both transects in terms of the models of Sarmiento (Chapter 30).

THE PHYSICAL ENVIRONMENT

Climate

The nearest long-term weather station to Maracá Island is in Boa Vista (Fig. 18.2), 130 km to the south-east. Records from 1923 (Instituto

Nacional de Meteorologia/Delegacia Federal de Agricultura em Roraima, unpublished) show a seasonal climate with a wet season between April and September. From 1923 until 1988, the mean annual precipitation was 1696 mm. Rainfall was recorded daily in a clearing next to the Maracá Island Field Station at 0700 hours from 20 March 1987 to 31 May 1988 in a standard rain gauge of which the upper rim was 110 cm above the ground. The total rainfall between 1 April 1987 and 31 March 1988 was 1822 mm. (The Boa Vista rainfall was 1299 mm.) Moskovits (1985) measured rainfall (2385 mm) for the same period in 1981–82 (cf. Boa Vista 1711 mm). During our study the wettest month was May 1987 with 371 mm and the driest was April 1988 with 0.1 mm. Occasional, but fairly regular, heavy rains were a feature of the Maracá dry season. Eleven days with more than 5 mm of rain occurred between 15 October 1987 and 24 March 1988 when a rainfall of 25 mm was followed by a very dry spell of 44 days, until 37 mm fell on 8 May 1988.

Temperature records were made at 0700 daily in a Stevenson screen in the Field Station clearing from 5 September 1987 to 31 May 1988. These records unfortunately do not span the wettest part of the year but they do include some very wet days. The monthly mean maxima ranged from 36.6 °C in September 1987 to 39.8 °C in April 1988, with the highest temperature of 42.0 °C recorded on 1 February 1988. There was a sharp decrease in May 1988, coinciding with the onset of the wet season, when the lowest daily maximum of 26.5 °C (9 May 1988) was recorded. Mean minimum temperatures were less variable than the maxima and ranged from 22.2 °C in March 1988 to 24.7 °C in April 1988. The lowest minimum recorded was 20.0 °C on 20, 22–24 and 28 March 1988 and the highest was 28.0 °C on 23 April 1988. Temperature data collected on Maracá by Moskovits (1985) for 1981–82 using a thermometer placed in the shade of the forest showed a similar pattern with mean monthly maxima ranging from 22.5 °C (July 1981) to 36.5 °C (December 1981) and mean monthly minima ranging from 20.5 °C (July 1981) to 26.5 °C (May 1982).

Soils

The soils of the transects are developed on drift deposits which overlie the Precambrian bedrock (Nortcliff and Robison, 1989). Eleven soil pits (1.4–2.8 m deep) were dug along transect A (Fig. 18.3) and eight (1.0–1.4 m deep) along transect B (Fig. 18.4). On both transects the pits were positioned in contrasting vegetation zones. The pits were used for the water depth measurements and those for transect B provided the only samples for the surface soils. The surface samples for transect A were collected differently, as described later. Since this chapter is largely concerned with surface samples no detailed soil profile descriptions will be

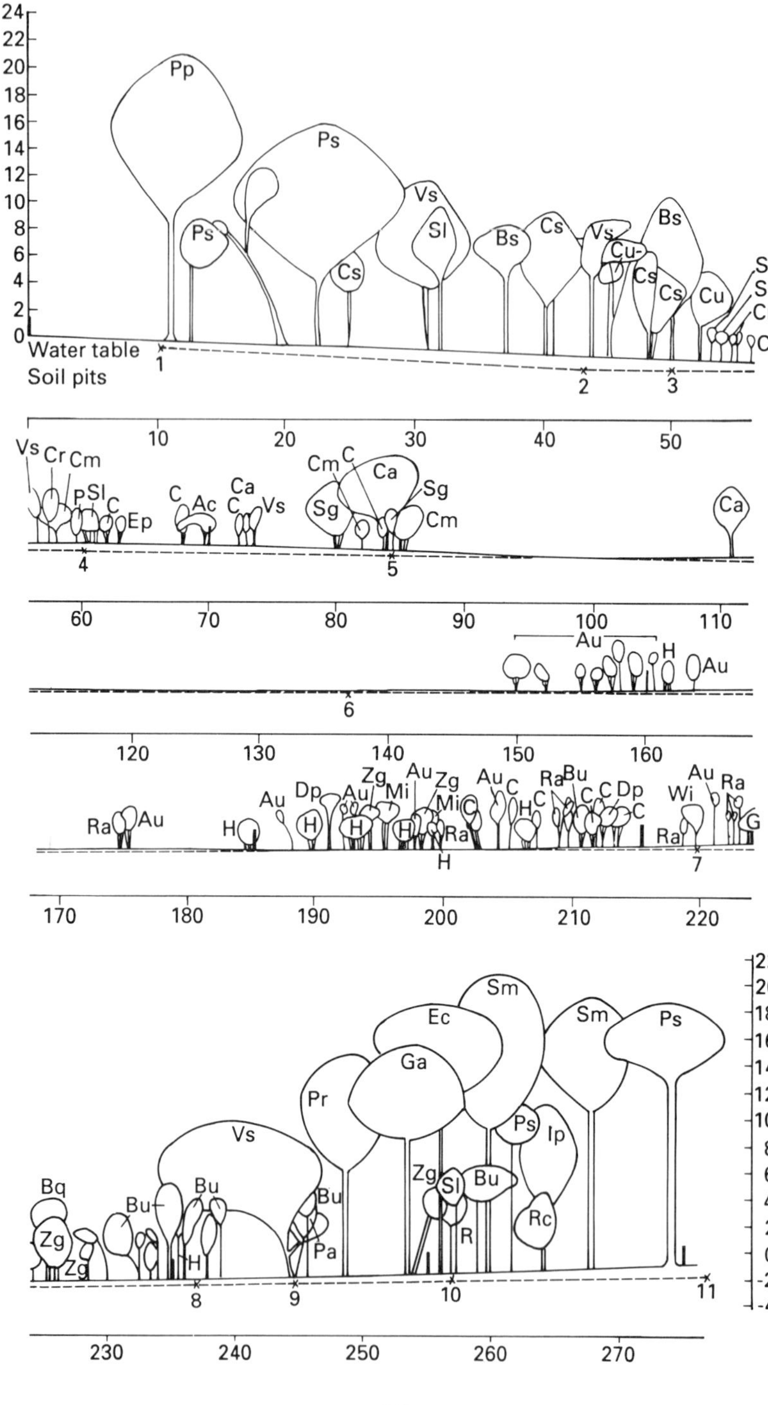
24
22
20
18
16
14
12
10
8
6
4
2
0
Pp
Ps
Ps
Vs
Sl
Cs
Bs
Cs
Vs
Cu-
Cs
Cs
Bs
Cu
Sl
Sg
Cr
Ca
Water table
Soil pits
1
2
3
10
20
30
40
50
Vs
Cr
Cm
P
Sl
C
Ep
C
Ac
Ca
C
Vs
Cm
C
Ca
Sg
Sg
Cm
Ca
4
5
60
70
80
90
100
110
Au
H
Au
6
120
130
140
150
160
Ra
Au
H
Au
Dp
Au
H
H
Zg
Mi
Au
Zg
H
Mi
C
Ra
H
Au
C
H
C
Ra
Bu
C
C
Dp
C
Au
Wi
Ra
Ra
G
7
170
180
190
200
210
220
22
20
18
16
14
12
10
8
6
4
2
0
-2
-4
Sm
Ec
Sm
Ps
Ga
Pr
Ps
Ip
Vs
Zg
Sl
Bu
Bq
Bu
Bu
Bu
Rc
R
Zg
Zg
H
Pa
8
9
10
11
230
240
250
260
270

given here. Furley (1989) described soil profiles for transect A and found that the soil textures ranged from sands to clays with the heavier soils in the middle zones of the transect. The profiles showed an angular blocky structure in the B-horizons and a poor granular structure in the surface A-horizons; the soils were relatively deep under the forests, dense thicket and marginal woodland and much more shallow under low scrub and savanna grassland. A stone-line occurred (at depths between 87 and 103 cm) under both forest zones and under the marginal woodland and low scrub zones (1); it was absent in the pits (up to 130 cm deep) under the savanna grassland, low scrub (2) and dense thicket. Gleying was more evident on the evergreen forest side where it occurred at progressively lower depths in the soil from the savanna into the forest. Profile descriptions for the pits on transect B have been made by S. Ross (unpublished). The soils have a predominantly sandy or loamy sand texture but become more silty at the wetter, southern end of the transect. They resemble those of transect A in having a weak angular blocky structure in the B-horizons and a poor granular structure in the A-horizons. Four profiles under tree/shrub savanna showed ochreous discolourations indicating iron depositions. This feature, along with distinct surface mottlings of organic matter, was absent from the other profiles.

Fig. 18.3 Profile (3 m wide) of the forest–savanna–forest transect A on Maracá Island. Woody plants ≥6 m tall or ≥3 cm *dbh* in the forest and ≥2 m tall in the savanna were recorded. The dashed line shows the water tables on 15 September 1987. The soil pits (1–11) were in the following vegetation zones: 1 and 2, evergreen forest (EF); 3, marginal woodland (MW); 4 and 5, low scrub (LS1); 6, savanna grassland (SG); 7 and 8, low scrub (LS2); 9, dense thicket (DT); 10 and 11, semi-evergreen forest (SEF). Symbols for the species are: Ac, *Arrabidaea cinnamomea* (DC.) Sandw.; Au, *Annona ulei* R.E. Fries; Bs, *Byrsonima schomburgkiana* Benth.; Bu, *Bauhinia ungulata* L.; Bq, *Bombacopsis quinata* (Jacq.) Dugand; C, *Chomelia* sp.; Ca, *Curatella americana* L.; Cm, *Cordia multispicata* Cham.; Cr, *Clusia renggerioides* Tr. & Pl.; Cs, *Casearia spinescens* (Sw.) Benth.; Cu, *Cupania rubiginosa* (Poir.) Radlk.; Dp, *Desmoncus polyacanthos* Mart.; Ec, *Enterolobium cyclocarpum* (Jacq.) Griseb.; Ep, *Eugenia punicifolia* (Kunth) DC.; G, *Gustavia augusta* L.; Ga, *Genipa americana* L. var. *caruto* (Kunth) Steyerm.; H, *Heteropterys*? sp.; Hg, *Helicteres guazumaefolia* Kunth.; Hp, *Henrietta patrisiana* DC.; Ip, *Isertia parviflora* Vahl; Mi, *Machaerium inundatum* (Mart. ex Benth.) Ducke; P, *Pavonia angustifolia* Benth.; Pp, *Peltogyne paniculata* Benth. ssp. *pubescens* (Benth.) M.F. da Silva; Pa, *Phyllanthus attenuatus* Miq.; Pr, *Pradosia surinamensis* (Eyma) Penn.; Ps, *Pouteria surumuensis* Baehni; R, *Rinorea* sp.; Ra, *Randia armata* (Sw.) DC.; Rc, *Rudgea crassiloba* (Benth.) Rob.; S, *Siparuna* sp. aff. *guianensis* Aubl.; Sg, *Swartzia grandifolia* Bong. ex Benth.; Sl, *Swartzia laurifolia* Benth.; Sm, *Spondias mombin* L.; Wi, *Waltheria involucrata* Benth.; Zg, *Zygia glomerata* (DC.) Pittier.

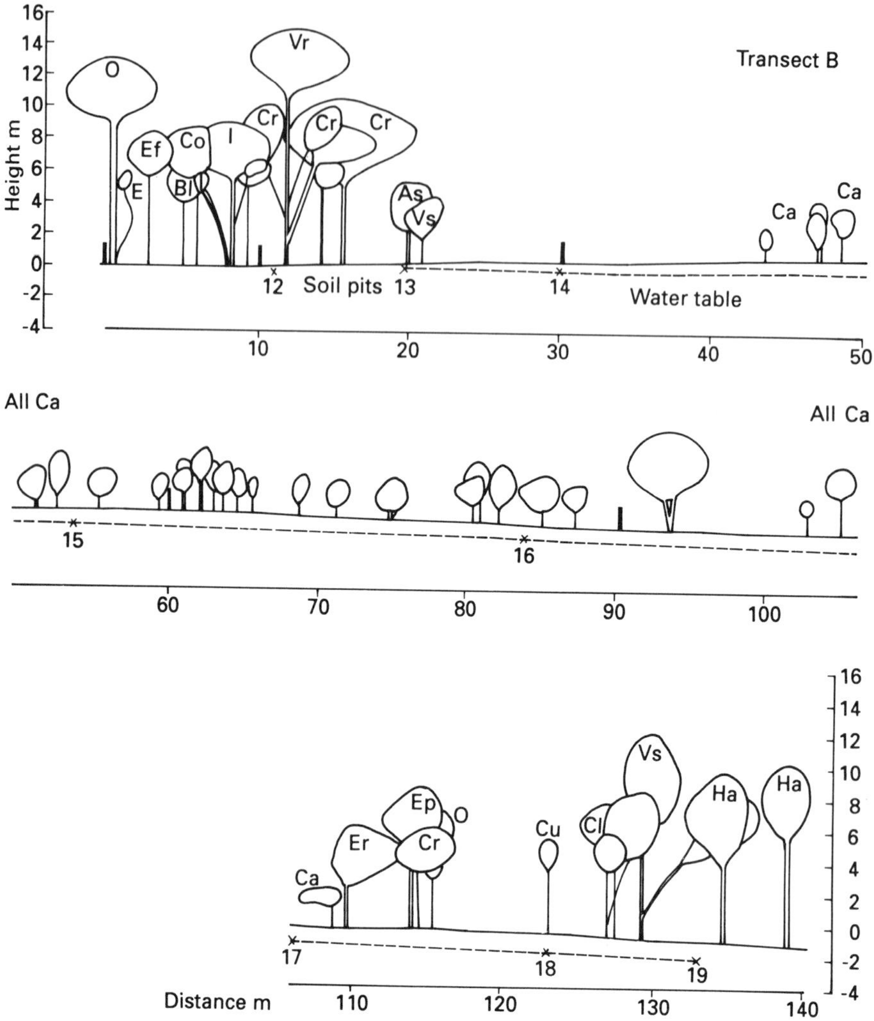

Fig. 18.4 Profile diagram (3 m wide) of the forest–savanna–forest transect B on Maracá Island. Woody plants ⩾6 m tall or ⩾3 cm *dbh* in the forests and ⩾2 m tall in the savanna were recorded. The dashed line shows the water tables on 15 September 1987. The soil pits (12–19) were in the following vegetation zones: 12, Swamp forest (SF); 13–17, tree/shrub/savanna (TS); 18, marginal woodland (MW); 19, evergreen forest (EF). Symbols for the species are: As, *Agonandra silvatica* Ducke; Bl, *Brosimum lactescens* (Moore) C.C. Berg; Ca, *Curatella americana* L.; Cl, *Casearia sylvestris* Sw. var. *sylvestris*; Co, *Cordia sellowiana* Cham.; Cr, *Clusia renggerioides* Tr. & Pl.; Cu, *Cupania rubiginosa* (Poir.) Radlk.; Dm, *Didymopanax morototoni* Decne. & Planch; E, *Erythroxylum anguifugum* Mart.; Ef, *Eugenia florida* DC. sens. lat.; Ep, *Eugenia punicifolia* (Kunth) DC.; Er, *Erythroxylum rufum* Cav.; Ha, *Himatanthus articulatus* (Vahl) Woods; I, *Inga* sp.; O, *Ormosia* sp.; Vr, *Virola surinamensis* (Rol.) Warb.; Vs, *Vitex schomburgkiana* Schauer.

Vegetation

The profile diagram (Fig. 18.3) shows that the appearance of the forest–savanna boundaries at each end of transect A differs. The evergreen forest (with trees up to about 20 m tall) adjoins a narrow zone of marginal woodland, which merges into savanna grassland through low scrub (1) which is *c*. 4 m tall and up to 15 m wide. The transition is more gradual at the next boundary, moving from savanna grassland (which has grasses up to 1 m tall at the end of the wet season) through a *c*. 90 m wide zone of low scrub (2) (*c*. 4 m tall) where the soil forms abundant small (*c* 50 cm high) hummocks. These hummocks result from the erosion associated with narrow water-runoff channels in the lowest parts of the savanna. About 15 m from the forest boundary there is a zone of dense thicket with taller (*c*. 7 m) shrubs and small trees. The semi-evergreen forest has trees up to 34 m tall. The proportion of deciduous species in the evergreen forest was not quantified, but it never had a substantially leafless appearance. It contrasted with the semi-evergreen forest which had leafless emergents in the dry season and had 41% (by basal area) of deciduous trees ($\geqslant$10 cm diameter at breast height, dbh) in a 50 × 50 m plot described by Thompson *et al*. 1992.

The forests at both ends of the transect have been described by the point centre quarter method of Cottam and Curtis (1956) by Milliken and Ratter (1989). The most important trees ($\geqslant$10 cm dbh) in the evergreen forest are: *Erythroxylum rufum* Cav. (importance value index, IVI, 31.8); *Spondias mombin* L. (IVI, 31.2); *Casearia sylvestris* Sw. var. *sylvestris* (IVI, 28.3); and *Peltogyne paniculata* Benth. ssp. *pubescens* (Benth) M.F. da Silva (IVI, 28.1). The most important trees in the semi-evergreen forest are *Apeiba schomburgkii* Szysz. (IVI, 54.1); *Ecclinusa guianensis* Eyma (IVI, 36.8); *Lecythis corrugata* Poit. ssp. *rosea* (Spruce ex Berg) Mori (IVI, 35.8) and *Guazuma ulmifolia* Lam. var. *tomentella* K. Schum (IVI, 22.8). The herbs of the savanna grassland are variable in height and density (depending on the time of year), ranging from a low covering of *Cuphea anti-syphylitica* Kunth, some species of Cyperaceae, a *Sauvagesia* species, and some *Xyris* species to a *c*. 1 m high sward of dense rank grasses. The dodder *Cassytha filiformis* L. is common. A full list of herbaceous species in the savanna grassland is given by Milliken and Ratter (1989). The grasses collected were: *Andropogon virgatus* Sesv., *Leptocoryphium lanatum* (Kunth) Nees, *Olyra longifolia* Kunth, *Panicum stenodes* Griseb., *Paspalum densum* Poir., and *Sorghastrum setosum* (Griseb.) Hitchc. The savanna grassland has scattered trees, mainly *Curatella americana* L. Floristic details of the low scrub, dense thicket, and marginal woodland are given by Milliken and Ratter (1989).

The forest–savanna boundaries on transect B (Fig. 18.4) are more abrupt, with a narrow and interrupted marginal woodland and no low scrub nor dense thicket. The grasses grow up to about 70 cm high in the tree/shrub savanna at the end of the wet season, and never reach the height of the tallest grasses on transect A. There is little floristic information available for transect B other than the profile diagram (Fig. 18.4). The swamp forest at the south end of the transect includes some 'buriti' palms (*Mauritia flexuosa* L. f.) which indicate a high water table close to their bases for much of the year. At the north end of the transect the evergreen forest is continuous with that of transect A. The tree/shrub savanna on transect B has *Curatella americana* L. as the sole woody species.

MATERIALS AND METHODS

Profile diagrams

Both transects were levelled using a clinometer and ranging pole, and profile diagrams were drawn of the vegetation in a 3 m wide strip along transect A (Fig. 18.3) and transect B (Fig. 18.4). The diagrams included all woody plants (⩾6 m high or ⩾3 cm dbh in the forest and savanna woodland and ⩾2 m tall in the other savanna types) and give an overall impression of the vegetation.

Water depth measurements

Water depths were measured on transect A at two-weekly intervals from 17 March to 18 December 1987 from the 11 soil pits (Figs 18.5 and 18.6). Soil-water depth was measured in seven pits on transect B (Fig. 18.5) from 10 September to 8 December 1987. The pit in the swamp forest of transect B was dug when this zone (which was inundated when the other pits were dug) had dried out. Water depth was, therefore, not recorded in this pit. On both transects, measurements ended when no pit had standing water. Frequent further observations until 1 April 1988 showed that all the pits remained dry.

Surface-soil sampling

On transect A, a 10 × 10 m plot was placed within 15 m of each side of the transect in a representative part of each of the seven vegetation zones (evergreen forest), marginal woodland, low scrub (1), savanna grassland, low scrub (2), dense thicket, and semi-evergreen forest. These plots are being used for a long-term monitoring of vegetation dynamics. At each of

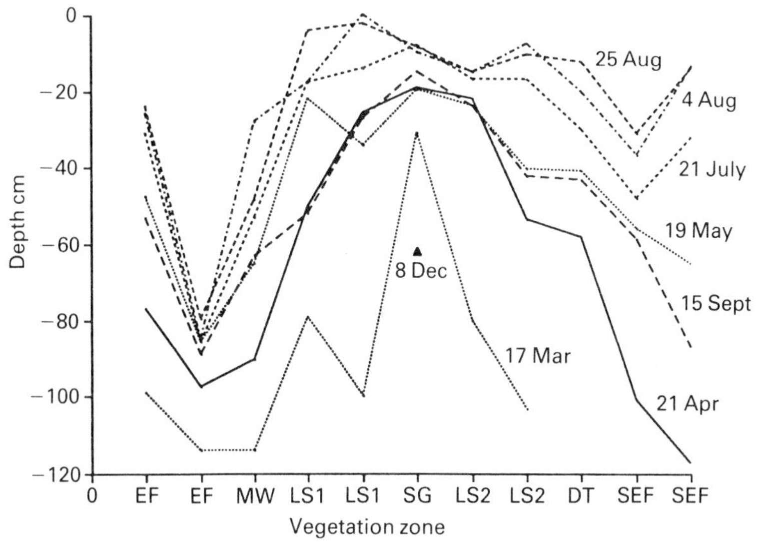

Fig. 18.5 Soil-water levels in 11 soil pits on transect A at selected dates between 7 March and 8 December 1987. The vegetation zone abbreviations of Fig. 18.3 apply.

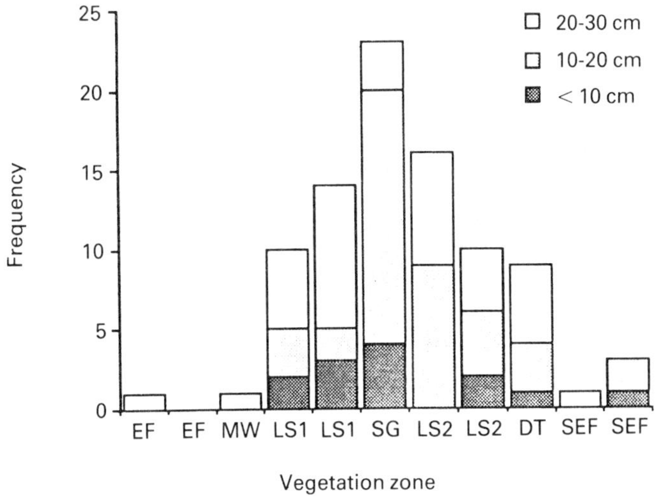

Fig. 18.6 The number of occasions (out of 24 two-weekly measurements) at which soil water was <10 cm, 10–20 cm, or 20–30 cm below the soil surface in 11 soil pits along transect A. The vegetation zone abbreviations of Fig. 18.3 apply.

the fourteen 10 × 10 m plots a 1 × 1 m quadrat for soil sampling was placed just outside the two plot corners nearest to the transect line. Soil samples (0–10 cm deep) were collected for chemical analysis from these quadrats on 29 June 1987 (mid wet season) and a second set on 5 January 1988 (mid dry season). Similar samples were collected for textural analysis from the quadrats on 25 April 1987 (wet season). A subsample of each soil was air dried and sieved through a 2 mm mesh in Brazil and 100 g returned to Stirling for analysis. Bulk density was measured by weighing the soil excavated from a 30 × 30 cm box inserted to a depth of about 10 cm at one point within each of the seven vegetation zones. For transect B the soil samples used for chemical and textural analyses were the surface horizons of the eight sampled pits (collected on 9 September 1987 for seven pits and on 4 February 1988 for the pit in the swamp forest (Fig. 18.4)). The soils were subsampled as described for transect A. The depths of the surface horizons were: pit 12, 0–24 cm; pit 13, 0–25 cm; pit 14, 0–15 cm; pit 15, 0–22 cm; pit 16, 0–11 cm; pit 17, 0–22 cm; pit 18, 0–20 cm; and pit 19, 0–7 cm.

Soil analyses

pH was measured in a 1:2 soil:deionized water mix, which was shaken for 2 h and allowed to stand for 1 h. Loss-on-ignition was measured on oven-dried subsamples after heating at 375 °C for 16 h. Total quantities of elements were extracted from 1.5 g subsamples of soil that had been digested in a mix of 6 ml concentrated sulphuric acid (containing 0.1% selenium as a catalyst) and 3 ml of '100 volume' hydrogen peroxide. Total nitrogen and total phosphorus were measured colorimetrically by Technicon auto-analyser. Nitrogen was determined using sodium salicylate and sodium nitroprusside with sodium dichloroisocyanurate as the chlorine source, and phosphorus using molybdenum blue. Available phosphorus was extracted from 10 g subsamples of soil by 50 ml Melich solution (a mix of 0.05 M hydrochloric acid and 0.0125 M sulphuric acid). It was analysed using an auto-analyser technique similar to that used for total phosphorus. Exchangeable cations were leached from 10 g subsamples of soil by ten successive additions of 10 ml of 1 M ammonium acetate solution adjusted (by the addition of acetic acid) to pH 4.8 (mean soil pH of the transect soils). They were analysed by atomic absorption spectrophotometry. For analyses of total acidity and exchangeable aluminium, 10 g subsamples of soil were leached by ten successive additions of 10 ml of 1 M potassium chloride solution. This was then treated with 0.0025 M sodium hydroxide solution and phenolphthalein indicator to measure total acidity. After adding 10 ml of 1 M potassium fluoride to the titrated solution, another titration with 0.005 M hydrochloric acid gave the values for exchangeable aluminium. Cation exchange capacity was calculated from the sum of total exchangeable cations plus total acidity. Particle-size analyses were made by the pipette method of Black *et al.* (1965).

RESULTS

Water depths

On transect A the water depths (Figs 18.5 and 18.6) show a relationship between the height of the wet-season soil water and vegetation zone with the water level in the savanna grassland > low scrub > dense thicket > marginal woodland > forest. The savanna grassland was waterlogged for much of the wet season. The measurement of standing soil water in pits is, however, a crude means of assessing the water table. The very low water depth in one of the evergreen forest pits is likely to be an artefact of the method since on several occasions water was observed flowing into the two pits adjacent to it through a connecting trench. However, there is little doubt that the vegetation/soil-water depth relationship is reasonably well represented by our data. On transect B the data are less complete and the situation is more complex since the water depth shows a gradient, from the swamp forest to the evergreen forest (Fig. 18.7). As on transect A, the water depths on transect B are higher in the savanna pits than in the evergreen forest (which is continuous with the evergreen forest of transect A). However, the raised shape of transect B means that the surface soils of most of the savanna are never waterlogged. On 15 September 1987, when the water table was very high (a stream was flowing at the south end) on transect B, the savanna pits (13–17) had water depths ranging from 29 to 106 cm below the surface (Fig. 18.4). On the same date, the savanna grassland pit (6) on transect A had water 14 cm below the surface (Fig. 18.3).

Soil analyses

The results of soil chemical analyses for transect A are shown in Fig. 18.8. There was a marked increase in soil percentage loss-on-ignition as one moved from the evergreen forest and low scrub (1) to the savanna grassland. The low scrub (2), dense thicket and semi-evergreen forest samples had a similarly high loss-on-ignition to the savanna grassland. Total nitrogen and phosphorus were at higher concentrations and pH values were lower in the soils with a higher loss-on-ignition. Available phosphorus was higher in the forest, dense thicket and marginal woodland

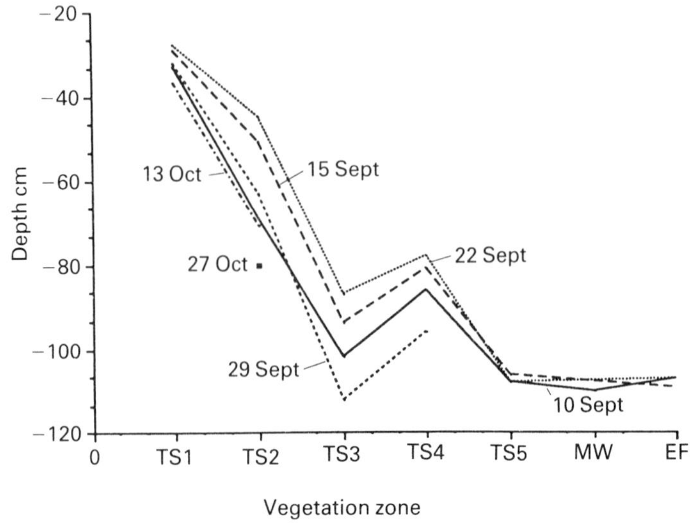

Fig. 18.7 Soil-water levels in seven soil pits on transect B between 10 September and 27 October 1987 (when pit 14 was the last with water). Pit 13 was only 40 cm deep at that time and the water depth in it (not shown in the figure) was probably just below that depth on 27 October. By the time pit 13 was fully excavated (to 124 cm) there was no standing water in it. No data are given for pit 12. The vegetation zone abbreviations of Fig. 18.4 apply.

samples than in the low scrub and savanna grassland. Cation exchange capacity and exchangeable acidity were least, and percentage base saturation highest, in the forest samples, but the exchangeable cations showed no clear patterns along the transect. All the soil chemical factors (except exchangeable calcium) showed highly significant differences between the vegetation zones (F test, $p < 0.001$) but were not related to them in any consistent way. In the wet season, total and available phosphorus, and pH were significantly higher ($p < 0.001$) than in the dry season, and exchangeable calcium, potassium and magnesium were generally higher ($p < 0.05$). There were no significant temporal variations for total nitrogen, exchangeable magnesium and loss-on-ignition. Only pH showed a significant interaction between time and vegetation zones ($p < 0.001$).

On transect B (Fig. 18.9) there was no sample replication and no statistical tests could be made. Loss-on-ignition was much higher towards the forest at the wetter south end and this was accompanied by high concentrations of total nitrogen, phosphorus and exchangeable cations. Nutrient concentrations in the tree/shrub savanna were generally low (except in the samples from the southernmost pit in this vegetation, close

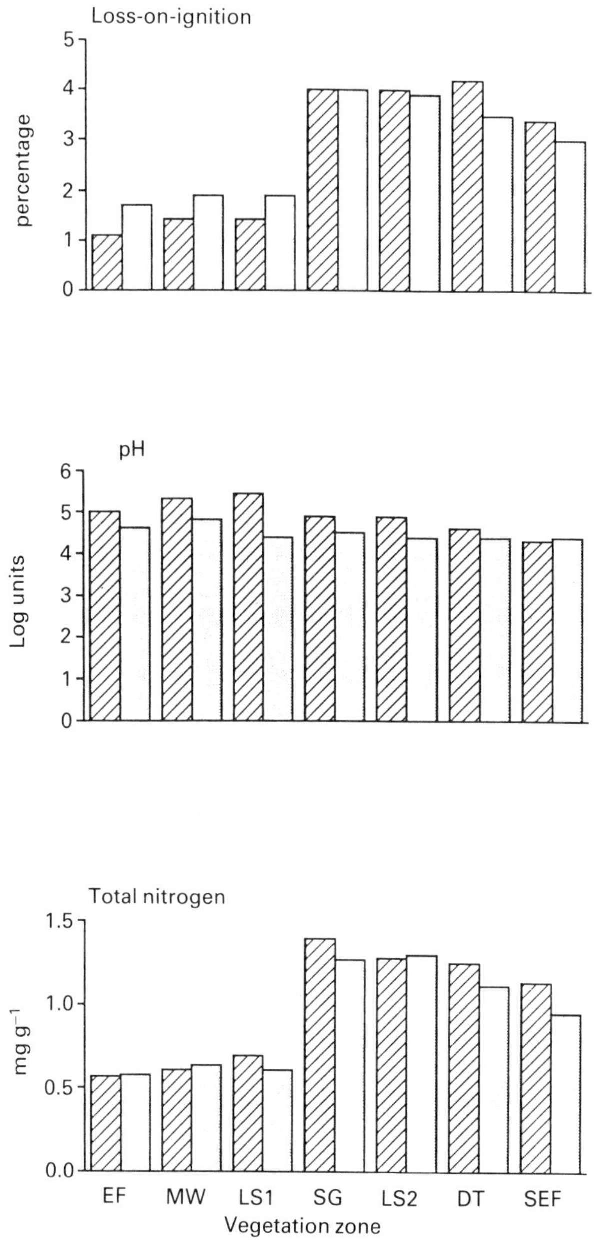

Fig. 18.8 Soil chemical factors for different vegetation zones along transect A. Where histograms are paired, those on the left are wet-season values, those on the right are dry-season values. The vegetation zone abbreviations of Fig. 18.3 apply.

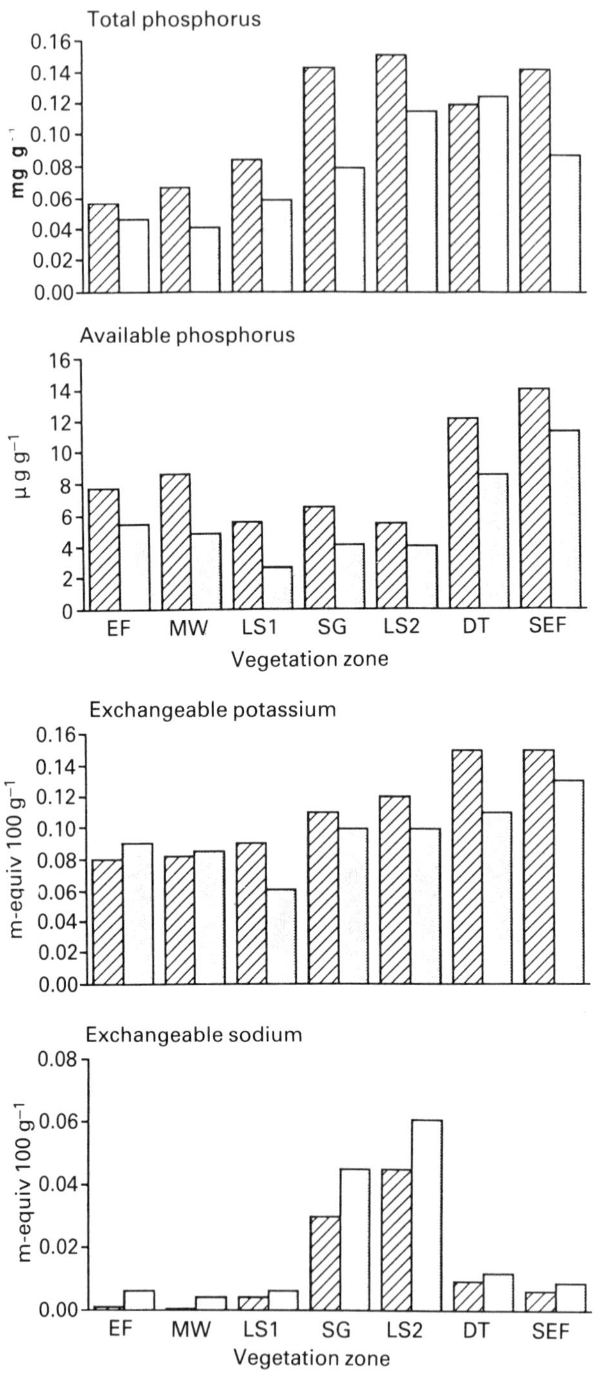
Total phosphorus
mg g⁻¹
0.16
0.14
0.12
0.10
0.08
0.06
0.04
0.02
0.00
Available phosphorus
µ g g⁻¹
16
14
12
10
8
6
4
2
0
EF
MW
LS1
SG
LS2
DT
SEF
Vegetation zone
Exchangeable potassium
m-equiv 100 g⁻¹
0.16
0.14
0.12
0.10
0.08
0.06
0.04
0.02
0.00
Exchangeable sodium
m-equiv 100 g⁻¹
0.08
0.06
0.04
0.02
0.00
EF
MW
LS1
SG
LS2
DT
SEF
Vegetation zone

Fig. 18.8

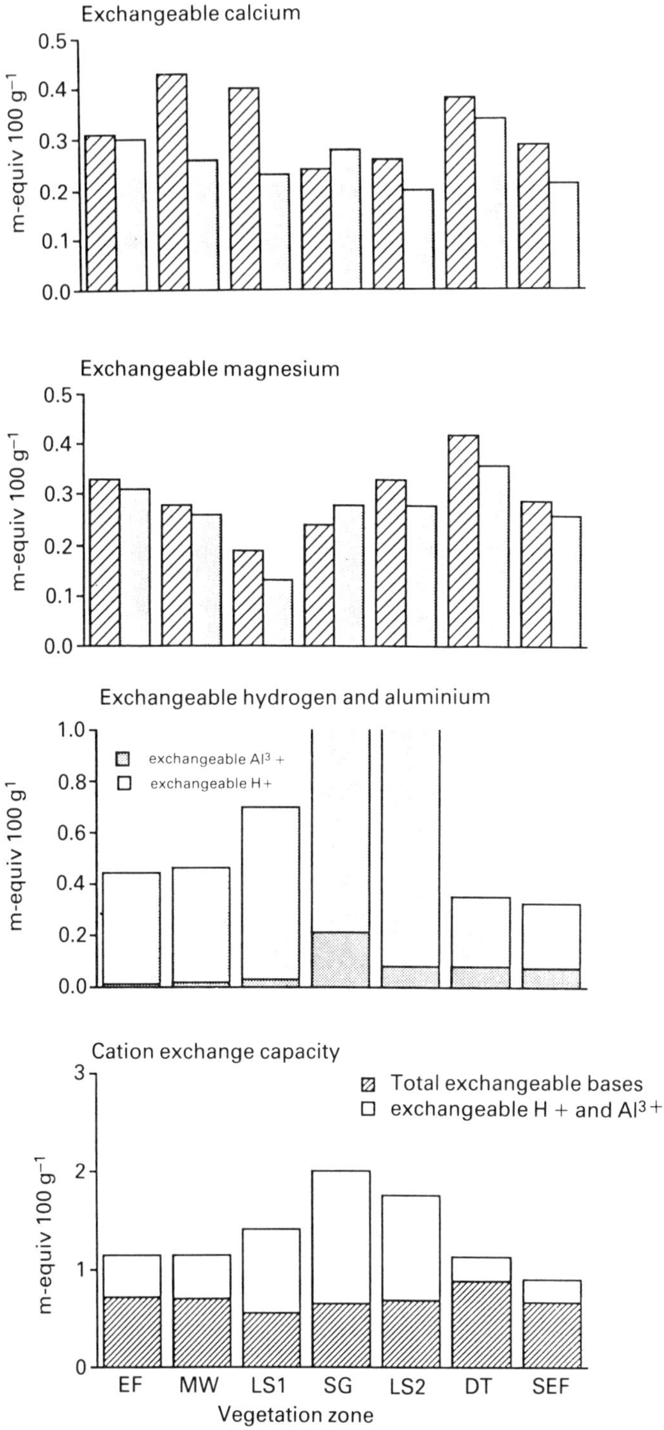
Exchangeable calcium
0.5
0.4
0.3
0.2
0.1
0.0
m-equiv 100 g⁻¹
Exchangeable magnesium
0.5
0.4
0.3
0.2
0.1
0.0
m-equiv 100 g⁻¹
Exchangeable hydrogen and aluminium
1.0
0.8
0.6
0.4
0.2
0.0
exchangeable Al3 +
exchangeable H+
m-equiv 100 g⁻¹
Cation exchange capacity
3
2
1
0
Total exchangeable bases
exchangeable H + and Al3+
m-equiv 100 g⁻¹
EF
MW
LS1
SG
LS2
DT
SEF
Vegetation zone

Fig. 18.8

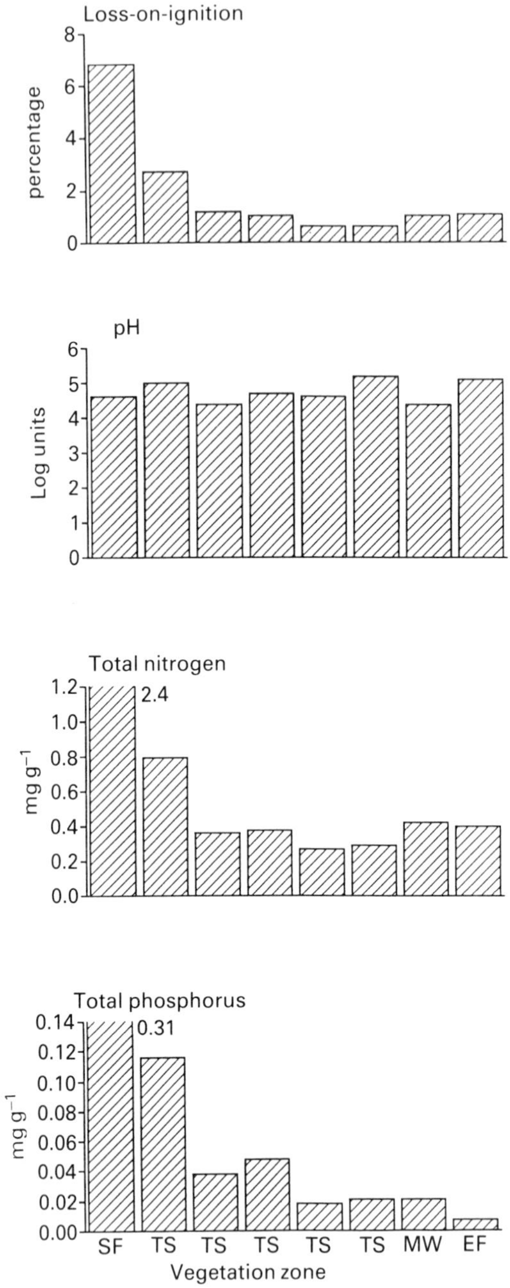

Fig. 18.9 Soil chemical factors for different vegetation zones along transect B. The vegetation zone abbreviations of Fig. 18.4 apply.

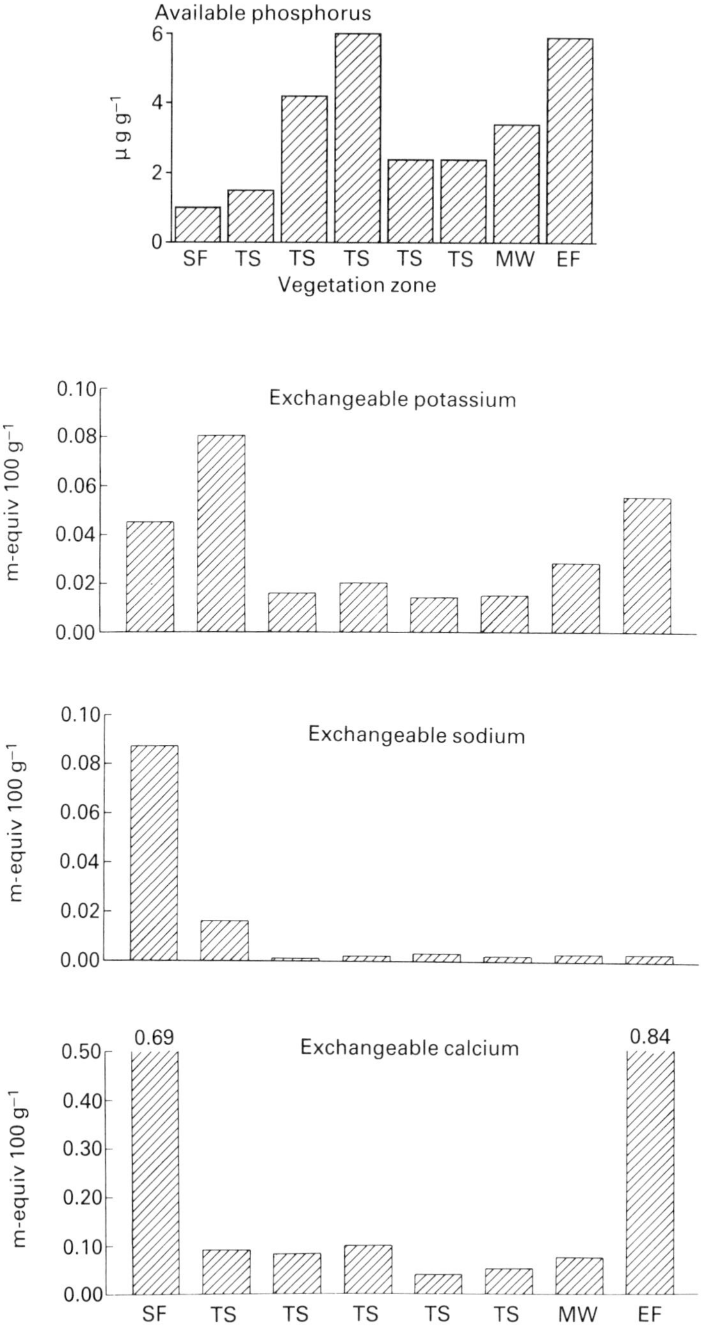
Available phosphorus
μg g⁻¹
0
2
4
6
SF TS TS TS TS TS MW EF
Vegetation zone
Exchangeable potassium
m-equiv 100 g⁻¹
0.00
0.02
0.04
0.06
0.08
0.10
Exchangeable sodium
m-equiv 100 g⁻¹
0.00
0.02
0.04
0.06
0.08
0.10
Exchangeable calcium
m-equiv 100 g⁻¹
0.00
0.10
0.20
0.30
0.40
0.50
0.69
0.84
SF TS TS TS TS TS MW EF

Fig. 18.9

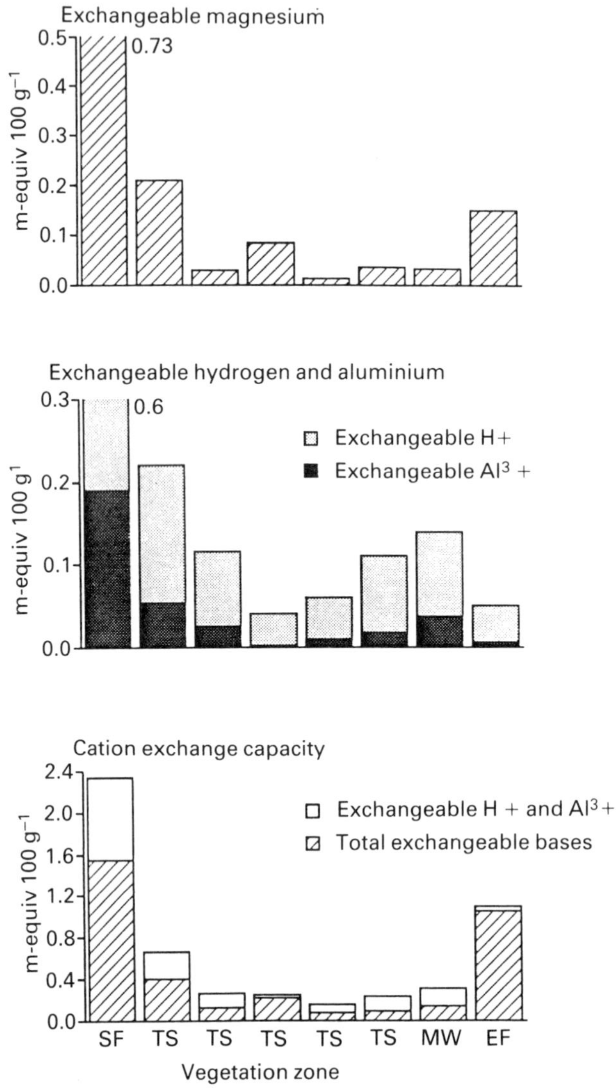

Fig. 18.9

to the swamp forest). Exchangeable potassium, calcium and magnesium were at relatively high concentrations in the evergreen forest while available phosphorus and pH showed no clear trends along the transect.

On transect A (Fig. 18.10) there was a decrease in percentage sand and

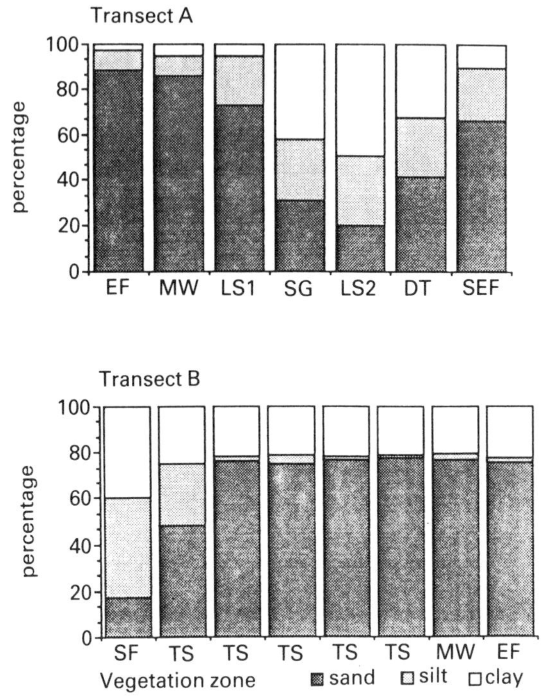

Fig. 18.10 Soil particle-size composition for different vegetation zones along (a) transect A (Fig. 18.3 for vegetation zone abbreviations), and (b) transect B (Fig. 18.4 for vegetation zone abbreviations).

a corresponding increase in silt and clay on moving from forest into savanna. By contrast, there was a large increase in percentage sand from swamp forest into the savanna at the south end of transect B. The transition from evergreen forest to savanna at the north end of transect B was not accompanied by any soil textural changes. Bulk density showed no clear patterns along transect A and was (in $g\,cm^{-3}$): 1.36, evergreen forest; 1.46, marginal woodland; 1.30, low scrub (1); 1.28, savanna grassland; 1.50, low scrub (2); 1.46, dense thicket; and 1.35, semi-evergreen forest. Bulk density was not measured along transect B.

Nutrient concentrations were generally higher on transect A and were associated with a higher loss-on-ignition and less sandy texture than in the soils of transect B. The main exception was the soil under the swamp forest of transect B which had a high loss-on-ignition and relatively high concentrations of several nutrients. The relatively high concentrations of calcium in the evergreen forest of transect B are surprising.

DISCUSSION

The alternating wet and dry climate of Maracá Island falls into Koppen's 'Aw' category which is more favourable for savanna than forest (Sarmiento and Monasterio, 1975). The east of Maracá has several vegetation types including lowland evergreen rainforest, semi-evergreen forest and savanna. There is a possibility that the areal extent of these vegetation types is changing, e.g. Furley and Ratter (1990) and Milliken and Ratter (1989) consider that there is evidence (the presence of relict savanna species within the forest) that forest is extending into savanna, at least locally. The line of rounded stones in the soil profiles at both ends of transect A suggests that the transect may cross a former river bed. This view is supported by the particle-size class distribution along the transect and lends weight to the view of the long-term dynamics of forest–savanna boundaries as a corollary of the evolution of landscapes (Sarmiento and Monasterio, 1975). The possibility of these long-term vegetation changes must be borne in mind when trying to relate present-day vegetation to soils, since they may change asynchronously. The data in the present paper show that at least two savanna types occur on Maracá and support the views of Furley and Ratter (1990) on the complexity of causes of this vegetation on the island.

Transect A fits Sarmiento's model of 'hyperseasonal savannas' (Chapter 30). These are savannas where periods of waterlogging and soil-water deficit reinforce the seasonality of the climate in causing alternating very wet and very dry conditions. Savanna, but not forest, can tolerate these fluctuations in soil water. It is noteworthy that, while accepting that the model is broadly true, the forest on transect A is able to withstand some short periods of waterlogging in alternation with the dry spells. On a subsequent visit to Maracá it was noticed (by one of us, J.T.) that during 1–14 June 1989 the land between soil pit 4 (low scrub, 1) (Fig. 18.3) to between soil pits 10 and 11 (semi-evergreen forest) was frequently under water but only for a few days at a time. The water levels in the River Uraricoera at this time were the highest seen by the local farmer, Zé Alves (personal communication), since he had arrived near Maracá in about 1958.

Most of the tree/shrub savanna of transect B fits the 'dry savanna' model of Sarmiento (Chapter 30). This contrasts with 'hyperseasonal savanna' since the raised topography and the coarse-textured soils mean that the soils are relatively dry even during the wet season (Fig. 18.7). The prolonged soil moisture deficits are held to inhibit forest colonization. At the south end the savanna changes sharply into a type of swamp forest which has a prolonged high water table, and often a stream, and much shorter periods (if any) of soil-water deficit. Over much of transect

B soil nutrients were low. One soil pit (14) at the south of the tree/shrub zone is in an area of rapid transition to swamp forests and is wetter and more nutrient-rich than the others in this vegetation.

The transect described by Ross *et al.* (Chapter 8) seems to combine features of transects A and B but shows some contrasts to them. The lower parts of their transect including the 'dry campo' and 'wet campo' are tree/shrub savannas of a hyperseasonal type which are, however, structurally different from any vegetation on transect A. The upper part of their transect including the 'transition zone' and the 'dry savanna' is savanna grassland of a dry savanna type but which is much less woody than any part of transect B. Their terra firme forest is similar to the evergreen forests of transects A and B. The upper part of the Ross *et al.* transect remained unflooded even during the exeptionally high water levels of 1989 and this confirms that this part does not fit the hyperseasonal model. Much of the Santa Rosa savanna south of transect A is similar to that described by Ross *et al.* (Chapter 8) and transect A is to be viewed as a local area where the hyperseasonal model fits particularly well.

In conclusion, the savanna models of Sarmiento (Chapter 30) formulated from elsewhere in tropical America seem to apply on Maracá. It is striking that adjacent and superficially similar savannas may have dissimilar causes. The presence of *Curatella americana* as an important species on both transects suggests either that it has a wide ecological amplitude or that it has evolved physiologically separate races. A similar occurrence of *C. americana* in contrasting hydrologic habitats (e.g. the so-called *lixeiro* community of the Mato Grosso Pantanal) and freely-drained cerrados (savanna woodlands) is well known in central Brazil.

The seasonal waterlogging has been shown (Fig. 18.8) to result in generally higher pH and supplies of nutrients. However, Marrs *et al.* (1991) showed that nitrification and nitrogen mineralization rates along transect A were much lower during the wet season (when they were often negative) (Fig. 18.11). Ross *et al.* (Chapter 8) showed that evergreen forest and savanna grassland adjacent to the forest boundary had higher available and potentially available nitrate nitrogen compared with their hyperseasonal savanna grassland and woodland savanna sites in the lower, wetter part of their transect. It is possible that the seasonally low supplies of nitrate are a key feature of hyperseasonal savannas, but experimental work is needed to confirm this.

There is no evidence that fire is of overriding importance on Maracá and Sarmiento and Monasterio (1975) consider 'that seasonal savannas could maintain themselves even with very low fire frequency as could be the case if only natural fires were considered'. It is impossible to be sure of the influence of the human disturbance on the forest–savanna

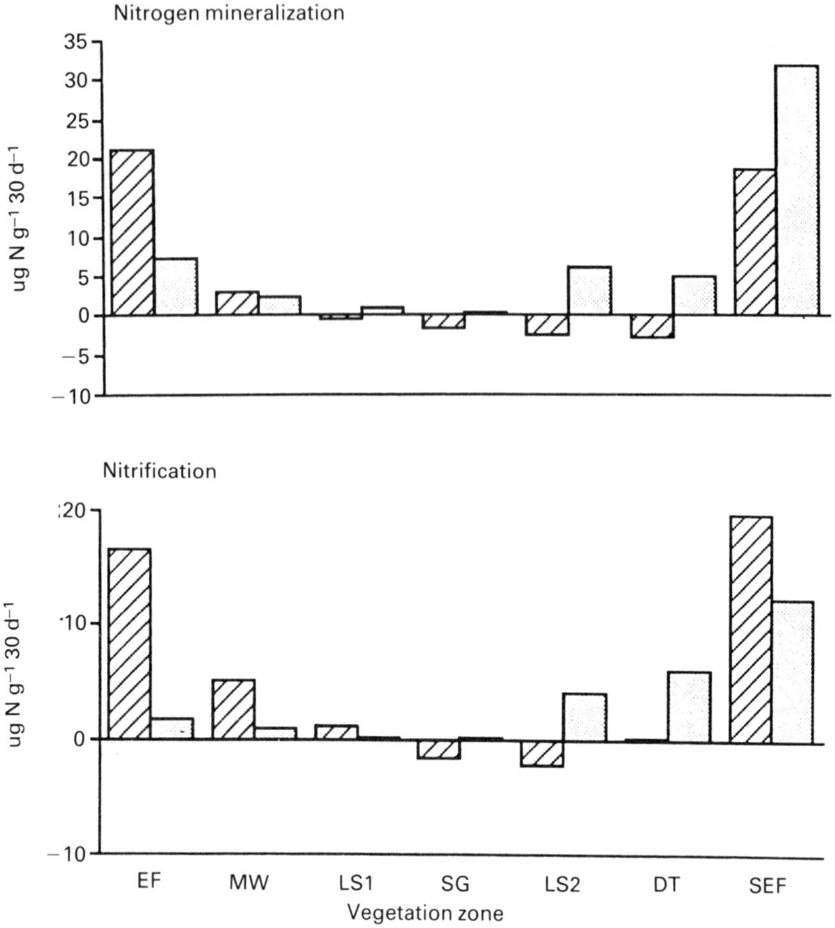

Fig. 18.11 Nitrogen mineralization (a) and nitrification rates (b) for different vegetation zones along transect A. The left histograms of each pair are wet-season values, the right hand are dry-season values. These data are from Marrs *et al.* (1991). The vegetation zone abbreviations of Fig. 18.3 apply.

boundaries on transects A and B. The boundaries do seem to reflect natural causes, since plantations (abandoned in 1980) within the forest area, a few hundred metres west of transect A, are reverting to forest. If the open areas of transects A and B were anthropogenic savannas then one would expect to see them being colonized throughout by forest, which is not the case. However, the possibility of the effect of forest removal raising the water table and locally favouring the development of hyperseasonal savanna should be further investigated.

ACKNOWLEDGEMENTS

We thank the Secretaria Especial do Meio Ambiente (SEMA) for their invitation to the Royal Geographical Society (RGS) to work on Maracá and Guttemberg Moreno de Oliveira and other SEMA staff for much help. The RGS, particularly Dr J. Hemming, Mr S. Bowles, and Miss F. Watson, gave support in many ways. We thank the Overseas Development Administration for providing funds for our work. Souza Cruz Ltd provided one of us (J.T.) with financial support. We also thank many other sponsors who supported the Maracá project. Dr J. Ferraz, Dr P.A. Furley, Miss S. Latham, Dr. R.H. Marrs, Mr R. Miller, W. Milliken, V. Viana and Dr S.M. Ross and technicians from Instituto Nacional de Pesquisas da Amazônia are also thanked for their help.

REFERENCES

Anonymous (1979) *Dados Climatológicos de Boa Vista Período: 1923 a 1978*, Instituto nacional de Meterologia/Delegacia Federal de Agricultura em Roraima.

Black, C.A., Evans, D.D., White, J.L. *et al.* (1965) *Methods of Soil Analysis*, American Society of Agronomy, Madison, USA.

Cottam, G. and Curtis, J.T. (1956) The use of distance measures in phytosociological sampling. *Ecology*, **37**, 451–60.

Furley, P.A. (1974a) Soil–slope–plant relationships in the northern Maya mountains, Belize, Central America. 1. The sequence over metamorphic sandstones and shales. *Journal of Biogeography*, **1**, 171–86.

Furley, P.A. (1974b) Soil–slope–plant relationships in the northern Maya mountains, Belize, Central America. 2. The sequence over phyllites and granites. *Journal of Biogeography*, **1**, 263–79.

Furley, P.A. (1976) Soil–slope–plant relationships in the northern Maya mountains, Belize, Central America. 3. Variations in the properties of soil profiles. *Journal of Biogeography*, **3**, 303–19.

Furley, P.A. (1989) The soils and soil–plant relationships of the eastern sector of Maracá Island. in *The Vegetation of the Ilha de Maracá* (ed. W. Milliken and J.A. Ratter), Royal Botanic Garden, Edinburgh and Royal Geographical Society, London, pp. 229–76.

Furley, P.A. and Ratter, J.A. (1989) Further observations on the nature of the savanna vegetation and soils in Belize in *Advances in Research in Belize* (ed. P.A. Furley), *Biogeographical Monographs*, **3**, pp. 1–23.

Furley, P.A. and Ratter, J.A. (1990) Pedological and botanical variations across the forest–savanna transition on Maracá Island. *Geographical Journal*, **156**, 251–66.

Furley, P.A., Ratter, J.A. and Gifford, D.R. (1988) Observations on the vegetation of eastern Mato Grosso, III. The woody vegetation and soils of the Morro de Fumaça, Torixoreu, Brazil. *Proceedings of the Royal Society of London*, **B235**, 259–80.

Marrs, R.H., Proctor, J., Thompson, J. and Scott, D.A. (1991) Soil nitrogen mineralization and nitrification in terra firme forest after felling and along forest–savanna boundaries on the Ilha de Maracá. *Journal of Tropical Ecology*, 7, 123–37.

Milliken, W. and Ratter, J.A. (1989) *The Vegetation of the Ilha de Maracá*, Royal Botanic Garden, Edinburgh and Royal Geographical Society, London.

Moskovits, D.K. (1985) The behaviour and ecology of two Amazonian tortoises *Geochelone carbonaria* and *G. denticulata* in northwestern Brazil. Unpublished Ph.D. thesis, University of Chicago, USA.

Nortcliff, S. and Robison, D. (1989) *The Soils and Geomorphology of the Ilha de Maracá: The Second Approximation*, Royal Geographical Society, London.

Proctor, J. (1992) The savannas of Maracá. In *Dynamics of a Brazilian Rainforest* (ed. J. Hemming), Manchester University Press (in press).

Proctor, J. and Miller, R.P. (1993) Human occupation on Maracá Island. Preliminary notes in *The Ecology of an Amazonian Rain Forest* (eds W. Milliken and J.A. Ratter), Manchester University Press, Manchester (in press).

Ratter, J.A., Richards, P.W., Argent, G. and Gifford, D.R. (1973) Observations on the vegetation of northeastern Mato Grosso I. The woody vegetation types of the Xavantina–Cachimbo Expedition area. *Philosophical Transactions of the Royal Society of London*, **B266**, 449–92.

Ratter, J.A., Askew, G.P., Montgomery, R.F. and Gifford, D.R. (1978) Observations on the vegetation of northeastern Mato Grosso II. Forests and soils of the Rio Suiá–Missu area. *Proceedings of the Royal Society of London*, **B203**, 191–208.

Sarmiento, G. (1983) The savannas of tropical America, in *Ecosystems of the World 13: Tropical Savannas* (ed. F. Bourlière), Elsevier, New York, pp. 245–88.

Sarmiento, G. and Monasterio, M. (1975) A critical consideration of the environmental conditions associated with the occurrence of savanna ecosystems in tropical America, in *Tropical Ecological Systems: Trends in Terrestrial and Aquatic Research* (eds F.B. Golley and E. Medina), Springer-Verlag, New York, pp. 223–50.

Thompson, J., Proctor, J. and Scott, D.A. (1992) A semi-evergreen forest on Maracá Island. I. Physical environment, forest structure and floristics. In Dynamics of a Brazilian Rainforest (ed. J. Hemming), Manchester University Press (in press).

19

Floristic composition, phytosociology and comparison of cerrado and gallery forests at Fazenda Água Limpa, Federal District, Brazil

JEANINE M. FELFILI and M.C. DA SILVA JR

SUMMARY

1. Permanent plots were established in *cerrado s.s.*, *cerradão* and two gallery forests as a basis for continuous monitoring.
2. In cerrado *s.s.* and cerradão all trees with a diameter ≥5 cm, 30 cm above ground level, were recorded. Sixty-one species in 50 genera and 30 families were found in the cerrado *s.s.*; the highest importance value indices (IVI) were recorded for *Qualea parviflora*, *Ouratea hexasperma* and *Q. grandiflora*. Sixty species in 46 genera and 29 families occurred in the cerradão; the highest IVIs were recorded for *Emmotum nitens*, *Blepharocalyx salicifolia* and *Didymopanax macrocarpum*.
3. The two gallery forests studied were both of the dry (i.e. not swampy) type. All trees ≥5 cm dbh were recorded. The plots on the Capetinga stream had 81 species belonging to 74 genera and 41 families with *Amaioua guianensis*, *Piptocarpha macropoda* and *Protium brasiliense* the species with highest IVI. The Gama stream was richer with 104 species in 93 genera and 45 families; here *Amaioua guianensis*, *Metrodorea pubescens*, *Licania apetala* and *Cheiloclinum cognatum* had the highest IVIs.
4. Shannon and Wienner diversity indices were calculated for each vegetation site and all were compared using Sørensen and Morisita similarity indices.

INTRODUCTION

Cerrado vegetation is categorized as savanna in the world classification and occupies $2 \times 10^6\,km^2$, representing about 23% of the vegetation cover of Brazil. It has its core area on the central Brazilian plateau in the states of Goiás, Mato Grosso, Minas Gerais, Bahia and in the Federal District, but also extends north into Maranhão and Piauí, west into central Mato Grosso and southern Rondônia and south-west to Mato Grosso do Sul and Paraguay. In southern Brazil it occurs in disjunct areas in São Paulo and Paraná (Eiten, 1972).

The site of the present study is in the Federal District where the predominant vegetation is cerrado. Cerrado is notable for great variation in physiognomy and many forms are found in the Federal District. The most common form there, and in central Brazil in general, is tree-scrub woodland with crown cover of 10–60%: this form is called cerrado *sensu strictu* (*s.s.*) (as opposed to cerrado *sensu lato* (*s.l.*) which is a generic term used for all forms of cerrado vegetation). A denser and taller forest-like form in which the trees often reach 15–18 m tall and produce a practically closed canopy is known as cerradão (Portuguese for big cerrado), while sparser forms with less crown cover than cerrado *s.s.* are known as *campo cerrado* and *campo sujo* (the latter a grassland with a sparse scattering of shrubs). In addition, the extreme form consisting of a herbaceous, principally grassy, vegetation devoid of trees and shrubs is called *campo limpo* (meaning clean field).

Evergreen gallery forests occur along stream- and river-sides in the cerrado landscape and are particularly well developed around headwaters (the so-called *cabeceiras*). They occur on both well drained and swampy sites on the valley floors.

Ratter (1980, 1985, 1986) studied the vegetation of Fazenda Água Limpa and described differences between floristic composition of cerrado and cerradão. He observed some characteristic tree species such as *Emmotum nitens*, *Ocotea spixiana*, *Callisthene major*, *Casearia grandiflora* and *Virola sebifera* were confined to cerradão and did not occur in more open forms of cerrado, while other species, for example *Plathymenia reticulata*, *Copaifera langsdorfii* and *Rapanea guianensis* were much more common in cerradão than cerrado. In contrast, species such as *Kielmeyera coriacea* were abundant in the cerrado but only occasional in cerradão.

The present study was carried out in the same area as Ratter's earlier work and was directed towards a detailed comparison of the structure and floristic composition of cerrado *s.s.*, cerradão and gallery forest.

MATERIALS AND METHODS

Fazenda Água Limpa is the experimental farm and ecological reserve of the University of Brasília. It is situated at 15°56′S 47°46′W and covers approximately 4000 ha. The climate belongs to Köppen's subtype Awi and the maximum average temperature is 28.5 °C and the minimum 12.0 °C. The predominant soils are Dystrophic Latosols (Furley, 1985).

The principal vegetation is cerrado *s.s.* but there are extensive areas covered with campo limpo and campo sujo, gallery forests along the streams, and two areas of cerradão (Ratter 1980, 1985, 1986).

In 1983 we began to establish permanent plots, as described below, in the cerrado *s.s.*, cerradão and gallery forests of the fazenda with the object of continuous monitoring over many years.

1. Cerrado *s.s.*: in a 148 ha area (400 × 3700 m), 21 permanent plots 1000 m^2 (20 × 50 m) were established randomly (Husch *et al.*, 1972). The first tree measurements were taken in 1985 and the second in 1988. The measurement procedure was:
   (a) all trees with *d* (base diameter at 0.30 m above soil level) ≥5 cm were identified and their *d* and total height measured;
   (b) the natural regeneration was identified in subplots, and the number of main branches occurring was also recorded.
2. Cerradão: a series of 100 point-centred quarter (PCQ) points (Cottam and Curtis, 1956) was established in an area of approximately 4 ha of cerradão occurring as a patch in cerrado *s.s.* The first tree measurements were made in 1987 using the same procedure as in the cerrado *s.s.*
3. Gallery forest of the Córrego Capetinga (Capetinga stream): 100 permanent plots 100 m^2 (10 × 10 m) were systematically established (Husch *et al.*, 1972) in the area of about 40 ha of dry gallery forests occurring at the headwaters of the Córrego Capetinga (a tributary of the Riberão do Gama). The plots were laid along four lines at right angles to the main stream bed. The first three lines were 150 m apart, and the fourth 300 m from the third. The first tree measurements were taken in 1983, the second in 1986 and the third in 1989. Ratter (1980, 1985, 1986) observed evidence of disturbance by fire and by man in the area. In 1987 a fire again entered the forest damaging mainly line 1. The measurement procedure was as follows:
   (a) all trees with dbh (diameter at breast height, i.e. 1.30 m above ground level) ≥5 cm were identified and their dbh, trunk height and total height recorded;
   (b) the positions of the trees in the plots were mapped;

(c) a qualitative classification was made of trunk and crown characteristics;
(d) natural regeneration was recorded in subplots.

4. Gallery forest of the Riberão de Gama (Gama stream): 151 permanent plots 200 m^2 (20 × 10 m) were systematically established in an area of approximately 80 ha of dry gallery forest along the course of the Riberão de Gama. The plots were located along ten lines at right angles to the stream bed with 100 m between lines. The same measurement procedure was used as for the Capetinga forest and the first measurements were made in 1985 and the second in 1988.

The phytosociological analyses were made on the 1985 measurements of the cerrado *s.s.*, the 1987 of the cerradão, the 1989 of the Capetinga gallery forest and the 1988 of the Gama gallery forest. Phytosociological parameters were calculated following Cain and Castro (1959) and Cottam and Curtis (1956). Floristic similarity indices (Mueller-Dombois and Ellenberg, 1974; Pielou, 1977) and the Shannon and Wienner floristic diversity index (Pielou, 1977) were calculated and species–area curves plotted.

RESULTS AND DISCUSSION

Floristic composition

In the area of cerrado *s.s.* 61 species occurred belonging to 30 families and 50 genera (Table 19.1). The Leguminosae had the greatest number of species (13) followed by the Vochysiaceae (5). The only genera with more that one species represented were *Aspidosperma* (Apocynaceae) (2 spp.), *Tabebuia* (Bignoniaceae) (2 spp.), *Erythroxylum* (Erythroxylaceae) (3 spp.), *Kielmeyera* (Guttiferae) (2 spp.), *Byrsonima* (Malpighiaceae) (3 spp.) *Miconia* (Melastomataceae) (2 spp.), *Qualea* (Vochysiaceae) (3 spp.) and *Vochysia* (Vochysiaceae) (2 spp.).

Sixty species in 29 families and 46 genera were recorded in the cerradão (Table 19.2). As in the cerrado, the families with the most species are the Leguminosae (13) and the Vochysiaceae (7). In addition to the genera listed in the cerrado as having more than one species, *Ocotea* (Lauraceae) had two, as had *Machaerium* (Leguminosae/Papilionoideae), while *Miconia* (Melastomataceae) is represented by four instead of two in the cerrado and *Vochysia* (Vochysiaceae) by three (two in the cerrado). In all, 40 species were common to the cerrado *s.s.* and the cerradão.

Eighty-one species of 41 families and 74 genera occurred in the Capetinga forest plots (Table 19.3). The Leguminosae, once more, was the family with the most species, followed by Myrtaceae, Melastomataceae,

Table 19.1 Species recorded in the cerrado *s.s.* at Fazenda Água Limpa, Federal District. Occurrence in other localities: C, Capetinga; Co, cerradão; G, Gama

Annonaceae
Annona crassiflora Mart.
Apocynaceae
Aspidosperma macrocarpon Mart., Co
A. tomentosum Mart., Co
Hancornia speciosa Muell. Arg., Co
Araliaceae
Didymopanax macrocarpum (C. & S.) Seem., Co
Bignoniaceae
Tabebuia caraiba (Mart.) Bur.
T. ochracea Cham., C, G
Bombacaceae
Eriotheca pubescens (Mart. & Zucc.) Schott. & Endl., Co
Caryocaraceae
Caryocar brasiliense Camb., Co
Celastraceae
Austroplenckia populnea (Reiss.) Lund
Compositae
Eremanthus glomerulatus Less., Co
Piptocarpha rotundifolia (Less.) Baker
Connaraceae
Connarus suberosus Planch. var. *fulvus* (Planch.) Forero
Rourea induta Planch
Dilleniaceae
Davilla elliptica St. Hil., Co
Ebenaceae
Diospyros burchellii Hiern
Erythroxylaceae
Erythroxylum deciduum St. Hil., Co
E. suberosum St. Hil., Co
E. tortuosum Mart., Co
Flacourtiaceae
Casearia sylvestris Sw., C, G
Guttiferae
Kielmeyera coriacea (Spr.) Mart., Co
K. speciosa St. Hil.
Hippocrateaceae
Salacia crassifolia (Mart.) Peyr., Co
Leguminosae/Caesalpinoideae
Hymenaea stigonocarpa [Mart. ex] Hayne
Sclerolobium paniculatum Vog. var. *subvelutinum*, C, Co, G
Leguminosae/Mimosoideae
Dimorphandra mollis Benth., Co
Enterolobium gummiferum (Mart.) Macbride, Co
Mimosa claussenii Benth., Co
Stryphnodendron adstringens (Mart.) Coville

Table 19.1 *Continued*

Leguminosae/Papilionoideae
Acosmium dasycarpum (Vog.) Yakovl., Co
Andira paniculata Benth., Co, G
Bowdichia virgilioides Kunth, Co
Dalbergia violacea (Vog.) Malme, Co
Machaerium opacum Vog., Co
Pterodon pubescens Benth., Co
Vatairea macrocarpa (Benth.) Ducke
Loganiaceae
Strychnos pseudoquina St. Hil.
Lythraceae
Lafoensia pacari St. Hil., Co
Malpighiaceae
Byrsonima coccolobifolia (Spr.) Kunth, Co
B. crassa Nied., Co
B. verbascifolia Adr. Juss., Co
Heteropterys byrsonimaefolia Adr. Juss.
Melastomataceae
Miconia ferruginata (DC.) Cogn., Co
M. pohliana Cogn., Co
Myrsinaceae
Rapanea guianensis Aubl., C, Co, G
Myrtaceae
Blepharocalyx salicifolia (Lamb.) Burr., Co
Myrcia rostrata DC
Psidium warmingianum Kiaersk., C, Co
Nyctaginaceae
Guapira noxia (Netto) Lund
Ochnaceae
Ouratea hexasperma (St. Hil.) Baill., Co
Proteaceae
Roupala montana Aubl., C, Co, G
Rubiaceae
Palicourea rigida Kunth.
Tocoyena formosa (C. & S.) K. Schum.
Sapotaceae
Pouteria ramiflora (Mart.) Radlk., C, Co, G
Symplocos rhamnifolia A. DC.
Styraceae
Styrax ferrugineus Nees & Mart.
Vochysiaceae
Qualea grandiflora Mart., Co
Q. multiflora Mart., Co
Q. parviflora Mart., Co
Vochysia elliptica (Spr.) Mart., Co
V. thyroidea Pohl, Co

Table 19.2 Species recorded in the cerradão at Fazenda Água Limpa, Federal District. Occurrence in other localities: C, Capetinga; Ce, cerrado; G, Gama (authorities for species omitted if given in Table 19.1)

Anacardiaceae
Tapirira guianensis Aubl., C, G
Apocynaceae
Aspidosperma macrocarpon, Ce
A. tomentosum, Ce
Hancornia speciosa, Ce
Aquifoliaceae
Ilex conocarpa Reiss.
Araliaceae
Didymopanax macrocarpum, Ce
Bignoniaceae
Tabebuia serratifolia (Vahl) Nichols.
Bombacaceae
Eriotheca pubescens, Ce
Caryocaraceae
Caryocar brasiliense, Ce
Compositae
Eremanthus glomerulatus, Ce
Dilleniaceae
Davilla elliptica, Ce
Erythroxylaceae
Erythroxylum deciduum, Ce
E. suberosum, Ce
E. tortuosum, Ce
Euphorbiaceae
Maprounea guianensis Aubl. C, G
Guttiferae
Kielmeyera coriacea, Ce
Hippocrateaceae
Salacia crassifolia, C, Ce, G
Icacinaceae
Emmotum nitens (Benth.) Miers, C, G
Lauraceae
Ocotea spixiana (Nees.) Mez
O. sp.
Leguminosae/Caesalpinoideae
Copaifera langsdorffii Desf., C, G
Sclerolobium paniculatum var. *subvelutinum*, C, Ce, G
Leguminosae/Mimosoideae
Dimorphandra mollis, Ce
Enterolobium gummiferum, Ce
Mimosa claussenii, Ce
Plathymenia reticulata Benth.

Table 19.2 *Continued*

Leguminosae/Papilionoideae
Acosmium dasycarpum, Ce
Andira paniculata, Ce, G
Bowdichia virgilioides, Ce
Dalbergia violacea, Ce
Machaerium acutifolium Vog., C, G
M. opacum, Ce
Pterodon pubescens, Ce
Lythraceae
Lafoensia pacari, Ce
Malpighiaceae
Byrsonima coccolobifolia, Ce
B. crassa, Ce
B. verbascifolia, Ce
Melastomataceae
Miconia albicans (SW.) Triana
M. ferruginata, Ce
M. pohliana, Ce
M. sellowiana Naud., C, G
Myrsinaceae
Rapanea guianensis, C, Ce, G
Myrtaceae
Blepharocalyx salicifolia, Ce
Gomidesia lindeniana Berg, C, G
Psidium warmingianum, Ce
Siphoneugena densiflora, Berg
Nyctaginaceae
Neea theifera Oerst.
Ochnaceae
Ouratea hexasperma, Ce
Proteaceae
Roupala montana, C, Ce, G
Rubiaceae
Alibertia edulis (L. Rich.) A. Rich., C, G
Rutaceae
Zanthoxylum rhoifolium Lam., G
Sapotaceae
Pouteria ramiflora, C, Ce, G
Simaroubaceae
Simarouba versicolor St. Hil.
Vochysiaceae
Callisthene major Mart., C, G
Qualea grandiflora, Ce
Q. multiflora, C, Ce, G
Q. parviflora, Ce
Vochysia elliptica, Ce
V. thyrsoidea, Ce
V. tucanorum (Spr.) Mart.

Table 19.3 Tree species recorded in the gallery forests at Fazenda Água Limpa, Federal District. C, Capetinga; G, Gama (authorities for species omitted if given Tables 19.1 and 19.2)

Anacardiaceae
  Astronium fraxinifolium Schott., C, G
  Tapirira guianensis, C, G
Annonaceae
  Guatteria sellowiana Schlecht., C, G
  Xylopia emarginata Mart., C, G
  X. sericea St. Hil., C, G
Apocynaceae
  Aspidosperma cylindrocarpum Muell. Arg., C, G
  A. olivaceum Muell. Arg., C, G
  A. subincanum Mart., C, G
  A. sp., G
  Indet., G
Araliaceae
  Didymopanax morototoni Decne. & Planch., G
  Gilibertia sp., C, G
Bignoniaceae
  Tabebuia ochracea, C, G
Bombacaceae
  Eriotheca gracilipes A. Robyns, G
Boraginaceae
  Cordia sp., C, G
Burseraceae
  Protium brasiliense Engl., C, G
Celastraceae
  Maytenus alaternoides Reiss., C, G
Chrysobalanaceae
  Hirtella glandulosa Spreng., C, G
  Licania apetala (E. Mey.) Fritsch., C, G
Combretaceae
  Terminalia argentea Mart., G
  T. brasiliensis L., C, G
Compositae
  Piptocarpha macropoda Bak., C, G
Cunoniaceae
  Lamanonia tomentosa Camb., G
Dichapetalaceae
  Tapura amazonica Poepp., C, G
Ebenaceae
  Diospyros hispida A. DC., G
Elaeocarpaceae
  Sloanea sp., G
Erythroxylaceae
  Erythroxylum sp., C, G

Table 19.3 *Continued*

Euphorbiaceae
Alchornea iricurana Casar. C, G
Maprounea guianensis Aubl., C, G
Flacourtiaceae
Casearia sylvestris, C, G
Guttiferae
Calophyllum brasiliense Camb., G
Rheedia brasiliensis Planch., C, G
Vismia sp., C, G
Hippocrateaceae
Cheiloclinum cognatum (Miers.) A.C. Smith, C, G
Salacia elliptica (Mart.) Peyr., C, G
Humiriaceae
Saccoglottis guianensis Benth., C, G
Icacinaceae
Emmotum nitens, C, G
Lauraceae
Cryptocarya aschersoniana Mez, C, G
Nectandra mollis., C, G
Ocotea sp., C, G
Indet., G
Lecythidaceae
Cariniana estrellensis (Raddi) O. Kuntze, G
Leguminosae/Caesalpinoideae
Apuleia molaris (Spruce) Benth., C, G
Bauhinia sp., G
Copaifera langsdorffii, C, G
Hymenaea courbaril L. var. *stilbocarpa* (Hayne) Lee & Langenheim, C, G
Sclerolobium paniculatum var. *rubiginosum*, C, G
Leguminosae/Mimosoideae
Inga alba Willd., C, G
Leguminosae/Papilionoideae
Andira paniculata, G
Dalbergia frutescens Britton, C
Machaerium acutifolium, C, G
Ormosia sp., C
Indet., C, G
Malpighiaceae
Byrsonima laxiflora Griseb., C, G
Melastomataceae
Miconia punctata D. Don., C, G
M. sellowiana, C, G
M. sp., C
M. sp., G
M. sp., G
M. sp., G
M. sp., G
M. sp., G

Table 19.3 *Continued*

Meliaceae
  Cabralea canjerana (Vell.) Mart., C, G
Monimiaceae
  Siparuna guianensis Aubl., C, G
Moraceae
  Cecropia pachystachya Tréc., C, G
  Pseudolmedia laevigata Tréc., C, G
  Sorocea ilicifolia Miq., C, G
  Indet., G
Myristicaceae
  Virola sebifera Aubl., C, G
Myrsinaceae
  Myrsine coriacea R. Br., C, G
  Rapanea guianensis, G
Myrtaceae
  Campomanesia sp., C, G
  Gomidesia lindeniana, C, G
  Myrcia deflexa Steud., C, G
  Psidium warmingianum, C
  Indet., C, G
  Indet., C, G
  Indet., C, G
  Indet., C, G
  Indet., G
Nyctaginaceae
  Guapira sp., C, G
Ochnaceae
  Ouratea castaneaefolia Engl., C, G
  O. sp., C, G
Proteaceae
  Euplassa inaequalis (Pohl) Engl., C, G
  Roupala montana, C, G
Rubiaceae
  Aliberta edulis, C, G
  Amaioua guianensis Aubl., C, G
  Ferdinandusa speciosa Pohl, G
  Guettarda sp., G
  Ixora warmingii Muell. Arg., C, G
  Psychotria sp., C, G
  Indet., C, G
Rutaceae
  Metrodorea pubescens St. Hil. & Tul., G
  Zanthoxylum rhoifolium, G
Sapindaceae
  Cupania vernalis Camb., C, G
  Matayba guianensis Aubl., C, G
Sapotaceae
  Chrysophyllum sp., C, G
  Micropholis venulosa (Mart. & Eichl.) Pierre, C, G

Table 19.3 *Continued*

Pouteria ramiflora, C, G
Solanaceae
Solanum sp., C, G
Styracaceae
Styrax guianensis A.DC., G
Tiliaceae
Apeiba tibourbou Aubl., G
Luehea paniculata Mart., C, G
Vochysiaceae
Callisthene major, C, G
Qualea dichotoma Warm., C, G
Q. multiflora, G
Vochysia pyramidalis Mart., G

Rubiaceae, Sapotaceae and Vochysiaceae. Only *Xylopia* (Annonaceae) (2 spp.), *Aspidosperma* (Apocynaceae) (3 spp.), *Miconia* (Melastomataceae) (3 spp.) and *Ouratea* (Ochnaceae) (2 spp.) had more than one species per genus.

One hundred and three species of 45 families and 93 genera occurred in the Gama forest plots (Table 19.3). The Leguminosae again was the family with the most species, followed by Myrtaceae, Rubiaceae, Apocynaceae, Lauraceae, Melastomataceae and Vochysiaceae. The following genera were represented by more than one species: *Xylopia* (Annonaceae) (2 spp.); *Aspidosperma* (Apocynaceae) (4 spp.); *Terminalia* (Combretaceae) (2 spp.); *Miconia* (Melastomataceae) (8 spp.); *Qualea* (Vochysiaceae) (2 spp.).

The total number of species recorded for gallery forest was 109 while for the cerrado *s.l.* (cerrado *s.s.* + cerradão) it was 81. The Gama forest is richer (103 spp.) than the Córrego Capitinga (80 spp.), although in part this may be explained by the larger sampled area of the former (see species–area curves, Fig. 19.1). About 70% of the total species recorded for these two forests were in common. As might be expected, the Gama forest had the highest Shannon and Wienner diversity index of all the areas studied (Table 19.4).

The similarity of the Gama and Capetinga forests is high as scored by Sørensen (1948) and Morisita (Horn, 1966) indices (Table 19.5): this is what one would expect from forests of the same type (dry gallery forest) only about 2 km apart. The similarity for cerrado *s.s.* and cerradão was high for the Sørensen index (which is based only on the presence or absence of species) but much lower for the Moristia index (which also considers relative abundance). Similarity between the cerrados and gallery forests is registered as low by both indices, although, predictably, cerradão is more similar to gallery forest than is cerrado *s.s.*

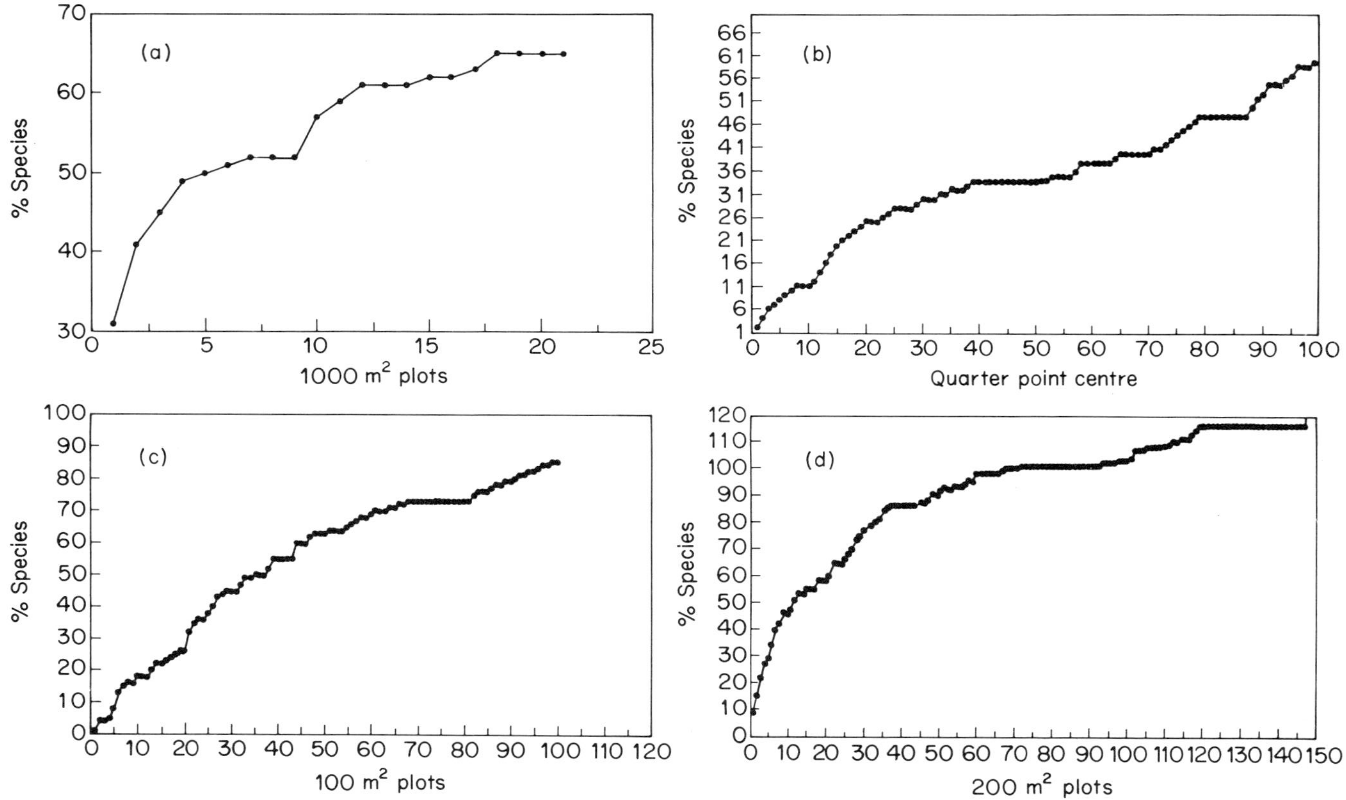

Fig. 19.1 Species–area curves for the study areas (a) cerrado *s.s.*, (b) cerradão, (c) Capetinga gallery forest, (d) Gama gallery forest.

Table 19.4 Diversity indices

Vegetation type	*Shannon and Wienner diversity index*
Cerrado *s.s.*	3.46
Cerradão	3.42
Capetinga forest	3.54
Gama forest	3.87

Shannon and Wienner diversity index: $H' = -\Sigma\, pi \ln pi$, where pi = the proportion of the total number of individuals which species i represents.

Table 19.5 Similarity indices of vegetation types Fazenda Água Limpa

			Sørensen index	
Morisita index	*Cerrado*	*Cerradão*	*Capetinga forest*	*Gama forest*
Cerrado		0.666	0.097	0.096
Cerradão	0.394		0.181	0.183
Capetinga forest	0.032	0.049		0.623
Gama forest	0.019	0.046	0.709	

Sørensen index: $I = \dfrac{2\,xa}{b + c}$

where a = no. of species in common between 1 and 2, b = no. of species in community 1, c = no. of species in community 2.

Morisita index: $\ln = 2\sum_{i=1}^{N} Xi \cdot Yi/(1 + 2)Nx \cdot Ny$

Where Xi = no. of individuals of species i in community X, Yi = no of individuals of species i in community Y, Nx = total number of individuals in community X, Ny = total number of individuals in community Y.

$$1 = \sum_{i=1}^{N} Xi(Xi - 1)/Nx(Nx - 1)$$

$$2 = \sum_{i=1}^{N} Yi(Yi - 1)/Ny(Ny - 1)$$

Phytosociology

The cerrado *s.s.* survey showed 958.10 trees ha^{-1} and 7.34 $m^2\,ha^{-1}$ total basal area. The highest IVIs (importance value indices) were for *Qualea parviflora, Ouratea hexasperma, Qualea grandiflora, Sclerolobium paniculatum, Caryocar brasiliense, Dalbergia violacea, Byrsonima crassa, Stryphnodendron adstringens, Palicourea rigida, Pouteria ramiflora and Miconia pohliana*. These 11 species represented about 53% of the total IVI value (Table 19.6).

Table 19.6 Phytosociological parameters for all trees and shrub ≥5 cm diam., 30 cm above ground level in an area of 2.1 ha of cerrado *sensu strictu*. Species in order of importance value index (IVI) (R.De., relative density; R.Do., relative dominance; RF, relative frequency; no. of trees ha^{-1} = 958.1; basal area ha^{-1} = 7.34 m^2)

	R.De.	*R.Do.*	*RF*	*IVI*
1. *Qualea parviflora*	8.95	19.22	3.41	31.58
2. *Ouratea hexasperma*	12.57	8.73	3.41	24.72
3. *Qualea grandiflora*	5.96	7.14	3.25	16.36
4. *Sclerolobium paniculatum*	5.37	6.30	3.41	15.08
5. *Caryocar brasiliense*	3.48	7.52	3.09	14.09
6. *Dalbergia violacea*	3.83	5.47	3.25	12.55
7. *Byrsonima crassa*	4.62	2.97	3.09	10.68
8. *Stryphnodendron adstringens*	3.68	2.96	3.09	9.73
9. *Palicourea rigida*	4.13	1.66	3.09	8.88
10. *Pouteria ramiflora*	2.34	3.14	2.60	8.07
11. *Miconia pohliana*	2.53	2.59	2.93	8.05
12. *Aspidosperma tomentosum*	2.73	2.05	2.76	7.54
13. *Byrsonima verbascifolia*	2.83	1.75	2.76	7.34
14. *Pterodon pubescens*	1.49	2.82	2.28	6.58
15. *Roupala montana*	2.09	1.34	3.09	6.51
16. *Didymopanax macrocarpum*	2.14	1.65	2.60	6.38
17. *Byrsonima coccolobifolia*	2.14	1.33	2.44	5.90
18. *Vochysia elliptica*	2.09	1.03	2.44	5.55
19. *Kielmeyera coriacea*	2.24	1.09	1.79	5.12
20. *Piptocarpha rotundifolia*	1.69	0.96	2.44	5.09
21. { *Erythroxylum suberosum*	1.94	0.61	2.44	4.98
{ *Qualea multiflora*	1.24	1.30	2.44	4.98
23. *Vochysia thyrsoidea*	1.34	2.00	1.63	4.97
24. *Styrax ferrugineus*	1.39	0.99	2.28	4.66
25. *Kielmeyera speciosa*	1.69	0.93	1.95	4.57
26. *Hymenaea stigonocarpa*	0.70	1.62	1.79	4.11
27. *Erythroxylum tortuosum*	1.34	0.63	2.11	4.08
28. *Rourea induta*	1.24	0.54	2.28	4.06
29. *Guapira noxia*	0.89	0.76	2.28	3.93
30. *Eremanthus glomerulatus*	1.49	0.57	1.79	3.85
31. *Acosmium dasycarpum*	0.84	0.54	1.95	3.34
32. *Miconia ferruginata*	0.55	0.44	1.63	2.62
33. *Blepharocalyx salicifolia*	0.50	1.13	0.81	2.44
34. *Salacia crassifolia*	0.55	0.31	1.46	2.32
35. *Diopyros burchellii*	0.75	0.34	1.14	2.22
36. *Symplocos rhamnifolia*	0.65	0.56	0.98	2.18
37. *Connarus suberosus*	0.84	0.45	0.81	2.11
38. *Rapanea guianensis*	0.40	0.73	0.98	2.10
39. *Mimosa claussenii*	0.50	0.26	1.14	1.89
40. *Eriotheca pubescens*	0.30	0.66	0.81	1.77
41. *Erythroxylum deciduum*	0.50	0.18	0.98	1.65
42. *Autroplenckia populnea*	0.35	0.28	0.98	1.61
43. *Enterolobium gummiferum*	0.25	0.66	0.65	1.56

Table 19.6 *(continued)*

	R.De.	*R.Do.*	*RF*	*IVI*
44. *Vatairea macrocarpa*	0.40	0.30	0.81	1.51
45. *Tabebuia caraiba*	0.35	0.13	0.98	1.45
46. *Tabebuia ochracea*	0.30	0.25	0.81	1.37
47. *Dimorphandra mollis*	0.25	0.08	0.81	1.15
48. *Davilla elliptica*	0.25	0.14	0.65	1.04
49. *Casearia sylvestris*	0.30	0.15	0.49	0.93
50. *Strychnos pseudoquina*	0.15	0.11	0.49	0.75
51. *Bowdichia virgilioides*	0.15	0.19	0.33	0.67
52. *Heteropterys byrsonimaefolia*	0.10	0.12	0.33	0.55
53. *Aspidosperma macrocarpon*	0.10	0.11	0.33	0.53
54. *Andira paniculata*	0.10	0.03	0.33	0.45
55. *Lafoensia pacari*	0.10	0.05	0.16	0.31
56. *Tocoyena formosa*	0.10	0.03	0.16	0.29
57. *Psidium warmingianum*	0.05	0.03	0.16	0.24
58. { *Annona crassiflora*	0.05	0.02	0.16	0.23
Machaerium opacum	0.05	0.02	0.16	0.23
60. { *Myrcia rostrata*	0.05	0.01	0.16	0.22
Hancornia speciosa	0.05	0.01	0.16	0.22

The 1987 cerradão survey recorded 815.00 trees ha^{-1} with 21.94 $m^2 ha^{-1}$ total basal area. The highest IVIs were for *Emmotum nitens, Blepharocalyx salicifolia, Didymopanax macrocarpum, Qualea grandiflora, Simarouba versicolor, Siphoneugena densiflora, Ocotea spixiana* and *Plathymenia reticulata*. These eight species represented about 57% of the total IVI value (Table 19.7). Five of them were not found in the cerrado *s.s.* The higher IVI values were concentrated in fewer species in cerradão than in cerrado (Table 19.7), because the dominant trees tend to be relatively much larger.

The Capetinga gallery forest had 982.00 trees ha^{-1} with 21.40 $m^2 ha^{-1}$ total basal area. The highest IVIs were for *Amaioua guianensis, Piptocarpha macropoda, Protium brasiliense, Cryptocarya aschersoniana, Licania apetala, Maytenus alaeternoides, Aspidosperma olivaceum* and *Tapirira guianensis*. These seven species represented about 50% of the total IVI (Table 19.8). Dead standing trees, classed together irrespective of species, had the second highest IVI, representing 16% of the total IVI for Capetinga forest trees. The pioneer species, *Piptocarpha macropoda* and *Cecropia pachystachya*, made up 10.80% of the total IVI. This probably is a consequence of disturbances in the past. At present there are areas of grasses, ferns and tangled lianas within the forest, giving evidence of relatively recent disturbance.

The Gama survey recorded 1350.55 trees ha^{-1} with 41.12 $m^2 ha^{-1}$ total basal area. The highest IVIs were for *Amaioua guianensis, Metrodorea*

Table 19.7 Phytosociological parameters for all trees and shrubs ≥5 cm diam., 30 cm above ground level on a 100 point point-centred-quarter transect in cerradão (abbreviations as in Table 19.6; no. of trees ha^{-1} = 815; basal area ha^{-1} = 21.94 m^2)

		R.De.	*R.Do.*	*RF*	*IVI*
1.	*Emmotum nitens*	6.35	23.97	8.26	38.58
2.	*Blepharocalyx salicifolia*	10.11	13.38	10.25	33.74
3.	*Didymopanax macrocarpum*	10.11	2.01	11.50	23.62
4.	*Qualea grandiflora*	8.50	4.36	7.22	20.08
5.	*Simarouba versicolor*	5.25	7.40	5.49	18.15
6.	*Siphoneugena densiflora*	3.50	6.02	3.18	12.70
7.	*Ocotea spixiana*	2.75	2.60	6.63	11.98
8.	*Plathymenia reticulata*	3.50	3.92	4.05	11.46
9.	*Sclerolobium paniculatum*	4.25	1.68	4.05	9.98
10.	*Bowdichia virgilioides*	2.50	2.57	2.89	7.96
11.	*Pouteria ramiflora*	2.75	2.11	2.89	7.75
12.	*Copaifera langsdorffii*	5.23	1.25	1.15	7.64
13.	*Caryocar brasiliense*	2.55	2.25	2.31	7.11
14.	*Aspidosperma macrocarpon*	2.50	1.38	2.89	6.77
15.	*Dalbergia violacea*	1.50	3.17	1.73	6.41
16.	*Qualea parviflora*	2.00	1.12	2.31	5.43
17.	*Roupala montana*	2.00	0.62	2.31	4.93
18.	*Miconia ferruginata*	2.00	0.40	2.31	4.71
19.	*Aspidosperma tomentosum*	2.00	0.59	2.02	4.61
20.	*Qualea multiflora*	1.75	0.37	1.73	3.85
21.	*Ouratea hexasperma*	1.50	0.48	1.73	3.71
22.	*Acosmium dasycarpum*	1.50	0.35	1.73	3.58
23.	*Pterodon pubescens*	1.00	1.27	1.15	3.42
24.	*Tapirira guianensis*	1.00	0.93	1.15	3.08
25.	*Salacia crassifolia*	1.00	0.65	1.16	2.80
26.	*Lafoensia pacari*	1.00	0.31	1.16	2.47
27.	*Miconia albicans*	0.75	0.35	0.87	1.97
28.	*Callisthene major*	0.25	1.28	0.29	1.82
29.	*Dimorphandra mollis*	0.75	0.39	0.58	1.72
30.	*Eremanthus glomerulatus*	0.75	0.08	0.87	1.70
	Erythroxylum suberosum	0.75	0.08	0.87	1.70
32.	*Andira paniculata*	0.50	0.57	0.57	1.65
33.	*Tabebuia serratifolia*	0.50	0.46	0.57	1.54
34.	*Vochysia thyrsoidea*	0.50	0.39	0.58	1.47
35.	*Byrsonima crassa*	0.75	0.15	0.58	1.48
36.	*Byrsonima coccolobifolia*	0.50	0.13	0.58	1.21
37.	*Eriotheca pubescens*	0.25	0.63	0.29	1.17
38.	*Ilex conocarpa*	0.50	0.07	0.59	1.15
	Psidium warmingianum	0.50	0.07	0.58	1.15
40.	*Kielmeyera coriacea*	0.50	0.03	0.58	1.12
41.	*Vochysia tucanorum*	0.25	0.21	0.29	0.75
42.	*Alibertia edulis*	0.25	0.14	0.29	0.68
43.	*Gomidesia lindeniana*	0.25	0.13	0.29	0.67
44.	*Ocotea* sp	0.25	0.13	0.28	0.66
45.	*Zanthoxylum rhoifolium*	0.25	0.09	0.29	0.63

Table 19.7 *(continued)*

		R.De.	R.Do.	RF	IVI
46.	*Erythroxylum deciduum*	0.25	0.06	0.29	0.60
	Mimosa claussenii	0.25	0.06	0.29	0.60
48.	*Miconia sellowiana*	0.25	0.04	0.29	0.58
	Miconia pohliana	0.25	0.04	0.28	0.58
50.	*Neea theifera*	0.25	0.03	0.29	0.57
	Rapanea guianensis	0.25	0.03	0.29	0.57
	Enterolobium gummiferum	0.25	0.03	0.29	0.57
	Machaerium acutifolium	0.25	0.03	0.29	0.57
	Hancornia speciosa	0.25	0.03	0.29	0.57
	Davilla elliptica	0.25	0.03	0.29	0.57
56.	*Maprounea guianensis*	0.25	0.03	0.29	0.57
57.	*Machaerium opacum*	0.25	0.02	0.29	0.56
	Byrsonima verbascifolia	0.25	0.02	0.29	0.56
	Vochysia elliptica	0.25	0.02	0.29	0.56
60.	*Erythroxylum tortuosum*	0.25	0.02	0.28	0.55

Table 19.8 Phytosociological parameters for all trees ≥5 cm d.b.h. in an area of 1 ha in the gallery forest of the Córrego Capetinga (abbreviations as in Table 19.6; no. of trees ha^{-1} = 982; basal area ha^{-1} = 21.40 m^2)

		R.De.	R.Do.	RF	IVI
1.	*Amaioua guianensis*	15.27	14.73	8.10	38.10
2.	Dead tree	16.50	7.72	8.10	32.32
3.	*Piptocarpha macropoda*	7.84	3.50	3.39	14.73
4.	*Protium brasiliense*	4.68	4.42	3.95	13.05
5.	*Cryptocarya aschersoniana*	2.04	5.70	2.82	10.56
6.	*Licania apetala*	3.36	4.04	2.64	10.08
7.	*Maytenus alaternoides*	3.16	3.06	3.39	9.61
8.	*Aspidosperma olivaceum*	2.34	5.21	1.69	9.24
9.	*Tapirira guianensis*	1.93	3.98	2.82	8.74
10.	*Pouteria ramiflora*	2.65	2.12	3.58	8.36
11.	*Guatteria sellowiana*	2.34	2.43	3.20	7.98
12.	*Nectandra mollis*	1.32	3.07	2.45	6.84
13.	*Gomidesia lindeniana*	2.34	1.47	2.82	6.64
14.	*Maprounea guianensis*	1.43	2.64	2.07	6.14
15.	*Copaifera langsdorffii*	1.32	2.55	2.26	6.13
16.	Myrtaceae indet.	2.04	1.02	3.01	6.07
17.	*Cecropia pachystachya*	2.95	0.83	2.07	5.86
18.	*Sclerolobium paniculatum*	1.43	1.65	2.26	5.33
19.	*Cabralea canjerana*	1.02	2.69	1.32	5.03
20.	*Euplassa inaequalis*	0.92	2.15	1.51	4.57
21.	*Saccoglottis guianensis*	0.81	2.01	1.32	4.14
22.	*Ocotea* sp.	0.81	1.62	1.51	3.94
23.	*Aspidosperma subincanum*	0.61	2.15	0.94	3.70
24.	*Inga alba*	1.22	0.89	1.51	3.62

Table 19.8 *(continued)*

		R.De.	*R.Do.*	*RF*	*IVI*
25.	*Terminalia brasiliensis*	0.41	2.18	0.56	3.15
26.	*Emmotum nitens*	0.51	1.66	0.94	3.12
27.	*Machaerium acutifolium*	0.01	1.15	1.13	3.09
28.	*Virola sebifera*	0.81	0.71	1.51	3.03
29.	*Ouratea castaneaefolia*	0.81	0.45	1.51	2.77
30.	*Ixora warmingii*	0.92	0.48	1.32	2.72
31.	*Miconia sellowiana*	0.92	0.47	1.32	2.70
32.	*Cheiloclinum cognatum*	1.02	0.68	0.94	2.64
33.	*Guapira* sp.	0.71	0.76	1.13	2.60
34.	*Callisthene major*	0.51	1.01	0.94	2.46
35.	*Hirtella glandulosa*	0.71	0.24	1.32	2.27
36.	*Casearia sylvestris*	0.71	0.33	1.13	2.18
37.	*Micropholis venulosa*	0.31	1.18	0.56	2.05
38.	*Pseudolmedia laevigata*	0.61	0.12	1.13	1.86
39.	*Alibertia edulis*	0.71	0.15	0.94	1.80
40.	*Myrsine coriacea*	0.20	1.21	0.38	1.79
41.	*Matayba guianensis*	0.41	0.47	0.75	1.63
42.	*Rheedia brasiliensis*	0.20	0.84	0.38	1.42
43.	*Sorocea illicifolia*	0.51	0.15	0.75	1.41
	Myrtaceae indet.	0.41	0.25	0.75	1.41
45.	*Salacia elliptica*	0.41	0.42	0.56	1.39
46.	*Ouratea* sp.	0.41	0.20	0.75	1.37
47.	*Myrsine coriaceae*	0.51	0.42	0.38	1.30
48.	Indet.	0.41	0.12	0.75	1.28
49.	*Chrysophyllum* sp.	0.31	0.28	0.56	1.15
50.	*Luehea paniculata*	0.41	0.15	0.56	1.13
	Qualea dichotoma	0.41	0.15	0.56	1.13
52.	*Xylopia emarginata*	0.41	0.11	0.56	1.08
53.	*Tapura amazonica*	0.41	0.16	0.38	0.94
54.	*Astronium fraxinifolium*	0.20	0.35	0.38	0.93
55.	*Siparuna guianensis*	0.31	0.03	0.56	0.91
56.	*Dalbergia frutescens*	0.20	0.03	0.38	0.61
	Gilibertia sp.	0.20	0.03	0.38	0.61
58.	*Cupania vernalis*	0.20	0.02	0.38	0.60
	Miconia sp.	0.20	0.02	0.38	0.60
60.	*Tabebuia ochracea*	0.10	0.24	0.19	0.53
	Cordia sp.	0.10	0.24	0.19	0.53
61.	*Xylopia sericea*	0.20	0.09	0.19	0.48
62.	*Apuleia molaris*	0.10	0.15	0.19	0.44
63.	*Aspidosperma cylindrocarpum*	0.10	0.13	0.19	0.42
64.	*Vismia* sp.	0.10	0.09	0.19	0.38
	Byrsonima laxiflora	0.10	0.05	0.19	0.34
65.	*Psychotria* sp.	0.10	0.05	0.19	0.34
	Myrtaceae indet.	0.10	0.05	0.19	0.34
68.	*Alchornea iricurana*	0.10	0.04	0.19	0.33
	Erythroxylum sp.	0.10	0.03	0.19	0.32
69.	*Solanum* sp.	0.10	0.03	0.19	0.32
	Myrcia deflexa	0.10	0.03	0.19	0.32

Table 19.8 *Continued*

		R.De.	R.Do.	RF	IVI
72.	Myrtaceae indet.	0.10	0.02	0.19	0.31
	Hymenaea courbaril var. *stilbocarpa*	0.10	0.02	0.19	0.31
	Ormosia sp.	0.10	0.02	0.19	0.31
	Miconia punctata	0.10	0.02	0.19	0.31
	Campomanesia sp.	0.10	0.01	0.19	0.31
77.	Myrtaceae indet.	0.10	0.01	0.19	0.30
	Psidium warmingianum	0.10	0.01	0.19	0.30
	Roupala montana	0.10	0.10	0.19	0.30
	Rubiaceae indet.	0.10	0.10	0.19	0.30
	Leguminosae indet.	0.10	0.10	0.19	0.30

pubescens, Licania apetala, Cheiloclinum cognatum, Tapirira guianensis, Lamanonia tomentosa, Guatteria sellowiana, Protium brasiliense, Copaifera langsdorffii, Maytenus alaeternoides, Nectandra mollis, Piptocarpha macropoda and *Cryptocarya aschersoniana*. These 13 species represented about 50% of the total IVI value (Table 19.9). In the Gama gallery forest the dominance of species was weaker than in the Capetinga forest: *Amaioua guianensis*, the main species, had only 6.1% of the total IVI in the former as opposed to 12.7% in the latter. Dead standing trees had the highest IVI, but made up only 5% of the total number of trees per hectare. The reason for their high IVI value was the large basal area: the forest is undisturbed and as expected there are many large dead standing trees.

Despite being close neighbours, the Capetinga and the Gama gallery forests had exclusive species: for example, *Metrodorea pubescens* had the second position in IVI in the Gama forest and was absent in the Capetinga gallery forest.

Some species such as *Roupala montana, Qualea multiflora* and *Sclerolobium paniculatum* show roughly similar numbers of trees per hectare in cerrado *s.s.* and in cerradão; but in general a number of exclusive, or almost exclusive, species predominate in the cerradão (*Emmotum nitens, Blepharocalyx salicifolia, Simarouba versicolor, Ocotea spixiana, Plathymenia reticulata, Copaifera langsdorfii*, etc.), while others such as *Qualea parviflora* and *Ouratea hexasperma* are very important in cerrado *s.s.* The occurrence of distinct cerradão elements in the flora of Fazenda Água Limpa is discussed by Ratter (1980, 1985, 1986). Only *Pouteria ramiflora, Sclerolobium paniculatum* and *Roupala montana* occurred in all vegetation types studied.

Table 19.9 Phytosociological parameters for all trees ≥5 cm d.b.h. in an area of 3.2 ha of the gallery forest of the Riberão do Gama (abbreviations as in Table 19.6; no. of trees ha^{-1} = 1350.5; basal area ha^{-1} = 60.1 m^2)

		R.De.	*R.Do.*	*RF*	*IVI*
1.	Dead tree	5.64	10.02	2.83	18.49
2.	*Amaioua guianensis*	8.66	6.33	3.33	18.32
3.	*Metrodorea pubescens*	5.72	4.47	3.17	13.36
4.	*Licania apetala*	4.59	3.97	3.17	11.75
5.	*Cheiloclinum cognatum*	5.04	3.40	3.17	11.61
6.	*Tapirira guianensis*	4.26	4.46	2.34	11.06
7.	*Lamanonia tomentosa*	2.28	5.65	1.84	9.77
8.	*Guatteria sellowiana*	3.91	3.38	2.34	9.63
9.	*Protium brasiliense*	4.16	3.78	1.50	9.44
10.	*Copaifera langsdorffii*	2.21	4.55	2.67	9.43
11.	*Maytenus alaternoides*	3.39	2.51	2.17	8.07
12.	*Nectandra mollis*	2.14	2.63	2.34	7.11
13.	*Piptocarpha macropoda*	2.68	2.03	2.17	6.88
14.	*Cryptocarya aschersoniana*	2.31	2.42	1.50	6.23
15.	*Aspidosperma* sp.	1.32	1.84	2.33	5.59
16.	*Callisthene major*	1.03	2.53	1.67	5.23
17.	*Matayba guianensis*	1.82	1.38	1.67	4.87
18.	*Aspidosperma olivaceum*	1.43	2.08	1.33	4.84
19.	*Ouratea castaneaefolia*	2.08	1.42	1.17	4.67
	Inga alba	1.72	1.58	1.17	4.47
21.	*Emmotum nitens*	1.05	1.02	1.84	3.91
22.	*Cupania vernalis*	1.03	0.68	1.84	3.55
23.	*Machaerium acutifolium*	1.09	1.21	2.33	4.63
24.	*Saccoglottis guianensis*	0.72	0.60	1.84	3.16
25.	*Miconia sellowiana*	1.29	0.91	0.83	3.03
26.	*Cabralea canjerana*	0.82	0.54	1.50	2.86
27.	*Virola sebifera*	0.95	0.66	1.17	2.78
28.	*Gomidesia lindeniana*	1.07	0.88	0.83	2.73
29.	*Sclerolobium paniculatum*	0.66	0.72	1.17	2.55
30.	Myrtaceae indet.	0.63	0.50	1.34	2.47
31.	*Siparuna guianensis*	0.72	0.52	1.12	2.43
32.	*Sorocea ilicifolia*	0.64	0.58	1.16	2.32
33.	*Terminalia brasiliensis*	0.41	1.06	0.83	2.30
	Roupala montana	0.58	0.38	1.34	2.30
35.	*Qualea multiflora*	0.48	1.27	0.50	2.25
36.	*Byrsonima laxiflora*	0.66	0.38	1.16	2.20
37.	*Maprounea guianensis*	0.43	0.59	1.17	2.19
38.	*Vochysia pyramidalis*	0.93	0.89	0.33	2.15
39.	*Myrsine coriacea*	0.43	0.70	1.00	2.13
40.	Leguminosae indet.	0.42	0.36	1.34	2.12
41.	*Guapira* sp.	0.56	0.37	1.16	2.09
42.	*Salacia elliptica*	0.42	0.16	1.50	2.08
43.	*Hirtella glandulosa*	0.50	0.37	1.17	2.04
44.	*Alchornea iricurana*	0.62	0.41	0.83	1.86
45.	*Astronium fraxinifolium*	0.41	0.43	1.00	1.84

Table 19.9 *(continued)*

		R.De.	R.Do.	RF	IVI
46.	*Aspidosperma cylindrocarpum*	0.55	0.27	1.00	1.82
	Pouteria ramiflora	0.56	0.43	0.83	1.82
48.	*Psychotria* sp.	0.54	0.39	0.83	1.76
49.	*Ocotea* sp.	0.56	0.42	0.67	1.65
50.	Myrtaceae indet.	0.44	0.29	0.83	1.56
51.	*Aspidosperma subincanum*	0.35	0.36	0.83	1.54
	Xylopia emarginata	0.41	0.30	0.83	1.54
53.	*Ixora warmingii*	0.29	0.23	1.00	1.52
54.	*Casearia sylvestris*	0.57	0.36	0.50	1.43
55.	*Diospyros hispida*	0.32	0.24	0.83	1.39
56.	*Miconia* sp.	0.37	0.33	0.67	1.37
57.	*Miconia* sp.	0.37	0.32	0.67	1.36
58.	*Xylopia sericea*	0.56	0.30	0.50	1.35
59.	*Terminalia argentea*	0.19	0.47	0.66	1.32
60.	*Didymopanax morototoni*	0.52	0.29	0.50	1.31
61.	*Erythroxylum* sp.	0.63	0.51	0.16	1.30
61.	Lauraceae indet.	0.24	0.22	0.83	1.29
62.	*Qualea dichotoma*	0.20	0.55	0.50	1.25
63.	*Chrysophyllum* sp.	0.45	0.39	0.33	1.17
64.	*Pseudolmedia laevigata*	0.49	0.33	0.33	1.15
65.	*Calophyllum brasiliense*	0.33	0.48	0.33	1.14
67.	*Miconia* sp.	0.27	0.16	0.67	1.10
68.	*Gilibertia* sp.	0.28	0.14	0.67	1.09
69.	*Styrax guianensis*	0.58	0.33	0.16	1.07
70.	*Ouratea* sp.	0.24	0.12	0.67	1.03
71.	*Cordia* sp.	0.17	0.18	0.67	1.02
72.	*Andira paniculata*	0.26	0.19	0.50	0.95
73.	*Zanthoxylum rhoifolium*	0.20	0.20	0.50	0.90
74.	*Tapura amazonica*	0.47	0.07	0.33	0.87
75.	Myrtaceae indet.	0.20	0.14	0.50	0.86
76.	*Eriotheca gracilipes*	0.07	0.43	0.33	0.83
77.	*Apuleia molaris*	0.21	0.16	0.45	0.82
78.	Apocynaceae indet.	0.13	0.27	0.41	0.81
79.	*Cecropia pachystachya*	0.31	0.10	0.33	0.74
80.	*Bauhinia* sp.	0.49	0.07	0.16	0.72
81.	*Luehea paniculata*	0.09	0.10	0.50	0.69
82.	*Sorocea* sp.	0.34	0.18	0.16	0.68
83.	Indet.	0.11	0.16	0.36	0.63
	Moraceae indet.	0.07	0.40	0.16	0.63
85.	*Miconia* sp.	0.11	0.16	0.34	0.61
	Hymenaea courbaril	0.07	0.21	0.33	0.61
87.	*Miconia punctata*	0.26	0.18	0.16	0.60
88.	*Rapanea guianensis*	0.24	0.19	0.16	0.59
89.	*Miconia* sp.	0.05	0.16	0.30	0.51
90.	*Guettarda* sp.	0.18	0.12	0.16	0.46
91.	*Tabebuia ochracea*	0.05	0.16	0.24	0.45
92.	*Rheedia brasiliensis*	0.13	0.15	0.16	0.44

Table 19.9 *(continued)*

		R.De.	*R.Do.*	*RF*	*IVI*
93.	*Apeiba tibourbou*	0.08	0.15	0.16	0.39
94.	Myrtaceae indet.	0.13	0.07	0.16	0.36
95.	*Euplassa inaequalis*	0.07	0.12	0.16	0.35
96.	*Myrcia deflexa*	0.10	0.06	0.16	0.32
	Vismia sp.	0.08	0.08	0.16	0.32
98.	Rubiaceae indet.	0.07	0.04	0.16	0.27
99.	Myrtaceae indet.	0.05	0.02	0.16	0.23
100.	*Solanum* sp.	0.02	0.03	0.16	0.21
101.	*Cariniana estrellensis*	0.02	0.01	0.16	0.19
	Campomanesia sp.	0.02	0.01	0.16	0.19
	Sloanea sp.	0.02	0.01	0.16	0.19
	Alibertia edulis	0.02	0.01	0.16	0.19
	Ferdinandusa speciosa	0.02	0.01	0.16	0.19

Species–area curves

Species–area curves are given for all vegetation types in Fig. 19.1. As can be seen, those for cerrado *s.s.* and the gallery forest of the Riberão de Gama flatten off above, whereas those of the cerradão and the Capetinga gallery forest are still climbing quite rapidly. As would be expected, in the latter two localities smaller areas were sampled. In all cases more than 60% of the total number of species were found in the first 50% of sampling.

ACKNOWLEDGEMENTS

We thank Mr Newton Rodrigues de Oliveira and some of our students and colleagues at the University of Brasília and CENARGEN-EMBRAPA for field assistance and Dr Timothy Synnott, Mr Howard Wright and Mr Gareth Phillips of the Oxford Forestry Institute for the valuable corrections and suggestions.

REFERENCES

Cain, S.A. and Castro, G.M.O. (1959) *Manual of Vegetation Analysis*, Harper & Brothers, New York.

Cottam, G. and Curtis, J.T. (1956) The use of distance measures in phytosociological sampling. *Ecology*, **37**, 451–60.

Eiten, G. (1972) The cerrado vegetation of Brazil. *Botanical Review*, **38**, 201–341.

Furley, P.A. (1985) *Notes on the Soils and Plant Communities of Fazenda Água Limpa (Brasília, DF, Brazil)*, Dept of Geography Occasional Publications, NS, No. 5, University of Edinburgh, Edinburgh, UK.

Horn, H.S. (1966) Measurement of 'overlap' in comparative ecological studies. *American Naturalist*, **100**, 419–24.

Husch, B., Miller, C.I. and Beers, T.W. (1972) *Forest Mensuration*, 2nd edn, John Wiley, New York.

Mueller-Dombois, D. and Ellenberg, M. (1974) *Aims and Methods in Vegetation Ecology*, John Wiley, New York.

Pielou, E.C. (1977) *Mathematical Ecology*, 2nd edn, John Wiley, New York.

Ratter, J.A. (1980) *Notes on The Vegetation of Fazenda Água Limpa (Brasília, DF, Brazil)*, Royal Botanic Garden, Edinburgh, UK.

Ratter, J.A. (1985) *Notas Sobre a Vegetação da Fazenda Água Limpa (Brasília, DF); com uma Chave para os Gêneros Lenhosos de Dicotyledôneas do Cerrado*, Royal Botanic Garden, Edinburgh, UK.

Ratter, J.A. (1986) *Notas Sobre a Vegetação da Fazenda Água Limpa (Brasília, DF); com uma Chave para os Gêneros Lenhosos de Dicotyledôneas do Cerrado*, Editora Universidade de Brasília, textos universitários no. 003. Brasília.

Sørensen, T.A. (1948) A method of establishing groups of equal amplitude in plant sociology based on similarity of species content, and its application to analyses of the vegetation on Danish commons. *Det Kongelige danske Videnskabernes Selskab Bioloqiske Skrifter*, **5**, 1–34.

20

Transitions between cerrado and forest vegetation in Brazil

J.A. RATTER

SUMMARY

1. The paper discusses transitions between *cerrado* vegetation and forest in central Brazil and southern Amazonia and between hydrologic savanna and forest in Roraima.
2. Transition from deciduous and semi-deciduous forest to cerrado is related to soil conditions: the forest occurs on more fertile soils with higher levels of calcium and magnesium and a characteristic *cerradão* community (termed mesotrophic facies cerradão or *Callisthene fasciculata/Magonia pubescens* cerradão) is found on the intermediate mesotrophic soils lying between the fertile forest soils and the dystrophic soils of the cerrado.
3. A succession from cerrado towards forest related to protection from fire was observed on dystrophic soils in some areas of the Federal District. After some years of protection, large numbers of saplings of fire-sensitive species became established and a dense cerradão tending towards forest was established. Similar observations were made in a southern cerrado area in São Paulo state.
4. A study of the area of contact of cerrado and Amazonian forest on dystrophic soils on the Serra do Roncador, east Mato Grosso, demonstrated a characteristic ecotonal cerradão lying as a band, sometimes 3–4 km broad, between the two vegetation types. Presence of giant, relict, cerrado trees 4–5 km within the Amazonian forest indicated that the forest was expanding into the cerrado, with the ecotonal cerradão, in which saplings of forest trees were establishing, representing its advancing face. Other observations on the Amazonian forest–cerrado transition at Iquê on the Mato Grosso border demonstrated a transitional vegetation dominated by the leguminous tree *Sclerolobium paniculatum*.

5. Investigation of the abrupt transition between Amazonian forest and hydrologic savanna on Maracá Island, Roraima, have demonstrated that it is related to fluctuations in groundwater: the hydrologic savanna being tolerant of a seasonally high water table while the forest is not.
6. One of the conclusions emerging from the study is that the cerrado–forest transition is often dynamic, with forest expanding into cerrado in many localities. This seems to be a continuation of the forest expansion which has been occurring in South America since the end of the last glaciation – a process which is often masked by destruction caused by man's agricultural activities.

INTRODUCTION

Cerrado *sensu lato* is the predominant natural vegetation of central Brazil and before the recent extensive destruction for agriculture covered approximately $2 \times 10^6\,km^2$, representing about 23% of the land surface of the country. In terms of area it is the second most important vegetation formation of Brazil, exceeded only by the Amazonian forest with $3.5 \times 10^6\,km^2$. It is a savanna vegetation which varies from treeless grassland to dense woodland, and represents the climax on the poor, freely drained soils which are so abundant over the central plateau. A brief summary of the characteristics of cerrado vegetation are given in Furley and Ratter (1988) and the distribution of cerrado and Amazonian forest in Brazil is given in Fig. 20.1.

Although the climate of the cerrado region varies considerably, there are a number of well defined characteristics: (a) a pronounced wet season generally from about late September to April and a dry season coincident with the coldest months of the year, annual precipitation averages approximately 1400 mm (<800 mm in NE to >2000 mm in NW extremes); (b) an average temperature range from 20 to 25 °C; and (c) an annual evapotranspiration deficit (Thornthwaite's method) of 4–491 mm (Lopes and Cox, 1977). Cerrado vegetation represents a climatic climax on dystrophic soils, and, as would be expected, precipitation figures for cerrado areas at the margin of the Amazonian forest are at about the maximum of the range given above – the climate in general is the transition of Köppen's Savanna (Aw) and Monsoon (Am) subtypes of the Tropical Rain Climate (A).

Like all savanna vegetation, the cerrado is very resistant to burning, and occasional natural fires must have been an environmental factor throughout its long history stretching back to the Miocene, Eocene or perhaps even earlier (Van der Hammen, 1983). Excessively frequent

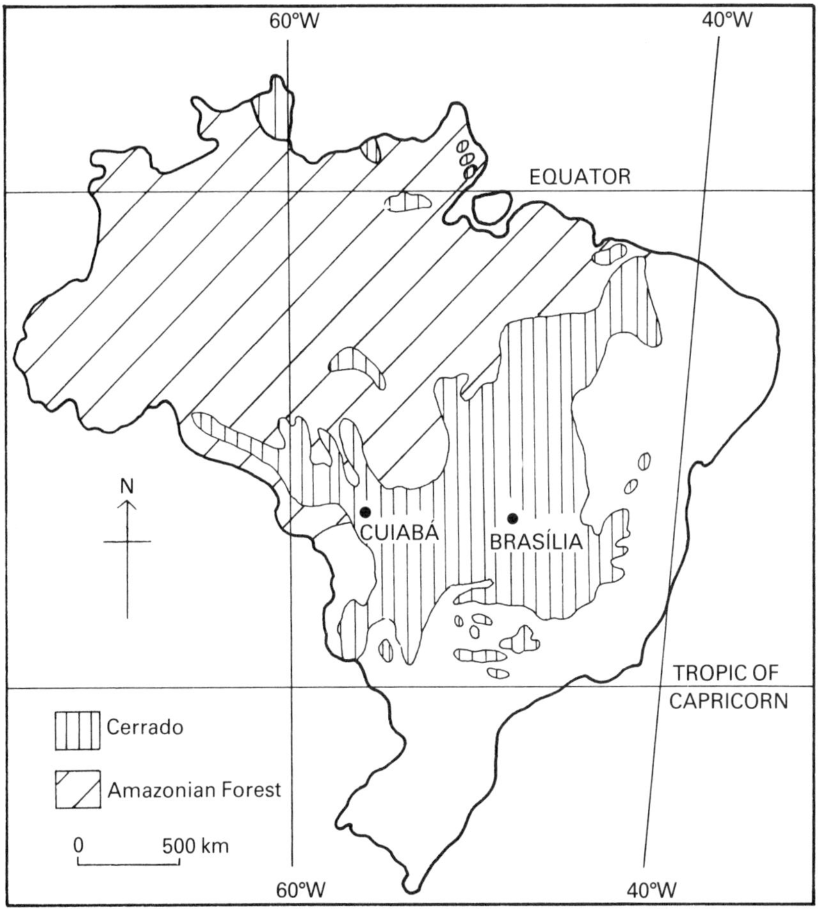

Fig. 20.1 Distribution of cerrado and Amazonian forest in Brazil.

burning, however, favours the herbaceous as opposed to the woody component of the vegetation and when this occurs woodlands are transformed into grasslands. Conversely, protection from fire often has the opposite effect and a dense woodland with considerable shade and only a very sparse ground layer is established. Fire is thus an important factor in determining the physiognomy of cerrado which forms a continuum from open grasslands to dense woodlands in which more or less recognizable nodes are given vernacular names. Dry grassland without shrubs or trees is called *campo limpo* (clean field); grassland with a scattering of shrubs and trees is known as *campo sujo* (dirty field); where there are numerous trees and shrubs but still a large area of grassland, the vegetation is termed *campo cerrado* (closed field); the next stage when the vegetation is obviously dominated by trees and shrubs but there is still a fair amount of herbaceous vegetation between them is known as cerrado* (closed, i.e. the vegetation has closed); the last stage is an almost closed woodland made up of trees, often of 8–12 m or even taller, casting a considerable shade so that ground vegetation is much reduced – this form is called *cerradão* (augmentative of cerrado). Clearly the dividing line between the last three forms is somewhat arbitrary, but workers in the field usually agree surprisingly well on the classification of the vegetation.

Within the cerrado region, however, the vegetation is by no means exclusively cerrado *sensu lato*. The typical landscape pattern consists of cerrado woodlands on the well drained interfluves and gallery forests following the watercourses. These two types of vegetation are often separated by campo grasslands which occur in valleys with a wide annual fluctuation in water table.

Cerrado is essentially a vegetation of poor dystrophic soils and where more fertile soils occur in the cerrado domain they are occupied by deciduous and semi-deciduous forests. Formerly, there were large expanses of such forests, but since the soils are ideal for agriculture they have been almost entirely cleared.

In addition to the typical cerrados of well drained soils, a relatively species-poor type of savanna is found in areas where seasonal water-logging occurs. This is hydrologic savanna (Denevan, 1968) which is widespread in Central America, South America north of the Amazonian forest, and in wet parts of Brazil's cerrado area, such as the Pantanal and the Ilha do Bananal.

For many years the author has worked in central Brazil and, to a lesser extent, in Roraima in northern Amazonia. The observations which follow relate to a number of different forest–cerrado and forest–hydrologic savanna transitions which occur in these areas.

TRANSITION BETWEEN DECIDUOUS FOREST AND CERRADO

As already mentioned, more fertile soils with higher levels of calcium and magnesium occurring in the cerrado landscape are covered with deciduous or semi-deciduous forest. Such forests are dominated by a very characteristic calciphilous community of tree species consisting of

*i.e. cerrado *sensu stricto* – it is unfortunate that, in common usage, the term should refer both to Brazilian savanna vegetation in its generic sense and to one particular subvariant of it.

Astronium urundeuva (Fr. All.) Engl., *A. fraxinifolium* Schott., *Anadenanthera* spp., *Aspidosperma subincanum* Mart., *Cedrela fissilis* Vell., *Ceiba* spp., *Sterculia striata* St. Hil. and Naud., *Tabebuia avellanedae* [Lor. ex] Griseb., etc. (Ratter *et al.*, 1973, 1978a).

We have followed the transition from deciduous forest to cerrado in over a dozen localities in the states of Goiás, Mato Grosso and Minas Gerais and in the Federal District. The transition is related to the change from mineral-rich meso- or eutrophic soils to poor dystrophic soils (Table 20.1). In all areas studied a floristically very characteristic cerradão is found on the transitional soils. It has a number of important marker species, such as *Callisthene fasciculata* (Spr.) Mart., *Magonia pubescens* St. Hil., *Luehea paniculata* Mart., *Pseudobombax tomentosum* (Mart. and Zucc.) A. Robyns, *Dilodendron bipinnatum* Radlk. and *Dipteryx alata* Vog., which do not grow on poor dystrophic soils. This vegetation is termed mesotrophic facies cerradão or *Callisthene fasciculata*/*Magonia pubescens* cerradão, and has been described in detail in a number of papers (e.g. Ratter, 1971; Ratter *et al.*, 1973, 1977). It is found on soils with surface calcium levels >2 mequiv 100 g^{-1} and covers large areas where such soils, intermediate between those characteristic of cerrado on the one hand and deciduous forest on the other, are extensive (e.g. in the Valley of the Rio Paranã in Goiás). Soil analyses for a number of mesotrophic facies cerradão–deciduous forest ecotones are given in Furley and Ratter (1988) and Furley *et al.* (1988). They show that there is a considerable overlap in the soil conditions characteristic of each vegetation type, i.e. a soil of the fertility that carries deciduous forest in one area may bear mesotrophic facies cerradão in another. This overlap seems to have sometimes allowed a succession from mesotrophic facies cerradão to forest to occur, as demonstrated by the occurrence of giant relict cerradão trees in the forest. It should, however, be pointed out that the two types of vegetation are closely related and that some of the species, such as *Dilodendron bipinnatum*, seem equally happy as cerradão or forest trees.

Table 20.1 Analysis of surface layers of soil of the Morro de Fumaça, Torixoreu, Mato Grosso, in areas of deciduous forest, mesotrophic facies cerradão and cerrado (surface sampling of the A-horizon)

	pH (H_2O)	*ex. Ca*	*ex. Mg* (*mequiv. 100 g^{-1}*)	*ex. K*	*Available P* (*ppm*)
Deciduous seasonal forest	6.7	16.9	4.1	0.49	60
Mesotrophic facies cerradão	6.1	5.8	3.6	0.49	8
Cerrado	5.3	0.9	0.5	0.37	2

From Furley *et al.*, 1988.

The deciduous and semi-deciduous forests, like the cerrado, seem to be highly fire-resistant and many of the trees such as *Anadenanthera* and *Erythrina* species are notable for their thick, corky barks. The transition from deciduous forest to cerrado therefore seems to be related entirely to edaphic factors with fire having little significance.

TRANSITIONS AND SUCCESSION ON DYSTROPHIC SOILS

Most of our observations on forest–cerrado transitions and successions on dystrophic soils have been made in the Federal District and on the margin of the Amazonian forest in eastern Mato Grosso. In both areas, forest and cerrado are on essentially similar nutrient-poor soils (Table 20.2).

Federal District

The chief study area was Fazenda Água Limpa, the University of Brasília Experimental Station at 15°57′S, 47°56′W, some 18 km south-west of the centre of Brasília. It has an area of 4062 ha and the principal vegetation is cerrado, thickening in areas to cerradão, with well developed gallery forests along the streams. A large part of the area is an ecological reserve and has been protected from fire, with varying degrees of effectiveness, for the past 14 years.

In one area of cerrado and cerradão a striking thickening of the woody vegetation occurred after five years' protection from fire and this was more pronounced after ten years. Great numbers of very fire-sensitive saplings of species such as *Siphoneugena densiflora* Berg, *Blepharocalyx salicifolia* Berg, *Emmotum nitens* (Benth.) Miers, *Tapirira guianensis* Aubl., *Virola sebifera* Aubl. and *Simarouba versicolor* St. Hil. were established. These species are not found in more open forms of cerrado but are characteristic of cerradão

Table 20.2 Analysis of surface layers of soil (0–15 cm) of adjoining Amazonian forest (*mata seca*), transitional (*Hirtella glandulosa*/dystrophic) cerradão and cerrado

	pH (H_2O)	*ex. Ca*	*ex. Mg* *(mequiv. 100 g^{-1})*	*ex. K*	*available P (ppm)*
Forest	4.7	0.04	0.03	0.09	22
Cerradão	4.1	0.07	0.01	0.04	27
Cerrado	4.8	0.05	0.19	0.07	14

From previously unpublished data, Askew *et al.*, *in litt.*

and some types of forest. We excavated the root systems of some of these young plants and discovered that *S. densiflora* and *B. salicifolia* were establishing from seed, whereas *S. versicolor* was spreading by woody stolons from parental trees (see also Raw and Hay, 1985, for observations on *Simarouba*). The vegetation was following a succession towards a type of forest occurring at an area a few kilometres from Fazenda Água Limpa (the Escola Fazendária) and in other places in the Federal District. Such forests have a floristic composition similar to that being established at Fazenda Água Limpa, but also include more exclusively forest species such as the pioneers *Jacaranda copaia* (Aubl.) D. Don, *Amaioua guianensis* Aubl. and *Piptocarpha macropoda* (DC.) Baker. The succession depends on protection from burning and for establishment of forest species probably provision of some shade. It can easily be reversed, at least in the early stages, as was demonstrated by a disastrous fire which entered the reserve in 1987 and destroyed thousands of saplings of the fire-sensitive species. Further information is given in Ratter (1980, 1985, 1986).

Expansion of gallery forest into fringing cerrados has also taken place at Fazenda Água Limpa during the last 14 years. This has occurred particularly at the headwaters of streams, the so-called *cabeceiras*, where there is often a considerable area of forest surrounded by a narrow belt of cerrado abutting on to damp campo. Young forest trees, such as *Cupania vernalis* Camb. and *Maprounea guianensis* Aubl., establish in the fringing cerrado, and there is evidence from the presence of large trees of cerrado species deep within the cabeceira forest that this process of forest expansion has continued for a long period.

Evidence of a similar succession was also obtained in a study of a small cerrado reserve at Angatuba in São Paulo state (Ratter *et al.*, 1988). Here there had been protection from fire for at least 19 and perhaps for as long as 35 years, and in places the cerrado vegetation had closed to produce a thicket with over 10 000 trees of $\geqslant$3 cm dbh per hectare and many fire-susceptible and forest species were establishing.

Transition to Amazonian forest

The transition between Amazonian forest and cerrado vegetation was studied in the Serra do Roncador region of Mato Grosso by the Royal Society/Royal Geographical Society Xavantina–Cachimbo Expedition of 1967–69 and in a subsequent expedition in 1972 (Ratter *et al.*, 1973, 1978b). The work was centred on a base camp at 12°49′S, 51°46′W which lay right in the transition.

The Amazonian forest in this area is the widespread marginal form known as *mata seca* or 'dry forest' (Pires, 1974; Pires and Prance, 1985). It is evergreen, with only a tiny percentage of the trees having a short deciduous period, and consists in the main of widespread Hylaean species. In most places there is a belt of ecotonal vegetation, sometimes 3–4 km broad, between the forest and cerrado. This is a cerradão containing the typical cerrado species of the area but dominated by a number of characteristic cerradão tree species generally not found in more open forms of cerrado, the most important of which are *Hirtella glandulosa* Spreng., *Emmotum nitens* (Benth.) Miers, *Sclerolobium paniculatum* Benth. and *Vochysia haenkeana* Mart. The nearby cerrado vegetation almost always showed charring of barks but there was seldom any evidence of fire in the transitional cerradão, and in fact many of the trees have thin, smooth bark, apparently ill-adapted to resisting fire, in contrast to the thick, corky barks of the cerrado trees.

Detailed soil studies showed that the soils of the forest, cerrado and transitional cerradão were essentially the same, apart from some differences in clay content (Askew *et al.*, 1970a,b; Table 20.2).

We concluded, principally on the basis of the occurrence of relict, usually giant, cerrado trees within the forest, that the forest was expanding into the cerrado and that the cerradão belt, in which there were many saplings of forest species, represented its advancing face. The cerrado trees found deepest in the forest were emergent specimens of *Pterodon pubescens* Benth., some specimens of which were 4–5 km from the nearest cerrado shown on aerial photographs, indicating a considerable expansion of forest into cerrado during recent times (Ratter *et al.*, 1973, 1978b).

Although the commonest transition of the Serra do Roncador was through a broad ecotonal cerradão as just described, other areas were encountered where the change was abrupt. Such abrupt transitions were usually found in valleys where forest occurred on one side and cerrado on the other, with riverine vegetation in between.

The only other area where the author has observed the transition between cerrado and Amazonian vegetation is at Iquê on the Rondônia–Mato Grosso border (*c.* 12°12′S, 59°40′W). Here, there is a vast belt of transitional vegetation dominated by *Sclerolobium paniculatum* about 10–15 m tall, sometimes in almost pure stand. *S. paniculatum* is a short-lived, abundantly fruiting tree, and many of the trees in the stand are dead, giving the landscape an untidy, depressing aspect. The vernacular name used for *S. paniculatum* in the region is ecologically significant. It is called '*apaga-apaga*' (Portuguese = it extinguishes, it extinguishes), apparently because cerrado fires go out when they reach the transitional vegetation dominated by *S. paniculatum*. In this region there seems to be a complex mosaic of patches of Amazonian forest and transition vegetation, and much study would be required to derive a detailed interpretation.

TRANSITION TO HYDROLOGIC SAVANNA

Hydrologic savannas (Denevan, 1968) are widespread in South and Central America (Beard, 1953; Chapter 30). They are known by a variety of terms which include edaphic and topographical savannas and correspond to Sarmiento's hyperseasonal and semiseasonal savannas (Chapter 30). They consist of campo dominated by grasses, sedges, xyrids, etc., and their presence is determined by a fluctuating water table that is very high during the rainy season (often producing flooding) and low during the dry season. This contrasts with true cerrado vegetation where soils are always well drained (Eiten, 1972) but a few tree and shrub species found in cerrado are to some extent tolerant of hydrologic savanna conditions. Most important of such species are the very widely distributed *Curatella americana* L. and *Byrsonima crassifolia* (L.) Kunth *sens. lat.* which form campo sujo in less extremely inundated areas in, for example, Roraima, the Pantanal and the Ilha do Bananal – as they also do in other parts of South and Central America.

The transition of forest and hydrologic savanna is usually very abrupt. We conducted detailed studies on the Ilha de Maracá, Roraima, where the pattern is often for the forest margin to form a wall abutting on to a campo sujo dominated by *Curatella americana* and *Byrsonima crassifolia* (Milliken and Ratter, 1989; Furley and Ratter, 1990). Studies of groundwater changes throughout the year in relation to effective rooting depth seem to demonstrate clearly that the principal causal factor controlling the vegetation boundary is the upper limit of seasonal flooding – the forest, like true cerrado, requiring a well drained soil (Furley and Ratter, 1990).

DISCUSSION

An interesting feature of the observations described here is that in the main they indicate a dynamic relationship between forest and savanna vegetation. Only in the case of the hydrologic savanna–forest boundary, where the causal factor of vegetation change is water-table level, does the situation appear to be static. Even in the transitions in the mesotrophic soil landscapes between deciduous (or semi-deciduous) forest and mesotrophic facies cerradão, where the change is related to nutrient levels, there is some evidence of forest encroaching on cerrado vegetation. In all the other areas we have studied there seems to be clear evidence of expansion of forest into cerrado.

Our observations on the Amazonian forest–cerrado transition on the Serra do Roncador, east Mato Grosso, point to a particularly rapid expansion of the forest; while evidence from the Federal District indicates expansion of the galleries and a succession from cerrado to cerradão and towards forest when there is protection from fire. Similar indications of encroachment of forest into savanna were recorded in some areas on the Ilha de Maracá, Roraima (Furley and Ratter, 1990). Such forest extension probably represents a continuation of the process which has been occurring ever since the end of the last glaciation when the Amazonian forest began to expand once more after its peak contraction into isolated refugia. No doubt this phenomenon would have been observed much more widely, were it not for man's effect in preventing forest extension by agricultural activities involving the use of fire, grazing, etc.

It is, however, important to stress that not all cerrado areas protected from fire, show rapid transition towards cerradão and forest. Such closing of vegetation seems to be very uneven and for instance at Angatuba, São Paulo state, while some areas protected for a long period from fire were occupied by an almost impenetrable thicket, others, which had enjoyed equal protection, carried campo cerrado or even sparse campo sujo. Similar observations were made at Fazenda Água Limpa and in other localities in the Federal District. Soil analyses of these contrasting areas fail to provide clues to the vegetational differences and, in fact, areas clothed with cerradão/forest are often slightly more oligotrophic than those bearing sparser vegetation (Ribeiro, 1983). Ferri (1973) speculates on such transformation of cerrado into forest, based principally on observations made on an area at Emas (São Paulo state) protected from fire and other disturbance for 30 years.

Ackerley *et al.* (1989) carried out a very interesting study of the forest–cerrado transition zone in Mato Grosso. In total they studied some 25 sites, varying from pure Amazonian terra firme forest to typical Planaltine cerrado, and characterized the vegetation in terms of the numbers of species belonging to the following floristic elements: Planaltine, Transitional, Amazonian, Widespread and Other. They described transitional cerradão of a type similar to that of the Serra do Roncador at Sinop, some 500 km north of Cuiabá, but consider on the evidence of Projeto RADAM (1981, 1982) that such vegetation is probably limited to the headwaters of the Xingu basin and may not be a feature of the transition in other areas. Their study indicates that there are probably only very small areas of this genuine ecotonal community within the broad belt in which Amazonian and cerrado vegetation interdigitate.

The areas of forest–cerrado contact are under great pressure from agriculture, particularly in southern Amazonia, and Ackerley *et al.* (1989) emphasize that observations may soon only have historic interest. Whitmore (1987) points out 'that in past times of drier climate transition forest became much more extensive . . . yet this forest formation is poorly known and is today the one most rapidly disappearing at the hand of man'. This makes further studies in the ecotonal vegetation, and the establishment of a reserve to conserve an area of it, an urgent priority. At present, perhaps only 1.5% of the great cerrado domain is in preserved areas: clearly much more is necessary if the biodiversity of this ancient and heterogeneous vegetation is to be preserved, and reserves in the contact zone with the Amazonian forest are of vital interest.

REFERENCES

Ackerley, D.A., Wayt Thomas, W., Cid Ferreira, C.A. and Pirani, J.R. (1989) The forest–cerrado transition zone in southern Amazonia: results of the 1985 Projeto Flora Amazônica expedition to Mato Grosso. *Brittonia*, **41**, 113–28.

Askew, G.P., Moffat, D.J., Montgomery, R.F. and Searl, P.L. (1970a) Soil landscapes in north eastern Mato Grosso. *Geographical Journal*, **136**, 211–27.

Askew, G.P., Moffat, D.J., Montgomery, R.F. and Searl, P.L. (1970b) Interrelationships of soils and vegetation in the savanna–forest boundary zone of north-eastern Mato Grosso. *Geographical Journal*, **136**, 370–6.

Beard, J.S. (1953) The savanna vegetation of northern tropical America. *Ecological Monographs*, **23**, 149–215.

Denevan, W.M. (1968) *The Ecology of the Forest/Savanna Boundary – Proceedings of the IGU Humid Tropics Commission Symposium, Venezuela, 1964*, (eds T.L. Hills and R.E. Randall), McGill University, Montreal, Canada, pp. 45–9.

Eiten, G. (1972) The cerrado vegetation of Brazil. *Botanical Review*, **38**, 201–341.

Ferri, M.G. (1973) Sobre a origem, a manutenção e a transformação dos cerrados, tipos de savana do Brasil. *Revista Biologia*, **9**, 1–13.

Furley, P.A. and Ratter, J.A. (1988) Soil resources and plant communities of the central Brazilian cerrado and their development. *Journal of Biogeography*, **15**, 97–108.

Furley, P.A. and Ratter, J.A. (1990) Pedological and botanical variations across the forest–savanna transition on Maracá Island. *Geographical Journal*, **156**, 251–66.

Furley, P.A., Ratter, J.A. and Gifford, D.R. (1988) Observations on the vegetation of eastern Mato Grosso. III. The woody vegetation and soils of the Morro de Fumaça, Torixoreu. *Proceedings of the Royal Society of London*, **B235**, 259–80.

Lopes, A.S. and Cox, F.R. (1977) A survey of the fertility status of surface soils under 'cerrado' vegetation in Brazil. *Soil Science Society of America Journal*, **41**, 742–7.

Milliken, W. and Ratter, J.A. (1989) *The Vegetation of the Ilha de Maracá*, Royal Botanic Garden, Edinburgh/Royal Geographical Society, London.

Pires, J.M. (1974) Tipos de Vegetação da Amazônia. *Brasil Florestal* no. 17, 48–58.

Pires, J.M. and Prance, G.T. (1985) The vegetation types of the Brazilian Amazon in *Amazonia – Key Environments*, (eds G.T. Prance and T.E. Lovejoy), Pergamon Press, Oxford, pp. 109–45.

Projeto RADAM (1981) *Levantamento de Recursos Naturais*, **25**, Goiás, Departamento Nacional da Produção Mineral, Rio de Janeiro, Brazil.

Projeto RADAM (1982) *Levantamento de Recursos Naturais*, **25**, Goiás & *Ibid*. **26**: *Cuiabá*, Departamento Nacional da Produção Mineral, Rio de Janeiro, Brazil.

Ratter, J.A. (1971) Some notes on two types of cerradão occurring in northeastern Mato Grosso, in *III Simpósio sobre o Cerrado* (ed. M.G. Ferri), Universidade de São Paulo, Brazil, pp. 100–102.

Ratter, J.A. (1980) *Notes on the vegetation of Fazenda Água Limpa (Brasília, DF, Brazil). Including a Key to the Woody Genera of Dicotyledons of the Cerrado*, Royal Botanic Garden, Edinburgh, UK.

Ratter, J.A. (1985) *Notas Sobre a Vegetação da Fazenda Água Limpa (Brasília, DF); Com uma Chave para os Gêneros Lenhosos de Dicotyledôneas do Cerrado*, Royal Botanic Garden, Edinburgh, UK.

Ratter, J.A. (1986) *Notas Sobre a Vegetação da Fazenda Água Limpa (Brasília, DF); Com uma Chave para os Gêneros Lenhosos de Dicotyledôneas do Cerrado*, Editora Universidade de Brasília, textos universitários no. 003. Brasília.

Ratter, J.A., Askew, G.P., Montgomery, R.F. and Gifford, D.R. (1977) Observaçoes adicionais sobre o cerradão de solos mesotróficos no Brasil Central, in *IV Simpósio sobre o Cerrado*, (ed. M.G. Ferri), Universidade de São Paulo, Brazil, pp. 303–16.

Ratter, J.A., Askew, G.P., Montgomery, R.F. and Gifford, D.R. (1978a) Observations on forests of some mesotrophic soils in central Brazil. *Revista brasileira de Botânica*, **1**, 47–58.

Ratter, J.A., Askew, G.P., Montgomery, R.F. and Gifford, D.R. (1978b) Observations on the vegetation of northeastern Mato Grosso. II. Forests and soils of the Rio Suiá-Missu area. *Proceedings of the Royal Society of London*, **B203**, 191–208.

Ratter, J.A., Leitão Filho, H. de F., Argent, G. *et al.* (1988) Floristic composition and community structure of a southern cerrado area in Brazil. *Notes RBG Edinburgh*, **45**, 137–51.

Ratter, J.A., Richards, P.W., Argent, G. and Gifford, D.R. (1973)

Observations on the vegetation of northeastern Mato Grosso. I. The woody vegetation types of the Xavantina–Cachimbo Expedition area. *Philosophical Transactions of the Royal Society of London*, **B266**, 449–92.

Raw, A. and Hay, J. (1985) Fire and other factors affecting a population of Simarouba amara in 'cerradão' near Brasília, Brazil. *Revista brasileira de Botânica*, **8**, 101–7.

Ribeiro, J.F. (1983) *Comparação da concentração de nutrientes na vegetação arbórea e nos solos de um cerrado e um cerradão no Distrito Federal, Brasil*, Master's thesis, University of Brasília.

Van der Hammen, T. (1983) The palaeoecology and palaeogeography of savannas, in *Tropical Savannas*, (ed. F. Bourlière), Elsevier, Amsterdam, pp. 19–35.

Whitmore, T.C. (1987) Introduction, in *Biogeography and Quaternary History in Tropical America*, (eds T.C. Whitmore and G.T. Prance), Clarendon Press, Oxford, pp. ix–x.

21

Observations on the differentiation of woodland and wet savanna habitats in the Pantanal of Mato Grosso, Brazil

B. DUBS

SUMMARY

1. Semi-deciduous forest and *cerradão* were studied on two 1 ha plots in the Pantanal do Rio Negro (19°30′S, 56°10′W). The semi-deciduous forest had a top-storey dominated by the trees which typify this community in central Brazil and a dense understorey of the palm *Attalea phalerata*. The cerradão is of the mesotrophic facies type and contains many of the characteristic marker species of this community. It is speculated that soil differences, particularly calcium levels, probably control whether semi-deciduous forest or cerradão is present at a particular site.
2. Root systems of 17 trees belonging to different species were excavated. As a general rule the main lateral roots were confined to the upper soil layer of *c*. 40 cm which is not reached by the water table.
3. A study was made of forest-islands occurring on earthmounds surrounded by wet savanna. Such islands provide a habitat for both cerrado and forest trees since they provide patches of unflooded soil. They are associated with the activities of a number of species of termites and, more rarely, with leaf-cutter ants.

INTRODUCTION

The Pantanal of Mato Grosso is a 150 000 km^2 alluvial area in the upper Paraguay basin (Fig. 21.1). This basin is part of a subsidence zone formed

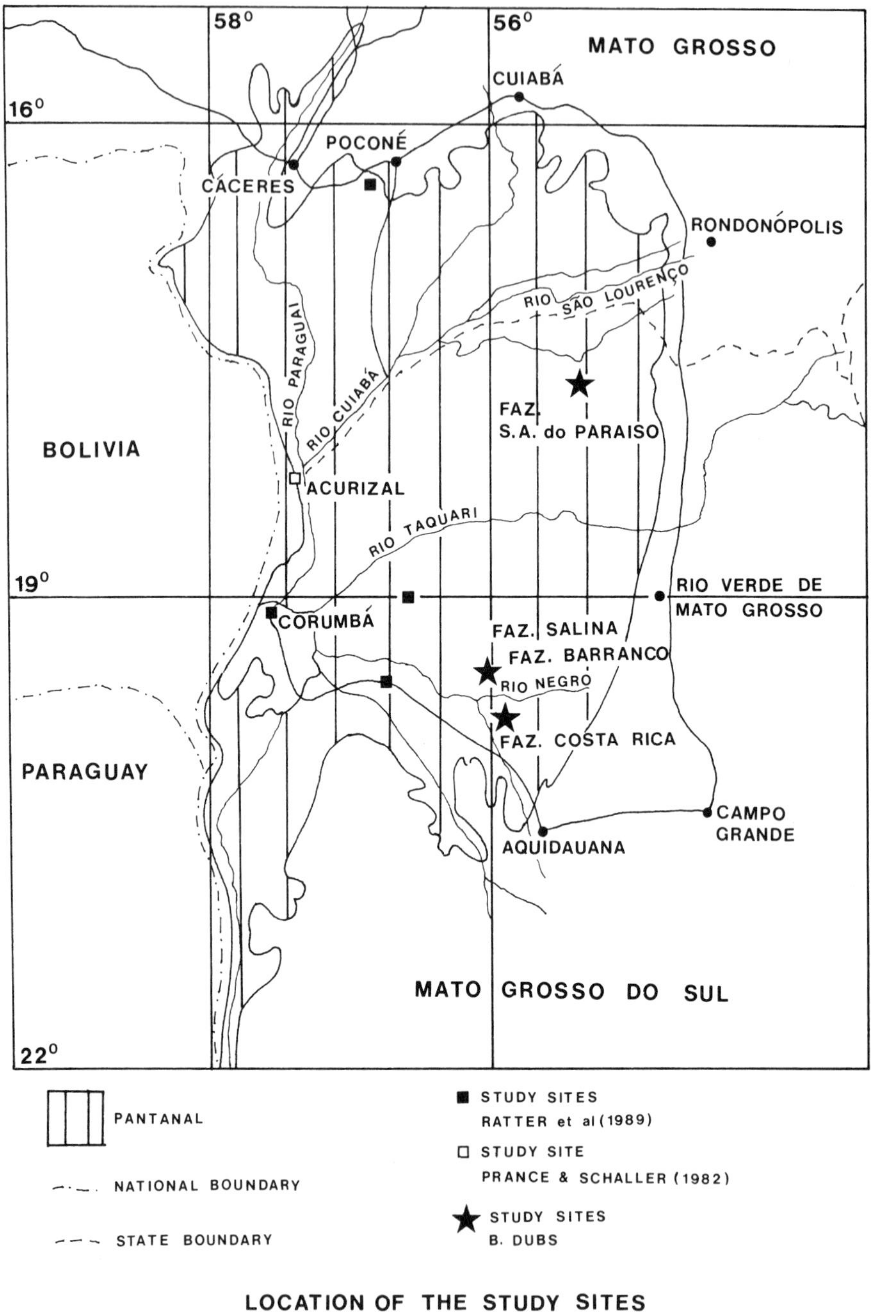

Fig. 21.1 Location of the study sites.

at the time of the Andes orogenesis stretching east of the Andes from the llanos in Columbia right into the Chaco area. The slope of the Pantanal plain is extremely slight: only 25 cm km^{-1} in an east–west direction and 2 cm km^{-1} in a north–south direction (Oliveira Carvalho, 1986). In the north and east, the Pantanal is bordered by the central Brazilian Planaltos, in the south it ends in the hill area of Bodoquena, and in the west the Rio Paraguay forms a natural border. Isolated mountain ranges of Precambrian origin, flanking the Rio Paraguay in the Corumbá region and further north at Amolar, are the only elevations of any note in the wide Pantanal plain. The region has a seasonal climate with three to four arid months while during the rainy season (November–April) there is a rainfall of 1000–1400 mm. The rivers, which spring from the borders of the Planalto and run in a south-westerly or westerly direction to the Rio Paraguay, are not able to cope with the amount of rainfall, and each year, by the end of the rainy season, overflow their banks and deposit their sediments. The changing drainage pattern of the river system has continually altered the topography of the Pantanal landscape, and has caused a mosaic of varying types of scenery from small and large rivers, ponds, swamps and floodplains to wooded areas which remain dry all year long.

Owing to the seclusion and sparse colonization of the area until recently, the Pantanal is a superb natural landscape, rich in flora and fauna. Because of its size and unspoiled nature, it is the most significant wetlands area in the world.

This chapter describes the composition of forest types occurring in the area of Pantanal studied and considers how they differ in soil conditions from the neighbouring wet savanna. The formation of soil mounds on wet savannas by termites and ants is also investigated.

STUDY SITE

Since 1984 the vegetation of the Pantanal do Rio Negro, Mato Grosso do Sul, on Fazendas Salina, Barranco and Costa Rica (19°30′S, 56°10′W) (Fig. 21.1) have been studied on behalf of SODEPAN*. This area takes its name from the Rio Negro, which, rising at the western side of the Planalto, meanders through the Pantanal in a westerly direction and finally disappears into a large swamp area without reaching the Rio Paraguay. North of the Rio Negro the landscape (Fig. 21.2) is marked by numerous, mostly circular, lakes and ponds (*baias*) which are separated by strips of higher-lying forested land (known locally in the Pantanal as

*SODEPAN, Sociedade de Defesa do Pantanal, Campo Grande.

Fig. 21.2 Typical landscape with ponds and forest-bearing strips of dry land, Pantanal do Rio Negro, May 1989.

cordilheiras). South of the Rio Negro, there are broad *campos* (grasslands) which are inundated for a large part of the year. The remarkably sharp boundary between these two types of landscape reflects the pattern of deposition of the sediments of the Rio Taquari lying to the north. Radar shots from satellites have shown that these sediments were deposited in the form of a giant alluvial fan (Moreira Franco and Pinheiro, 1982; Cardoso da Silva, 1986). This caused the Rio Negro, which now bounds the southern part of this alluvial fan, to be deflected to the south. The ponds of the landscape to its north are therefore the remnants of former riverbeds (cut-off meander-loops, oxbow lakes), whereas the ridges represent former river banks.

METHODS

Characterization of the vegetation

A north–south transect of approximately 5 km was run from the Rio Negro inland through a forested area that also contained wet savanna and ponds and their shore zones. The transect areas were divided into

10 × 10 m squares in which all trees with diameter at breast height (dbh) ≥10 cm were recorded. The transect included two distinct forest types, in each of which a hectare study plot was set up for more detailed analysis. Relative frequency, relative density, relative dominance and importance value for each of the tree species in these plots were calculated.

Structure of root systems

The soil was excavated within a radius of 1 m around the bases of 17 individual trees of different species, in an area of forest situated between two ponds. Root structures were described, and the textural and hydrological characteristics of the surrounding soil were noted.

Forest islands and termite and ant activity

Relationships between the occurrence of forest islands within wet savanna and the presence of ant and termite mounds were investigated on the southern side of the Rio Negro. Further studies were carried out at Fazenda SA do Paraíso (17°36′S, 55°24′W), further to the north.

RESULTS

Types of forest

Semi-deciduous forest with *Attalea* palm understorey

In some areas, where the soil is darker and more fertile, we find a forest with an upper, incomplete canopy 18–21 m high. Here, the characteristic trees are *Tabebuia impetiginosa**, *Anadenanthera colubrina* var. *cebil*, *Sterculia striata*, *Aspidosperma* sp., etc., with trunk diameters of 30–50 cm. The density of trees, however, is relatively low and the degree of canopy coverage of the upper storey is estimated at only about 50%. The trees of this strorey are completely deciduous during the dry season. The lower storey provides almost complete cover and is made up of mixed evergreen and deciduous trees, mostly 8–10 m high, and a large number of somewhat lower *Attalea phalerata* palms. Just a few of these palms have been drawn in the profile diagram to illustrate the physiognomy of the forest (Fig. 21.3). After the *Attalea* palms, the most numerous species of this layer are *Tabebuia roseo-alba*, *Combretum leprosum*, *Casearia gossypiosperma* and *Tabebuia ochracea*. The undergrowth is made up of palm-

*A full list of species mentioned in the text with authorities is given in the appendix on page 448.

Fig. 21.3 Profile (60 × 10 m) through semi-deciduous forest at Fazenda Salina. Note understorey of *Attalea phalerata* palms.

Fig. 21.4 Profile (60 × 10 m) through cerradão at Fazenda Salina.

seedlings, *Bromelia balansae,* shrubs and some sparse grasses. Where *Attalea* palms are predominant, shade is extremely dense and there is often no undergrowth. Lianas are common but epiphytes are rare. From a physiognomic aspect, this type of forest can be described as the *Attalea*-facies of a semi-deciduous forest. Analyses of the trees of two areas of this forest are given in Tables 21.1 and 21.2.

Savanna forest (cerradão)

This type of forest (Fig. 21.4) is much more common than the semi-deciduous forest. The trees are (8–)10–12(–14) m, often branching low in the middle third of the trunk and with a tendency to twisted growth. There is no distinct stratification in the tree-layer. Only the low-growing but abundant *Alibertia sessilis* grows in the shadow of taller trees. In the undergrowth, there is a shrub layer and often a dense growth of terrestrial bromeliads but practically no other herbaceous plants. Vegetation analysis for three areas of cerradão are given in Tables 21.3–21.5.

Table 21.1 Trees with dbh ⩾10 cm in a 0.42 ha transect of semi-deciduous forest at Fazenda Salina, ranked by importance value (seven sub-units of 10 × 60 m; BA, basal area cm^2; IVI, importance value index; no., number of trees present; R.De., relative density; R.Do., relative dominance; RF, relative frequency; SU, occurrence in sub-units)

	SU	*No.*	*BA*	*RF*	*R.De.*	*R.Do.*	*IVI*
1. *Attalea phalerata*	7	160	50 240	8.75	53.51	40.67	102.93
2. *Tabebuia roseo-alba*	7	29	8 516	8.75	9.70	6.89	25.34
3. *Tabebuia impetiginosa*	5	9	11 782	6.25	3.01	9.54	18.80
4. *Combretum leprosum*	5	15	9 105	6.25	5.02	7.37	18.64
5. *Casearia gossypiosperma*	5	9	1 563	6.25	3.01	1.27	10.53
6. *Tabebuia ochracea*	5	7	2 197	6.25	2.34	1.78	10.37
7. *Anadenanthera colubrina*	4	5	3 543	5.00	1.67	2.87	9.54
8. *Sterculia striata*	2	2	7 682	2.50	0.67	6.22	9.39
9. *Rhamnidium elaeocarpum*	4	8	1 516	5.00	2.68	1.23	8.90
10. *Aspidosperma tomentosum*	3	6	3 338	3.75	2.01	2.70	8.46
11. *Trichilia elegans*	3	7	673	3.75	2.34	0.54	6.64
12. *Cordia trichotoma*	2	5	2 646	2.50	1.67	2.14	6.31
13. *Astronium urundeuva*	2	4	2 982	2.50	1.34	2.41	6.25
14. *Sapium longifolium*	3	4	998	3.75	1.34	0.81	5.90
15. *Acacia paniculata*	3	3	341	3.75	1.00	0.28	5.03
16. *Vitex cymosa*	1	1	3 848	1.25	0.33	3.12	4.70
17. *Ocotea suaveolens*	2	3	1 112	2.50	1.00	0.90	4.40
18. *Dilodendron bipinnatum*	2	2	1 480	2.50	0.67	1.20	4.37
19. *Licania parvifolia*	1	4	1 879	1.25	1.34	1.52	4.11
20. *Platypodium grandiflorum*	2	2	1 075	2.50	0.67	0.87	4.04
21. *Hymenaea* sp. 'Jatobá-mirim'	1	2	2 470	1.25	0.67	2.00	3.92
22. *Casearia decandra*	2	2	409	2.50	0.67	0.33	3.50
23. *Guarea trichilioides*	1	1	1 385	1.25	0.33	1.12	2.71
24. *Albizia niopoides*	1	1	804	1.25	0.33	0.65	2.24
25. *Casearia sylvestris*	1	2	314	1.25	0.67	0.25	2.17
26. *Salacia elliptica*	1	1	452	1.25	0.33	0.37	1.95
27. *Pseudobombax marginatum*	1	1	380	1.25	0.33	0.31	1.89
28. *Fagara hassleriana*	1	1	314	1.25	0.33	0.25	1.84
29. *Agonandra brasiliensis*	1	1	201	1.25	0.33	0.16	1.75
30. 'Quatro folhas'	1	1	201	1.25	0.33	0.16	1.75
31. *Erythrina dominguezii*	1	1	78	1.25	0.33	0.06	1.65
	80	299	123 524	100.00	100.00	100.00	300.00

Groundwater and the rooting structure of forest trees

Soil profile

A soil profile sited at the highest point of a wooded ridge lying between two ponds 160 m apart was studied to determine the relationship between

Table 21.2 Trees with dbh ⩾10 cm in a 1 ha plot of semi-deciduous forest at Fazenda Barranco, ranked by importance value (25 sub-units of 20 × 20 m; abbreviations as in Table 21.1)

	SU	*No.*	*BA*	*RF*	*R.De.*	*R.Do.*	*IVI*
1. *Attalea phalerata*	23	108	33 912	13.69	29.92	18.02	61.63
2. *Anadenanthera colubrina*	20	39	57 839	11.90	10.80	30.74	53.44
3. *Tabebuia impetiginosa*	16	26	18 206	9.52	7.20	9.68	26.40
4. *Aspidosperma tomentosum*	16	41	8 944	9.52	11.36	4.75	25.63
5. *Astronium urundeuva*	14	32	13 777	8.33	8.86	7.32	24.52
6. *Casearia gossypiosperma*	13	27	3 911	7.74	7.48	2.08	17.30
7. *Pterogyne nitens*	8	10	13 793	4.76	2.77	7.33	14.86
8. *Combretum leprosum*	11	18	3 904	6.55	4.99	2.07	13.61
9. *Pseudobombax marginatum*	7	7	9 526	4.17	1.94	5.06	11.17
10. *Guarea trichilioides*	7	11	3 448	4.17	3.05	1.83	9.05
11. *Tabebuia ochracea*	4	8	2 659	2.38	2.22	1.41	6.01
12. *Dilodendron bipinnatum*	4	4	3 007	2.38	1.11	1.60	5.09
13. *Swartzia jorori*	4	5	1 275	2.38	1.39	0.68	4.44
14. *Tabebuia roseo-alba*	4	5	833	2.38	1.39	0.44	4.21
15. *Cordia glabrata*	3	4	2 239	1.79	1.11	1.19	4.08
16. *Bumelia obtusifolia*	2	3	3 685	1.19	0.83	1.96	3.98
17. Sp. indet. Snr. 938	2	2	2 277	1.19	0.55	1.21	2.95
18. *Cordia trichotoma*	2	3	540	1.19	0.83	0.29	2.31
19. *Rhamnidium elaeocarpum*	2	2	174	1.19	0.55	0.09	1.84
20. *Acacia paniculata*	2	2	174	1.19	0.55	0.09	1.84
21. *Aspidosperma australe*	1	1	1 810	0.60	0.28	0.96	1.83
22. *Sterculia striata*	1	1	1 590	0.60	0.28	0.84	1.72
23. *Platypodium grandiflorum*	1	1	573	0.60	0.28	0.30	1.18
24. Sp. indet. Snr. 941	1	1	79	0.60	0.28	0.04	0.91
	168	361	188 175	100.00	100.00	100.00	300.00

soil characteristics and tree rooting. The profile is described below and further data are given in Table 21.6.

The top 40 cm of soil appeared somewhat lighter in colour than the next horizon, probably since it was drier. From 45–85 cm, the soil was uniformly brown (6E5) – the soil colours and code follow Kornerup and Wanscher (1978). At 85–115 cm deep, the soil was somewhat lighter (6E4) and finely speckled with rust-red and greyish-black, 6–12 mm, mottles. In this layer there were also a few round, brown concretions, 5–7 mm in diameter. From 115 to 147 cm, the colour remained the same but with more greyish-black patches. At 147–180 cm, the main colour was orange-brown (6C3) with rusty red and greyish-black patches. At a depth of 180 cm groundwater appeared. The size of the grains indicates an even mixed, fine-sandy sediment. Clay content is very low, as is that of organic carbon.

Table 21.3 Trees with dbh ≥10 cm in a 0.18 ha transect of cerradão at Fazenda Salina, ranked by importance value (three sub-units of 10 × 60 m; abbreviations as in Table 21.1)

	SU	*No.*	*BA*	*RF*	*R.De.*	*R.Do.*	*IVI*
1. *Lafoensia pacari*	2	10	2545	4.88	12.66	10.59	28.13
2. *Eriotheca gracilipes*	2	6	2912	4.88	7.59	12.12	24.59
3. *Terminalia argentea*	3	6	2262	7.32	7.59	9.41	24.32
4. *Astronium fraxinifolium*	3	8	1097	7.32	10.13	4.56	22.01
5. *Luehea paniculata*	1	3	2859	2.44	3.80	11.90	18.13
6. *Caryocar brasiliense*	2	3	2030	4.88	3.80	8.45	17.12
7. *Magonia pubescens*	2	5	1302	4.88	6.33	5.42	16.62
8. *Kielmeyera coriacea*	2	5	1057	4.88	6.33	4.40	15.60
9. *Tabebuia ochracea*	2	5	1014	4.88	6.33	4.22	15.43
10. *Byrsonima coccolobifolia*	2	5	882	4.88	6.33	3.67	14.88
11. *Pouteria ramiflora*	2	4	957	4.88	5.06	3.98	13.92
12. *Tabebuia aurea*	2	3	776	4.88	3.80	3.23	11.90
13. *Curatella americana*	2	2	872	4.88	2.53	3.63	11.04
14. *Hymenaea stigonocarpa*	2	2	458	4.88	2.53	1.91	9.32
15. *Casearia sylvestris*	2	2	156	4.88	2.53	0.65	8.06
16. *Cecropia pachystachya*	1	1	531	2.44	1.27	2.21	5.91
17. *Buchenavia tomentosa*	1	1	531	2.44	1.27	2.21	5.91
18. *Guettarda viburnoides*	1	1	380	2.44	1.27	1.58	5.29
19. *Platypodium grandiflorum*	1	1	346	2.44	1.27	1.44	5.14
20. *Andira* sp. 'Morcego'	1	1	346	2.44	1.27	1.44	5.14
21. *Qualea grandiflora*	1	1	201	2.44	1.27	0.84	4.54
22. *Connarus suberosus*	1	1	177	2.44	1.27	0.74	4.44
23. *Diptychandra aurantiaca*	1	1	154	2.44	1.27	0.64	4.35
24. *Sclerolobium aureum*	1	1	95	2.44	1.27	0.40	4.10
25. *Alibertia sessilis*	1	1	95	2.44	1.27	0.40	4.10
	41	79	24035	100.00	100.00	100.00	300.00

The soils appear to be of medium fertility with low levels of organic matter, silt and clay and with mildly acidic horizons. The soil of the forested area of this cordilheira demonstrated a Bg-horizon penetrated by the fluctuating water table, appearing beneath an 85 cm thick A-horizon. The upper level of this Bg-horizon is at about the same height as the level of the adjacent pond during the wet season. The Bg-horizon, more than a metre thick, shows a strong amplitude in fluctuation of the groundwater level, from which one can conclude that the groundwater level is related to the water level of the nearby pond.

Root system examination

Seventeen individuals of different tree species were excavated as described in the methods section. With very few exceptions the trees had

Table 21.4 Trees with dbh ⩾10 cm in a 1 ha plot of cerradão at Fazenda Salina, ranked by importance value (five sub-units of 20 × 100 m; abbreviations as in Table 21.1)

	SU	*No.*	*BA*	*RF*	*R.De.*	*R.Do.*	*IVI*
1. *Alibertia sessilis*	5	45	6 221	4.10	14.71	5.61	24.42
2. *Protium heptaphyllum*	5	28	11 737	4.10	9.15	10.59	23.84
3. *Plathymenia reticulata*	5	19	12 379	4.10	6.21	11.17	21.47
4. *Buchenavia tomentosa*	4	13	10 949	3.28	4.25	9.88	17.40
5. *Dipteryx alata*	5	17	7 832	4.10	5.56	7.07	16.72
6. *Caryocar brasiliense*	5	10	8 892	4.10	3.27	8.02	15.39
7. *Ocotea suaveolens*	5	17	5 243	4.10	5.56	4.73	14.38
8. *Erythrina dominguezii*	5	14	4 346	4.10	4.58	3.92	12.59
9. *Luehea paniculata*	4	12	3 252	3.28	3.92	2.93	10.13
10. *Vitex cymosa*	5	7	3 930	4.10	2.29	3.55	9.93
11. *Astronium fraxinifolium*	4	14	2 194	3.28	4.58	1.98	9.83
12. *Curatella americana*	4	8	3 564	3.28	2.61	3.21	9.11
13. *Tabebuia aurea*	5	8	1 713	4.10	2.61	1.55	8.26
14. *Fagara hassleriana*	4	9	1 836	3.28	2.94	1.66	7.88
15. *Terminalia argentea*	3	8	2 659	2.46	2.61	2.40	7.47
16. *Tabebuia impetiginosa*	4	6	1 614	3.28	1.96	1.46	6.70
17. *Eugenia* sp.	4	6	1 287	3.28	1.96	1.16	6.40
18. *Agonandra brasiliensis*	4	6	762	3.28	1.96	0.69	5.93
19. *Dimorphandra mollis*	3	4	2 128	2.46	1.31	1.92	5.69
20. *Dilodendron bipinnatum*	3	4	1 625	2.46	1.31	1.47	5.23
21. *Tabebuia roseo-alba*	3	6	849	2.46	1.96	0.77	5.19
22. *Acosmium subelegans*	2	4	1 966	1.64	1.31	1.77	4.72
23. *Mouriri elliptica*	3	3	1 280	2.46	0.98	1.15	4.59
24. *Eriotheca gracilipes*	2	3	1 769	1.64	0.98	1.60	4.22
25. *Astronium urundeuva*	2	2	1 508	1.64	0.65	1.36	3.65
26. *Lafoensia pacari*	2	3	953	1.64	0.98	0.86	3.48
27. *Qualea grandiflora*	2	3	839	1.64	0.98	0.76	3.38
28. *Tabebuia ochracea*	2	3	449	1.64	0.98	0.41	3.02
29. *Sterculia striata*	1	2	1 685	0.82	0.65	1.52	2.99
30. *Stryphnodendron obovatum*	2	2	381	1.64	0.65	0.34	2.64
31. *Platypodium grandiflorum*	2	2	272	1.64	0.65	0.25	2.54
32. *Pouteria ramiflora*	2	2	255	1.64	0.65	0.23	2.52
33. *Bauhinia* sp.	2	2	173	1.64	0.65	0.16	2.45
34. *Hymenaea stigonocarpa*	1	2	1 062	0.82	0.65	0.96	2.43
35. *Aspidosperma tomentosum*	1	1	855	0.82	0.33	0.77	1.92
36. *Pseudobombax marginatum*	1	1	755	0.82	0.33	0.68	1.83
37. *Andira* sp. 'Morcego'	1	2	355	0.82	0.65	0.32	1.79
38. *Casearia decandra*	1	2	322	0.82	0.65	0.29	1.76
39. *Magonia pubescens*	1	2	266	0.82	0.65	0.24	1.71
40. *Casearia sylvestris*	1	2	208	0.82	0.65	0.19	1.66
41. *Byrsonima coccolobifolia*	1	1	314	0.82	0.33	0.28	1.43
42. *Sclerolobium aureum*	1	1	177	0.82	0.33	0.16	1.31
	122	306	110 856	100.00	100.00	100.00	300.00

Table 21.5 Trees with dbh ≥10 cm in a 0.18 ha transect of cerradão at Fazenda Salina, ranked by importance value (three sub-units of 10 × 60 m; abbreviations as in Table 21.1)

	SU	*No.*	*BA*	*RF*	*R.De.*	*R.Do.*	*IVI*
1. *Sapium longifolium*	3	11	2755	6.38	11.83	9.79	28.00
2. *Caryocar brasiliense*	2	6	4453	4.26	6.45	15.82	26.53
3. *Vitex cymosa*	3	4	4398	6.38	4.30	15.62	26.31
4. *Casearia sylvestris*	3	10	1450	6.38	10.75	5.15	22.29
5. *Ocotea suaveolens*	3	7	1482	6.38	7.53	5.26	19.17
6. *Fagara hassleriana*	3	5	1568	6.38	5.38	5.57	17.33
7. *Rhamnidium elaeocarpum*	2	8	903	4.26	8.60	3.21	16.07
8. *Curatella americana*	3	4	907	6.38	4.30	3.22	13.91
9. *Protium heptaphyllum*	2	5	894	4.26	5.38	3.18	12.81
10. *Tabebuia aurea*	2	3	1327	4.26	3.23	4.71	12.20
11. *Astronium urundeuva*	1	4	1362	2.13	4.30	4.84	11.27
12. *Luehea paniculata*	1	1	1963	2.13	1.08	6.97	10.18
13. *Qualea grandiflora*	2	3	570	4.26	3.23	2.02	9.51
14. *Diospyros hispida*	1	5	477	2.13	5.38	1.69	9.20
15. *Plathymenia reticulata*	1	1	707	2.13	1.08	2.51	5.71
16. *Astronium fraxinifolium*	1	1	572	2.13	1.08	2.03	5.23
17. *Terminalia argentea*	1	2	228	2.13	2.15	0.81	5.09
18. *Lafoensia pacari*	1	1	415	2.13	1.08	1.47	4.68
19. *Sclerolobium aureum*	1	1	227	2.13	1.08	0.81	4.01
20. *Licania minutiflora*	1	1	201	2.13	1.08	0.71	3.92
21. *Dipteryx alata*	1	1	201	2.13	1.08	0.71	3.92
22. *Albizia niopoides*	1	1	154	2.13	1.08	0.55	3.75
23. *Guazuma ulmifolia*	1	1	154	2.13	1.08	0.55	3.75
24. *Casearia gossypiosperma*	1	1	154	2.13	1.08	0.55	3.75
25. *Eriotheca gracilipes*	1	1	133	2.13	1.08	0.47	3.68
26. *Rheedia brasiliensis*	1	1	113	2.13	1.08	0.40	3.60
27. *Bowdichia virgilioides*	1	1	113	2.13	1.08	0.40	3.60
28. *Buchenavia tomentosa*	1	1	95	2.13	1.08	0.34	3.54
29. *Magonia pubescens*	1	1	95	2.13	1.08	0.34	3.54
30. *Agonandra brasiliensis*	1	1	78	2.13	1.08	0.28	3.48
	47	93	28149	100.00	100.00	100.00	300.00

Table 21.6

Depth of sample (cm)	*pH ($CaCl_2$)*	*Org. C (%)*	*Fe (%) (HNO_3)*	*Soil texture classes (mm)* 2.0–1.0	1.0–0.5	0.5–0.25	0.25–0.125	0.125–0.063	<0.063
25	4.5	0.2	0.79	0	0.3	7.1	45.8	32.6	14.2
60	3.9	0	0.82	0	0.5	8.5	46.6	33.7	10.7
130	4.2	0	1.00	0	0.2	6.7	49.7	32.1	11.3
170	5	0	0.70	0.1	0.5	7.7	43.5	37	11.2

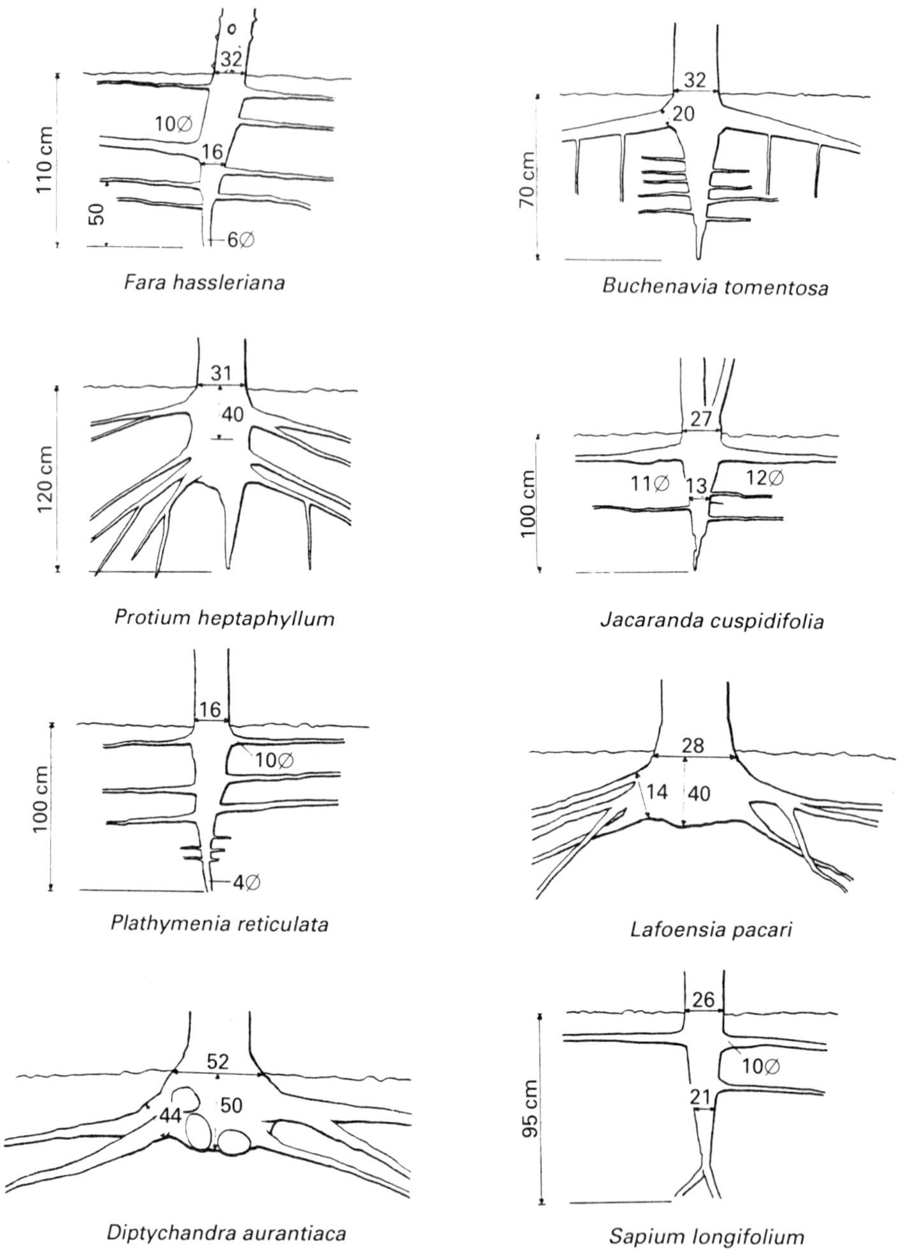

Fig. 21.5 Root systems of trees at Fazenda Barranco.

strong tap-roots which reached the upper layer of the Bg-horizon. As a general rule (Fig. 21.5), the main lateral roots were confined to the upper soil layer which is not penetrated by groundwater. Close to a relatively dense disk of surface lateral roots, further lateral roots branching from the tap-roots were often encountered, at varying depths. Obliquely growing roots and secondary tap-roots were also present, especially when the tap-root was not strongly developed. These observations may be interpreted as follows: (a) a strong conglomeration of surface lateral roots facilitates the intake of the accumulated nutrients from the upper soil layer during the rainy season, while (b) the network of lateral roots, which branch out from the tap-root at varying depths, enables the production of fine roots to adapt continually to the seasonally changing depth of the capillary fringe.

Forest islands associated with termites and ants

Areas of forest of varying size can also be found on patches of elevated ground in typical wet savanna. Usually these areas take the form of small woods, which are scattered about the flood savanna like islands (Fig. 21.6). This type of forest-island can frequently be found in the floodplain on the southern side of the Rio Negro, and during the rainy season these

Fig. 21.6 Wet savanna with forest-islands on termite mounds, halfway between Aquidauana and the Rio Negro, May 1989.

Fig. 21.7 Forest-island, Fazenda Costa Rica, September 1989.

forest-islands are surrounded by large sheets of water on the flooded savanna. The wet savanna in which they appear is a very sparse tree and shrub savanna with few species of woody plants, of which *Curatella americana* and *Byrsonima orbignyana,* species noted for their tolerance of flooding, are most common. In complete contrast, however, the forest-islands are covered with a variety of tree species of the cerrado and others characteristic of semi-deciduous forest (Fig. 21.7). Measurements of these forest-islands showed that their surfaces were higher than the surrounding wet savanna, sometimes by 40–60 cm but in extreme cases by as much as 140 cm. These mounds are therefore out of reach of the wet-season floodwaters, so that trees can become established on them.

In the centre of these forest-islands there is often a termite nest or, more rarely, a nest of leaf-cutter ants; the termitaria include those of both *Cornitermes snyderi* (Emerson, 1952) and *Rotunditermes bragantinus* (Roonwal and Rathore, 1976). In wet savanna grassland cone-shaped, 35 cm high soil nests of *Armitermes euamignathus* Silvestri 1901 have been found, which occur locally in great numbers. Earth-mounds in wet savanna near Corumbá, with isolated trees of *Tabebuia aurea* growing on them, may also prove to be termite nests of this species, although I do not have specimens collected from this area.

On several occasions, I studied small mounds in the wet savanna which carried bushes or shrubs. Small eruptions, varying in size from that of a fist to a child's head, contained the nests of *Paratrechina fulva* (Mayr). Small mounds up to 60 cm high and with growth of small bushes, contained the ant-nest of a *Camponotus* species.

A 22 m diameter forest-island south of the Rio Negro, which at its centre was 139 cm higher than the surrounding wet savanna, was studied. The elevation was made by an occupied nest of leaf-cutter ants, *Atta sexdens* (L.). It bore 29 dicotyledonous trees belonging to 10 species and also 17 *Attalea phalerata* palms. One of these trees, a *Tabebuia impetiginosa*, was rooted at the highest point of the *Atta* nest. Another forest-island, 30–50 m wide and several hundred metres long, contained a large number of nests, some occupied, some not, of leaf-cutter ants. The topography of the inner part of the island was undulating and the highest points were always occupied by the *Atta* nests. The mean difference in height between this type of ant-nest and the level of the surrounding wet savanna was 138 cm. The vegetation of this large forest-island was strongly dominated by *Attalea* palms which were overtopped by individual deciduous trees and, therefore, showed a strong resemblance to the aforementioned semi-deciduous forest on the northern side of the Rio Negro.

DISCUSSION

Extremely few detailed studies of the forests of the Pantanal cordilheiras have been made and the only papers with information comparable to that presented here are Prance and Schaller (1982) and Ratter *et al.* (1989). The semi-deciduous forests (Tables 21.1 and 21.2) are similar to those described in central Brazil by Ratter *et al.* (1978) and show a close floristic resemblance to those occurring on the cordilheiras of Fazenda Nhumirim in the Pantanal of Nhecolândia (Ratter *et al.*, 1989). The cerradão (savanna forest, Tables 21.3–21.5) is of the mesotrophic type described by Ratter *et al.* (1977) and contains various characteristic marker species of this community such as *Astronium fraxinifolium, Luehea paniculata, Magonia pubescens, Terminalia argentea, Platypodium grandiflorum, Dipteryx alata* and *Dilodendron bipinnatum*, while some calcicolous species more typical of forest are also present, e.g. *Albizia niopoides, Astronium urundeuva, Rhamnidium elaeocarpum, Sterculia striata* and *Vitex cymosa*. Ratter *et al.* (1973, 1977) have demonstrated that semi-deciduous forest soils usually have higher calcium levels in the surface soil horizons than those of mesotrophic facies cerradão, and this is probably occurring here.

The flat topography of the floodplains of the Pantanal does not, at first sight, offer very favourable conditions for trees to thrive, since only very few species can tolerate the varying effects of flooding and drought. However, the root systems of the tree species examined suggest that they are well adapted to habitats where quite a shallow surface layer of soil (40 cm or more) remains unflooded for most of the year. Thus relatively slight elevations of the terrain, which protect the trees from contact with groundwater, permit the growth of certain tree species, particularly those of the cerrado. In this sense, the assertion of Eiten (1972) that cerrado can only develop on well drained land with a low groundwater level, becomes questionable.

The observations made in this study show that various ant and termite species build nests in the flood areas which, as small elevations in the otherwise completely flat land, are more or less out of reach of groundwater. It is reasonable to suppose that nests which are built at lower level are flooded in years with exceptionally high water, and that their occupants are destroyed. As Matthews (1977) has stated, 'the creation of the mound may indeed be because a termite or ant colony dies off occasionally and their mud erections erode and crumble; anteaters (*Myrmecophaga tridactyla*) who damage nests which then collapse, could also contribute to a build-up in the landscape. If a nest is damaged or destroyed, it is repaired or a new colony of the same or another species builds a new nest in the same place'. Construction and disintegration of termite hills, over a long period, lead to the formation of mounds which provide suitable places for colonization by trees. These vegetated mounds, in turn, become sheltering places for animals which forage in the surrounding grasslands and thus further accumulate material. Individual forest islands probably can, in time, merge to form larger forest parcels (so-called 'capões'), which are often several hundred metres in diameter.

Two completely different processes provide the conditions necessary for the establishment of forest areas: on the one hand, the purely mechanically caused elevation of the banks of meandering lowland rivers, and on the other, the biological influence of the termite or ant constructions producing higher mounds in the landscape. Cerrado and/or semideciduous forest colonize the riverbanks in the flood areas and, by this means, forest corridors come into being, whereas the termite hills provide the only land elevations in the otherwise flat floodplain. Together with wind-dispersed diaspores, seeds which are carried by birds also reach these islands of forest. The fruit of savanna shrubs, e.g. *Byrsonima orbignyana*, attract many frugivorous birds which, while flying through the Pantanal, sometimes come to rest in the trees of the termite forest islands and thus contribute to the spreading of species from the cerrado and/or forest riverbanks.

ACKNOWLEDGEMENTS

The work in the Pantanal formed part of a vegetation survey which was carried out for SODEPAN (Sociedade de Defesa do Pantanal, Campo Grande). I am indebted to the landowner of Fazendas Salina and Barranco, Dr Jorge Schweizer, and the owner of Fazenda SA do Paraíso, João Carlos Marinho Lutz for giving me free access to their land. Many specialists in plant taxonomy have contributed to the identification of collected plant specimens. Dr Ernst, Allschwil, has identified the termites. Dr Bolton, Department of Entomology, British Museum (Nat. Hist.), identified the ant species. The soil samples were analysed by Prof. Fitze, Geographisches Institut, University of Zurich. To all these specialists I am greatly indebted.

REFERENCES

Cardoso da Silva, T. (1986) Contribuição da geomorfologia para o conhecimento e valorização do Pantanal, in *Anais do I°. Simpósio sobre recursos naturais e socio-econômicos do Pantanal*, EMBRAPA, Brasília.

Eiten, G. (1972) The Cerrado vegetation of Brazil. *Botanical Review*, **38**, 201–341.

Kornerup, A. and Wanscher, J.H. (1978) *Methuen Handbook of Colour*, 3rd ed., Methuen, London.

Matthews, A.G.A. (1977) *Studies on termites from the Mato Grosso State, Brazil*. Academia Brasileira de Ciências, Rio de Janeiro.

Moreira Franco, M. and Pinheiro, R. (1982) Geomorfologia, in *Projeto Radambrasil*, Vol. 27, Ministério das Minas e Energia, Secretaria-General, Rio de Janeiro.

Oliveira Carvalho, N. (1986) Hidrologia da Bacia do Alto Paraguai, in *Anais do I°. Simpósio sobre recursos naturais e socio-econômicos do Pantanal*, EMBRAPA, Brasília.

Prance, G.T. and Schaller, G.P. (1982) Preliminary study of some vegetation types of the Pantanal, Mato Grosso, Brazil. *Brittonia*, **34**, 228–51.

Ratter, J.A., Askew, G.P., Montgomery, R.F. and Gifford, D.R. (1977) Observações adicionais sobre o cerradão de solos mesotróficos no Brasil Central, in *IV Simpósio sobre o Cerrado*, Universidade de São Paulo, Brazil, pp. 303–16.

Ratter, J.A. Askew, G.P., Montgomery, R.F. and Gifford, D.R. (1978) Observations on forests of some mesotrophic soils in central Brazil. *Revista Brasileira de Botânica*, **1**, 47–58.

Ratter, J.A., Pott, A., Pott, V.J. *et al.* (1989) Observations on woody vegetation types in the Pantanal and at Corumbá, Brazil. *Notes RBG Edinburgh*, **45**, 503–25.

Ratter, J.A., Richards, P.W., Argent, G. and Gifford, D.R. (1973) Observations on the vegetation of northeastern Mato Grosso. 1. The woody vegetation types of the Xavantina–Cachimbo Expedition area. *Philosophical Transactions of the Royal Society of London*, **B266**, 449–92.

APPENDIX

List of plant species mentioned in the text

Anacardiaceae
Astronium fraxinifolium Schott
A. urundeuva Schott

Apocynaceae
Aspidosperma australe M. Arg.
A. tomentosum Mart.

Bignoniaceae
Jacaranda cuspidifolia Mart.
Tabebuia aurea (Manso) B. & H. ex S. Moore
T. impetiginosa (Mart.) Standl.
T. ochracea (Cham.) Melch.
T. roseo-alba (Ridley) Sandw.

Bombacaceae
Eriotheca gracilipes (K. Schum.) A. Robyns
Pseudobombax marginatum (St. Hil.) A. Robyns

Boraginaceae
Cordia glabrata (Mart.) A.DC.
C. trichotoma (Vell.) Arrab. ex Steud.

Bromeliaceae
Bromelia balansae Mez

Burseraceae
Protium heptaphyllum (Aubl.) March.

Caryocaraceae
Caryocar brasiliensea Cambess.

Celastraceae
Salacia elliptica (Mart.) G.Don

Chrysobalanaceae
Licania minutiflora (Sagot) Fritsch
L. parvifolia Huber

Combretaceae
Buchenavia tomentosa (Mart.) Eichl.
Combretum leprosum Mart.
Terminalia argentea Mart.

Connaraceae
Connarus suberosus Planchon

Dilleniaceae
Curatella americana L.

Ebenaceae
Diospyros hispida A.DC.

Euphorbiaceae
Sapium longifolium (M.Arg.) Huber

Flacourtiaceae
Casearia decandra Jacquin
C. gossypiosperma Briquet
C. sylvestris Sw.

Guttiferae
Kielmeyera coriacea Mart.
Rheedia brasiliensis (Mart.) Pl. & Tr.

Lauraceae
Ocotea suaveolens (Meissn.) Hassler

Leg.-Mimosoideae
Acacia paniculata Willd.
Albizia niopoides (Spruce ex Benth.) Burk.
Anadenanthera colubrina (Vell.) Brenan
Plathymenia reticulata Benth.
Stryphnodendron obovatum Benth.

Leg.–Caesalpinioideae
Dimorphandra mollis Benth.
Hymenaea stigonocarpa Mart. ex Hayne

Pterogyne nitens Tul.
Sclerolobium aureum (Tul.) Benth.
Swartzia jorori Harms

Leg.-Papilionoideae
Dipteryx alata Vog.
Acosmium subelegans (Mohlenbr.) Yakovlev
Bowdichia virgilioides H.B.K.
Diptychandra aurantiaca Tul.
Erythrina dominguezii Hassler
Platypodium grandiflorum Benth.

Lythraceae
Lafoensia pacari St.Hil.

Malpighiaceae
Byrsonima coccolobifolia Kunth
B. orbignyana Adr. Juss.

Melastomataceae
Mouriria elliptica Mart.

Meliaceae
Guarea guidonia (L.) Sleumer (*Guarea trichiloides* L.)
Trichilia elegans Adr. Juss.

Moraceae
Cecropia pachystachya Trecul.

Opiliaceae
Agonandra brasiliensis Miers

Palmae
Attalae phalerata Mart.

Rhamnaceae
Rhamnidium elaeocarpum Reiss.

Rubiaceae
Alibertia sessilis Sch.
Guettarda viburnoides Cham. & Schlecht.

Rutaceae
Fagara hassleriana Chod.

Sapindaceae
Dilodendron bipinnatum Radlk.
Magonia pubescens St.Hil.

Sapotaceae
Bumelia obtusifolia Roem. & Schult.
Pouteria ramiflora (Mart.) Radlk.

Sterculiaceae
Guazuma ulmifolia Lam.
Sterculia striata St.Hil. & Naud.

Tiliaceae
Luehea paniculata Mart.

Verbenaceae
Vitex cymosa Bert.

Vochysiaceae
Qualea grandiflora Mart.

22

The Chaco-Pantanal transition in southern Mato Grosso, Brazil

D.E. PRADO, P.E. GIBBS, A. POTT and V.J. POTT

SUMMARY

1. Phytogeographical accounts and maps of South America depict a wedge of so-called Chaco vegetation in south-west Mato Grosso do Sul state, Brazil.
2. A survey undertaken in the area of Corumbá to Porto Murtinho revealed the presence of at least five different woody communities with varying degrees of relationship with typical Chaco; the physiognomy is sometimes very similar, but with one exception, their floristic composition is quite distinct.
3. The *Diplokeleba–Tabebuia–Capparis* scrublands, the *Schinopsis balansae* parklands and the *Aspidosperma–Mimosa*–cactus scrublands are transitional formations occupying low-lying areas. Their flora is related mainly to the Pantanal and *cerrados* with a few Chaco elements present.
4. The Chaco forests close to Porto Murtinho, dominated by *Schinopsis balansae,* are the only ones with an unmistakable chaquenian character, and apparently occur in a very small area.
5. The more widespread *Schinopsis brasiliensis*-dominated calcareous woodlands on high, stony, mesotrophic ground, do not contain any Chaco elements, but are linked to similar formations in the Pantanal and on calcareous outcrops within the cerrados.
6. In all likelihood, the dominant tree of the last community has been previously mistaken for *Schinopsis quebracho-colorado,* which does not occur in Brazil.
7. Thus the Brazilian 'Chaco' area is a meeting place of two separate elements – chaquenian and mesotrophic-type forest – which scarcely mingle but rather occur as distinct woodland types. The extent of this Chaco ingression into Brazil seems likely to be far smaller than previously thought.

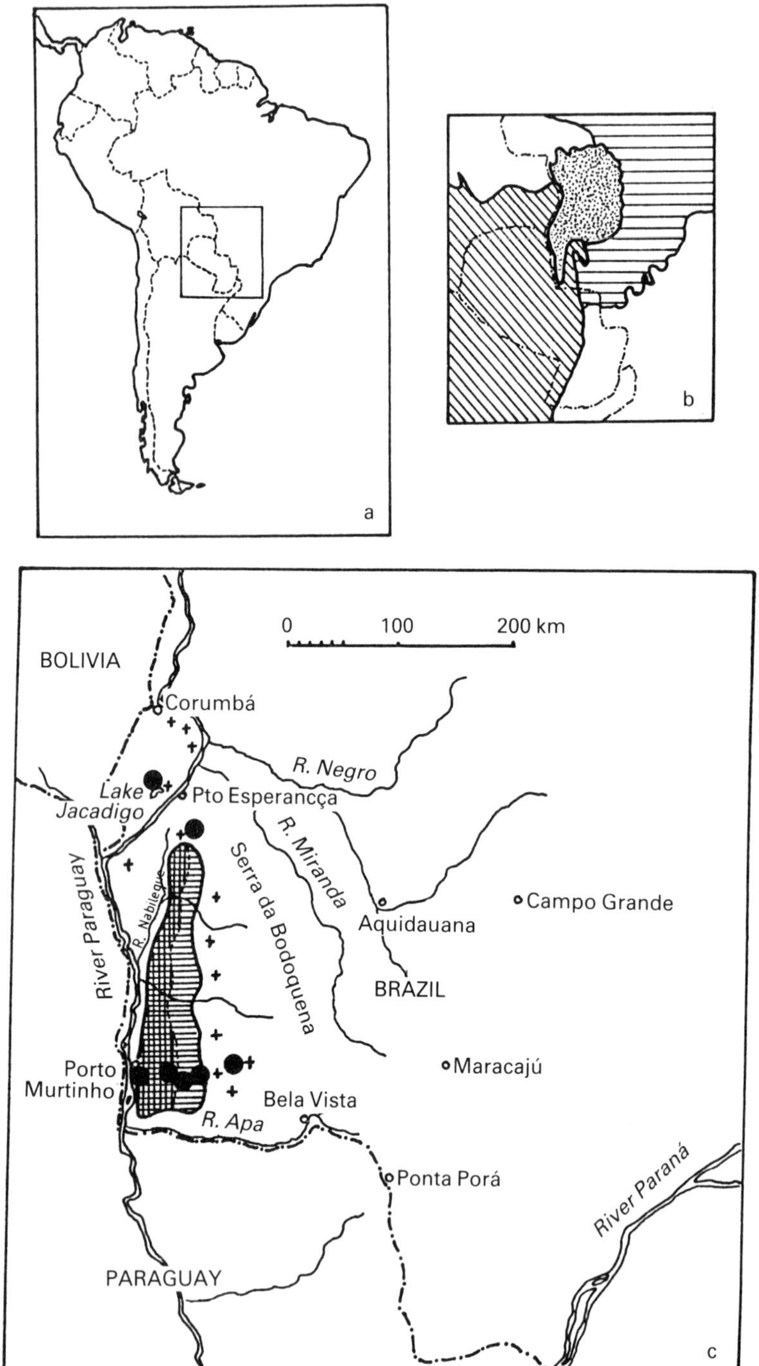
a
b
0
100
200 km
BOLIVIA
Corumbá
R. Negro
Lake
Jacadigo
Pto Esperancça
R. Miranda
Serra da Bodoquena
River Paraguay
R. Nabileque
Aquidauana
Campo Grande
BRAZIL
Porto
Murtinho
Maracajú
Bela Vista
R. Apa
Ponta Porá
River Paraná
PARAGUAY
c

INTRODUCTION

A number of phytogeographical accounts and vegetation maps of South America depict a narrow wedge of so-called Chaco vegetation, entering into Brazil in south-west Mato Grosso do Sul state as a continuation of the main area in Argentina, Paraguay and Bolivia. Figure 22.1b, modified from Cabrera and Willink's biogeographical provinces of South America (1980), is just one example of the currently prevailing concept of the phytogeography of the continent in general and of the Chaco in particular; other maps show approximately the same spur of the great Chaco extending into Brazil, as an intermediate zone separating the cerrados from the River Paraguay Pantanal. The presence of Chaco forests in that area has been reported by diverse authors, e.g. Hueck (1955, 1959, 1972), Ragonese and Castiglioni (1968), Morello and Adámoli (1968), Cabrera and Willink (1980), Hueck and Seibert (1981), Bucher (1982), Prance and Schaller (1982), Adámoli (1982), Eiten (1983), Ratter (1984), Allem and Valls (1987), and Ratter *et al.* (1988).

Veloso (1947) appears to have been one of the first to make a comprehensive study of the vegetation of this area, but although he cited the presence of the genera *Schinopsis* and *Bulnesia* he did not use the word Chaco. Hueck (1955) was the first phytogeographer to report *'bosques chaquenhos'* within Brazil, and he included a map showing the extent of the Chaco intrusion in the area (Fig. 22.1c), with three different formations: (a) *Copernicia* palm thickets, (b) *Celtis-Ziziphus* scrubland and (c) *'quebracho'* forests. According to Hueck (1955) the 'quebracho' type consists of formations dominated by either *Schinopsis balansae, Schinopsis quebracho-colorado* or *Aspidosperma quebracho-blanco*. No comment was made as to whether the first two species, *S. balansae* and *S. quebracho-colorado,* occur in the same forests, but he stated that the latter species was nine times more frequent than the former in the Brazilian Chaco while the proportion was two to eight for the same species in neighbouring Paraguay. Although Hueck did not provide floristic lists for his types of vegetation, we have made a list based on his citations (Table 22.1).

Fig. 22.1 (a) Map of South America. *Inset*: area covered by map b. (b) Biogeographical provinces in central South America (modified from Cabrera and Willink, 1980). Oblique hatching, Chaco; dotted, Pantanal; horizontal hatching, cerrado; blank areas, Amazonian Dominium and mesophilous forest. (c) The Chaco ingression into Brazil according to Hueck, 1955 (modified). Horizontal hatching, forests of 'quebracho' type (*Aspidosperma quebracho-blanco, Schinopsis quebracho-colorado, S. balansae*); cross-hatching, *Celtis-Ziziphus* forest type; crosses, isolated elements of chaquenian forests; solid black dots, localities at which our observations were made.

Table 22.1 Comparison of some floristic lists for the vegetation of south-west Mato Grosso do Sul

Hueck (1955)			*Fiebrig (1933)*
Schinopsis balansae	!	R	***Schinopsis balansae***
Schinopsis quebracho-colorado			*Schinopsis quebracho-colorado*
Aspidosperma quebracho-blanco	!	R	*Aspidosperma* spp.
Caesalpinia paraguariensis[a]	!	R	***Caesalpinia paraguariensis***[a]
Acacia sp.	!	R	
Celtis sp.	!	R	
Ziziphus sp.	!	R	*Ziziphus* sp.
Copernicia australis	!	R	***Copernicia australis***
Peltophorum dubium	!	R	
Anadenanthera colubrina v. *cebil*[a]	!	R	***Anadenanthera colubrina*** v. ***cebil***[a]
Calycophyllum multiflorum	!	R	*Calycophyllum multiflorum*
Ruprechtia sp.	!	R	*Ruprechtia* sp.
Maclura tinctoria[a]	!	R	*Maclura* sp.[a]
Prosopis ruscifolia[a]	!	R	*Prosopis* spp.
Astronium urundeuva	!	R	*Astronium urundeuva*
Tabebuia sp. (lapacho)	!	R	***Tabebuia impetiginosa***[a]
Ceiba insignis agg.[a]	!	R	several Bombacaceae: ***Ceiba***[a], etc.
Trithrinax sp.			
Cereus sp.	!	R	
	!	R	*Fagara* sp.
	!	R	*Machaerium* sp.
		R	*Luehea* sp.
		R	*Pithecollobium saman*
	!		*Diatenopterix sorbifolia*
	!	R	*Pterogyne nitens*
		R	*Jacaranda cuspidifolia*
	!		*Phyllostylon rhamnoides*
	!	R	*Rhamnidium* sp.
		R	*Xylosma venosum*
			Carica quercifolia
			Cochlospermum sp.

!: taxa seen in the field, or as exsiccata collected in this area.
R: exsiccata Ratter *et al.* (1987), or taxa cited for this area in Ratter *et al.* (1988).
bold: cited by Fiebrig but for other communities.
[a] under synonym in the original.

Fiebrig (1933) describes a community similar to Hueck's 'quebracho' type for the opposite bank of the River Paraguay on high ground; he calls this formation *'campos-parque'* (parkland–grassland) reporting that *Elionurus muticus* covers the open spaces in almost pure populations, while the forested areas are dominated by *Schinopsis quebracho-colorado,* which replaces *S. balansae* since the former is 'less exigent in salinity'. His floristic list is very similar to that of Hueck (Table 22.1) but since Fiebrig provided lists for each community, that for the 'campos-parque' covers only part of Hueck's, and the rest of the species (in bold type) listed by Fiebrig are of different communities in the same area of northern Paraguay.

The predominance of *Schinopsis quebracho-colorado* over *S. balansae* in Mato Grosso do Sul as reported by Hueck seems very odd since this area has a rainfall of *c.* 1250 mm yr^{-1} (Ratter *et al.*, 1988), far more than the 850–900 mm yr^{-1} to 500 mm yr^{-1} in the Chaco area of Argentina where *S. quebracho-colorado* is dominant. On the other hand, *S. balansae* is found in areas of the Great Chaco with 1250 to around 800 mm yr^{-1}, so that there is a zone of contact in the centre of the Argentinian Chaco where both species are at opposite extremes of their ecological amplitudes. In general, the environmental conditions favourable to one species exclude the other, and even within the area where both species can be found together they occupy different topographical positions in the soil gradient (Morello and Adámoli, 1974, p. 46). Moreover, *S. balansae* dominates the forests in the eastern Chaco (Cabrera, 1976; Lewis and Pire, 1981) whereas *S. quebracho-colorado* is virtually absent from this sector in Argentina (Morello and Adámoli, 1974, Fig. 8). On a rainfall-climatic basis one would expect the vegetation of this area in south-west Brazil should relate to the eastern Chaco.

For background reference, Table 22.2 gives an abridged list of the main woody species of the Chaco phytogeographic province (Ragonese and Castiglioni, 1968; Morello and Adámoli, 1968; Cabrera, 1976). The Chaco province consists of forests dominated by species of the arboreal genus *Schinopsis*. In the frequently waterlogged eastern plains the predominant species is the simple-leaved *S. balansae*; in the dry and flat plains of central and western Chaco it is *S. quebracho-colorado*; with numerous linear-lanceolate leaflets. Also characteristic are the forests of *Bulnesia sarmientoi* in west Formosa and east Salta, and the woodlands dominated by *Aspidosperma quebracho-blanco* in the south-west Argentinian Chaco (Ragonese and Castiglioni, 1968). Savannas are also a very important component of the landscape: both *Spartina argentinensis* and *Elionurus muticus* grasslands with scattered trees of *Prosopis* spp. are common throughout the region. Palm tree woodlands of *Copernicia australis* are very important from the Brazilian Pantanal to north-east Santa Fe in Argentina, always on alluvial, seasonally flooded ground. Both from a floristic and physiognomic point of view there is little contact between the Chaco and the cerrados, and they have been placed in widely different floristic dominia in South America (Cabrera and Willink, 1980). The Chaco–Pantanal links have remained obscure until Adámoli's (1982) discussion on the general phytogeographic relationships of the Pantanal, but the matter is still far from settled.

Table 22.2 Some characteristic species of the Chaco province

Schinopsis quebracho-colorado (Schlecht.) Bark. & Meyer
Schinopsis balansae Engler
Aspidosperma quebracho-blanco Schlecht.
Prosopis alba Gris.
Prosopis nigra (Gris.) Hieron.
Prosopis ruscifolia Gris.
Prosopis spp.
Bulnesia sarmientoi Lorentz ex Gris.
Ziziphus mistol Gris.
Caesalpinia paraguariensis (D. Parodi) Burkart
Geoffroea decorticans (Hook. & Arn.) Burkart
Acacia praecox Gris.
Acacia spp.
Cercidium praecox (Ruiz & Pavón) Harms
Schinus spp.
Capparis spp.
Celtis spp.
Bougainvillea spp.
Opuntia spp.
Cereus spp.
Eriocereus spp.

RESULTS

In February 1989 three of the authors (D.E.P., A.P. and V.J.P.) carried out a survey of the area (Fig. 22.1c), on the basis of which we have been able to segregate the vegetation into the following major woody communities (floristic composition and frequency estimation of their elements are listed in Table 22.3).

Diplokeleba–Tabebuia–Capparis scrubland

To the south of Corumbá, in the Fazenda Coqueiro area, close to the Jacadigo Lake within the Pantanal of Nabileque (Adámoli, 1982), there is a scrubland up to 8 m, but generally not more than 5–6 m tall. Trees and shrubs are usually clustered in small polyspecific groups of three to four individuals on slight elevations of the otherwise flat, low-lying terrain which was waterlogged at the time of observation. The treelet-shrub layer is dominated by *Diplokeleba floribunda* and *Tabebuia nodosa,* with *Machaerium hirtum* and *Capparis* spp. also common; the herbaceous layer is either dominated by Gramineae or pure populations of *Selaginella sellowii*. The physiognomy of this community seems to match well with Hueck's *Celtis–Ziziphus* type but although both taxa are present here we

Table 22.3 Occurrence of woody species in the Mato Grosso do Sul study area (*Frequency*: a, abundant; c, common; f, frequent; o, occasional; r, rare. *Occurrence in other formations*; ca, calciphilous species (Ratter *et al.*, 1988); ce, cerradão species (Heringer *et al.*, 1977, Ratter *et al.*, 1988);* Chaco species *s.s.*)

		Schinopsis brasiliensis woodland	*Schinopsis balansae parkland*	*Aspidosperma –Mimosa scrubland*	*Diplokeleba– Tabeb.–Capp. scrubland*	*Chaco Porto Murtinho*	*Floristic groups*
1 *Schinopsis brasiliensis*		a					
2 *Amburana cearensis*		c					
3 *Phyllostylon rhamnoides*		c					
4 *Cnidoscolus vitifolius* var. *cnicodendron*	ca/ce	c					I
5 *Helicteres lhotskyana*		c					
6 *Anadenanthera colubrina* var. *cebil*	ca/ce	f					
7 *Sapium* sp.		f					
8 *Croton* sp. I		f					
9 *Calycophyllum multiflorum*	ce	o					
10 *Anemopaegna velutinum*		r					
11 [Myrtaceae]		r					
12 *Sterculia striata*	ca/ce	r					
13 *Astronium urundeuva*	ca/ce	c	f	o			
14 *Acrocomia totai*	ca/ce	f	r	o			
15 *Cordia glabrata*	ca/ce	o	r	f	r		
16 *Tabebuia impetiginosa*	ca/ce	f	r		r		
17 *Machaerium hirtum*	ce	o	f		c		II
18 *Celtis* cf. *pubescens*	ce	f		f	f		
19 *Ziziphus oblongifolius*			c	f	f		
20 *Cereus* cf. *peruvianus*		r		f	o		
21 *Randia armata*	ce	o			r		
22 *Chomelia obtusa*	ce	f	r				
23 *Pseudobombax* sp.	ce	o	r				III
24 *Rhamnidium elaeocarpum*	ce	o	r				
25 *Maytenus ilicifolia*		r	r				

Table 22.3 (continued)

		Schinopsis brasiliensis woodland	*Schinopsis balansae parkland*	*Aspidosperma –Mimosa scrubland*	*Diplokeleba–Tabeb.–Capp. scrubland*	*Chaco Porto Murtinho*	*Floristic groups*
26 *Tocoyena formosa*	ce		o				
27 *Pisonia* cf. *zapallo*			r				
28 *Tabebuia ochracea*	ce		r				IV
29 *Terminalia argentea*	ca/ce		r				
30 *Schinopsis balansae*	*		a			a	
31 *Erythroxylum* cf. *cuneifolium*			o			c	V
32 *Fagara* cf. *rhoifolia*	ce		o			o	
33 *Mimosa hexandra*			c	c			
34 *Mimosa* sp. (Pott 4591)			c	c			VI
35 *Aspidosperma quebracho-blanco*	*		f	a	r	c	
36 *Cercidium praecox*	*			o		r	VII
37 *Stachytarpheta* sp.				f			
38 *Cordia* cf. *curassavica*				f			
39 *Jatropha ribifolia*				o			VIII
40 *Prosopis rubriflora*				o			
41 *Diplokeleba floribunda*					a		
42 [Mimosoidea (*Prosopis* sp.?)]					o		
43 *Bougainvillea* sp.					o		
44 *Goldmania paraguensis*					r		
45 *Crateva tapia*					r		IX
46 *Tabebuia caraiba*	ce				r		
47 *Rauwolfia mollis*					r		
48 *Casearia aculeata*	ce				r		
49 *Paullinia pinnata*					r		

50	*Tabebuia nodosa*	*				a	o	
51	*Capparis* cf. *retusa*	*				c	o	
52	*Coccoloba* sp.					f	f	
53	*Capparis tweediana*	*				f	r	X
54	*Capparis speciosa*	*				f	r	
55	*Aporosella chacoensis*					r	o	
56	*Acacia praecox*	*					c	
57	*Sideroxylon obtusifolium* ssp. *obtusifolium*	*					f	
58	*Copernicia australis*	*					f	
59	*Prosopis ruscifolia*	*					f	
60	*Croton* sp. II						f	XI
61	*Aspidosperma triternatum*	*					o	
62	*Ziziphus mistol*	*					o	
63	*Geoffroea striata*						o	
64	*Celtis spinosa*	*					o	
65	*Ruprechtia* cf. *triflora*	*					r	
66	*Bulnesia sarmientoi*	*					r	
67	*Rollinia emarginata*						r	
	Number of species		24	20	14	23	23	
	Number of Chaco *s.s.* (*) species		–	2	2	5	16	
	Percentage of Chaco *s.s.* species		–	10.00	14.30	21.70	69.60	

have been unable to find vegetation stands dominated by them in southwest Mato Grosso do Sul.

The subsequent survey comprised a transect running east–west from the Serra da Bodoquena ranges to Porto Murtinho on the River Paraguay. Leaving aside the cerrado vegetation *sensu lato* which covers most of the eastern half of this sector, the first Chaco-related formation to appear is the following.

Schinopsis balansae parklands ('quebrachal')

These consist of small to large islets or groves of trees, up to 50 m in diameter, usually dominated by *S. balansae* (sometimes forming nearly pure populations). The wooded areas are enclosed within a matrix of a grassland consisting primarily of *Elionurus muticus* (Fig. 22.2), which closely resembles that described by Allem and Valls (1987) for the Pantanal region. The tallest trees reach 20 m or more, but most fall within a range of 10–14 m. The physiognomy is similar to that of equivalent communities further south in the eastern Chaco of Argentina, but details of the floristic composition reveal substantial differences (Table 22.3), and only two species (albeit one of them the dominant) are strictly chaquenian

Fig. 22.2 *Schinopsis balansae* parkland within a matrix of *Elionurus muticus* grassland. In the centre a small group of cerrado elements. Mato Grosso do Sul, 50 km east of Porto Murtinho.

in our view. Within the area occupied by the grassland there are some isolated individuals, or usually very small clumps, of shrubs and trees belonging to the *'cerradão'* or *'mata calcárea'* communities described by Ratter *et al.* (1988), such as *Tabebuia ochracea, Terminalia argentea, Cordia glabrata, Chomelia obtusa* or *Tabebuia impetiginosa* (*'piuva'*). The understorey of these *Schinopsis* woods consists sometimes of pure populations of *Bromelia serra*, although patches of *Selaginella sellowii* can be found. Other relevant elements of this layer are *Microchloa indica, Gouinia paraguariensis, Setaria geniculata, Eupatorium christieanum, Tournefortia* sp., *Cuphea* sp., *Stachytarpheta* sp., etc.

Aspidosperma–Mimosa scrubland

Some 18 km west of Córrego Capivara and 30 km east of Porto Murtinho there is an open woodland and cactus scrub on sandy, alkaline soils. This community is dominated by *Aspidosperma quebracho-blanco* accompanied by two species of *Mimosa*, and a great candelabra-like *Cereus* (Fig. 22.3). Again, the physiognomy of this community strikingly resembles certain stands of vegetation in the Paraguayan or Argentinian Chaco, such as the *'cardonales'* of *Stetsonia coryne* (Lewis and Pire, 1981; Esser, 1982), but once more its floristic composition shows only two chaquenian species *sensu stricto*.

Fig. 22.3 *Aspidosperma–Mimosa*–cactus scrubland, 18 km west of Córrego Capivara and 30 km east of Porto Murtinho, Mato Grosso do Sul.

Table 22.4 Diagnostic characters of *Schinopsis quebracho-colorado* and *S. brasiliensis*

	S. quebracho-colorado	*S. brasiliensis*
Leaf		
petiole (mm)	20–30	18–20
colour	discolorous	concolorous
Leaflets		
number	13–31	11–21
width (mm)	4–7	7–11
shape	lanceolate	obovate
apex	acute	obtuse, emarginate
Flower		
sepals	ciliate	glabrous
stamens (mm)	2	2.5
Fruit		
length (mm)	26–30	34–40
width (mm)	7–9	*c.* 13
seed (mm)	5 × 7	12 × 17

Calcareous woodlands

Around 20 km east of Porto Murtinho there is a different kind of forest which is widespread in the region and occurs on high, stony ground with isolated rock outcrops and bearing a calcareous, mesotrophic soil. This comprises a 15–20 m tall forest which is dominated by another species of the genus *Schinopsis*, locally called *'coronillo'* (*'chamucoco'* in the Corumbá area). According to Hueck's survey this species should be *S. quebracho-colorado* but is in fact *S. brasiliensis*. These two species are readily distinguished in the field on account of the 'willow-like' habit of *S. quebracho-colorado* with its pendent, discolorous leaves, while *S. brasiliensis* has erect, concolorous leaves, and there are other differential characters (Table 22.4). The floristic composition of this community (Table 22.3), which according to rural workers extends to the north and south of the main road BR267, in parallel to the general direction of the River Paraguay, closely resembles the calcareous woodlands near Corumbá (Ratter *et al.*, 1988). Some stands are rather open (Fig. 22.4) with groves of *S. brasiliensis* in *Elionurus muticus* grasslands. However, the grasslands here were not as extensive as in the previous *S. balansae* formation and the woodlands were more open.

Chaquenian forests

Within the urban area of Porto Murtinho, surrounded by a dike which prevents the waters of the River Paraguay from regularly flooding the

Fig. 22.4 *Schinopsis brasiliensis* woodland, 22 km east of Porto Murtinho, Mato Grosso do Sul.

city, there are still some enclaves of '*mata chaquenha*', and other stands outside the dike were also visited. This community, on a depressed, floodable topographic position, with fine-grain loamy alkaline soils, is essentially a *Schinopsis balansae* 'quebrachal' which resembles typical Chaco vegetation in both physiognomy and floristics (Fig. 22.5). Of the species of this community, 70% are true Chaco elements, while the rest are mainly riparian elements of the Paraguay and Paraná river valleys, such as *Geoffroea striata, Aporosella chacoensis* and *Rollinia emarginata*. In our view, and also according to ex-employees of the now-closed tannin company of Porto Murtinho, these 'quebracho' woodlands only occur in a small area around the town.

DISCUSSION

Contrary to Hueck (1955), it seems very unlikely that the species *Schinopsis quebracho-colorado* extends its range into Brazil. No examples of this species were encountered in our survey, and rural workers only know one type of 'coronillo', which is *S. brasiliensis*. Furthermore, we have seen and studied collections of this genus from a number of

Fig. 22.5 The Chaco forest at Porto Murtinho, Mato Grosso do Sul. *Schinopsis balansae* is the dominant tree in this community.

herbaria, and have never encountered any exsiccata of *S. quebracho-colorado* from this area. Moreover, in the monograph of the genus (Meyer and Barkley, 1973) *S. quebracho-colorado* is not cited for Brazil. There are, however, numerous collections of *S. brasiliensis* from these localities. As noted previously, rainfall in Mato Grosso do Sul is far in excess of that for the central Chaco of Paraguay and Argentina where *S. quebracho-colorado* is dominant. It seems most likely that Hueck (1955) confused the identity of these two *Schinopsis* species, particularly since most of the accompanying species cited by him appear in the *S. brasiliensis* community described here (see Table 22.3), or have been seen by us either in the field or as herbarium specimens collected in this area (Table 22.1). All this evidence offers an explanation for the paradox of a more xeric species (*S. quebracho-colorado*) being apparently more abundant in a relatively wetter area than the congeneric species known to be adapted to more humid localities (*S. balansae*).

This case of mistaken identity by Hueck (1955) seems to have led to subsequent confusions by other workers. Thus, Cabrera (1970) stated that the 'mata chaquenha' enters into Brazil on high calcareous ground, whereas we consider it is restricted to saline and seasonally flooded areas. In this context, it is of interest that the soil map reproduced by Allem and

Valls (1987) indicates solodized-solonetz soils for those sectors of Mato Grosso do Sul which we consider to have a higher chaquenian influence. It is also of interest that Fiebrig (1933) described a community for the area of Paraguay opposite to Porto Murtinho which according to him is dominated by *S. quebracho-colorado*. However, the accompanying woody species, and the grasslands, are those we have recognized for the *S. brasiliensis* community. Although no further studies have been undertaken in Fiebrig's area, the likelihood that the woodlands in this sector of Paraguay are indeed also dominated by *S. brasiliensis* rather than *S. quebracho-colorado* finds support in Tortorelli's (1967) overview of Paraguayan forests. According to this work, *S. quebracho-colorado* only occurs further to the west, in a different kind of chaquenian community, whilst closer to the river only *S. balansae* is cited. The species list for the latter community (Tortorelli, 1967) is in fact very similar to that for *S. brasiliensis* woodlands in Brazil.

The five woody communities described here show different degrees of relationship with the main core of Chaco vegetation in Bolivia, Paraguay and Argentina. The calcareous woodlands have no Chaco species at all (Table 22.3), and we regard this community as absolutely non-chaquenian in character, both in physiognomy and floristic composition. Interestingly, some of its elements can be found elsewhere in South America in formations such as the caatingas in north-eastern Brazil, the yungas of Bolivia and Argentina, or the subtropical forests at both banks of the river Paraná (D.E. Prado, unpublished). At least seven species of this community have been considered calciphilous (Table 22.3) and are indicators of mesotrophic forests (Ratter, 1987; Ratter *et al.*, 1978, 1988; Furley *et al.*, 1988), which occur on calcium-rich soils in the cerrados from Bahia, north of Minas Gerais to south-west Brazil. Ratter *et al.* (1988) described a very similar deciduous forest near Corumbá with *Tabebuia impetiginosa* as dominant, and although *Schinopsis brasiliensis* was not cited, we have seen this species on calcareous hills near Corumbá in an area very close to the one studied by Ratter *et al.*

The *Schinopsis balansae* parklands contain elements of several floristic stocks, of which the true chaquenian species, although including the dominant, are otherwise rather exceptional (only 10%), whereas other taxa of semi-deciduous tropical and subtropical forests such as *Astronium urundeuva, Tabebuia impetiginosa, Pseudobombax* sp., etc., make up the floristic bulk of the community. These last elements, which are important, for example, in the Paranaense phytogeographical province (Cabrera and Willink, 1980), co-occur in these parklands with some cerrado species (*Tabebuia ochracea, Terminalia argentea, Chomelia obtusa*, etc.), thus resulting in a perplexing mosaic of at least three different lineages.

The *Aspidosperma–Mimosa* scrubland contains only 14.3% of chaquenian species, but again one of them (*Aspidosperma quebracho-blanco*) is dominant; the predominant floristic stock is the semi-deciduous forest elements and some endemic species of the Pantanal and north-east Paraguay region such as *Ziziphus oblongifolius* and *Prosopis rubriflora*, the latter known only from the border area between Paraguay and Brazil close to the river Paraguay (Burkart, 1976). This community might very well be the remains of stands of *Schinopsis balansae* parklands subjected to selective felling of valuable timber in the past, since it shares 50% of its members with the latter. Although we did not find any *S. balansae* stumps or seedlings, it has been reported that even remaining rootstocks of this particular species were removed for tannin extraction in the Argentinian eastern Chaco (Parodi, 1934). There were clear signs of moderate erosion, perhaps because of overgrazing in this area, and it has also been shown that *Aspidosperma quebracho-blanco* (through its gemmiferous roots) and *Cercidium praecox* can easily pervade in areas felled or overgrazed (Morello and Saravia Toledo, 1959), thus explaining the dominance of the former species and the presence of the latter in this community.

The *Diplokeleba–Tabebuia–Capparis* scrubland is closer to a true Chaco community with nearly 22% of Chaco species. This community also includes some cerrado elements (*Tabebuia caraiba, Cordia glabrata*), riparian species (*Crateva tapia, Aporosella chacoensis, Coccoloba* sp.) and subtropical forest elements which extend along the Paraguay–Paraná River systems such as *Goldmania paraguensis* and *Diplokeleba floribunda*.

These three foregoing communities are regarded here as transitional, as none of them can be fully ascribed to any of the major phytogeographical units that converge in the Pantanal, i.e. cerrados, Amazonian, Paranaense and Chaco provinces (Adámoli, 1982). The remaining community, the Chaco forest at Porto Murtinho, is the only one from south-west Mato Grosso do Sul that can be accepted without reserve as Chaco vegetation, since both dominants and accompanying species are Chaco elements *sensu stricto*, and similar communities can be found elsewhere in Paraguay or as far south as the north of Santa Fe province in Argentina (Lewis and Pire, 1981).

The floristic lists of the five major communities have been worked out as a phytosociological table (Braun-Blanquet, 1979), and 11 different floristic groups (Mueller-Dombois and Ellenberg, 1974) characterize each community or set of communities (Table 22.3). Floristic groups I, IV, VIII, IX and XI comprise those species exclusive to a single community in the study area, whereas floristic groups III, V, IV and X are common to pairs of communities. All five communities are reasonably well defined, but the most distinctive are the *Schinopsis brasiliensis* forests and the Chaco. The latter is in fact quite isolated, and the highest concentration of Chaco species occurs in floristic group XI; more than half of the taxa bearing an asterisk in Table 22.3 are in this single group. Although the *Diplokeleba–Tabebuia–Capparis* scrubland contains some Chaco elements, it actually includes more species of the Pantanal or cerrado areas, e.g. group II comprises most of the calciphilous and cerrado trees and shrubs, but even more relevant, it contains precisely those species that are absent from the forests at Porto Murtinho. Thus, these group II species give a well defined, non-chaquenian character to the other four communities. Group VII, which comprises two chaquenian trees (*Aspidosperma quebracho-blanco* and *Cercidium praecox*), occurs in four of the five communities; these two chaquenian trees are presumably at the extreme of their ecological range here. They can be found from the driest corner of the Chaco in the Argentina, the Salinas Grandes, to the Mato Grosso do Sul in Brazil, and they are particularly abundant in human-disturbed ecosystems.

If Hueck's proportion of 9:1 of *Schinopsis brasiliensis* (mistaken for *S. quebracho-colorado*) to *S. balansae* still holds true (as does seem to be the case), the main consequence is that around 80–90% of what is shown in Fig. 22.1c as Chaco vegetation is wrong. Thus the Brazilian 'Chaco' area is a meeting place of two distinct forest elements – chaquenian and mesotrophic-type forest, and it is noteworthy that they scarcely mingle but rather occur as distinct woodland types. Further studies are needed to assess the exact extent of the Chaco ingression into Brazil, but this seems likely to be far less than previously thought, and hardly mappable on a continental scale.

REFERENCES

Adámoli, J. (1982) O pantanal e suas relações fitogeográficas com os Cerrados. Discussão sobre o conceito de 'Complexo do Pantanal'. *XXXII Congresso Nacional de Botânica, Teresina, PI*, 109–19.

Allem, A.A. and Valls, J.F.M. (1987) *Recursos forrageiros nativos do Pantanal mato-grossense*, Empresa Brasileira de Pesquisa Agropecuária, CENARGEN, Brasília.

Braun-Blanquet, J. (1979) *Fitosociología*, H. Blume Ediciones, Madrid.

Bucher, E.H. (1982) Chaco and caatinga. South American arid savannas, woodlands and thickets, in *Ecology of Tropical Savannas*, (eds B.J. Huntley and B.H. Walker), Springer-Verlag, Berlin, pp. 48–79.

Burkart, A. (1976) A monograph of the genus *Prosopis* (Leguminosae Subfam. Mimosoideae). *Journal of the Arnold Arboretum*. **57**, 450–525.

Cabrera, A.L. (1979) La vegetación del Paraguay en el cuadro fitogeográfico de América del Sur. *Boletín de la Sociedad Argentina de Botánica*, **11 (Suppl.)**, 121–31.

Cabrera, A.L. (1976) Regiones fitogeográficas argentinas. *Enciclopedia Argentina de Agricultura y Jardinería*, 2nd edn, Vol. II, p. 1.
Cabrera, A.L. and Willink, A. (1980) *Biogeografía de América Latina*, Secretaría General de la Organización de los Estados Americanos, Washington, DC, USA.
Eiten, G. (1983) *Classificação da Vegetação do Brasil*, Conselho Nacional de Desenvolvimento Científico e Tecnológico, Brasília.
Esser, G. (1982) Vegetationsgliederung und Kakteen-vegetation von Paraguay. *Tropische und Subtropische Pflanzenwelt*, **38**, 365–471.
Fiebrig, C. (1933) Ensayo fitogeográfico sobre el Chaco Boreal. *Revista del Jardín Botánico y Museo de Historia Natural del Paraguay*, **3**, 3–87.
Furley, P.A., Ratter, J.A. and Gifford, D.R. (1988) Observations on the vegetation of eastern Mato Grosso, Brazil. III. The woody vegetation and soils of the Morro de Fumaça, Torixoreu. *Proceedings of the Royal Society, London*,. **B235**, 259–80.
Heringer, E.P., Barroso, G.M., Rizzo, J.A. and Rizzini, C.T. (1977) A flora do cerrado, in *IV Simposio sobre o Cerrado*, (ed. M.G. Ferri), Editora Universidade de São Paulo, São Paulo, pp. 211–32.
Hueck, K. (1955) Bosques chaquenhos e extração de tanino no Brasil. *Revista Brasileira de Geografia*, **17**, 343–6.
Hueck, K. (1959) Bosques secos de la zona tropical y subtropical de la América del Sur. *Boletín del Instituto Forestal Latinoamericano*, **4**, 1–49.
Hueck, K. (1972) *As Florestas da América do Sul. Ecologia, Composição e Importancia Econômica*, Editora Polígono & Editora da Universidade de Brasília, São Paulo.
Hueck, K. and Seibert, P. (1981) *Vegetationskarte von Südamerika*, G. Fischer Verlag, Stuttgart.
Lewis, J.P. and Pire, E.F. (1981) *Reseña sobre la vegetación del Chaco santafesino*, Instituto Nacional de Tecnología Agropecuaria, Serie Fitogeográfica, 18, Buenos Aires.
Meyer, T. and Barkley, F.A. (1973) Revisión del género *Schinopsis* (Anacardiaceae). *Lilloa*, **33**, 207–58.
Morello, J. and Adámoli, J. (1968) *Las Grandes Unidades de Vegetación y Ambiente del Chaco Argentino. Primera parte: Objetivos y Metodología*, Instituto Nacional de Tecnología Agropecuaria, Serie Fitogeográfica No. 10, Buenos Aires.
Morello, J. and Adámoli, J. (1974) *Las Grandes Unidades de Vegetación y Ambiente del Chaco Argentino. Segunda Parte: Vegetación y Ambiente de la Provincia del Chaco*, Instituto Nacional de Tecnología Agropecuaria, Serie Fitogeográfica, 13, Buenos Aires.
Morello, J. and Saravia Toledo, C. (1959) El bosque chaqueño. I: Paisaje primitivo, paisaje natural y paisaje cultural en el oriente de Salta. *Revista de Agronomía del Noroeste Argentino*, **3**, 5–81.

Mueller-Dombois D. and Ellenberg, H. (1974) *Aims and Methods of Vegetation Ecology*, John Wiley, New York.
Parodi, L.R. (1934) La vegetación de Reconquista (Provincia de Santa Fe). *Revista Geográfica Americana*, **1**, 389–407.
Prance, C.T. and Schaller, G.B. (1982) Preliminary study of some vegetation types of the Pantanal, Mato Grosso, Brazil. *Brittonia*, **34**, 228–51.
Ragonese, A.R. and Castiglioni, J.A. (1968) *La vegetación del Parque Chaqueño*, Instituto Nacional de Tecnología Agropecuaria, Serie Fitogeográfica, 12, Buenos Aires.
Ratter, J.A. (1984) *Report of a visit to the Unidade de Execução de Pesquisa de Ambito Estadual (UEPAE) de EMBRAPA (Pantanal Station), Corumbá, MS. 28th August–2nd September, 1984*, Mimeograph, Royal Botanic Garden, Edinburgh.
Ratter, J.A. (1987) Notes on the vegetation of the Parque Nacional do Araguaia (Brazil). *Notes of the Royal Botanic Garden, Edinburgh*, **44**, 311–42.
Ratter, J.A., Askew, G.P., Montgomery, R. and Gifford, D.R. (1978) Observations on forests of some mesotrophic soils in central Brazil. *Revista Brasileira de Botânica*, **1**, 47–58.
Ratter, J.A., Pott, A., Pott, V.J. *et al.* (1988) Observations on woody vegetation types in the Pantanal and at Corumbá, Brazil. *Notes of the Royal Botanic Garden, Edinburgh*, **45**, 503–25.
Tortorelli, L.A. (1967) Formaciones forestales y madereras del Paraguay. *Boletín del Instituto Forestal Latinoamericano*, **24**, 3–34.
Veloso, H.P. (1947) Considerações gerais sôbre a vegetação do Estado de Mato Grosso. II – Notas preliminares sôbre o pantanal e zonas de transição. *Memórias do Instituto Oswaldo Cruz*, **45**, 253–72.

APPENDIX: SPECIES RECORDED IN THE STUDY AREAS

Acacia praecox Gris.
Acrocomia totai Hoehne
Amburana cearensis (Fr. All.) A.C. Smith
Anadenanthera colubrina (Vell.) Bren. var. *cebil* (Gris.) Altsch.
Anemopaegma velutinum Mart. ex DC.
Aporosella chacoensis (Morong) Speg.
Aspidosperma quebracho-blanco Schlecht.
Aspidosperma triternatum Rojas Acosta
Astronium urundeuva (Fr. All.) Engler

Bougainvillea sp.
Bromelia balansae Mez
Bulnesia sarmientoi Lorentz ex Gris.

Caesalpinia paraguariensis (D. Parodi) Burkart
Calycophyllum multiflorum Gris.
Capparis cf. *retusa* Gris.
Capparis speciosa Gris.
Capparis tweediana Eichler
Casearia aculeata Jacq.
Ceiba insignis (Kunth) Gibbs & Semir sp. agg.

Celtis cf. *pubescens* Kunth
Celtis spinosa Spreng.
Cercidium praecox (Ruiz & Pavón) Harms
Cereus cf. *peruvianus* (L.) Miller
Chomelia obtusa Cham. & Schlecht.
Cnidoscolus vitifolius Mill. ex Pohl var. *cnicodendron* (Gris.) Pax
Coccoloba sp.
Copernicia australis Becc.
Cordia cf. *curassavica* (Jacq.) Roem. & Sch.
Cordia glabrata (Mart.) A. DC.
Crateva tapia L.
Croton sp. 1
Croton sp. 11
Cuphea sp.

Diatenopterix sorbifolia Radlk.
Diplokeleba floribunda N.E. Brown

Elionurus muticus (Spreng.) Kuntze
Erythroxylum cf. *cuneifolium* (Mart.) Schulz
Eupatorium christieanum Baker

Fagara cf. *rhoifolia* (Lam.) Engler

Geoffroea striata (Willd.) Morong
Goldmania paraguensis (Benth.) Brenan
Gouinia paraguariensis (O.K.) L.R. Parodi

Helicteres lhotskyana (Schott) K. Schum.

Jatropha ribifolia (Pohl) Baill.

Machaerium hirtum (Vell.) Stellf.
Maclura tinctoria (L.) Don ex Steud.
Maytenus ilicifolia Mart. ex Reiss.
Microchloa indica (L.f.) P. Beauv.
Mimosa hexandra Micheli
Mimosa sp. (Pott 4591)

Paullinia pinnata L.
Peltophorum dubium (Spreng.) Taub.
Phyllostylon rhamnoides (Poiss.) Taub.
Pisonia cf. *zapallo* Gris.
Prosopis rubriflora Hassler
Prosopis ruscifolia Gris.
Pseudobombax sp.
Pterogyne nitens Tulasne

Randia armata (Sw.) DC.
Rauwolfia mollis S. Moore
Rhamnidium elaeocarpum Reiss.
Rollinia emarginata Schlecht.
Ruprechtia cf. *triflora* Gris.
Ruprechtia sp.

Sapium sp.
Schinopsis balansae Engler
Schinopsis brasiliensis Engler
Selaginella sellowii Hieron.
Setaria geniculata (Willd.) P. Beauv.
Sideroxylon obtusifolium (Roem. & Sch.) Penn. ssp. *obtusifolium*
Stachytarpheta sp.
Sterculia striata St. Hil. & Naud.

Tabebuia caraiba (Mart.) Bur.
Tabebuia impetiginosa (Mart.) Standley
Tabebuia nodosa (Gris.) Gris.
Tabebuia ochracea (Cham.) Standl.
Terminalia argentea Mart. & Zucc.
Tocoyena formosa (Cham. & Schl.) Schum.
Tournefortia sp.

Ziziphus mistol Gris.
Ziziphus oblongifolius S. Moore

[Mimosoideae (*Prosopis* sp.?)]
[Myrtaceae]

PART FIVE

Animals at the Forest–Savanna Boundary

23

Soil fauna at the forest–savanna boundary: role of termite mounds in nutrient cycling

LUC ABBADIE, MICHAEL LEPAGE and XAVIER LE ROUX

SUMMARY

1. In a preforest savanna (Lamto, Côte d'Ivoire), termites locally modify soil physio-chemical properties through their building activities. Their action is studied both at the ecosystem and soil profile levels.
2. Large termitaries are built by epigeous species of Macrotermitinae, then abandoned and colonized by other species. Their bases cover about 9% of the total ground area in the savanna and they have a volume of about 300 $m^3 ha^{-1}$. They show higher contents of organic carbon and nitrogen than the surrounding savanna soil, probably as a result of clay enrichment in mound soil. A higher density of trees was observed on the mounds than in the surrounding savanna.
3. Underground Macrotermitinae build fungus comb chambers within the soil profile. The walls of such chambers show a very high potential microbial activity which results in a strong nitrogen mineralization. These chambers are particularly abundant within mounds and are sites of intense and permanent activity of the soil fauna.

We hypothesize that nutrient cycling in relation to termite nests in this savanna could have two aspects. First, immobilization of nutrients in the soil brought about by the primary termite builders. Second, mineralization activated by secondary termite colonizers that develop their nest within the abandoned mound.

INTRODUCTION

Termites of the Macrotermitinae are a prominent component of most African ecosystems. Through their feeding and building activities, they make a major contribution to the structure and the functioning of savannas.

Owing to their special exosymbiosis with a fungus (*Termitomyces* sp.), they withdraw large quantities of dead vegetable matter from ordinary litter decomposition (Wood and Sands, 1978), often more than $1\,ha^{-1}\,yr^{-1}$ (Lepage, 1981). Large epigeous nest systems and associated galleries and runways modify soil profiles (Boyer, 1975) and occupy a large percentage of the land area. This action has a large effect on the chemical and physical properties of soils (Lee and Wood, 1971), and consequently on the vegetation. Some savannas have even been described as '*Termitensavannen*' (Troll, 1936).

The aim of this chapter is to give a preliminary view of the role of Macrotermitinae on nutrient cycling in a savanna within a landscape of forest–savanna mosaic, regarding the dynamics of storage and mineralization of the organic matter.

MATERIALS AND METHODS

Location of the study

The savanna studied is located at Lamto in the Côte d'Ivoire (West Africa) at the edge of the rainforest (5°02′W, 6°13′N). The sampling was made in an open shrub facies on well drained oxisols with a low organic matter concentration (total about 0.5%). The mean annual temperature is 27 °C and rainfall averages 1200 mm per year. More than 50% of the aerial primary production is burnt each year in the dry season.

Field sampling and laboratory measurements

Soil samples were taken between 5 and 10 cm depth with a soil corer (8 cm diameter) on two mounds randomly selected. For each mound, samples were taken each metre from the centre to the exterior, air dried and stored. Field capacity was measured by centrifugation (1000 g). Total carbon was determined by potassium dichromate oxidation and titration of dichromate remaining with ammonium iron II sulfate (Anne, 1945). Total nitrogen was measured by the Kjedldahl method (Kjeltec apparatus, Tecator). Size-class distribution of soil particles was estimated according to Feller's (1979) methodology. Potential biological activity was estimated by respirometry on air-dried samples (15 g) held at 28 °C and adjusted to 80% of the field capacity; the CO_2 produced was absorbed in 0.2 M NaOH and titrated with 0.1 M HCl, using phenolphthalein indicator after adding 2 ml of a $200\,g\,l^{-1}$ $BaCl_2$ solution (Dommergues, 1960); the NaOH solution was renewed on each occasion. Mineral nitrogen was extracted by mixing the incubated soil samples with 60 ml of a 0.2 M KCl solution for 30 min. Ammonium was measured with a specific electrode (ORION). Nitrates were reduced to nitrites with cadmium and their concentrations measured by colorimetry after adding sulphanilamide and *n*-(napthyl 1) ethylenediamine (Morris and Riley, 1963). Isotopic composition of soil carbon was measured according to Balesdent *et al.* (1987).

The species studied and their nests

The study deals with two species of Macrotermitinae: *Macrotermes bellicosus* and *Ancistrotermes cavithorax*. *Macrotermes* is a builder of large termitaries, a typical feature of most African landscapes. In Lamto, most of these structures have been abandoned by termites probably more than 50 years ago for unclear reasons and exposed to erosion (Vuattoux, 1968; Menaut, 1971). Forty-five mounds were counted on a 3.5 ha area (Fig. 23.1). The average density was 13 mounds ha^{-1}, a higher value than that of 3.5 mounds ha^{-1} in a savanna on hydromorphic soil recorded by Kouassi (1987). The mean area of the mound bases was 70 m^2 and their mean height 0.9 m. The mound bases cover about 9% of the total surface of the studied zone and they represent a soil volume of more than 300 m^3 above the ground (which would correspond to a 3 cm soil layer if completely eroded).

The abandoned mounds are colonized by a great number of soil animals, especially earthworms and hypogeous Macrotermitinae. Among termites, *Ancistrotermes cavithorax* has been well studied in the soil conditions where its nests average a density of 150 ha^{-1} (Baroni-Urbani *et al.*, 1978). Underground termitaries (Fig. 23.2) are made of fungus comb chambers, mostly found between 5 and 25 cm deep, which reach a density of about 100 000 units ha^{-1}, connected by galleries (several kilometres per hectare). Chambers and gallery walls are covered by a fine layer of reddish clay material (Fig. 23.3).

The vegetation of the mounds

In contrast to other parts of Africa (Malaisse, 1976), there is not a specific flora associated with the mounds. Tree density is increased, however, especially for the palm *Borassus aethiopum*, *Piliostigma thonringii* and *Bridelia ferruginea* (Table 23.1). The mounds are patches of typical woody savanna vegetation (Menaut *et al.*, 1985).

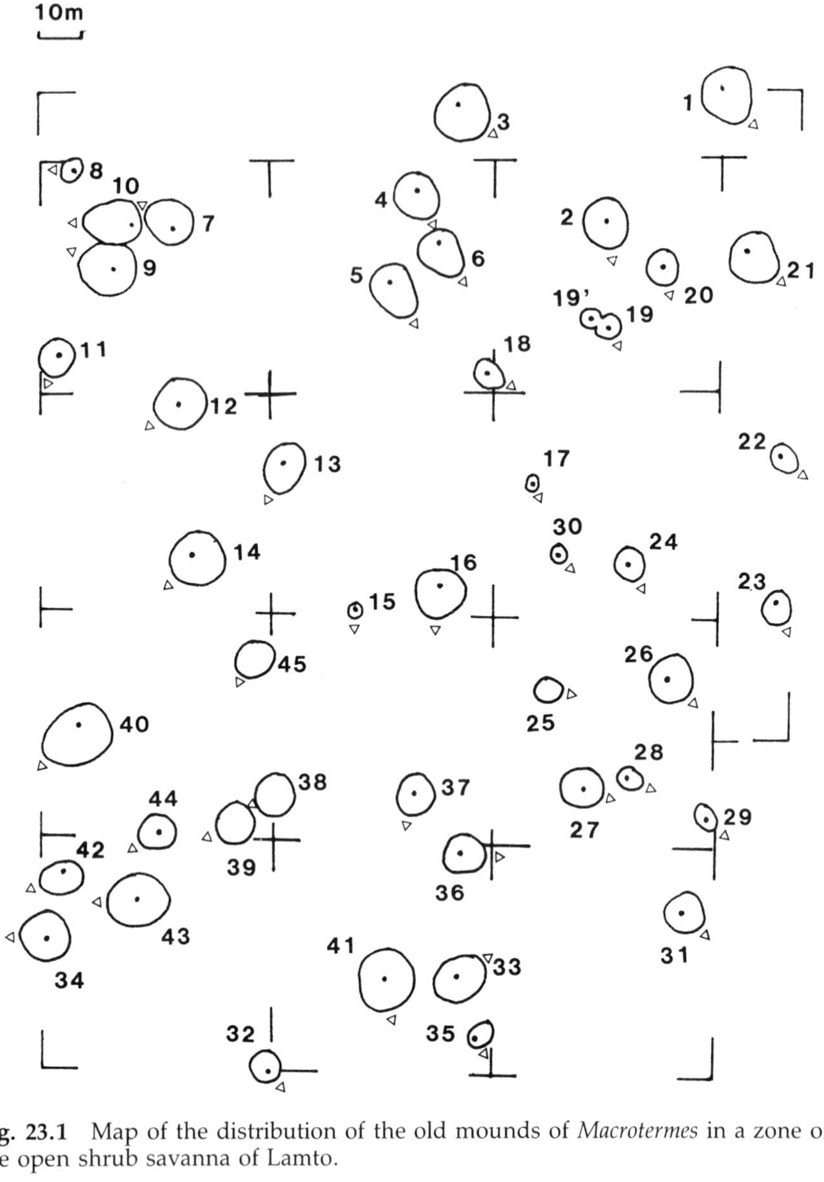

Fig. 23.1 Map of the distribution of the old mounds of *Macrotermes* in a zone of the open shrub savanna of Lamto.

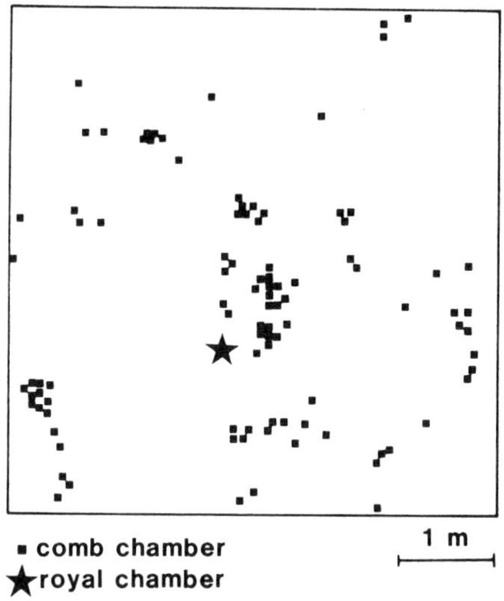

Fig. 23.2 Map of the distribution of the underground chambers of *Ancistrotermes cavithorax* in a sample area of the open shrub savanna (from Josens, 1972).

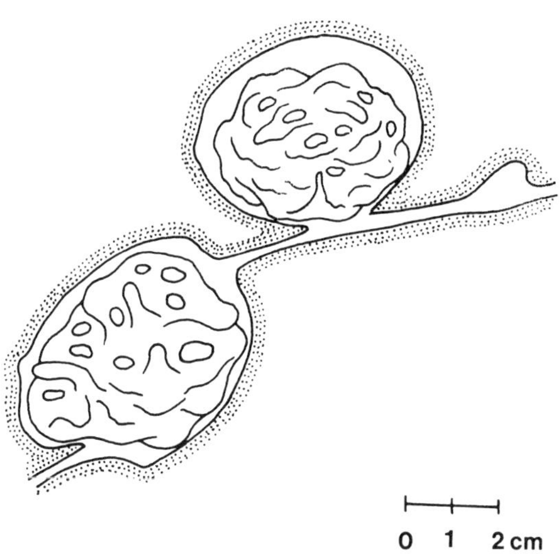

Fig. 23.3 Underground fungus comb chambers of *Ancistrotermes cavithorax* (from Josens, 1972).

Table 23.1 Comparison of tree abundance on and off mounds

Tree species	*Adults*	*On mounds (2585 m²) seedlings*	n/*1000 m²*	*Adults*	*Off mounds (2500 m²) seedlings*	n/*1000 m²*
Borassus aethiopum	10	195	79.3	7	22	11.6
Piliostigma thoningii	78	43	46.8	13	35	19.2
Bridelia ferruginea	13	20	12.8	8	6	5.6
Cussonia barteri	11	5	6.2	4	6	4.0
Crossopteryx febrifuga	6	6	4.6	11	6	6.8
Annona senegalensis	6	3	3.5	6	8	5.6
Other species	7	18	9.7	13	11	9.6
Total	131	290	162.9	62	94	62.4

Table 23.2 Size-class distribution (as % dry weight) of soil particles at several distances from the centres of two typical termite mounds

Distance from top (m)	*Particle size class (μm)*					
	2000–250	*250–100*	*100–50*	*50–20*	*20–2*	*2–0*
0	21.4–23.2	16.6–19.1	14.9–11.6	13.1–11.8	20.0–13.5	13.9–20.8
3	23.8–24.7	23.7–21.0	12.5–12.8	8.7–8.4	20.4–22.7	10.9–10.5
8	32.6–27.2	31.6–28.4	11.8–13.3	7.3–9.5	10.5–13.1	6.2–8.1
14	37.9–36.2	31.4–27.5	12.3–12.0	7.7–9.4	5.2–6.4	5.6–8.4

Mound mean diameter is 7.5 m (mounds $n = 2$).

MOUND SOIL CHARACTERISTICS

The chemical and physical properties of savanna mound soil are different from those of the non-mound soil. Mounds are three or four times richer in clay and, consequently, have less sand (Table 23.2). This has consequences on water dynamics. The field capacity on the top of the mound is four times higher than in the savanna soil around. As a consequence, the seasonality of soil functioning is modified: the growing period of the vegetation (grasses and trees) probably does not occur at the same time in both situations.

Table 23.3 Characteristics of soil organic matter on two termite mounds (a and b) and in the surrounding savanna: mean total organic carbon (as per cent of dry soil weight, ±s.e.; $n = 4$) and nitrogen (as $mg\,g^{-1}$ of dry soil weight, ±s.e.; $n = 3$), and isotopic composition of soil organic carbon (as $\delta\,^{13}C$ as per thousand; $n = 1$) at several distances from the centre (mound a has a diameter of 7 m, mound b has a diameter of 8 m)

	Mound a, distance from centre (m)				*Mound b, distance from centre (m)*			
	0	3	8	14	0	3	8	13
C (%)	1.75	1.88	1.01	0.96	2.24	1.81	1.28	1.17
	(0.06)	(0.06)	(0.03)	(0.06)	(0.12)	(0.04)	(0.02)	(0.02)
N ($mg\,g^{-1}$)	1.45	1.17	0.63	0.61	1.59	1.23	0.89	0.74
	(0.01)	(0.02)	(0.01)	(0.01)	(0.1)	(0.1)	(0.1)	(0.2)
$\delta\,^{13}C$ (‰)	−21.8	−14.4	−13.4	−13.2	−20.6	−18.4	−14.8	−13.4

Mounds can be considered as patches of soil, characterized by an alteration of strong dry and wet periods. This results in a strong periodic drought stress, especially in the upper horizons where the grasses have their roots. This could account for the low productivity of the grass layer on the mounds compared with the surrounding savanna. Probably, trees are less sensitive to the drought stress because they take up a substantial part of the water they need from the deeper layers of the soil.

The mounds are also richer in organic matter. Total organic carbon and nitrogen on the top are twice as high as in the surrounding soil. The concentration in organic matter, for carbon as for nitrogen, decreases rapidly from the centre of the mound to the bottom (Table 23.3).

Three explanations can be given for this observation. First, it is well known that clays protect organic matter against mineralization (McLaren and Peterson, 1965). Secondly, termites collect litter on the ground and concentrate this material in their nests. Thirdly, the mounds support a large plant biomass (mostly trees), and annual bush fires do not completely destroy the surface litter. As a result, the proportion of litter incorporated is higher on mound soils and the storage of organic matter is increased.

It is not only that the quantity of soil organic matter increases by *Macrotermes*' building activities, but its origin is also changed. The $\delta^{13}C$ (Table 23.3) of the non-mound savanna soil organic matter is around −13 per thousand, i.e. characteristic of C_4 grasses. The mound soil carbon exhibits an isotopic composition tending to that of C_3 plants. Obviously, the contribution of the trees to the soil carbon stock is negligible in the surrounding savanna, but very important in the mound soil, a consequence of the high density of the trees on the mounds and of the foraging behaviour of Macrotermitinae species, which collect large quantities of tree litter in their underground nests.

Because of their richness in organic matter and assimilable nutrients released, the mounds seem to be favourable sites for tree development especially for seedlings (Table 23.1). Trees may generally need nutrient-enriched soils for their establishment (Kellman, 1979) although there is evidence to suggest that they can grow on very poor soils (Proctor *et al.*, 1983). Mound patches certainly play a key role in tree dynamics. Further observations have shown that large termitaria closer to the forest boundary constitute favourable environments for tree seedlings from the forest (Kouassi, 1987).

THE POSSIBLE ROLE OF SOIL FAUNA

The old mounds of *Macrotermes* have been colonized by numerous soil animals, such as earthworms and termites. Table 23.4 is a comparison of the hypogeous termite community on the mounds and in the surrounding savanna. The densities of living fungus comb chambers reach 39 units m^{-2} in the mounds versus 8 units m^{-2} in the savanna. Food stored in fungus combs is 241.1 $g\,m^{-2}$ (dry weight) on mounds, as compared to 16.4 $g^{-2}\,m^{-2}$ off. These underground termite species feed on litter and are not supposed to attack living trees in natural conditions, but some of them may act as pests in agricultural and pastoral ecosystems (Wood and Johnson, 1978).

Fungus comb chambers of hypogeous termites are sites of strong microbial activity as shown by Abbadie and Lepage (1989) for the species *Ancistrotermes cavithorax*. The more the soil has been reworked by the termites, the higher the CO_2 production and NH_4^+ accumulation, and to a lesser extent NO_3^- (Table 23.5).

Table 23.4 Density of living chambers and mass of living combs (dry weight) of underground termites

Species	*Inside mounds*		*Outside mounds*[a]	
	Chambers (Number m^{-2})	*Combs ($g\,m^{-2}$)*	*Chambers (Number m^{-2})*	*Combs ($g\,m^{-2}$)*
Ancistrotermes cavithorax	12.4	67.8	3.5	13.2
Microtermes toumodiensis	3.1	1.7	3.3	2.2
Odontotermes sp.	16.9	122.8	0.1	0.4
Pseudacanthotermes militaris	6.2	48.8	1.0	0.6

[a] Data from Josens (1972).

Table 23.5 Characteristics of organic matter mineralization in the soil reworked by *Ancistrotermes cavithorax*: quantities of carbon dioxide, ammonium and nitrate produced by 15 g of soil incubated 30 days (data from Abbadie and Lepage, 1989) and proportion of carbon mineralized (in per cent of total initial organic carbon). The samples were taken outside the mounds (control) and 2–5 and 0–2 mm from the walls of the chambers

Soil sampled	*mg of C (CO_2)*	*% of C mineralized*	*μg of N (NH_4^+)*	*μg of N (NO_4^-)*
Control soil	2.4	2.2	817	7
2–5 mm from walls	3.9	3.8	1185	13
0–2 mm from walls	5.2	4.6	2088	5

The mineralization rates of carbon (percentage of total initial organic carbon mineralized in 30 days) varied with sampling location: 2.2% in the control soil, 3.8% at a distance of 2–5 mm from the chamber walls and 4.6% in the first 2 mm from the chamber wall. It is clear that the availability of carbon is improved by the action of the termites through a biochemical transformation of the soil organic matter (saliva, pugging, etc.), or by the supply of easily assimilable carbon (saliva, digestive products, etc.) resulting in a priming effect (Jenkinson, 1966). Whatever it may be, it is clear that the fungus comb chambers are sites of strong mineralization.

The impact of this activation on nutrient supply to plants is not easy to assess in spite of the large number of the structures built by savanna termite species. A realistic quantification of the role of soil fauna supposes the elaboration of complex models to integrate, at ecosystem level, both spatial and temporal variations of the microbial activities of a number of small fugacious entities.

CONCLUSIONS

Termite mounds constitute interesting models for studying mechanisms of nutrient cycling through soil animal activities in Lamto because they exhibit antagonistic processes: storage and release.

These two processes are directly related to two groups of termites of different behaviour. The first group (*Macrotermes*) is the primary builder of epigeous termitaries, creating a particular soil rich in clay compared to the surrounding savanna soil. Foraging activities of termites and the accumulation of small mineral particles lead to an increase of organic matter in the soil. Modification of water dynamics and organic matter status results in a local increase in tree density.

After the termitaries have been abandoned by their builders, earthworms and hypogeous termites colonize them, often in high densities. These animals (Barois *et al.*, 1987) are well known to cause sites of intense organic matter mineralization, and their feeding and building activities probably result in a strong production of assimilable nutrients.

By the combination of a strong organic matter storage, and a strong mineral nutrient production, mound soils are fertile places compared with those of the surroundings. In many sites, soil fauna are known to stimulate mineralization (Anderson, 1988). On the mounds, not only is mineralization increased but also the storage of organic matter (i.e. mineralizable substrate), as a result of the succession of the two termite communities.

ACKNOWLEDGEMENTS

We thank R. Vuottoux, Director of the Lamto Research Station (National University of Côte d'Ivoire) for the facilities in the field, A. Mariotti who performed $\delta^{13}C$ measurements, F. Kouame, J.B. Kouassi and G. Guillaume for their technical assistance, and Centre National de la Recherche Scientifique (Ecotrop and Piren) for grants.

REFERENCES

Abbadie, L. and Lepage, M. (1989) The role of subterranean fungus comb chambers (Isoptera, Macrotermitinae) in soil nitrogen cycling in a preforest savanna (Côte d'Ivoire). *Soil Biology and Biochemistry*, **21**, 1067–71.

Anderson, J.M. (1988) Spatiotemporal effects of invertebrates on soil processes. *Biology and Fertility of Soils*, **6**, 216–27.

Anne, P. (1945) Dosage rapide de carbone organique des sols. *Annales d'Agronomie*, **2**, 162–72.

Balesdent, J., Mariotti, A. and Guillet, B. (1987) Natural ^{13}C abundance as a tracer for soil organic matter dynamics studies. *Soil Biology and Biochemistry*, **19**, 25–30.

Barois, I., Verdier, B., Kaiser, P. *et al.* (1987) Influence of the tropical earthworm *Pontoscolex corethrurus* (Glossoscolecidae) on the fixation and mineralization of nitrogen, in *Selected Symposia and Monographs, Unione Zoologica Italiana*, (ed. A.M. Bonvicini and P. Omodeo), Mucchi, Modena, pp. 151–8.

Baroni-Urbani, C., Josens, G. and Peakin, G.K. (1978) Empirical data and demographic parameters, in *Production Ecology of Ants and Termites*, (ed. M.V. Brian), Cambridge University Press, Cambridge, pp. 5–44.

Boyer, Ph. (1975) Étude patriculière de trois termitières de *Bellicositermes* et de leur action sur les sols tropicaux. *Annales des Sciences Naturelles, Zoologie, Paris,* **12e série 17**, 273–446.

Dommergues, Y. (1960) La notion de coefficient de minéralisation du carbone dans les sols. *L'agronomie Tropicale,* **15**, 54–60.

Feller, C. (1979) Un méthode de fractionnement granulométrique de la matière organique des sols: application aux sols tropicaux a texture grossière, très pauvres en humus. *Cahiers ORSTOM, série Pédologie,* **17**, 339–46.

Jenkinson, D.S. (1966) The priming action, in *Isotopes in Soil Organic Matter Studies (Journal of Applied Radiation and Isotopes,* Suppl.), Pergamon Press, Oxford, pp. 199–208.

Josens, G. (1972) *Études biologique et écologique des termites (Isoptera) de la savane de Lamto-Pakobo (Côte d'Ivoire),* Doctoral Thesis, University of Bruxelles.

Kellman, M. (1979) Soil enrichment by neotropical savanna trees. *Journal of Ecology,* **67**, 565–77.

Kouassi, P.K. (1987) *Étude Comparative de la Macrofaune Endogée d'Écosystèmes Guinéens Naturels et Transformés de Côte d'Ivoire,* Thèse de 3e cycle, Université Nationale de Côte d'Ivoire.

Lee, K.E. and Wood, T.G. (1971) *Termites and Soils,* Academic Press, New York.

Lepage, M. (1981) L'impact des populations récoltantes de *Macrotermes michaelseni* (Sjöstedt) (Isoptera: Macrotermitinae) dans un écosysteme semi-aride (Kakjiado–Kenya). II–La nourriture récoltée, comparaison avec les grands herbivores. *Insectes Sociaux,* **28**, 309–19.

McLaren, A.D. and Peterson, G.H. (1965) Physical chemistry and biological chemistry of clay mineral–organic complexes, in *Soil Nitrogen,* (eds W.V. Bartholomew and F.E. Clark), American Society of Agronomy, Madison, Wisconsin, pp. 261–84.

Malaisse, F. (1976) De l'origine de la flore termitophile du Haut-Shaba (Zaire). Origine des flores africaines et malagaches. *Boissiera,* **24**, 505–13.

Menaut, J.C. (1971) *Étude de Quelques Peuplements Ligneux d'une Savane Guinéenne de Côte d'Ivoire,* These de Doctorat de 3e cycle, Université de Paris.

Menaut, J.C., Barbault, R., Lavelle, P. and Lepage, M. (1985) African savannas: biological systems of humification and mineralization, in *Ecology and Management of the World's Savannas,* (eds J.C. Tothill and J.J. Mott), Australian Academy of Sciences, Canberra, pp. 14–33.

Morris, A.W. and Riley, J.P. (1963) Determination of nitrate in sea water. *Analytica Chimica Acta,* **29**, 272–9.

Proctor, J., Anderson, J.M., Chai, P. and Vallack, H.W. (1983) Ecological studies in four contrasting lowland rain forests in Gunung Mulu

National Park, Sarawak I. Forest environment, structure and floristics. *Journal of Ecology*, **71**, 237–60.
Troll, C. (1936) Termitensavannen, in *Landekundliche Forschrift für Norbert Krebs*, Englehorn, Stuttgart, pp. 275–312.
Vuattoux, R. (1968) Le peuplement du palmier rônier (*Borassus aethiopum*) d'une savane de Côte d'Ivoire. *Annales de l'Université d'Abidjan*, ser. E, **1**, 1–138.
Wood, T.G. and Johnson, R.A. (1978) Abundance and vertical distribution in soil of *Microtermes* (Isoptera, Termitidae) in savanna woodland and agricultural ecosystems at Mokwa Nigeria. *Memorabilia Zoologica*, **29**, 203–13.
Wood, T.G. and Sands, W.A. (1978) The role of termites in ecosystems, in *Production Ecology of Ants and Termites*, (ed. M.V. Brian), Cambridge University Press, Cambridge, pp. 245–92.

24

The chemical ecology of forest and savanna termites

P.E. HOWSE

SUMMARY

1. There is typically a large termite biomass in tropical forest and savanna regions, and there are large mound densities in savannas. Termites aid nutrient cycling, nitrogen fixation and soil displacement.
2. The most conspicuous mound-builders are members of the Old World genus *Macrotermes* and of the New World genera *Syntermes* and *Cornitermes*. Colony defence depends largely upon the thick-walled mounds, which may incorporate relatively mineral-rich material from the subsoil, and upon the soldiers which typically have powerful mandibles and fight most effectively within the nest galleries.
3. Within forest and at the forest edge, trees supply above-ground nesting sites for many species of the tropicopolitan sub-family Nasutitermitinae. Defence of the colony rests mainly with the soldiers, which commonly use terpenic chemical sprays that are toxic or repellent to intruders. Their nests usually have thin friable covers, which rupture easily, allowing the soldiers to fight outside the nest.
4. Predation on colonies of mound-building termites by specialized mammalian and ant species ultimately results in mound decay and release of mineral salts into the soil. The enriched soil may then support the germination of seedlings and the regeneration or spread of forest, although the extent to which this occurs is little understood.

INTRODUCTION

The important role of termites in the maintenance of tropical ecosystems by nutrient cycling, nitrogen fixation and soil displacement has rarely been fully appreciated. A few figures will serve to illustrate their ecological importance. The population density in Nigerian riparian forest is $2646\,m^{-2}$, and $3163\,m^{-2}$ in semi-deciduous forest, corresponding to a biomass of $6.9\,g\,m^{-2}$ and $8.0\,g\,m^{-2}$ respectively (Wood *et al.*, 1982), while the biomass in Malaysian rain forest rises to $8.69–10.13\,g\,m^{-2}$ (Abe, 1979). Mound densities of *Amitermes laurensis* in savannas reach $217\,ha^{-1}$ in northern Australia (Lee and Wood, 1971), and the soil-feeding *Cubitermes exiguus* has nest densities up to $650\,ha^{-1}$ in Zaïrean savanna and a biomass of $19–28.5\,g\,m^{-2}$ (Bouillon and Mathot, 1964). Mounds of *Macrotermes* species are common features of African savannas. The large epigeal mounds of *M. bellicosus* in Nigerian southern Guinea savannas attain a density of $6.45\,ha^{-1}$ (Collins, 1981), but subterranean nests of *M. michaelseni* exist at around $150\,ha^{-1}$ in Kenya (Abe and Darlington, 1985). Nevertheless, the density of above-ground nests gives only a poor idea of the total biomass, of which the subterranean genera form by far the greater part (Wood *et al.*, 1982).

The mound-builder *Odontotermes* translocates 1059 kg of soil to the surface per hectare per year (Bagine, 1984) in northern Kenya. Small islands of woodland in Africa and South America which are confined to large termites (or leaf-cutting ant) mounds are commonly referred to as 'termite savannas'. It has been suggested that forest in various parts of Africa has spread by coalescence of termite savanna islands (see Harris, 1971), although there is little concrete evidence for this. Harris notes that the main large mound-builders in Africa (*Macrotermes bellicosus*, *M. natalensis* and *M. subhyalinus*, for example) do not occur in primary tropical forest and so empty old mounds in forest may provide evidence for the recent spread of the forest.

Fewer data are available for neotropical termites. Mound-builders are common in *cerrado* although there are no Macrotermitinae in the New World and the fungus-growing niche is occupied by leaf-cutting ants (Attini). In Brazilian cerrado of Goiás State, Redford (1984) found 55 *Cornitermes cumulans* mounds per hectare. All mound-builders (*Orthognathotermes gibberorum*, *Velocitermes heteropterus*, *Syntermes dirus* and *Armitermes euamignathus* included) amounted to $323\,ha^{-1}$. Over 50% of *Cornitermes* mounds had no *Cornitermes* within them but were secondarily occupied by other species.

Salik *et al.* (1983) found that termitaria (*Syntermes* and *Cornitermes*?) along the Rio Negro were up to 20 times richer in exchangeable nutrients than the surrounding soils and that the high turnover of mounds at $11\,yr^{-1}$ contributed to seedling establishment in the enriched soil. Amazonian *Nasutitermes* species fix nitrogen at high levels (Sylvester-Bradley *et al.*, 1983) but the importance of this to the general nitrogen economy is unknown.

Termite species include some that are found only in forest or only in savanna, although forest species may spill over into wetter savannas. Food sources in forest may include soil, decomposing wood, bark, lichens and leaf litter, and in savanna, grasses, soil or sound wood. In general terms, nest structures differ between forest and savanna and the differences may be seen not only as adaptations to the differing conditions of climate and soils but also as adaptations to resist predation by ants and termitophagous mammals. In this respect, the role of termites in relation to the dynamics of tropical ecosystems cannot be considered apart from the impact of such predators.

MACROTERMITINAE

The mounds of the old-world genus *Macrotermes* are among the most impressive animal artefacts. Their form varies with soils and regions; the cathedral-type mounds of Nigerian southern Guinea savanna at Mokwa reaching heights of over 7 m. Such thick-walled mounds protect the colony and the symbiotic fungus against the effects of fire, and the entry of predatory ants and vertebrates. Typically, mounds are elaborate in structure to provide a homeostatic internal environment. Lüscher (1955) was the first to describe the air circulation system within the nests of *Macrotermes bellicosus* from the Ivory Coast (Fig. 24.1). A central 'hive' of thin-walled clay chambers is enclosed within the thick outer wall. Here, the metabolism of the brood and the fungus gardens results in convection of air to cavities above the hive from whence it is ducted to narrow galleries within the outer wall, lying close to the outer surface. Gas exchange takes place here and the cooled air descends in ducts to the base of the hive, so completing a circuit. In other types of *M. bellicosus* mound from other regions of Africa it is unlikely that a true thermosiphon operates: air may simply leave through chimney openings in mound turrets or may accumulate in large reservoirs above the hive. Hive temperature, however, remains very stable at around 30 °C in mounds from the Ivory Coast (Lüscher, 1955) and from Zaïre (Ruelle, 1964).

As the outer wall of the mound, which extends below soil level, is made of fine clay particles cemented together with a salivary secretion, it is extremely hard and difficult to penetrate, especially during the dry season. New construction during the wet season remains soft for several days, and the termitaria may be vulnerable then. Opening of the mound for emergence of alates, for example, or by anteaters, may permit sporadic ant predation, but otherwise only a few species are able to prey on *Macrotermes* mounds on a regular basis. These tend to make use of the foraging galleries of the termites which extend underground from the

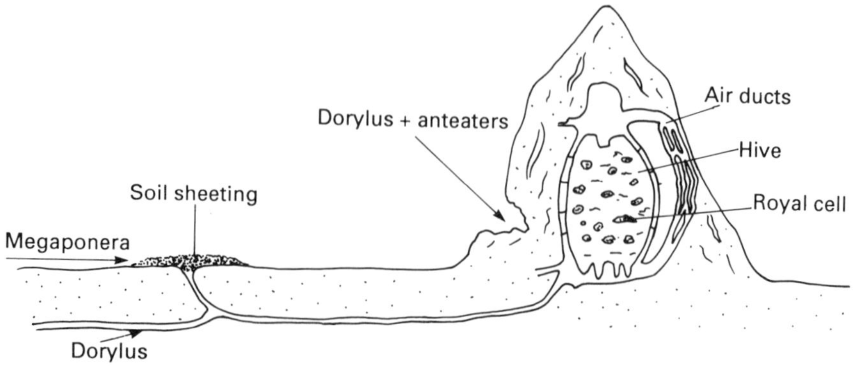

Fig. 24.1 Diagram of a *Macrotermes* nest to show the internal structure and the points of attack by ants (*Dorylus dentifrons, Dorylus* sp. and *Megaponera foetens*) and anteaters.

nest and open to the surface often many metres away where the workers build a protective sheeting of soil particles cemented together with saliva (Fig. 24.1).

In the Ivory Coast, *Dorylus dentifrons* have been observed to kill large numbers of *M. bellicosus* colonies (Bodot, 1967) estimated at 26% yr^{-1} (Collins, 1981). Here they appear to enter pre-existing holes in the mound, while other *Dorylus* species foraging on *M. michaelseni* in Kenya enter through the subterranean galleries of the termites (Abe and Darlington, 1985). In the latter case, dorylines are an important cause of mortality of mounds.

Another major predator of *Macrotermes* is the ponerine ant *Megaponera foetens*. Scout ants detect the presence of termites by means of a salivary kairomone in the soil particles at feeding sites (Longhurst and Howse, 1978). They then recruit a file of workers which attack and sting the termites before they have time to escape underground. The defences of the termites are also partly chemical, in that the soldiers have a cephalic frontal gland. *M. foetens*, however, is immune to the toxins in this secretion and to the secretion of the related mound-builder *Odontotermes smeathani*. Indeed, the worker ants are strongly attracted to filter paper on which soldiers of these species have been squashed (Longhurst, 1978).

The predation rates on *M. bellicosus* in Nigerian Guinea savanna have been estimated by Longhurst and Howse (1978) at $142\,m^{-2}\,yr^{-1}$ from a population density estimated at $37\,m^{-2}$ by Wood and Sands (1977).

Predation on mound-building termites has also been examined in the neotropics. There is heavy predation on *Cornitermes* mounds in Goiás, Brazil, by armadillos and giant anteaters (Redford, 1984). There were armadillo holes in 35% of the mounds. Approximately 90% of these had no *Cornitermes* colony compared with 40% of those without a hole. *Myrmecophaga* predation was more difficult to assess because the animals scrape the surface of mounds to open them and the damage is quickly repaired.

The ponerine ant *Pacycondyla commutata* is a specialist predator on mound-building *Syntermes* in South America, and has behaviour remarkably similar to that of *Magaponera foetens*. In semi-deciduous forest in Roraima (Maracá), Mill (1984) found a population density of 49.7 m^{-2} of three species of *Syntermes* which forage in leaf litter. The predation rate was calculated at 147 $m^{-2} yr^{-1}$. Although this study was of short duration, it shows that, as in African savannas, specialized ant predators take a heavy toll of termite populations (approximately 50% per year, assuming a worker lives for two months).

NASUTITERMITINAE

Syntermes, *Armitermes* and *Cornitermes* are prominent among the genera of neotropical mound-builders that replace the Macrotermitinae and certain Termitinae of the Ethiopian region. Their soldiers, however, are only partially 'nasute' and they possess powerful mandibles. The opening of the frontal gland is prolonged forwards into a nozzle, from which a droplet of frontal gland secretion can be exuded. The soldiers in genera such as *Nasutitermes* have vestigial mandibles and nests constructed by such species are common in forest and woodland throughout the tropics. They are typically subspherical, made of carton, and placed on tree trunks or larger branches.

The contrast between typically arboreal nests of *Nasutitermes* species and those of *Macrotermes* is extreme. The former have nest walls which are thin and fragile, so they can often be broken with finger pressure. They are generally made of carton, consisting of numerous small interconnected chambers. In a forest environment they are not normally exposed to fire and are not subject to wide variations of temperature and humidity. The differences can be ascribed in a large degree to the differing defence strategies of the species (Howse, 1984). Those which rely upon chemical defence are, in general, using mixtures of monoterpenes and sticky diterpenes (Prestwich, 1983; Baker and Walmsley, 1982) but are nevertheless susceptible to the effects of their own secretions. The discharge of numerous sticky threads within the galleries and chambers of the nest, together with the accumulation of volatile toxic monoterpenes within the nest atmosphere would mean that the nest inhabitants would perish. In contrast, mandibulate soldiers can fight within the nest galleries and block them with their snapping jaws. Wherever possible, therefore, nasute species must fight outside the nest, and the fragility of the nest wall can be viewed as an adaptation to facilitate this.

The chemical defences of nasutes are highly efficient. They appear to be repellent to army ants (Ecitonini) in S. America. Rettenmeyer *et al.* (1983) record that all *Eciton* species they have observed were repelled by *Nasutitermes* defence secretions. Harvester termites of the genus *Trinervitermes* which forage above ground for dry grasses in African savannas are highly inimical to *Megaponera foetens* which are repelled and killed by their secretions.

The soldier secretions of *Trinervitermes* species contain mixtures of monoterpene hydrocarbons such as pinenes, limonene, α-phellandrene, etc., together with characteristic diterpenes (Prestwich and Chen, 1981), α- and β-pinenes are highly repellent to *M. foetens* in laboratory assays, and Longhurst (1978) found that 20–63% of ants never returned to the nest after encounters with *T. geminatus*, and in one case 42 out of a column of 471 were dead at the site of the encounter.

The South American counterparts of *M. foetens* are *Pachycondyla* species. Traniello (1981) noted that *P. laevigata* in Panama never initiated a raid after encountering *Nasutitermes* soldiers. It avoided contact with *N. costalis* soldiers in 99 out or 147 encounters and retreated and groomed after being sprayed on the remaining 48 encounters. In 860 encounters with nine species of predatory ant, the ant attack was successful on only 62 occasions (7.2%).

In observations of Tamandua anteaters in Panama, Lubin and Montgomery (1981) found evidence that workers and alates of *Nasutitermes* species were eaten but the soldiers were not. In feeding choice tests, a captive *Tamandua mexicana* frequently rejected nasute soldiers from a feeding dish and ate very few although it accepted workers and both castes of *Microcerotermes exiguus* which is not chemically defended. The rejection was attributed to α- and β-pinenes which irritate mucous membranes of the anteaters, which tend to feed in short bouts and so crop the termite population, doing considerable damage to nests without destroying them.

The giant anteater *Myrmecophaga tridactyla* also tends to avoid nasutes (Coles, 1980; Coles de Negret and Howse, 1982). Feeding experiments again showed that samples of *Nasutitermes* with soldiers were eaten significantly less than those without, and that synthetic monoterpenes presented with food inhibited feeding; the strongest repellent activity was induced by the odour of limonene and β-pinene.

Syntermes, *Cornitermes* and *Armitermes* are mainly builders of relatively small mounds in cerrado and have soldiers with both well developed mandibles and non-squirting frontal glands. Evidence that chemical defence makes above-ground harvesting of grasses possible comes from a study of three species of *Syntermes* (Coles de Negret and Howse, 1982; Coles, 1980). The *S. grandis* soldier has the largest mandibles and smallest frontal glands, while *S. molestus* is smaller and has a relatively larger frontal gland and smaller mandibles. *S. dirus* is intermediate, and along with *S. molestus* has a frontal gland secretion rich in *cis*-β-ocimene (Baker *et al.*, 1980). *S. grandis,* however, produces only small above-ground foraging parties over an average distance of 12.4 m accompanied by few soldiers. *S. molestus* has the largest trails (42 cm) and the highest soldier: worker ratios on the trails (1:1.56 against 1:12.11 for *S. grandis*). Over 80% of above-ground foragers studied by Coles (1980) had monoterpenes in their soldier secretions and these had high soldier:worker ratios. This was confirmed by Mill (1983) who found that foraging parties of subterranean and cryptic foragers from a variety of Brazilian cerrado and tropical forest habitats had about 12% soldiers, while covered-way and open-air foragers had approximately 25% (Table 24.1).

Cis-β-ocimene is toxic to camponotine ants, causing paralysis and death (Coles, 1980), but it is not sufficient to prevent heavy predation of *Syntermes* by *Pachycondyla commutata* (Mill, 1984), and is not as repellent to giant anteaters as limonene and β-pinene (Coles de Negret and Howse, 1982).

Table 24.1 Correlations between foraging habits (subterranean and cryptic or open-air and covered way), percentage of soldiers in surface-foraging parties or in samples collected away from the nest (first two columns) for 186 species of Brazilian termites studies by Mill (1983). Column 1: soldiers with biting and cutting mandibles, column 2: soldiers with snapping mandibles, column 3: soldiers with mandibles and small defensive glands, column 4: soldiers with mandibles and large defensive glands, column 5: fully nasute soldiers. Gland volumes for *Syntermes calvus**, *Armitermes holmgreni* and *Nasutitermes* sp. # are calculated as a percentage of body volume. (Data from Mill, 1983)

	Biting and cutting	*Snapping*	*Syntermes type*	*Armitermes type*	*Fully Nasute*
Cryptic foraging (N = 102) and volume of frontal gland	25	24	27		25
Open foraging (N = 84)	0	3	40		56
% soldiers	3–4	5	20	35–40	30–40
Gland volume	0	0	8.5*	18.5**	20.5#

DEFENCE STRATEGIES

Syntermes, Cornitermes and *Armitermes* species are predominantly genera that occur in cerrado but which are sometimes found in forest (e.g. *S. molestus,* and *S. calcus* which cuts dry leaves in forests). Their nesting habits and nest structure are intimately related to the form of soldier defence and these in turn to foraging habits and consequent impact on the ecosystem. This mechanical defence is most efficient in a hard nest opening to the exterior by few narrow galleries: a fortress type of defence system. Energy is invested in building at the expense of soldier-production, few soldiers are found with foraging parties, which must move under litter or within the soil to escape predation; the workers cooperate in rapid repair of any breaks in the nest and the soldiers are large and heavy with limited mobility. Soldier mandibles hence become modified in various ways for cutting, slicing or crushing intruders within the confines of nest galleries.

In chemically defended genera (Fig. 24.2), energy is invested in biosynthesis at the expense of building. The danger of toxic chemicals accumulating within the nest means that the soldiers must defend within the open air. Consequently, colonies have many highly mobile soldiers, the nest cover is fragile and nest repair by workers is delayed.

Another form of compromise in defence is represented by the soldierless termites (Apricotermitinae) in which workers are able to release liquid toxins contained in abdominal glands by rupture of the body wall (*autothysis*). By this highly altruistic behaviour, intruders can become glued together with the unfortunate defender blocking the galleries of fortress-type nests, as in the cerrado species *Grigiotermes metoecus,* or of the diffuse subterranean nests commonly built by African Apicotermitinae.

Various measures of the differing strategies are shown in Tables 24.1 and 24.2. Frontal gland volume, the percentage of soldiers in foraging parties and the ratios of open:cryptic foragers are lowest in species such as *Labiotermes* and *Spinitermes,* highest in nasutes such as *Velocitermes,* and intermediate in *Syntermes* and *Armitermes* (Table 24.1). Nest hardness (measured by a penetrometer) is inversely correlated with the number of soldiers emerging, the latency of nest closure and the time taken for closure (Table 24.2). The percentage of nests with holes believed to have been made by armadillos in cerrado near Brasília was lowest for *Nasutitermes* and *Velocitermes,* and the average time for which *Myrmecophaga* anteaters spent feeding on these species was far lower than for subnasute species for *Grigiotermes* (Coles, 1980).

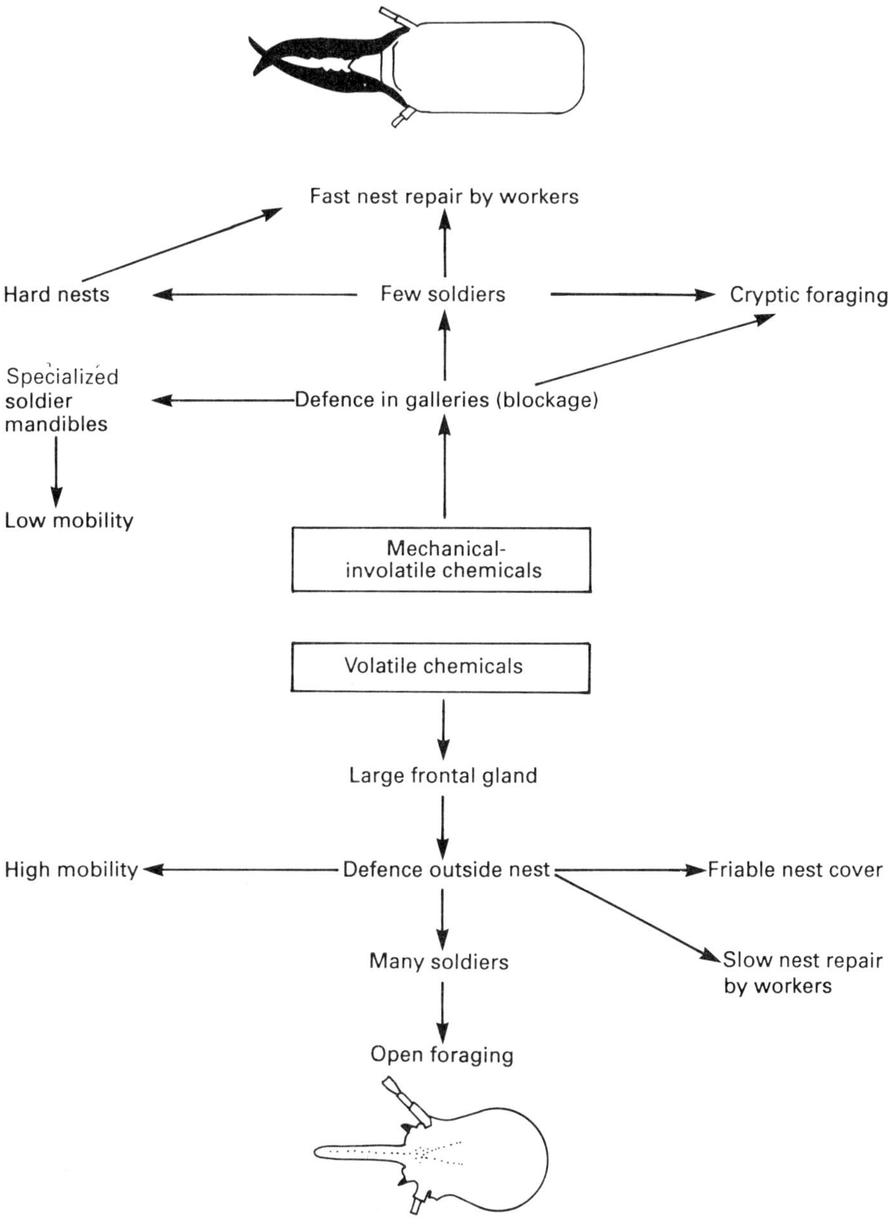

Fig. 24.2 Some of the possible consequences of termite species investing in soldiers with powerful mandibles (mechanical defence) or with chemical defence secretions.

Table 24.2 Relationships between nest hardness (measured with a penetrometer), number of soldiers emerging from the nest during the first three minutes after a breach in the wall, time to the beginning of nest repair, time taken for closure of the breach, percentage of nests with armadillo holes (*n* in parentheses), and time captive *Myrmecophaga tridactyla* spent eating termites in food-choice experiments. Species: *Grigiotemes metoecus* (soldierless), *Cornitermes* sp. nov., *Armitermes euamignathus*, *Nasutitermes* sp. nov. and *Velocitermes paucipilis*. Data from Coles (1980)

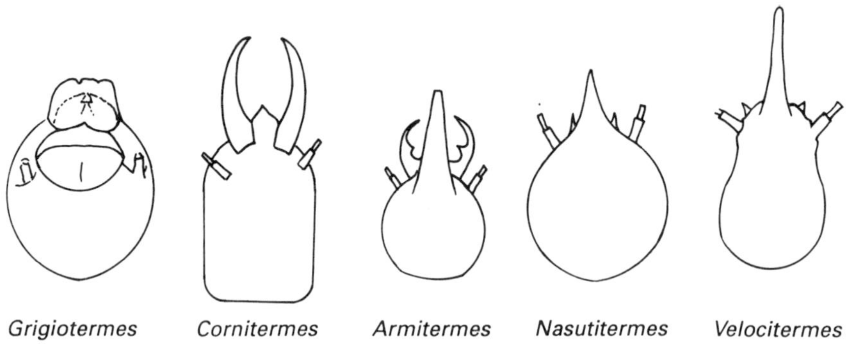

	Grigiotermes	*Cornitermes*	*Armitermes*	*Nasutitermes*	*Velocitermes*
Nest hardness	25	15.2	4.66	0.25	0.11
Soldiers emerging in 3 minutes	–	6	8	275	230
Latency of nest closure (s)	5	276	43	653	772
Time to complete closure (s)	39	–	423	1479	1341
% Nests with armadillo holes	31(100)	32(60)	19(100)	16(73)	11(100)
Time anteater spent eating (s)	949	1437	1086	309	138

CONCLUSIONS

Lubin and Montgomery (1981) suggest that anteater (Tamandua) predation has been a strong selective force in the evolution of the defence system of *Nasutitermes*. Evidence is reviewed here to indicate that the predation by ponerine ants, *Myrmecophaga* and armadillos in the New World and by ponerine and doryline ants in the Old World is also considerable.

The establishment of nests in savanna depends upon mound-building behaviour and soil excavation which make possible nests which are fire-resistant and which provide homeostasis of the nest environment – this may then reduce the energy available to such termites (e.g. *Macrotermes* and *Cornitermes*) for extensive chemical synthesis. Notable exceptions to this are the African *Trinervitermes*, which are above-ground harvesters, and the Australian *Nasutitermes* mound-builders of dry savannas. Mound densities are often very high (*T. geminatus* = $232\,ha^{-1}$ at Mokwa, Nigeria; Ohiagu, 1979) and the termites vary their location in the mound according to the temperature, often going below ground where temperatures are more stable (Josens, 1971). The soldier to worker ratio is 1:6 in *T. geminatus* against 1:15 in *Macrotermes bellicosus* (Collins, 1981). Haverty (1977) found that the proportion of soldiers in colonies with mandibulate soldiers was generally 1–5% while it is much higher (up to 30% or more) in nasute species. The terpenoid secretion of *Trinervitermes* appears to be a very efficient defence that also protects against vertebrates. There is little information on vertebrate predators of African termites, but the aardwolf, *Proteles cristatus*, specializes on *T. bettonianus* (Kruuk and Sands, 1972) although it ceases feeding when the soldiers get too high a density at a given point. The niche of *Macrotermes* and *Trinervitermes* has been mainly occupied by certain Nasutitermitinae in South America which are subnasute. Even here, within the genus *Syntermes*, an evolutionary trend towards the use of chemical defence can be identified, making possible more above-ground activity for harvesting grasses. The impact of vertebrate predators on *Cornitermes* has been investigated by Redford (1984) who noted that anteaters and armadillos that damaged mounds of *C. cumulans* laid them open to ant predation.

When chemical defence of any form exists in termites its effect appears to be to limit predation to a few specialists that have (like the mammalian predators) a degree of immunity to the chemicals. The ultimate effect of predation is to maintain colony mortality at a certain level, thereby exposing the mounds to increased weathering and liberating mineral salts and other nutrients that encourage plant growth and possible spread of forest vegetation. Lee and Wood (1971) found that at one site in northern Australia, 62 500 kg of subsoil per hectare was translocated to the surface by mound-building termites. A little less than a third of this, which is rich in organic material and nutrients, is in abandoned mounds and becomes available to plants as a result of mound erosion. Along the Rio Negro in Venezuela the turnover rate of termitaria on laterite soils has been estimated at $165\,nests\,ha^{-1}\,yr^{-1}$ (Salik *et al.*, 1983), corresponding to a mean of $0.78\,tons\,ha^{-1}\,yr^{-1}$. Specialized termite predators may thus play a key role in the control of natural ecosystems.

Mounds, however, may not simply decay, but may be re-occupied by numerous other species of social insect. Multiple secondary occupation is particularly common in mounds of *Cornitermes cumulans* (Redford, 1984) and *Syntermes* species (Coles de Negret and Howse, 1982; Coles, 1980). The effect of this is to produce a guild of species occupying different parts of the mound and possessing a variety of different defence mechanisms. Secondary or multiple occupation of this type may be important in restricting plant growth on mounds and in stabilizing the savanna plant communities.

ACKNOWLEDGEMENTS

This paper is based on the pioneering field studies of Helen Coles de Negret, Chris Longhurst and Alan Mill. The following organizations kindly provided help and facilities: Laboratório de Ecologia (University of Brasília), the CNPq, British Council and SEMA (Brazil).

REFERENCES

Abe, T. (1979) Studies on the distribution and ecological role of termites in a lowland rain forest on West Malaysia (2) Food and feeding habits of termites in Pasoh Forest Reserve. *Japanese Journal of Ecology*, **29**, 121–35.

Abe, T. and Darlington, J.P.E.C. (1985) Distribution and abundance of a mound-building termite, *Macrotermes michaelseni*, with special reference to its subterranean colonies and ant predators. *Physiological Ecology Japan*, **22**, 59–74.

Bagine, R.K.N. (1984) Soil translocation by termites of the genus *Odontotermes* (Holmgren) (Isoptera: Macrotermitinae) in an arid area of Northern Kenya. *Oecologia (Berlin)*, **64**, 263–6.

Baker, R. and Walmsley, S. (1982) Soldier defense secretions of the South American termites *Cortaritermes silvestri*, *Nasutitermes* sp. N.D. and *Nasutitermes kemneri*. *Tetrahedron*, **38**, 1899–1910.

Baker, R., Coles, H.R., Evans, D.A. *et al.* (1980) Chemical composition of the frontal gland secretion of *Syntermes* soldiers (Isoptera, Termitidae). *Journal of Chemical Ecology*, **7**, 135–45.

Bodot, P. (1967) Etude écologique des termites des savannas de basse Côte d'Ivoire. *Insectes Sociaux*, **14**, 229–58.

Bouillon, A. and Mathot, G. (1964) Observations sur l'ecologie et le nid de *Cubitermes exiguus* Mathot. Description de nymphes-soldats et d'un pseudoimago. In *Etudes sur les Termites Africains*, (ed. A. Bouillon), Masson, Paris, pp. 215–30.

Coles, H.M. (1980) *Defensive Strategies in the Ecology of Neotropical Termites*, Unpublished Ph.D thesis, University of Southampton.

Coles de Negret, H. and Howse, P.E. (1982) Termite defensive secretions: Ecological aspects, in *Social Insects in the Tropics* (International Union for the Study of Social Insects Symposium, (ed. P. Jaisson), Université Paris Nord, Paris.

Collins, N.M. (1981) Populations, age structure and survivorship of colonies of *Macrotermes bellicosus* (Isoptera: Macrotermitinae). *Journal of Animal Ecology*, **50**, 293–311.

Harris, W.V. (1971) *Termites, Their Recognition and Control*, Longman, London.

Haverty, M.I. (1977) The proportion of soldiers in termite colonies; a list and bibliography. *Sociobiology*, **2**, 199–216.

Howse, P.E. (1984) Sociochemicals of termites, in *Chemical Ecology of Insects*, (eds W.J. Bell and R.Y. Carde), Chapman & Hall, London, pp. 475–519.

Josens, G. (1971) Variations thermiques dans les nids de *Trinervitermes geminatus* Wasmann, en relation avec le milieu exterieur dans la savanne de Lamto (Côte d'Ivoire). *Insectes Sociaux*, **18**, 1–14.

Kruuk, H. and Sands, W.A. (1972) The aardwolf (*Proteles cristatus* Sparrman 1783) as a predator of termites. *East African Wildlife Journal*, **10**, 211–27.

Lee, K.E. and Wood, T.G. (1971) *Termites and Soils*, Academic Press, New York.

Longhurst, C. (1978) *Behavioural, Chemical and Ecological Interactions between West African Ants and Termites*, Ph.D. thesis, University of Southampton.

Longhurst, C. and Howse, P.E. (1978) The use of kairomones by *Megaponera foetens* (Fab.) (Hymenoptera: Formicidae) in the detection of its termite prey. *Animal Behaviour*, **26**, 1213–18.

Lubin, Y.D. and Montgomery, G.G. (1981) Defence of *Nasutitermes* termites (Isoptera, Termitidae) against Tamandua anteaters (Edentata, Myrmicophagidae). *Biotropica*, **13**, 66–76.

Lüscher, M. (1955) Der Sauerstoffverbrauch bei Termiten und die Ventilation de Nestes bei *Macrotermes natalensis* (Haviland). *Acta Tropica*, **12**, 289–307.

Mill, A.E. (1983). *Foraging and defence in Neotropical termites (Insecta, Isoptera)*. Unpublished Ph.D. thesis, University of Southampton.

Mill, A.E. (1984) Predation by the ponerine ant *Pachycondyla commutata* on termites of the genus *Syntermes* in Amazonian rain forest. *Journal of Natural History*, **18**, 405–10.

Ohiagu, C.E. (1979) Nest and soil populations of *Trinervitermes* spp. with particular reference to *T. geminatus* (Wasmann), (Isoptera), in Southern

Guinea savanna near Mokwa, Nigeria. *Oecologia (Berlin)*, **40**, 167–79.

Prestwich, G.D. (1983) Chemical systematics of termite exocrine secretions. *Annual Review of Ecology and Systematics*, **14**, 287–311.

Prestwich, G.D. and Chen, D. (1981) Soldier defense secretions of *Trinervitermes bettonianus* (Isoptera, Nasutitermitinae): chemical variation in allopatric populations. *Journal of Chemical Ecology*, **7**, 147–57.

Redford, K.H. (1984) The termitaria of *Cornitermes* and their role in determining a potential keystone species. *Biotropica*, **16**(2), 112–19.

Rettenmeyer, C.W., Chadab-Crepet, R. and Naumann, M.G. (1983) Comparative foraging by neotropical army ants, in *Social Insects in the Tropics* (ed. P. Jaisson), Université Paris-Nord, Paris, pp. 59–73.

Ruelle, J.E. (1964) L'architecture du nid de *Macrotermes natalensis* en son sens fonctionnel. In *Etudes sur les Termites Africains* (ed. A. Bouillon), Masson, Paris, 327–63.

Salick, J., Herrera, R. and Jordan, C.F. (1983) Termitaria nutrient patchiness in nutrient deficient rainforests. *Biotropica*, **15**(1), 1–7.

Sylvester-Bradley, R., Bandeira, A.G. and Oliveira, L.A. (1978) Fixação de nitrogênio (redução de acetilino) em cupins (insecta: Isoptera) da Amazônia Central. *Acta Amazônica*, **8**, 621–7.

Traniello, J.F.A. (1981) Enemy deterrence in the recruitment strategy of a termite: soldier-organized foraging in *Nasutitermes costalis*. *Proceedings of the National Academy of Sciences, USA*, **78**, 1976–79.

Wood, T.G. and Sands, W.A. (1977) The role of termites in ecosystems, in *Production Ecology of Ants and Termites*, (ed. M.V. Brian), Cambridge University Press, Cambridge, pp. 245–92.

Wood, T.G., Johnson, R.A. and Anderson, J.M. (1982) Abundance and distribution of termites (Isoptera) in a riparian forest in the Southern Guinea Savanna vegetation zone of Nigeria. *Biotropica*, **14**, 25–39.

25

The forest-savanna boundary and habitat selection by Brazilian social wasps

ANTHONY RAW

SUMMARY

1. Social wasps (Hymenoptera, Vespidae) were used to illustrate the use of the forest–savanna boundary by animals in Brazil. Wasps and their nests were sought in forest, at its edge, in nearby savanna and man-made clearings. Information is presented on a total of 83 species and four subspecies: 40 species from Brasília, 43 from Goiás Velho in Goiás state, 29 from Maracá in Roraima and 30 from three localities in eastern Rondônia.
2. The wasps in central Brazil show clear habitat preferences and are placed in 10 categories based on where they nested and searched for food. Preliminary data collected in Amazonia suggest the same pattern exists among social wasps there as well. Of the 83 species studied 78 were recorded nesting in only one type of habitat. The forest boundary was the most frequented habitat with greater numbers of individuals, colonies and species occurring there than elsewhere. In the four study areas, 53% to 58% of the species nested at the edge, while 72% to 86% of them searched for food there.

INTRODUCTION

The wall of trees and shrubs abutting open savanna is the characteristic ecotone where forest and savanna meet. Often canopy trees, 6 m to 20 m tall, overhang the ground plants of the open vegetation and even where the transition is more gradual, it is rarely more than 3 m or 4 m wide. Many animals frequent the forest–savanna boundary (personal observation). They are drawn from the forest and the open savanna, while some are confined to the edge. Among the more commonly encountered are insectivorous birds and lizards and many species of insects, notably butterflies. Likewise, the edge of forest bordering artificial clearings is colonized by species of animals and plants typical of the natural forest edge.

Social wasps (Hymenoptera, Vespidae) are a convenient group to illustrate the use of the forest boundary by neotropical animals. They are abundant and many are known to be restricted to particular habitats (Raw, 1988; Raw, 1989a). In particular, they have preferences for the types of vegetation formation where they build their nests and where they search for food. These preferences proved useful in the present investigation.

BIOLOGY OF THE WASPS

The 320 species of social wasps occurring in Brazil (Richards, 1978; Raw, 1985; Raw, 1989b; Richards, unpublished; Raw, unpublished) vary greatly in the number of occupants per nest and in colony longevity. A colony of *Mischocyttarus* or of *Polistes* is founded by an individual female. Most of these nests support only one or two dozen wasps at their maximum and the colony lasts only a few weeks or months. All the other neotropical species initiate new colonies by swarming. The colony may comprise several hundreds or thousands of wasps and persists for several months or years.

The nests of most species are attached to vegetation; suspended from twigs and branches or fixed to a leaf or a trunk. However, some species of *Polybia* and *Stelopolybia* nest in hollow trunks and in holes in the ground.

The normal hunting range of the wasps varies from 50 m to more than 500 m, with the members of larger colonies hunting further afield than those of smaller ones irrespective of the sizes of the adult wasps (Raw, unpublished). The wasp species studied obtain all their protein from the prey they catch which is mostly defoliating caterpillars and beetle larvae, winged termites and some spiders. Although small, these wasps are so numerous that they comprise a major group of predators in neotropical ecosystems.

STUDY SITES

Most of the information presented here was collected from 1976 to 1984 within a 15 km radius of Brasília airport (47°53′W, 15°52′S) and an area of similar size around the town of Goiás Velho (50°10′W, 15°55′S), 200 km west of Brasília. During the nine-year period all the habitats mentioned were sampled many times. Additional data collected on short visits to localities in the north and west of the Amazon basin are also presented. One was a two-week visit to Maracá, Roraima (61°50′W, 3°10′N), 90 km from the Venezuelan border. Two visits of one week each to Rondônia were to Ouro Preto do Oeste (62°18′W, 10°50′S), 250 km south-east of Porto Velho (1977), and to the Suruí (61°30′W, 11°30′S) and the Cinta Larga (60°05′W, 11°25′S) Amerindian communities, 360 km and 400 km south-east of Porto Velho (1986).

Around Brasília and Goiás Velho and at Maracá collections were made at both natural and artificial forest margins, while in Rondônia only artificial ones bordering deforested areas were available. Around Brasília and Goiás Velho the gallery forest gives way sharply to natural open savanna. At Maracá there is also a sudden change from evergreen forest to natural grassland. The deforested areas near Brasília and Goiás Velho were pasture and market gardens and that on Maracá was the clearing occupied by the field station. All three open areas in Rondônia were artificially cleared lands occupied by cultivated fields. Data were also collected in the permanent village clearings of the Suruí and Cinta Larga Amerindians. The three localities in the eastern part of Rondônia stretched over 100 km.

Numerous species of herbs, shrubs and vines are largely confined to the forest boundary. The more common genera include *Arrabidaea, Eupatorium, Passiflora, Piper* and *Pteridium*. For the present study only forest boundaries of less than 4 m width were sampled. In places where the boundary was sharp and often with forest trees overhanging the low vegetation a narrow band (2 m wide) of the latter was included as boundary.

Artificial forest boundaries often appear similar to natural ones. Immediately a patch of forest is felled the edge of the remaining forest is opened and bears little vegetation. However, it is soon colonized by plants and within a few years a wall of shrubs and small trees occupies the open space and the artificial edge resembles a natural forest boundary.

METHODS

In social wasps, where hundreds or thousands of individuals may occupy a single colony, the distributions of many species are very patchy. The species occurring at a particular site depend on which have colonies there and often a species is extremely abundant at one place and 200 m away is totally replaced by another. In order to examine the general patterns of the wasps' habitat preferences, the information presented here was compiled from general collecting in numerous habitats and localities on many occasions. This method of collecting seems to be appropriate for the sampling of species with very patchy distributions. All the wasps were collected, although most of the individuals of the more common species were released after being identified.

The nests were discovered by making meticulous searches in places where it was known that the wasps generally nest. Only nests containing colonies of live wasps were included in the analyses. Some larger colonies were located accidentally by colleagues and local people. Individual wasps were encountered when they were hunting. The more common species were identifid in the field and then released, while less common ones were collected. The nomenclature used is that of Richards (1978) with additions of Carpenter and Wenzel (1989) and Raw (1985, 1989b).

A simple technique was employed to identify the species occupying a large nest without alerting the entire colony. The nest envelope was touched very gently with a long straw or twig. When a few wasps had walked onto the straw it was carefully moved a few metres from the nest and the wasps caught in a net.

It should also be mentioned that it was not possible to collect in the forest canopy and any species which might be restricted to that habitat was/were not sampled.

RESULTS

During the course of this study a total of 83 species and four subspecies were investigated. These are 40 species from Brasília, 43 from Goiás Velho, 29 from Maracá and 30 from the three localities in Rondônia. The total numbers of colonies located and wasps observed were 930 and 5173 respectively. (The names of the species and the numbers of nests and wasps recorded are given in Appendix 25A, 1–4.)

Three types of habitat: forest, forest boundary and savanna (or clearing) were easily distinguished and the species which occupied them could be placed in 10 categories according to where they nested and searched for food (see Table 25.1). Fewer data were collected in the localities in Roraima and Rondônia because less time was spent there than in Brasília and Goiás Velho. The former data are preliminary and are included to compare Amazon with central Brazilian localities. In particular, there are few records of wasps from the forests of Rondônia because very little time was/were not sampled.

Compiling the data of the four study areas, 78 of the 83 species nested

Table 25.1 The numbers of species of social wasps which nested and hunted in three habitats and in four study areas in Brazil

Category of nesting and hunting	*Brasília*	*Goiás Velho*	*Maracá*	*Rondônia*
a	6	1	4	–
b	1	6	7	4
c	1	5	1	4
d	1	–	–	–
e	17	10	13	6
f	5	11	3	9
g	3	1	–	1
h	5	5	1	6
i	–	1	–	–
j	2	3	–	1
Total species using forest	10	13	12	8
Total species using margin	29	37	24	24
Total species using savanna	14	25	5	20
Total species in study area	41	43	29	31

The species which occupied the three types of habitat: forest, margin and savanna (or clearing) were placed in nine categories according to where they nested and searched for food. They are:
a) confined to the forest,
b) nesting in the forest and hunting there and at the forest edge,
c) nesting in the forest and hunting in all three habitats,
d) nesting at margin and hunting there and in forest,
e) confined to the forest edge,
f) nesting at edge and hunting there and in savanna,
g) nesting in savanna and hunting there and at forest edge,
h) confined to the savanna,
i) nesting and hunting in forest and at its edge,
j) nesting and hunting at forest edge and in savanna.
The names of the species and the numbers of colonies and wasps recorded are given in the appendices.

in one type of habitat and only five (6%) nested in two types. Around Brasília and Goiás Velho *P. occidentalis* and *Brachygastra lecheguana* nested and hunted both at the forest margin and in the savanna. *Chartergellus communis* also did so around Goiás Velho as did *Polistes versicolor* in Rondônia. *Metapolybia cingulata* was the only species found nesting both inside the forest and at its edge. However, 18 species (22%) were found hunting in only one type of habitat, 60 (72%) hunted in two types while only five (6%) hunted in all three. No species nested in the savanna and hunted in the forest and none nested in all three habitats.

Five of the species which nested in the forest were encountered in all three habitats. All have long-lived colonies comprising thousands of individuals. The four around Goiás Velho and in Rondônia were *Polybia liliacea* and *P. rejecta* which nest in the canopy, *P. dimidiata* which builds its nest close to the ground around the stem of a sapling and *P. ignobilis* which nests in hollow trees and holes in the ground. Likewise, *P. ignobilis* hunted in three habitats in Brasília, while the fifth species, *Stelopolybia pallipes* did so around Goiás Velho, but was encountered only in gallery forest near Brasília. *P. liliacea* was found in all three habitats in all four study areas.

In all four study areas, greater numbers occurred at the forest boundary than elsewhere, with 53% to 58% of the species nesting at the edge and 72% to 86% of them searching for food there. Of the total numbers recorded, 58% of the individual wasps and 73% of the colonies were at the forest boundary. Similarly, greater numbers of species were confined to the forest margin than to either of the other habitats, with 17 species in Brasília, 10 around Goiás Velho, 13 on Maracá and 6 in Rondônia found nesting and searching for food only at the forest edge. Fewer species were restricted to the forest or to open vegetation (see Table 25.1).

DISCUSSION

Based on where they nested and searched for food, the social wasps in central Brazil showed clear habitat preferences and the preliminary data collected in the Amazon suggest the same situation exists there as well. The following generalizations on the use of habitat can be drawn from this investigation.

Firstly, wasps which have large colonies, like some *Polybia* species, hunt over wide areas and often cover more than one type of habitat. In contrast, those with smaller colonies, notably species of *Mischocyttarus* and *Polistes*, restrict their hunting mostly to the habitat where they nest. The latter situation is illustrated by the three closely related species of *Mischocyttarus* (subgenus *Mischocyttarus*) from central Brazil. The adults of these species are difficult to identify from their morphology. However, *M. rotundicollis* is restricted to forest, *M. drewseni* occurs at the forest edge and sometimes in disturbed vegetation, while *M. matogrossoensis* is a species of the open cerrado.

Secondly, a given species generally nests in only one of the three types of habitat studied. Five of the 83 species investigated were exceptional. There was only one case of a species which nested in both the forest and at its edge (*Metapolybia cingulata* near Goiás Velho). Four species nested at both the boundary and in the open vegetation. One of them was *Polybia occidentalis* (subspecies *occidentalis*) which nested at the forest edge and occasionally in the campo and cerrado vegetation around Goiás Velho and Brasília. The situation in Roraima is more difficult to interpret. *P. occidentalis* (subspecies *venezuelana*) nested at the forest edge while, what appears to be an undescribed subspecies of *P. occidentalis*, nested only 1.5 km away in the forest. Identification of the 27 species and subspecies which comprise the group of *P. occidentalis* (Richards, 1978: 63) is very difficult. The close proximity of nests of two supposed subspecies clearly requires further study before a satisfactory decision can be made on the taxonomic status of these wasps. In the present study it was found simplest to analyse the data on these taxa as separate species.

Thirdly, several species hunt only in the habitat where they nest and, although many hunt in an adjacent habitat, few hunt in all three types.

Fourthly, the species which nest at the forest edge, the savanna or clearings do not enter the forest in search of food. Only one exception was recorded when a few individuals of *Polybia chrysothorax* were found in gallery forest near Brasília. Hence, some species may search for food in more than one type of habitat, but the ones encountered in the forest also nest there. Several explanations might be examined to account for this apparent preference for nesting in the forest or at its boundary. Among them are the availability of supports for the nest, a more easily controlled nest microclimate and reduced predation pressure.

Finally, the forest edge is a favoured habitat of many species of social wasps in various parts of Brazil. In all four areas, greater numbers of species nested at the margin than elsewhere (53% to 60%) and numerous species that nested in the two other habitats also searched for food at the forest edge. Thus 70% to 86% of the species recorded at the various localities, were found at the forest edge.

The observation that greater numbers of species are often found at the forest edge is not new. Leopold (1933) wrote that 'two major habitat types intergrade' at the boundaries of North American forests and the higher species richness is, therefore, due to use as a partial habitat by animals occupying the two adjacent habitats. This explanation accounts for only part of the Brazilian situation. The forest edge there is a particular habitat and many species are largely restricted to it.

The social wasps which nest in the forest and hunt over wide areas reflect the habitat use of other animals. The forest provides shelter for many species which seek their food at the edge and on the open savanna. Among the better known animals that do so are bats, raptors and cats. In central Brazil the gallery forest provides shelter for several species including social bees, like *Eulaema nigrita* and *Trigona amalthea*, which forage in the cerrados, and the maned wolf and puma which hunt there.

The sharp border of forest and savanna is also the particular habitat of many species of animals. Several insectivorous species are largely restricted to forest margins and wooded river banks. Among the birds are the rufous-tailed jacamar (*Galbula ruficauda*), white-eared puffbird (*Nystalus chacuru*), long-tailed tyrant (*Colonia colonus*) and other species of tyrant flycatchers. The bird perches on a high branch and flies after passing insects. Another insectivore typical of the forest edge of central Brazil is the arboreal lizard *Tropidurus torquatus* which is rarely seen in forest or cerrado. All four species catch social wasps (Raw, unpublished; Bamberg, personal communication).

Some artificial habitats are similar to the natural forest edge and many forest edge species are found there. The edges of Amerindian fields are probably the man-made environment most similar to the forest boundary. The sides of highways function in a similar manner, serving as lines of communication for the forest boundary species.

Some of the species associated with the forest boundary have successfully invaded suburban gardens which offer similar environmental conditions for mobile animals like insects and small birds. Others also live along the wooded sides of streams and several are successional species, being among the more typical colonizers of forest clearings.

Although the total area that constitutes the forest edge is relatively small, it stretches over great distances and the geographical distributions of the species inhabiting the boundary are, likewise, narrow, sinuous bands of populations. Numerous species that are restricted to this distinctive habitat have distributions that are often only a few metres wide yet stretching over hundreds of thousands of kilometres. The long, more or less continuous distributions of some of the edge species extend from Central America to Bolivia, Paraguay and southern Brazil. The distributions of these species may be likened to those of aquatic species confined to the littoral zone. For this reason, the term 'margin' may be more suitable than 'boundary' or 'edge' for the place where the forest meets the savanna.

In summary, although often a narrow strip of vegetation forming the interface between two major tropical ecosystems, the forest edge is a distinctive habitat which is permanently or temporarily occupied by many species of animals for which its existence is crucial.

ACKNOWLEDGEMENTS

I thank the University of Brasília for funding field work around Brasília, the Supirentendência do Desenvolvimento do Centro-Oeste and the Bank of International Reconstruction and Development of Washington for the same in Rondônia. The Fundação Nacional do Índio of Brasília kindly allowed me to visit the two groups of indians. The Instituto Brasileiro do Meio-Ambiente of Brasília [previously the Secretaria Especial do Meio-Ambiente] and the Royal Geographical Society of London invited me to visit Maracá in Roraima. Professora I. Rezende Rocha supplied me with information on *Polistes geminatus* and *P. subsericeus*. Dr S. Simpson and Dr C. O'Toole of the University Museum, Oxford, Dr John Proctor of the University of Stirling and an anonymous reviewer are thanked for their comments on the manuscript.

REFERENCES

Carpenter, J.M. and Wenzel, J.W. (1989) Synonymy of the genera *Protopolybia* and *Pseudochartergus* (Hymenoptera: Vespidae; Polistinae). *Psyche*, **96**, 177–86.

Leopold, A. (1933) *Game Management*. Schribners, New York.

Raw, A. (1985) Two new species of *Mischocyttarus* (Vespidae, Hymenoptera) from Brazil. *Revista Brasileira de Entomologia*, **29**, 107–12.

Raw, A. (1988) Social wasps (Hymenoptera: Vespidae) and insect pests of crops of the Suruí and Cinta Larga indians in Rondônia, Brazil. *The Entomologist*, **107**, 104–9.

Raw, A. (1989a) Suggested preferences for habitat and clearing size among social wasps (Hymenoptera, Vespidae) on the Ilha de Maracá, Roraima, Brazil, in *Maracá Rainforest Project, Invertebrates & Limnology, Preliminary Report*, (eds W. Milliken and J.A. Ratter) 39–45.

Raw, A. (1989b) Social wasps (Hymenoptera, Vespidae) of the Ilha de Maracá, Roraima, Brazil with comments on their biogeography, in *Maracá Rainforest Project, Invertebrates & Limnology, Preliminary Report*, (eds W. Milliken and J.A. Ratter) 27–38.

Richards, O.W. (1978) *The Social Wasps of the Americas*. British Museum (Natural History), London.

Richards, O.W. (1979) Additions and corrections to the 1978 publication *The Social Wasps of the Americas*. Unpublished manuscript, Entomology Library, British Museum (Natural History), London.

Richards, O.W. and Richards, M.J. (1951) Observations on the social wasps of South America (Hymenoptera, Vespidae). *Transactions of the Royal Entomological Society of London*, **102**, p. 107.

APPENDIX: HABITATS WHERE SOCIAL WASPS NEST AND SEARCH FOR FOOD

Table 25A.1 Habitats where social wasps nest and search for food within a 15 km radius of Goiás Velho, Goiás State, Central Brazil

	Type of habitat					
	Forest		*Margin*		*Savanna*	
	N	*H*	*N*	*H*	*N*	*H*
Mischocyttarus rotundicollis	X	4				
Polybia striata	2	14		15		
Polybia tinctipennis	1	2		8		
Stelopolybia multipicta	7	12		58		
Stelopolybia flavipennis	7	9		18		
Stelopolybia vicina	9	24		17		
Angiopolybia pallens	2	16		12		
Ploybia rejecta	4	13		17		11
Polybia ignobilis	11	68		160		38
Polybia dimidiata	6	86		56		21
Polybia liliacea	6	77		70		10
Stelopolybia pallipes	8	18		27		14
Metapolybia cingulata	14	28	4	20		
Metapolybia suffusa			3	11		
Epipona tatua			21	48		
Polybia fastidiosuscula			1	8		
Protopolybia chartergoides			2	12		
Protopolybia sedula			18	39		
Parachartergus smithii			1	6		
Pseudopolybia vespiceps			4	13		
Mischocyttarus flavicornis			2	8		
Mischocyttarus melanoxanthus			2	5		
Polistes versicolor			12	29		
Synoeca surinama			15	70		27
Chartergus chartius			15	29		6
Chartergus globiventris			35	36		14
Brachygastra augusti			2	12		15
Brachygastra moebiana			6	24		22
Polybia quadricincta			8	22		9
Polybia chrysothorax			6	18		5
Polybia jurinei			4	40		12
Polybia emaciata			14	25		16
Protopolybia exigua			39	60		18
Parachartergus fraternus			44	60		12
Brachygastra lecheguana			12	70	1	80
Chartergellus communis			8	28	1	5
Polybia occidentalis			14	152	24	250
Polybia sericea				20	14	129
Polistes subsericeus					2	25
Polistes davillae					1	8
Mischocyttarus matogrossoensis					3	7
Polistes billardieri					N	10
Polistes geminatus					N	6
Total number of colonies	77		292		46	
Total number of wasps		371		1323		770

N = Number of nests recorded.
H = Number of wasps recorded in flight and presumed to be hunting.
X = Nesting habitat recorded by Richards and Richards (1951) and O.W. Richards (unpublished).

Table 25A.2 Habitats where social wasps nest and search for food within a 15 km radius of Brasília airport (abbreviations as Table 25A.1)

	Type of habitat					
	Forest		*Margin*		*Savanna*	
	N	*H*	*N*	*H*	*N*	*H*
Mischocyttarus rotundicollis	2	9				
Mischocyttarus occipitalis	X	2				
Mischocyttarus goeldii	X	2				
Mischocyttarus giffordi	1	2				
Angiopolybia pallens	2	12				
Polybia ignobilis	6	34		184		49
Polybia dimidiata	3	16		28		
Polybia liliacea	X				4	1
Stelopolybia pallipes	2	12				
Polybia chrysothorax		6	3	26		
Clypearia angustior			5	18		
Epipona tatua			2	8		
Epipona quadrituberculata			1	8		
Polybia fastidiosuscula			7	35		
Protopolybia sedula			2	12		
Pseudopolybia vespiceps			4	12		
Mischocyttarus cassununga			87	96		
Mischocyttarus ceberus			94	87		
Mischocyttarus lecointei			2	2		
Mischocyttarus goyanus			23	22		
Polistes cinerascens			8	25		
Synoeca surinama			4	18		
Polybia emaciata			4	12		
Parachartergus fraternus			7	28		
Mischocyttarus drewseni			4	11		6
Brachygastra moebeiana			1	9		
Protopolybia exigua			34	58		
Protonectarina sylveirae			2	26		24
Polistes versicolor			6	190		12
Chartergellus communis			1	5		
Brachygastra lecheguana			1	24	1	31
Polybia occidentalis			10	160	32	182
Polybia sericea				42	28	165
Mischocyttarus marginatus				8	4	14
Mischocyttarus campestris					1	6
Polistes geminatus				6	3	14
Polistes subsericeus					1	18
Polistes davillae					1	6
Mischocyttarus matogrossoensis					9	21
Polistes billardieri					3	18
Total number of colonies	16		312		83	
Total number of wasps		95		1164		567

Table 25A.3 Habitats where social wasps nest and search for food on the Ilha de Maracá, Roraima, northern Brazil (abbreviations as Table 25A.1)

	Type of habitat					
	Forest		*Margin*		*Savanna*	
	N	*H*	*N*	*H*	*N*	*H*
Polybia occidentalis subsp. *AR1*	3	18				
Stelopolybia angulata	X	3				
Mischocyttarus carbonarius	X	1				
Mischocyttarus alboniger	3	6				
Stelopolybia fulvofasciata	X			9		
Stelopolybia testacea	X			1		
Stelopolybia ornata	X			2		
Angiopolybia pallens	1			16		
Polybia rejecta	X			8		
Polybia ignobilis	X	2		7		
Polybia dimidiata	1	8		21		
Polybia liliacea	2	2		35		11
Metapolybia unilineata			2	12		
Brachygastra smithii			5	28		
Stelopolybia multipicta			X	6		
Stelopolybia cajennensis			2	25		
Mischocyttarus labiatus			X	2		
Mischocyttarus injuncundus			8	15		
Mischocyttarus granadaensis			1	2		
Mischocyttarus maracaensis			1	6		
Polistes versicolor kaieteurensis			8	27		
Synoeca virginea			X	32		
Chartergus chartarius			1	1		
Polybia dimorpha			1	19		
Protopolybia exigua			1	2		
Synoeca surinama			X	38		6
Polybia occidentalis venezuelana			1	24		15
Polistes brevifissus			25	18		7
Polybia sericea					X	4
Total number of colonies	10		56		–	
Total number of wasps		40		356		43

Table 25A.4 Habitats where social wasps nest and search for food in three localities in central Rondônia, Brazil (abbreviations as Table 25A.1)

	Type of habitat					
	Forest		*Margin*		*Savanna*	
	N	*H*	*N*	*H*	*N*	*H*
Polybia striata	X	1		5		
Stelopolybia fulvofasciata	X	2		5		
Stelopolybia testacea	X	1		6		
Angiopolybia pallens	1	3		13		
Polybia rejecta	1	2		6		15
Polybia ignobilis	2	2		15		18
Polybia dimidiata	1	5		12		15
Polybia liliacea	2	1		11		8
Metapolybia unilineata			4	18		
Charterginus fulvus			1	2		
Protopolybia sedula			X	3		
Mischocyttarus labiatus			X	1		
Mischocyttarus surinamensis			2	4		
Mischocyttarus injucundus			2	5		
Brachygastra bilineolata			1	5		9
Stelopolybia cajennensis			2	6		8
Protopolybia exigua			X	6		8
Protopolybia holoxantha			1	8		6
Parachartergus smithii			X	4		2
Synoeca surinama			1	14		25
Epipona tatua			1	8		18
Polybia occidentalis			2	15		29
Polybia quadricincta			1	2		8
Polistes versicolor			2	6	4	19
Polybia bistriata					2	20
Polybia spinifex					1	5
Polybia sericea					X	7
Polistes testaceicolor					1	8
Polistes biguttatus					2	11
Polistes lanio					1	8
Total number of colonies	7		20		11	
Total number of wasps		17		180		247

26

The importance of forest edges in the ecology of open country cerrado birds

ROBERTO B. CAVALCANTI

SUMMARY

1. Forest edges contain important resources for birds of the open country in the highly seasonal *cerrado* regions of central Brazil. During the rainy season, several species of insectivorous/frugivorous tyrannid flycatchers breed in the savanna-like cerrado. In Brasília, at the end of the rainy season, some species (e.g. *Tyrannus savanna*) migrate, while others, such as *Elaenia chiriquensis*, disappear from cerrado, but are found in low numbers at the edge of the gallery forests, feeding on fruit during the dry season.
2. The seasonal pattern of forest edge use seen in flycatchers is also present in other families. Hummingbirds are common in the open *campo sujo* during the rainy season and when plants are flowering, but use the gallery forest edge heavily at other times of the year.
3. Birds of the forest, such as *Turdus* spp. and *Saltator similis*, use the edge and may forage in the adjoining campo during the early rainy season. At this time, the alates of termites and ants are an abundant food source in cerrado and campo.
4. The food resources provided at the forest edge may be critical for the survival of some cerrado birds during the dry season. In addition, after the grassland fires common during the mid and late dry season, many open habitat species seek cover at the forest edge.
5. Certain widespread species occur in cerrado typically at the forest edge. These include the migratory fork-tailed flycatcher (*Tyrannus savanna*), the shiny cowbird (*Molothrus bonariensis*), and the rufous-collared sparrow (*Zonotrichia capensis*).

INTRODUCTION

In many places in the cerrado region of central Brazil, the landscape shows a mosaic pattern, with different vegetation types occurring next to each other. In particular, long narrow 'corridors' of gallery forest are interspersed in areas of cerrado tree and scrub woodland (Eiten, 1984).

The rainfall in Brasília, Brazil, in the core of the cerrados, is highly seasonal. The dry season lasts from May until September, and rains start in late September or October. Fruit production and insect abundances also vary with time of the year. Both are most available to birds in the first half of the rainy season. Several migratory insectivorous birds, such as the fork-tailed flycatcher (*Tyrannus savanna*) arrive in July or August, breed between August and December, and leave by the following April.

The purpose of this study is to investigate the seasonal use of the cerrado and the forest edge by birds of open habitats. We studied primarily *Elaenia* flycatchers, a group of frugivorous/insectivorous birds.

STUDY SITE AND METHODS

This work was conducted at the University of Brasília Ecological Station, 20 km south-west of Brasília, Brazil. The area includes over 2000 ha of preserved habitat, with the major cerrado vegetation types (Ratter, 1980). Data on *Elaenia* were obtained by mist-netting, banding and visual censuses from 1983 to 1989. Observations on the effects of fire were made at the Ecological Station and in burned areas nearby.

OBSERVATIONS

Use of habitat by *Elaenia* flycatchers

Of the four common species of *Elaenia* in the area, three were seen both in cerrado and at the edge of the gallery forest (Table 26.1). The pattern of use differed in each case, however.

Elaenia chiriquensis arrived in large numbers in the cerrado from August onward. It was also seen at the edge of the gallery forest. Banding of this species in cerrado showed low frequencies of recapture (15 recaptures out of 52) at one site throughout the breeding season, suggesting extensive movements even after breeding sites are established. After February, *E. chiriquensis* became scarce in cerrado, but continued to be recorded in the edge of the gallery forest as late as May. It probably persists in low numbers throughout the year in this habitat, where we saw it feeding on fruit with other species in mixed flocks.

Table 26.1 Seasonal habitat use in four species of *Elaenia*

Species	*Habitat*	
	Cerrado	*Forest edge*
Elaenia cristata	All year and spring migrants	All year
Elaenia chiriquensis	Spring and summer (very abundant)	Most of the year (low numbers)
Elaenia flavogaster	Spring and summer (low numbers)	All year (abundant)
Elaenia obscura	Absent	All year

Elaenia cristata is a year-round resident of cerrado. However, during the rainy season its numbers increase considerably, probably through migration, since the new birds arriving are adults. It is also common year-round in the open *campo sujo* habitat beside the edges of the gallery forests. We also saw it in mixed-species flocks with *Elaenia chiriquensis* at fruiting trees at the forest edge during the dry season.

Elaenia flavogaster is most common at the forest edge. During the rainy season, it is also seen in the cerrado, in low numbers. It prefers the more humid *campo de murundus* habitat just outside the gallery forest.

Elaenia obscura is basically restricted to the forest edge, throughout the year.

Use of forest edge by cerrado birds

The hummingbird (*Colibri serrirostris*) defends feeding territories, and has a characteristic vocal display. During the rainy season it is extremely common in cerrado. It defends territories in the months November–December around plants in flower such as *Palicourea rigida*. During the dry season it may defend territories around large flowering trees in cerrado, but it is generally scarce in this habitat. However, at this time it is common along the edges and canopies of fairly small gallery forests, where it feeds on flowers from vines. In this situation, the individuals give vocal displays fairly continuously throughout the day.

The curl-crested jay (*Cyanocorax cristatellus*) travels in noisy flocks through the canopy of cerrado trees, and frequently uses the forest edge. Raptors of the cerrado, such as the crested caracara (*Polyborus plancus*) and the roadside hawk (*Buteo magnirostris*), also perch regularly at the forest edge.

The grassland fires which burn cerrado and campo habitats are usually stopped at the moist gallery forest edge. Many birds which survive the fires are deprived of cover in the burned-over habitat. Insects which survive fire by fleeing into holes are easily captured in the burned areas, and the soil exposure facilitates foraging for seeds. In these situations, birds of cerrado may use the burned areas for foraging, but choose the cover at the forest edge for roosting. These include the chalk-browed mockingbird (*Mimus saturninus*) and the picazuro pigeon (*Columba picazuro*).

Use of cerrado by forest birds

The gallery forests support a distinct avifauna as compared to cerrado, with a larger number of permanent residents. A few may use cerrado on a temporary basis, especially during the wet season when termite alates and fruit are produced in abundance in cerrado. Species which forage in cerrado adjoining gallery forests are the green winged saltator (*Saltator similis*), the pale-breasted thrush (*Turdus leucomelas*) and the rufous-bellied thrush (*Turdus rufiventris*).

Other birds of gallery forest use cerrado on a more extensive and permanent basis. They may be found over 1 km away from the nearest forest, and possibly set up territories in the denser areas of cerrado. These include the rufous-tailed jacamar (*Galbula ruficauda*), the burnished-buff tanager (*Tangara cayana*) and the guira tanager (*Hemithraupis guira*).

Forest edge birds

Several otherwise widespread species are fairly restricted to the forest edge in undisturbed cerrado. One is the shiny cowbird (*Molothrus bonariensis*), a brood parasite. As a rule, many species of the forest edge adapt fairly well to disturbed habitats. Other such species include the ruddy ground-dove (*Columbina talpacoti*), the rufous-browed peppershrike (*Cyclarhis gujanensis*), the rufous-collared sparrow (*Zonotrichia capensis*) and the fork-tailed flycatcher (*Tyrannus savana*).

As the cerrado habitats become increasingly disturbed and fragmented by human interference, these forest edge species may well increase in numbers, and interact more extensively with species of undisturbed cerrado. For instance, the shiny cowbird shows high rates of parasitism on the white-banded tanager (*Neothraupis fasciata*), in disturbed habitats (Cavalcanti and Pimentel, 1988). Further increases in the cowbird population may result in decline in the numbers of cerrado birds which presently have experienced little parasitism.

DISCUSSION

In this work some fairly specific aspects of the role of forest edges in bird communities of the cerrado were examined. Our evidence so far suggests that forest edges are important in maintaining year-round populations of flycatchers, and possibly hummingbirds, which would otherwise be wholly migratory in our study areas. Birds that depend on seasonal resources such as fruit and flowers are often quite mobile, and are able to take advantage of the different patterns of availability in the distinct but adjoining cerrado and forest edge habitats. The migratory species are an important component of the bird fauna in the cerrado region, in nearly all habitats (Negret and Negret, 1981; Negret, 1983).

The forest edge acts as a corridor for birds of successional habitats, which occur in low density in native cerrado. In areas disturbed by human activity such as farming, many of these species invade and establish populations in fragments of cerrado.

The interior of the forest also supports a bird fauna very distinct from the neighbouring cerrado. Fry (1970) found few birds common to both habitats at a site in Mato Grosso, with the woodpecker (*Dryocopus lineatus*) one of the few species to forage in cerrado and gallery forest. On a geographical scale, the gallery forests form a link between the Atlantic forest and the Amazon forest, across the open expanses of the cerrado region. Migratory species, such as *Turdus nigriceps*, travel through the gallery forests of Goias and the Federal District en route to their breeding grounds in the forests of south-central Brazil (Antas and Valle, 1987).

Gallery forests are a key habitat in the cerrado landscape, owing to their unique fauna and the presence of food and water resources utilized by animals of open habitats during the dry season and after grassland fires.

ACKNOWLEDGEMENTS

I am grateful to Dr Edwin Willis, Álvaro Negret and Dr Colin Johnson for advice in the field and for sharing their extensive knowledge of birds of cerrado. Miguel A. Marini, Maria Alice dos S. Alves, Thais Martins and Rita de Cassia S. Medeiros helped obtain data in the field. Field work was supported by the University of Brasília and by the Conselho Nacional de Desenvolvimento Cientifíco e Tecnológico (CNPq).

REFERENCES

Antas, P.T.Z. and Valle, M.P. (1987) Dados preliminares sobre *Turdus nigriceps* no Distrito Federal, in *Anais do II Encontro Nacional de Anilhadores de Aves,* Universidade Federal do Rio de Janeiro, Rio de Janeiro, Brasil, pp. 213–20.

Cavalcanti, R.B. and Pimentel, T.M. (1988) Shiny cowbird parasitism in central Brazil. *Condor,* **90**, 40–3.

Eiten, G. (1984) Vegetation of Brasília. *Phytocoenologia,* **12**, 271–92.

Fry, C.H. (1970) Ecological distribution of birds in north-eastern Mato Grosso State, Brazil. *Anais de Academia Brasileira de Ciências,* **42**, 275–318.

Negret, A.J. (1983) *Diversidade e Abundância da Avifauna da Reserva Ecológica de IBGE,* Unpublished M.Sc. thesis, Universidade de Brasília, Brasília, DF.

Negret, A.J. and Negret, R. (1981) *As Aves Migratórias do Distrito Federal,* Boletim Téchno no. 6, Instituto Brasileiro de Desenvolvimento Florestal, Brasília, Brasil.

Ratter, J.A. (1980) *Notes on the Vegetation of Fazenda Água Limpa (Brasilia, DF, Brazil),* Royal Botanic Garden, Edinburgh, UK.

27

The role of mammals in neotropical forest–savanna boundaries

RODRIGO A. MEDELLIN and KENT H. REDFORD

SUMMARY

1. Mammal assemblages in three areas of the neotropics were compared in relation to their distribution and ecological roles. The three areas (Venezuelan *llanos*, southern Veracruz State, Mexico, and Brazilian *cerrado*) represent the centre, and northern and southern extremes of the neotropics, and all are covered by mixed forest and savanna-like habitats.
2. Mammal species were examined in relation to dependence on the forest, vagility, type of interaction established with plants, and size of the species. In each of these categories, each species was assigned a ranked value.
3. Veracruz contained 124 mammal species, the llanos 122 and the cerrado 143. The majority of these species (93.6% in Veracruz, 93.5% in the llanos and 81.2% in the cerrado) were either partially or completely dependent on the forest, while few species depended on the savanna.
4. The general structure of assemblages per variable showed similarities between Veracruz and the llanos, but were different from the cerrado. This may be due to biogeographic, phytogenetic and climatic factors; the presence of the Sierra Madre Oriental and the Neovolcanic Belt may reduce the influence of temperate North American faunal elements in Veracruz, while the cerrado is open to the influence of extensive adjacent flat temperate grasslands in the south.
5. The influence of the bat fauna also affected the composition of the cerrado mammal fauna; bats were under-represented in the cerrado, probably because of the dominance of xeric vegetation types there and the association of bats with forested regions, mainly subhumid or

humid forests, available there only in the form of riparian gallery forests.

6. Mammal effects on the forest–savanna boundary are strong and can be perceived in their co-evolved association with many plants for their pollination and dispersal, playing an active role in modification of the boundary. Additionally, herbivores and seed predators affect the capabilities of forest plants to colonize savannas and vice versa. Carnivores and insectivores exert an indirect but strong influence by preying on insects and vertebrates which affect the forest–savanna boundary directly.

INTRODUCTION

Much ecological research has been done in tropical rainforests, to the neglect, particularly in the neotropics, of the large and diverse open vegetation ecosystems. Ranging from the deserts of central Mexico through the grasslands of Central America to the pampas of Argentina and Uruguay, these ecosystems cover vast areas of tropical and subtropical Central and South America. Many of these open ecosystems have a frequent and intimate association with forests, so that large ($>500\,km^2$) areas of open habitat without forests are rarely found.

Almost all the research that has focused on these ecosystems has been done by geographers, soil scientists and plant ecologists who have been particularly interested in the factors that accompany the forest–savanna boundaries. The point that very little work has been done on the animal component of this forest–savanna interface is witnessed by the fact that less than 15% of the chapters in this book deal with animals.

The purpose of this chapter is to examine and discuss the mammalian fauna of the forest–savanna boundary. Animal studies in this area are scarce, yet mammals are important because they account for a significant proportion of the animal biomass in the tropics. Their interactions with the environment range broadly and spread through most ecosystems, as some species are habitat specialists while others range across many different habitats. They play important roles in many ecological processes, including predation, pollination and seed dispersal. We examine the patterns of distribution and ecological roles of mammals in forest–savanna areas of three relatively well studied neotropical regions. Attempting to give the study representativity, we have included areas located at the extremes and the centre of the neotropics.

METHODS

The areas

Three areas in the neotropics were selected for the present study, located at about the centre, and the northern and southern extremes of the region. These are southern lowland Veracruz State, Mexico, the Venezuelan llanos and the Brazilian cerrado. We selected these areas because of their geographic position, the fact that they are covered by a mixture of forest and savanna-like habitats, and the degree of knowledge of the mammal faunas inhabiting them.

Southern lowland Veracruz State was defined as the area below 500 m altitude, south of 21 degrees of latitude and east of the Sierra Madre Oriental, encompassing roughly 75% of the state, an area of about 55 000 km^2 (García, 1986). It includes, among other vegetation types, natural and anthropogenic grasslands, tropical rainforest, and tropical semi-deciduous forest (Pennington and Sarukhán, 1968; Rzedowski, 1978; García, 1986). The habitats are intricately interdigitated and their boundaries shift continuously because of density-independent factors, secondary succession and human influence. The area has an annual precipitation ranging from about 800 to over 3200 mm, with forest–savanna boundaries running throughout the area, its dominant plant species varying accordingly (Pennington and Sarukhán, 1968; Rzedowski, 1978).

The second area, the Venezuelan llanos, is defined following Troth (1979) and Eisenberg and Redford (1979) as an area of about 200 000 km^2 in the central lowlands of the country, north and west of the Orinoco River, and south and east of the coastal mountain ranges and the Andes. There are grasslands, sometimes sparsely forested, forest patches and humid gallery forests; the altitude is below 500 m, and the annual precipitation ranges from 800 to 1600 mm (World Meteorological Organization/Unesco, 1975).

The third area, the cerrado, is the second largest plant association in Brazil, delimited to the north by the Amazonian forest, to the east by the caatinga and the Atlantic forest, to the west by the Pantanal, and to the south by the Chaco (Mares *et al.*, 1989). The cerrado is characterized by mainly xerophytic associations, ranging from open grasslands to closed woodlands, with ribbons of mesophytic, evergreen gallery forest following water courses across an extension of over $1.4 \times 10^6 km^2$. The precipitation ranges from 1200 to 2800 mm (World Meteorological Organization/Unesco, 1975).

The mammals

Lists of mammal species were compiled for each of the three selected areas. For southern Veracruz, we used the work by Hall and Dalquest (1963), Hall (1981), Navarro (1982), and Ramírez-P. *et al.*, (1986). For the Venezuelan llanos we consulted Eisenberg *et al.* (1979), Linares (1986) and Eisenberg (1989). For cerrado lists we used Fonseca and Redford (1986), Redford and Fonseca (1984), J.S. Marinho Filho and M. Lima (unpublished) and Mares *et al.* (1989). All species reported at least once from any one area were included, and information on ecological variables examined here was obtained from either those same sources or more comprehensive references such as Eisenberg (1981) and Nowak and Paradiso (1983).

Four variables were used in classifying the role of each mammal species: its dependence on the forest, its degree of vagility, its type of interaction with plants and its size category (Table 27.1). We assigned mammal species to one of three forest dependence levels:

1. those that do not go into the forest, strictly inhabiting open areas and depending on the savannas for their existence;
2. those that occur in the forest, but do not depend exclusively on it, moving between forest and savanna;
3. species relatively dependent on the forest.

Vagility (other ecological synonyms: motility or mobility; Pianka, 1978; Colwell and Winkler, 1984) was quantified in five levels:

1. fossorial species;
2. terrestrial and arboreal species with no cursorial capabilities (i.e. not relying on running to escape; Eisenberg (1981), defines cursorial as 'strongly adapted for running across open terrain. This adaptation typifies certain ungulates and carnivores');
3. terrestrial and arboreal species with cursorial capabilities or otherwise increasing their capabilities to move, but not having very large home ranges;
4. ungulates and large cats with large home ranges;
5. volant species.

We decided to use these five levels because they divide mammal species into well defined groups; only the three intermediate categories need an explanation: category 2 includes most mice and small, non-volant or fossorial species; category 3 includes most medium-sized species which clearly have increased mobility capabilities when compared to mice, but reduced when compared to ungulates and large cats; category 4 contains terrestrial mammals with the highest movement capabilities, and although frequently large cats have the largest home ranges, some

Table 27.1 Four variables quantified in this study and the definition of levels within each variable

Forest dependence levels	
1	Never occurs; these are savanna-dependent
2	Occurs but not dependent
3	Dependent on the forest; without forest they are not present
Vagility levels	
1	Fossorial species
2	Terrestrial and arboreal species with no cursorial capabilities
3	Terrestrial and arboreal species with cursorial capabilities or otherwise increasing their mobility; not very large home ranges
4	Ungulates and large cats with very large home ranges
5	Volant species
Plant–Mammal Interaction categories	
0	Seed predators
1	Grazers/browsers
2	Species partially dependent on plants, taking seeds and leaves and sometimes fruit
3	Species that do not include plant material in their diets
4	Species partially dependent on plants, taking only fruits
5	Frugivorous species
6	Nectar/pollinivorous species
Size categories	
1	Below 50 g
2	Between 50 g and 5 kg
3	Between 5 and 20 kg
4	Above 20 kg

ungulates (e.g. white-lipped peccaries) carry out equally extensive movements.

The third variable evaluated the nature of plant–mammal interactions: we have seven categories, ranked from the most negative to the most positive interaction from the plant's perspective, as follows:

0. seed predators;
1. grazers/browsers;
2. species that partially depend on plants, taking seeds and leaves and sometimes fruit;
3. species that do not include plant material in their diets;
4. species that partially depend on plants, taking exclusively fruits;
5. frugivorous species;
6. nectar- and pollen-eating species.

It is evident that each group differs from the others: frugivores are assumed to function primarily as seed dispersers, given that fruits have been generally recognized as baits for seed dispersers, frequently having specific characteristics (such as the presence of high acidic contents, tannins, strychnine, etc.) that seem to function as deterrents of wrong dispersers (Janzen, 1977, 1983; Herrera, 1982); species partially dependent on plants may spend long periods of time with no plant interaction, thus considerably reducing their impact.

Finally, we considered four size categories:

1. <50 g;
2. >50 g–5 kg;
3. >5–20 kg;
4. >20 kg.

Mammal faunas in each locality were compared with each other to show variability throughout the neotropical region.

RESULTS

A total of 275 species was found in the comprehensive list for all three areas. Of these, 90% were either totally (122 species) or relatively (118) dependent on the forest, while only 12.7% (35) were found to be associated solely with savannas. Southern lowland Veracruz contains 124 mammalian species, the Venezuelan llanos 122 species, and the Brazilian cerrado 143 species (see Appendix). Figure 27.1 shows that the great majority of mammals are either partially dependent (level 2) or completely dependent (level 3) on the forest (respectively, 46.8% and 46.8% in Veracruz, 53.3% and 40.2% in the llanos, and 37.8% and 45.4% in the cerrado). There are relatively few savanna-dependent species: 8 species in Veracruz (6.4%), 8 in the llanos (6.5%) and 24 in the cerrado (16.8%).

In Veracruz, vagility histograms are dominated by levels 3 (31 species or 25%) and 5 (68 species or 54.8%; Fig. 27.2a). Both columns roughly divide in half into partially and totally dependent forest species, while savanna-dependent species are practically absent, except in vagility level 2, where four species represent 23.6%. The only fossorial species in Veracruz is partially dependent on the forest. Plant–mammal interactions in Veracruz (Fig. 27.3a), are dominated by level 3, with 44.3% (55 species), roughly three times as many species as the next three columns. Plant interaction categories 2, 4 and 5 contain 14% (17, 18 and 17 species respectively). Levels 2–5 are roughly divided in half by species partially or completely dependent on the forest, with some deviation to one or the other. Remaining levels are represented by seven or less species. Most (62% or 77 species) mammal species in Veracruz fall in size category 1 (Fig. 27.4a), probably an effect of the very diverse bat fauna, and those are divided in roughly 45.4% and 45.4% (35 species each) being partially

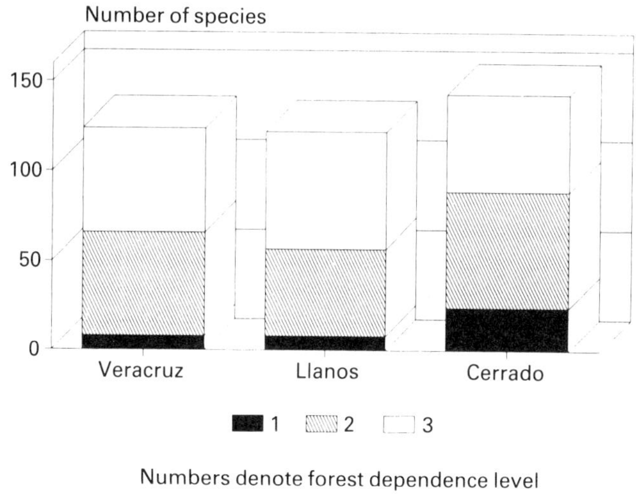

Fig. 27.1 Mammal species per forest dependence level in three neotropical localities. Level 1 includes species that do not go into the forest, strictly inhabiting open areas and depending on the savannas for their existence; level 2 indicates those species that occur in the forest, but do not depend on it, moving from forest to savanna and back; level 3 includes species relatively dependent on the forest, without which they could not survive.

or totally dependent on the forest, while only 7 species (9%) are dependent on the savannas. Size category 2 contains 28 species (22.6%), again roughly equally divided into totally and partially dependent on the forest, with only one species being dependent on the savannas.

Vagility in the Venezuelan llanos (Fig. 27.2b) is also dominated by level 5, with 70 species (58.2%) of bats, evidencing again the large influence of these mammals on neotropical ecosystems. Level 3 includes 31 species (25.4%), of which 19 (61.3%) were totally dependent on the forest, the remaining being partially dependent. In those two columns the fact is again evident that they are roughly divided in halves between totally and partially forest-dependent species. Only one species is in vagility level 1, and it is savanna-dependent. Plant–mammal interactions (Fig. 27.3b) in the llanos are dominated by level 3 (55 species, 45.1%); 25 species (45.4%) are totally dependent on the forest, 26 (47.3%) are partially dependent on the forest, and 4 (7.3%) are savanna-dependent. Interaction levels 2, 4 and 5 have 14, 18 and 17 species each (11.5, 14.7 and 13.9%), again divided in half between forest dependence levels 2 and 3, except for interaction level 4, dominated by forest-dependent species. One species is savanna-dependent. Four of the five seed predator species are totally dependent on the forest. Most mammals (57.4% or 70 species) in the

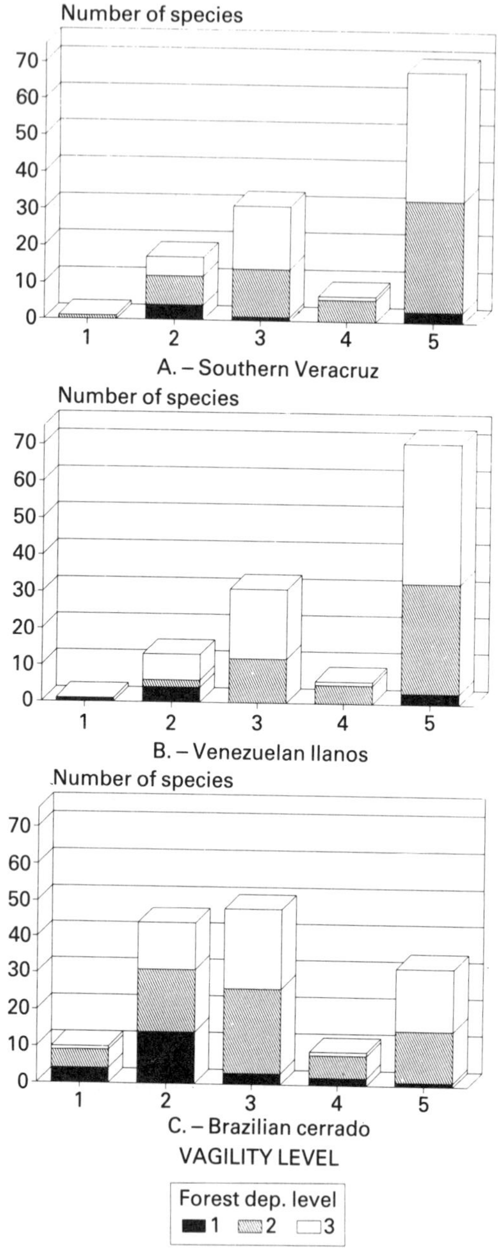

Fig. 27.2 Frequency distribution of mammal species per vagility level and per forest dependence level, defined in Fig. 27.1 and Table 27.1.

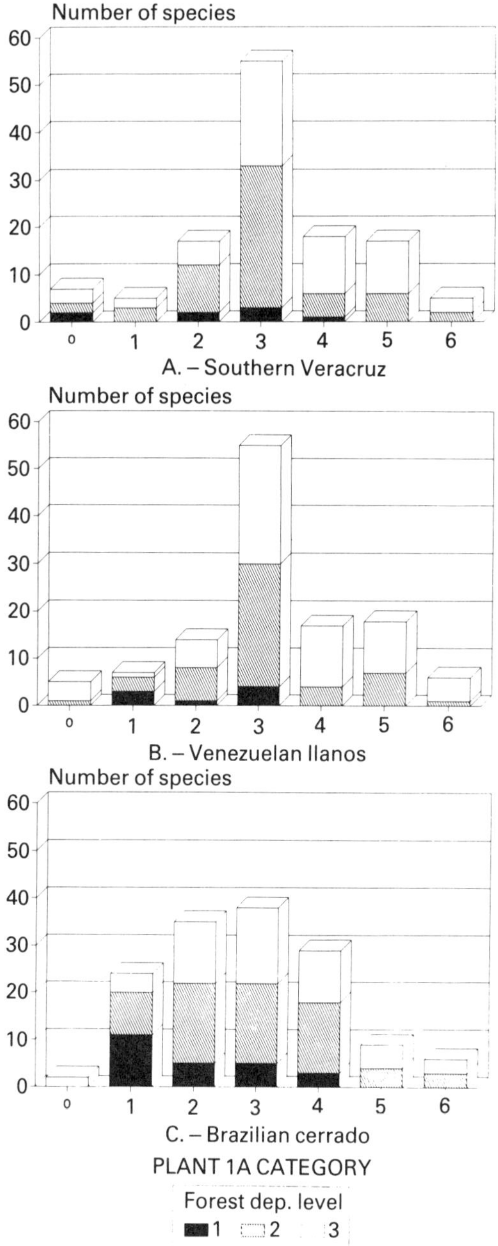

Fig. 27.3 Frequency distribution of mammal species per plant interaction category and per forest dependence level, defined in Fig. 27.1 and Table 27.1.

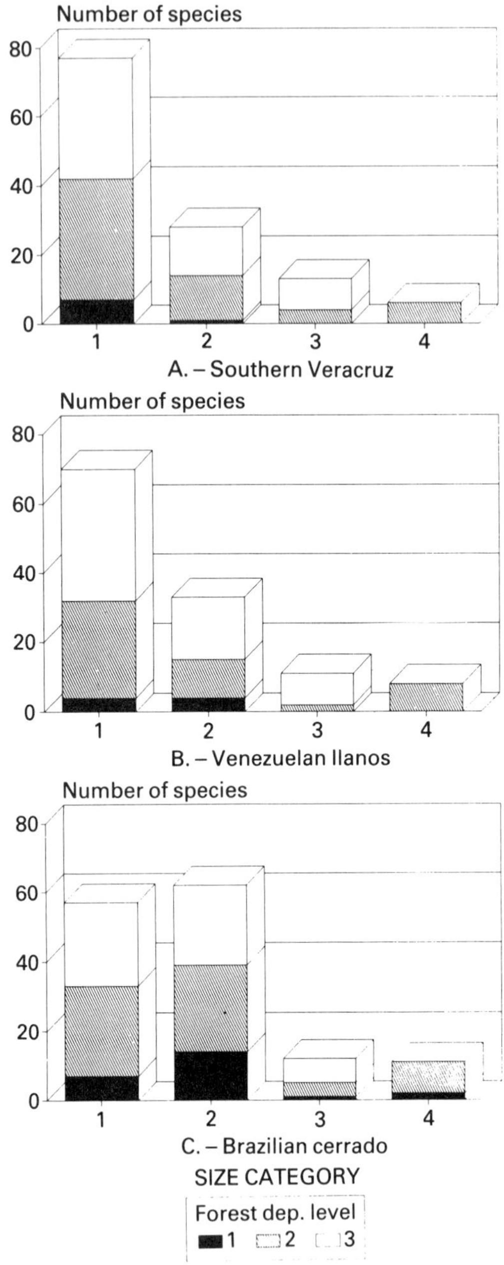

Fig. 27.4 Frequency distribution of mammal species per size category and per forest dependence level, defined in Fig. 27.1. and Table 27.1.

llanos fall in size category 1 (Fig. 27.4b), probably related to Pleistocene extinctions of large mammals with resulting dominance by bats and mice. About half are totally forest-dependent and the other half are partially dependent, with only four savanna-dependent species. In size category, 2, 35 species (18.7%) are included; again most are partially or completely forest-dependent. All eight (6.5%) species over 20 kg are partially dependent on the forest.

In the cerrado, vagility categories 2, 3 and 5 dominate, contrasting with the other two localities. Level 2 (44 species, 36.1%) is divided in three with each forest dependence level equally represented. Levels 3 and 5 (48 species, 33.6%, and 32 species, 26% respectively) are dominated by an almost equal proportion of partially and completely forest-dependent species. Plant–mammal interactions in the cerrado (Fig. 27.3c) show categories 2, 3 and 4 as dominant, with 35, 38 and 39 species (24.5, 26.6 and 20.2% respectively), again sharply contrasting with the other two localities. This is due to larger numbers of grazers/browsers and species that are partially dependent on plants. A distribution resembling normality can be perceived. In levels 2 through 4, species are nearly equally divided into partially and completely dependent on forest, and only level 1 shows dominance by savanna-dependent species. Only the cerrado size level 2 includes more species than level 1. The two represent over 80% of the total, size 1 with almost equal numbers of forest dependence levels 2 and 3 species, while a larger number of level 1 is present in size category 2.

DISCUSSION

Southern Veracruz and the llanos contain about the same proportions of species independent, partially dependent and dependent on the forest, while the cerrado has three times as many mammal species that are savanna-dependent (Fig. 27.1). It is surprising to find such similarities between the llanos and Veracruz, particularly since they share less than half their mammal species (56). We cannot identify the reason for such a similarlity, but we suggest an ecological adjustment mechanism in which competition, predation and other intrinsic, density-dependent factors have moulded such assemblages from the species pool determined by biogeographic, phylogenetic and climatic factors.

If all three localities were subjected to similar influences of both the density-dependent and the density-independent types, we would not expect any difference in species composition per variable level in each variable, but the differences observed in the cerrado warrant some explanation. The llanos lie in the middle of the neotropics, thus being exposed only to neotropical influences from all around. Veracruz lies in the northern extreme of the neotropical region and on a map it is evident that it is very close to the temperate Nearctic region; nevertheless, it does not show any significant influence of temperate faunal elements. This is probably due to the geographic isolation of Veracruz by the Sierra Madre Oriental and the Neovolcanic Belt, which represent an effective barrier for temperate colonizers. The cerrado is at the other extreme of the neotropics, and although the neotropical region extends as far south as Patagonia, it includes a vast temperate portion. This temperate area, represented by the Chaco and the pampas, is not separated from the cerrado by any mountain range of importance, and at the same time has evolved its own typical temperate grassland mammal fauna. This has allowed several temperate grassland-adapted species to colonize or to evolve forms capable of inhabiting tropical or subtropical areas. Thus the cerrado is enriched by 20 additional species when compared with the llanos or Veracruz, and consistently shows a larger representation of savanna-dependent species, in all levels of all variables. An alternative explanation is offered by the extension of the cerrado when compared to the other two localities; it is nearly one order of magnitude larger than either of the northern localities. Nevertheless, its uniformity casts some doubts on this argument.

In the vagility analysis (Fig. 27.2), the similarities between Veracruz and the llanos are evident; these two areas almost lack fossorial species, while they are more frequent in the cerrado, probably due to primary productivity occurring closer to the ground. A large proportion of the species in level 2 present only in the cerrado is independent or relatively independent of the forest, while the composition and size of level 4 (ungulates and large cats) is very constant throughout the localities examined. Veracruz differs from the llanos mainly in the number of species relatively dependent on the forest in vagility level 2, with Veracruz having eight while the llanos has only two. Temperate grassland influence may have affected the larger representation of vagility level 2 in the cerrado, and given the near universal dependence of bats on forest, vagility level 5 is reduced both in percentage and in number of species. An additional factor that may explain reduced representation of bats may be the dominance of xerophytic plant associations in the cerrado, leaving only riparian gallery forests to be packed with the dominant wet tropical bat fauna component. This is further supported by the fact that only one bat species in the cerrado is savanna-dependent.

The larger representation of grazers/browers in the cerrado may also be associated with a temperate open grassland influence, but other dissimilarities in plant interaction composition, such as the increased number of species partially dependent on plants and the reduced number of carnivores/insectivores, remain to be explained. The most distinctive characteristic of the interaction of mammals with plants (Fig. 27.3) is a unimodal distribution. In Veracruz and the llanos the composition is very similar, with species in level 3 representing an overwhelmingly important proportion of the total number of species. The pattern in the cerrado is markedly different from that in the llanos and Veracruz, with a more uniform distribution of species within interaction categories. The smaller size of category 3 when compared with Veracruz and the llanos is because there are fewer species relatively or completely forest-dependent, which may be due to a reason similar to that offered for reduced bat representation. The cerrado lacks some frugivorous forest-dependent species, possibly because of the dominance of open habitats and the reduced availability of forested regions.

Dividing the mammalian faunas into size levels (Fig. 27.4) again shows the similarities between Veracruz and the llanos, with most of the species falling in the smallest size level (below 50 g), and the whole distribution being similar to a log-normal; about half as many species in level one are present in level 2, about half as many as those in level 2 are in level 3; species in level 4 are fewer than any other column. Composition by forest dependence category is also remarkably similar between these two sites; the vast majority of species in size categories 1 and 2 divide equally among relatively and completely forest-dependent. By contrast with the other two sites, most mammals in the cerrado are equally distributed between size categories 1 and 2. The remaining size categories are similar in size and composition to those in Veracruz and the llanos. The reason for the reduced histogram of level 1 is an under-representation of forest-dependent species in the cerrado, while the increase in size level 2 is due to a large number of species that are savanna-dependent and relatively dependent from the forest. This may be associated with bat presence again, as most bats are forest-dependent, and with the presence of some medium-sized grassland adapted species, such as some armadillos (*Tolypeutes* sp.), cavimorph rodents (*Galea, Kerodon*) and a few mice (*Oxymycterus roberti, Euryzygomatomys guiara*, etc).

Given that such a large proportion of mammal species depend on the forest to one degree or another, and taking into account that many of these species frequently move between the forest and the savanna, it is evident that these species have important effects on the forest–savanna boundary. The appendix shows a summary of the results for the 240 mammal species that are either partially or totally dependent on the forest, classifying all species according to their vagility, type of interaction with plants, and size. Several conclusions are evident from an examination of the appendix. It is obvious that most species occur in vagility levels 2, 3 and 5, and in plant interactions 2 through 5. There are representatives for all vagility levels and plant interaction categories, but not all cells are occupied. The pattern of occupancy of vagility levels indicates a dominance of mice, bats and cursorial-not-hoofed species. Only a few species are seed predators, grazers/browers, or even pollinivorous. Most mammals fall in size categories 1 or 2 (below 5 kg), with only 31 species weighing over 5 kg, reflecting the dominance of small mammals in the neotropics. Also, there is a suggested trend where lower plant interaction categories co-occur with low vagility levels 3 and 5.

Examining the effects of the species in each vagility level helps explain the influence of each group on the forest–savanna boundary. There are very few fossorial species that are partially or totally dependent on the forest; these all are small, and are primarily affecting the forest as grazers/browsers. The reduced number of fossorial species may be due to the reduced nutrient contents of the forest soil or reduced ground-level primary productivity (Barbour *et al.*, 1987). Fossorial species may have a regulating effect on the advance of the forest into the savanna; in Arizona, pocket gophers accounted for 90% of the deaths of aspen, preventing aspen colonization of open areas (Cantor and Witham, 1989).

Among the species with no cursorial or volant capabilities such as mice (vagility level 2) there is considerable variation in the type of interaction with plants. A few species are either exclusively seed predators or grazers or browsers, while many species are only partially dependent on plants for their subsistence. In plant interaction level 2 there are 14 species, and in level 4 there are 10 species. Given that many of these species in vagility level 2 and in plant interaction levels 0, 1, 2 and 4 can reach very high population densities (e.g. *Sigmodon hispidus, Nectomys squamipes, Oryzomys* spp.), we can predict that this group will have a strong influence on the forest–savanna boundary processes, particularly since about half of these species customarily move between the forest and the savanna (level 2 of forest dependence). These species function mostly as seed predators and grazers/browsers, but also as seed dispersal agents, and there are two species that act as pollinators.

In vagility level 3 (arboreal and cursorial, not-hoofed species), we find a similar situation to that of species in vagility level 2: a few strict seed predators and grazers/browsers, and large numbers of partial grazers/browsers and partial frugivores. Some of the species (e.g. *Nasua nasua*, the coati, Kauffman, 1962; capuchin monkeys (*Cebus* sp.) and other primates, Terborgh, 1983; Robinson, 1986) included in this vagility level are only small to medium-sized, but they live in large groups that substantially affect local areas. The largest concentration of species within this vagility category occurs between 50 g and 20 kg, thus being able to process larger amounts of food per individual. Also in this category there are many large species that may affect plant species not influenced by any other animal. It is noteworthy that species whose plant interaction is type 3 (insectivores or carnivores) increase in numbers as vagility category increases. Examining vagility level 4, we see that ungulates are all above 5 kg and either obligate grazers/browsers, or partially frugivorous. Here we find species whose effects are among the most conspicuous (not necessarily the most important) on the forest, not only because of their size, but also because some species travel in groups that exploit vast areas. Peccaries tend to disturb the forest floor, digging and trampling on seedlings and saplings, feeding on large quantities of fruits and seeds, and thus removing them from near the parent plant and dispersing and preying on seeds, and feeding on foliage and on a variety of other food items (Kiltie, 1981; Kiltie and Terborgh, 1983; Sowls, 1984). In a similar role, but at a lower rate since these are not gregarious species, the tapirs feed (not exclusively) on large amounts of fruits whose seeds are frequently passed through their gut and deposited away from the parent plant (Terwilliger, 1978; Janzen, 1981, 1982).

The last vagility category includes bats, the only volant mammals. The effects of bats on the forest–savanna boundary are important because of the high densities of many species, and they influence the forest in a variety of ways. Their direct interactions with plants involve either frugivory (and seed dispersal) or pollen/nectarivory. A large percentage of the pioneer light-demanding plants are either bat-dispersed or bat-pollinated, among them *Bauhinia* sp., *Cecropia* sp., *Parkia*. sp. and *Piper* sp. At least in the case of one species of bat (*Carollia perspicillata*), the number of seeds dispersed amounts to many thousands per night per bat (Fleming, 1988). Although a large proportion of these seeds are deposited within 50 m of the parent plant, the bats can carry them for up to 2.5 km. Since bats defecate in flight, seeds are actually dispersed to a variety of microsites, and gut passage increases germination rates (Fleming, 1988).

Plant interaction category 3 (no direct interaction; carnivores and insectivores) produces some of the most important effects of mammals on the forest–savanna boundary by preying upon other species that do interact directly with plants. It is well known that rodents and ungulates are the main food item of a variety of predators, such as cats, bush dogs, weasels and grisons (Langguth, 1975; Svendsen, 1982; Emmons, 1987; Eisenberg, 1989) thus buffering the effects that those species could have on vegetation. Apparently, absence of large seed predators in forests leads to dramatically increased abundance of large-seeded tree species, and a superabundance of seed predators (e.g. pacas, peccaries), due to the absence of large carnivorous predators (e.g. pumas and jaguars), may cause reduced recruitment of large seeded plants in the forest (Terborgh, 1988). A similar effect is probably caused by insectivorous bats; over 50% of the bat species dependent or partially dependent on the forest are completely insectivorous or carnivorous, and given their large colonies (some into at least hundreds of thousands) and the fact that one 7 g bat can eat as much as 3 g of insects in one night (Kunz, 1980), they exert an enormous impact on insect populations (Fenton, 1982) and indirectly on the vegetation.

A comprehensive picture of the effects of mammals on the neotropical forest–savanna boundary is difficult to define, but this boundary is affected by at least the following components.

1. Mammalian herbivory at and below ground level may affect the quantitative and qualitative composition of savannas; herbivores frequently remove only particular species of grasses or other plants, and their effects are more evident around the forest–savanna boundary, as here even those mammalian species strictly dependent on the forest can affect the vegetation. Herbivory may effectively stop recessions or advances of one vegetation type into another, and may affect the type of dominance that is established in the forest or savanna. Fossorial species may apply a strong herbivory pressure on trees in open areas, sometimes preventing all colonization of savannas by forest species (Cantor and Witham, 1989).
2. Arboreal herbivores and browsers may reduce or inhibit the expansion of tree limbs of saplings into the savanna, 'sharpening' the limits between forest and grasslands. When fire is applied to the savannas, hardly any transitional zone is noticeable between the two habitats.
3. There is a continuous, heavy invasion of seeds from the forest into the savannas dispersed by many mammals, especially large ungulates, bats and some rodents. Frequently the first forest seeds dispersed into the savanna come in this fashion, owing to the fact that bats defecate in flight (Fleming, 1988), and that ungulates move through open areas depositing their forest-ingested seeds. The reverse is also true, where many species that feed in the savannas come during the night to the forest to roost. They defecate here, and thus bring savanna seeds into the forest.
4. Proposed coevolutionary schemes show that many species of plants in the Bombacaceae, Bignoniaceae, Cactaceae, Leguminosae, Sapotaceae and others depend on bats for their pollination (Butanda-Cervera *et al.*, 1978; Heithaus, 1982), and non-volant mammals are also known to serve as important pollinators (Janson *et al.*, 1981; Gribel, 1988), suggesting that mammals have actually moulded the structure of forests known today. It has been estimated that about 42% of the 320 tree species at La Selva, Costa Rica, depend on mammals for their dispersal (Hartshorn, 1983).
5. The effects of the carnivorous and insectivorous mammals are critical

for the continuation of the ecological processes both in the savanna and the forest, by exerting predation pressure on the populations of all herbivorous and seed-eating species, from insects to mammals.

6. Although beyond the scope of this chapter, it is important to note the remarkable resemblance between Veracruz and the Venezuelan llanos, in the composition of all categories in all variables studied. Total number of species, as well as per level per variable, are very similar in the two localities. When compared to the cerrado, similarities are still evident, and the most striking differences are that the savanna species are more frequent than in the other localities, there are several more fossorial species, and fewer bat species. This is probably associated with biogeographic, phylogenetic and climatic factors. Veracruz and the llanos lie close enough to have influenced each other with their faunas after the emergence of the Panama Isthmus, even though they share less than half (56 species) of the mammals. Some groups have undergone an impressive radiation in or near each of the localities, with a secondary local radiation that reduces the number of species in common, but since many of them share an ancestor, they are ecological homologues. Lastly, Veracruz is isolated from the adjacent temperate Nearctic region by intricate mountain ranges including the Sierra Madre Oriental and Neovolcanic Belt, thus reducing the influence from the temperate faunas, while the cerrado lies in a flatter region and is a continuum with the Chaco as far south as latitude 40°, thus allowing a much larger influence of grassland mammals.

CONCLUSIONS

Fully 90% of the mammal species present in the three areas selected for this study are dependent on the forest at least partially. Many species of plants are affected in a variety of ways by mammals; in many cases these play a key role for the successful reproduction and dispersal of the plants, while in others they exert a strong negative effect. Mammals affect vegetation in a variety of ways, not only pollinating flowers and moving seeds to suitable areas, but also increasing germination percentages (Janzen, 1983). Additionally, many species function as seed predators, grazers, browsers and root-eaters, further altering the potential for the forest–savanna boundary to modify in one direction or the other. Examining the detailed species composition of the mammal faunas that move between forests and savannas and their ecological roles may help explain the shifts of the forest–savanna boundary. Certainly, other animal taxa such as birds and ants, as well as abiotic factors such as fire and soils, affect the forest–savanna boundary but the role played by mammals is undoubtedly larger than anticipated. Many of the mammalian faunas found in forest–savanna complexes are being altered by habitat destruction, cattle grazing, agriculture and hunting. Given the large number of plants dependent on mammals for their reproduction or dispersal, such local or complete extinctions may quantitatively and qualitatively affect the myriad ecological processes in which mammals are vitally involved, destroying both neotropical forests and savannas as we know them today.

ACKNOWLEDGEMENTS

We thank J.S. Marinho Filho for his unpublished data for the cerrado mammals, and J.F. Eisenberg and R. Dirzo for their comments and suggestions. This paper is a contribution of the Program for Studies in Tropical Conservation, and was partly possible thanks to Grant Number 52701 from Consejo Nacional de Ciencia y Tecnología, Mexico.

REFERENCES

Barbour, M.G., Burk, J.H. and Pitts, W.D. (1987) *Terrestrial Plant Ecology*, 2nd edn, Benjamin/Cummings, Menlo Park, California.

Butanda-Cervera, A., Vázquez-Yanes, C. and Trejo, L. (1978) La polinización quiropterófila: una revisión bibliográfica. *Biotica, México*, **3**, 29–35.

Cantor, L.F. and Whitham, T.G. (1989) Importance of belowground herbivory: pocket gophers may limit aspen to rock outcrop refugia. *Ecology*, **70**, 962–70.

Colwell, R.K. and Winkler, D.W. (1984) A null model for null models in biogeography, in *Ecological Communities Conceptual Issues and the Evidence* (eds D.R. Strong Jr, D. Simberloff, L.G. Abele and A.B. Thistle), Princeton University Press, Princeton, pp. 344–59.

Eisenberg, J.F. (1981) *The Mammalian Radiations. An Analysis of Trends, Evolution, Adaptation and Behaviour*, University of Chicago Press, Chicago.

Eisenberg, J.F. (1989) *Mammals of the Neotropics*, Volume 1, University of Chicago Press, Chicago.

Eisenberg, J.F. and Redford, K.H. (1979) A biogeographic analysis of the mammalian fauna of Venezuela, in *Vertebrate Ecology in the Northern Neotropics* (ed. J.F. Eisenberg), Smithsonian Institution Press, Washington, DC, pp. 31–41.

Eisenberg, J.F., O'Connell, M.A. and August, P.V. (1979) Density, productivity and distribution of mammals in two Venezuelan habitats, in *Vertebrate Ecology in the Northern Neotropics*, (ed. J.F. Eisenberg), Smithsonian Institution Press, Washington, DC, pp. 187–207.

Emmons, L. (1987) Comparative feeding ecology of fields in a neotropical rainforest. *Behavioral Ecology and Sociobiology*, **20**, 271–83.

Fenton, M.B. (1982) Echolocation, insect hearing, and feeding ecology of insectivorous bats. in *Ecology of Bats* (ed. T.H. Kunz), Plenum Press, New York, pp. 261–85.

Fleming, T.H. (1988) *The Short-tailed Fruit Bat, a Study in Plant–Animal Interactions*, University of Chicago Press, Chicago.

Fonseca, G.A.B. and Redford, K.H. (1984) The mammals of IBGE's ecological reserve, Brasilia, and an analysis of the role of gallery forests in increasing diversity. *Revista Brasileira de Biologia*, **44**, 517–23.

Garcia de Miranda, E. (1986) *Nuevo Atlas Porrúa de la República Mexicana*, Ed. Porrúa, México.

Gribel, R. (1988) Visits of *Caluromys lanatus* (Didelphides) to flowers of *Pseudobombax tomentosum* (Bombacaceae): a probable case of pollination by marsupials in central Brazil. *Biotropica*, **20**, 344–7.

Hall, E.R. (1981) *The Mammals of North America*, 2nd edn, John Wiley, New York.

Hall, E.R. and Dalquest, W.W. (1963) The mammals of Veracruz. *University of Kansas Publications. Museum of Natural History*, **14**, 165–362.

Hartshorn, G.S. (1983) Plants. Introduction, in *Costa Rican Natural History* (ed. D.H. Janzen), University of Chicago Press, Chicago, pp. 118–57.

Heithaus, E.R. (1982) Coevolution between bats and plants, in *Ecology of Bats* (ed. T.H. Kunz), Plenum Press, New York, pp. 327–67.

Herrera, C.M. (1982) Defense of ripe fruits from pests: its significance in relation to plant disperser interactions. *American Naturalist*, **120**, 218–41.

Janson, C.H., Terborgh, J. and Emmons, L.H. (1981) Non-flying mammals as pollinating agents in the Amazonian forest. *Reproductive Botany*, **13**, 1–6.

Janzen, D.H. (1977) Why fruits rot, seeds mold, and meat spoils. *American Naturalist*, **111**, 691–713.

Janzen, D.H. (1981) Digestive seed predation by a Costa Rican Baird's tapir. *Biotropica*, **13**, 59–63.

Janzen, D.H. (1982) Seeds in tapir dung in Santa Rosa National Park, Costa Rica. *Brenesia*, **19/20**, 129–35.

Janzen, D.H. (1983) Dispersal of seeds by vertebrate guts, in *Coevolution* (eds D.J. Futuyma and M. Slatkin), Sinauer Associates, Sunderland, Massachusetts, pp. 232–62.

Kaufmann, J.H. (1962) Ecology and social behavior of the coati, *Nasua narica* on Barro Colorado Island, Panama. *University of California Publications in Zoology*, **60**, 95–222.

Kiltie, R.A. (1981) Stomach contents of rain forest peccaries (*Tayassu tajacu* and *T. pecari*). *Biotropica*, **13**, 234–6.

Kiltie, R.A. and Terborgh, J. (1983) Observations on the behavior of rain forest peccaries in Peru: why do whitelipped peccaries form herds? *Zeitschrift für Tierpsychologie*, **62**, 241–55.

Kunz, T.H. (1980) Daily energy budgets of free-living bats, in *Proceedings of the International Bat Research Conference*, (eds D.E. Wilson and A.L. Gardner), Texas Tech. Press, Lubbock, pp. 369–92.

Langguth, A. (1975) Ecology and evolution in the South American canids, in *The Wild Canids* (ed. M.W.Fox), Van Nostrand Reinhold, New York, pp. 192–206.

Linares, O.J. (1986) *Murciélagos de Venezuela*, Cuadernos Lagoven, Refolit, CA, Caracas.

Mares, M.A., Braun J.K. and Gettinger, D. (1989) Observations on the distribution and ecology of the mammals of the cerrado grasslands of central Brazil. *Annals of the Carnegie Museum*, **58**, 1–60.

Navarro, L.D. (1982) *Mamíferos de la Estación de Biologia Tropical 'Los Tuxtla', Veracruz.*, Bachelor's thesis, Facultad de Ciencias, Universidad Nacional Autónoma de México, Mexico.

Nowak, R. and Paradiso, R.L. (1983) *Walker's Mammals of the World*, Johns Hopkins University Press, Baltimore.

Pennington, T.D. and Sarukhán, J. (1968) *Manual Para la Identificación de Campo de los Principales Árboles Tropicales de México*, Instituto Nacional de Investigaciones Forestales/Food and Agriculture Organization, United Nations, Imprenta Benjamin Franklin, Mexico.

Pianka, E.R. (1978) *Evolutionary Ecology*, Harper & Row, New York.

Ramírez-P.J., Britton, M.C., Perdomo, A. and Castro, A. (1986) *Guía de los mamíferos de México* Universidad Autónoma Metropolitana Iztapalapa, Mexico, DF.

Redford, K.H. and da Fonseca, G.A.B. (1986) The role of gallery forests in the zoogeography of the cerrado's non-volant mammalian fauna. *Biotropica*, **18**, 126–35.

Robinson, J.G. (1986) Seasonal variation in use of time and space by the wedge-capped capuchin mnonkey *Cebus olivaceus*: implications for foraging theory, *Smithsonian Contributions in Zoology*, **431**, 1–60.

Rzedowski, J. (1978) *Vegetación de México*, Ed. Limusa, Mexico.

Sowls, L.K. (1984) *The Peccaries*, University of Arizona Press, Tucson.

Svendsen, G.E. (1982) Weasels (*Mustela* species), in *Wild Mammals of North America*, (eds J.A. Chapman and G.A. Feldhamer), John Hopkins University Press, Baltimore, pp. 613–28.

Terborgh, J. (1983) *Five New World Primates: a Study in Comparative Ecology*, Princeton University Press, Princeton.

Terborgh, J. (1988) The big things that run the world – A sequel to E.O.

Wilson. *Conservation Biology*, **2**, 402–3.
Terwilliger, V.J. (1978) Natural history of Baird's tapir on Barro Colorado Island, Panama Canal Zone. *Biotropica*, **10**, 211–20.
Troth, R.G. (1979) Vegetational types on a ranch in the central llanos of Venezuela, in *Verebrate Ecology in the Northern Neotropics* (ed. J.F. Eisenberg), Smithsonian Institution Press, Washington, DC, pp. 17–30.
World Meteorological Organization/Unesco (1975) *Climatic Atlas of South America*, Cartographia, Budapest.

APPENDIX

List of mammal species and values of forest dependence, vagility, interaction with plants and size

Table 27A.1 Mammals from lowland tropical Veracruz

Species	*Forest*	*Vagility*	*Plant*	*Size*
Order Marsupialia				
Didelphis marsupialis	2	3	4	2
Philander opossum	2	3	4	2
Marmosa mexicana	3	2	4	1
Caluromys derbianus	3	3	4	2
Order Insectivora				
Cryptotis mexicana	2	2	3	1
Cryptotis parva	2	2	3	1
Order Chiroptera				
Rhynchonycteris naso	3	5	3	1
Saccopteryx bilineata	3	5	3	1
Balantiopteryx plicata	2	5	3	1
Balaniotiopteryx	3	5	3	1
Peropteryx kappleri	3	5	3	1
Peropteryx macrotis	3	5	3	1
Centronycteris maximiliani	3	5	3	1
Diclidurus albus	2	5	3	1
Noctilio leporinus	2	5	3	2
Mormoops megalophylla	2	5	3	1
Pteronotus parnellii	2	5	3	1
Pteronotus personatus	2	5	3	1
Pteronotus davyi	2	5	3	1
Pteronotus gymnonotus	3	5	3	1
Micronycteris megalotis	3	5	4	1
Micronycteris sylvestris	3	5	4	1
Micronycteris brachyotis	3	5	4	1
Mimon cozumelae	3	5	3	1
Phyllostomus discolor	3	5	6	1
Trachops cirrhosus	3	5	3	1

Table 27A.1 *(continued)*

Species	Forest	Vagility	Plant	Size
Tonatia brasiliense	3	5	4	1
Tonatia evotis	3	5	4	1
Chrotopterus auritus	3	5	3	2
Vampyrum spectrum	3	5	3	2
Glossophaga soricina	2	5	6	1
Hylonycteris underwoodi	3	5	6	1
Choeroniscus godmani	3	5	6	1
Leptonycteris curasoae	2	5	6	1
Carollia perspicillata	2	5	5	1
Carollia brevicauda	2	5	5	1
Sturnira lilium	2	5	5	1
Sturnira ludovici	2	5	5	1
Chiroderma salvini	3	5	5	1
Chiroderma villosum	3	5	5	1
Artibeus watsoni	3	5	5	1
Artibeus jamaicensis	2	5	5	1
Artibeus lituratus	2	5	5	2
Artibeus phaeotis	3	5	5	1
Artibeus toltecus	3	5	5	1
Uroderma bilobatum	3	5	5	1
Vampyrops helleri	3	5	5	1
Vampyrodes major	3	5	5	1
Vampyressa pusilla	3	5	5	1
Centurio senex	3	5	5	1
Desmodus rotundus	2	5	3	1
Diphylla ecaudata	2	5	3	1
Natalus stramineus	2	5	3	1
Thyroptera tricolor	3	5	3	1
Eptesicus andinus	3	5	3	1
Eptesicus furinalis	2	5	3	1
Lasiurus borealis	2	5	3	1
Lasiurus cinereus	2	5	3	1
Lasiurus ega	2	5	3	1
Lasiurus intermedius	2	5	3	1
Myotis albescens	3	5	3	1
Myotis elegans	3	5	3	1
Myotis fortidens	2	5	3	1
Myotis keaysi	2	5	3	1
Myotis nigricans	2	5	3	1
Pipistrellus subflavus	3	5	3	1
Rhogeessa tumida	2	5	3	1
Bauerus dubaquercus	3	5	3	1
Eumops bonariensis	2	5	3	1
Eumops glaucinus	2	5	3	1
Molossus ater	2	5	3	1
Tadarida brasiliensis	1	5	3	1
Nyctinomops laticaudatus	1	5	3	1
Nyctinomops macrotis	1	5	3	1
Order Primates				
Alouatta palliata	3	3	2	3
Ateles geoffroyi	3	3	4	3
Order Edentata				
Cyclopes didactylus	3	3	3	2
Tamandua mexicana	2	3	3	2

Table 27A.1 *(continued)*

Species	*Forest*	*Vagility*	*Plant*	*Size*
Dasypus novemcinctus	2	3	2	2
Order Lagomorpha				
Sylvilagus brasiliensis	3	3	1	2
Sylvilagus floridanus	2	3	1	2
Order Rodentia				
Sciurus deppei	3	3	0	2
Sciurus aureogaster	3	3	0	2
Orthogeomys hispidus	2	1	2	2
Heteromys desmarestianus	3	2	0	2
Liomys irroratus	2	2	0	1
Liomys pictus	2	2	0	1
Baiomys musculus	1	2	0	1
Nyctomys sumichrasti	3	2	2	1
Oryzomys alfaroi	2	2	2	1
Oryzomys fulvescens	1	2	2	1
Oryzomys couesi	2	2	2	1
Oryzomys melanotis	2	2	2	1
Peromyscus leucopus	1	2	2	1
Peromyscus mexicanus	3	2	2	2
Reithrodontomys fulvescens	1	2	0	1
Sigmodon hispidus	2	2	1	2
Tylomys nudicaudus	3	2	5	2
Coendou mexicanus	3	3	2	2
Agouti paca	3	3	2	3
Dasyprocta punctata	2	3	2	2
Order Carnivora				
Canis latrans	2	3	3	3
Urocyon cinereoargenteus	2	3	3	3
Bassariscus sumichrasti	3	3	4	3
Potos flavus	3	3	4	3
Nasua nasua	2	3	4	3
Procyon lotor	2	3	4	3
Eira barbara	3	3	4	3
Galictis vittata	3	3	3	2
Mustela frenata	2	3	3	2
Conepatus leuconotus	2	3	4	2
Conepatus semistriatus	3	3	4	2
Mephitis macroura	1	3	4	2
Felis yagouaroundi	3	3	3	2
Felis wiedii	3	3	3	2
Felis pardalis	3	3	3	3
Felis concolor	2	4	3	4
Felis onca	2	4	3	4
Order Perissodactyla				
Tapirus bairdii	2	4	2	4
Order Artiodactyla				
Tayassu tajacu	2	4	2	4
Tayassu pecari	2	4	2	4
Odocoileus virginianus	2	4	1	4
Mazama americana	3	4	1	3

Table 27A.2 Mammals of the Venezuelan llanos

Species	*Forest*	*Vagility*	*Plant*	*Size*
Order Marsupiala				
Didelphis marsupialis	2	3	4	2
Caluromys lanatus	3	3	4	2
Chironectes minimus	3	3	3	2
Metachirus nudicaudatus	3	3	4	2
Monodelphis brevicaudata	2	3	4	2
Marmosa robinsoni	3	2	4	1
Order Chiroptera				
Rhynchonycteris naso	3	5	3	1
Saccopteryx bilineata	3	5	3	1
Saccopteryx canescens	3	5	3	1
Saccopteryx leptura	3	5	3	1
Cormura brevirostris	3	5	3	1
Peropteryx kappleri	3	5	3	1
Peropteryx macrotis	3	5	3	1
Centronycteris maximiliani	3	5	3	1
Noctilio leporinus	2	5	3	2
Noctilio albiventris	3	5	3	1
Mormoops megalophylla	2	5	3	1
Pteronotus parnellii	2	5	3	1
Pteronotus personatus	2	5	3	1
Micronycteris minuta	3	5	4	1
Micronycteris nicefori	3	5	4	1
Micronycteris megalotis	3	5	4	1
Lonchorhina orinocensis	1	5	3	1
Macrophyllum macrophyllum	3	5	3	1
Tonatia bidens	3	5	4	1
Tonatia brasiliense	3	5	4	1
Mimon crenulatum	3	5	3	1
Phyllostomus discolor	3	5	6	1
Phyllostomus hastatus	3	5	6	2
Phyllostomus elongatus	3	5	6	1
Phylloderma stenops	2	5	4	1
Trachops cirrhosus	3	5	3	1
Chrotopterus auritus	3	5	3	2
Vampyrum spectrum	3	5	3	2
Carollia perspicillata	2	5	5	1
Carollia brevicauda	2	5	5	1
Glossophaga soricina	2	5	6	1
Glossophaga longirostris	3	5	6	1
Choeroniscus godmani	3	5	6	1
Sturnira lilium	2	5	5	1
Sturnira ludovici	2	5	5	1
Artibeus cinereus	3	5	5	1
Artibeus fuliginosus	2	5	5	1
Artibeus lituratus	2	5	5	2
Artibeus jamaicensis	2	5	5	1

Table 27A.2 *(continued)*

Species	*Forest*	*Vagility*	*Plant*	*Size*
Sphaeronycteris toxophyllum	3	5	5	1
Ametrida centurio	3	5	5	1
Uroderma bilobatum	3	5	5	1
Uroderma magnirostrum	3	5	5	1
Vampyrops helleri	3	5	5	1
Vampyrops brachycephalus	3	5	5	1
Vampyressa pusilla	3	5	5	1
Ectophylla macconelli	3	5	5	1
Desmodus rotundus	2	5	3	1
Thyroptera tricolor	3	5	3	1
Natalus stramineus	2	5	3	1
Myotis nigricans	2	5	3	1
Myotis albescens	3	5	3	1
Eptesicus diminutus	2	5	3	1
Eptesicus furinalis	2	5	3	1
Eptesicus brasiliensis	3	5	3	1
Rhogeessa tumida	2	5	3	1
Lasiurus ega	2	5	3	1
Lasiurus borealis	2	5	3	1
Neoplatymops mattogrossensis	3	5	3	1
Molossops temminckii	2	5	3	1
Molossops greenhalli	2	5	3	1
Molossops planirostris	3	5	3	1
Molossus ater	2	5	3	1
Molossus molossus	2	5	3	1
Nyctinomops laticaudatus	1	5	3	1
Eumops perotis	1	5	3	1
Eumops bonariensis	2	5	3	1
Eumops dabbenei	2	5	3	1
Eumops glaucinus	2	5	3	1
Eumops auripendulus	2	5	3	1
Promops centralis	2	5	3	1
Order Primates				
Aotus brumbacki	3	3	4	2
Cebus olivaceus	3	3	4	2
Ateles belzebuth	3	3	4	3
Alouatta seniculus	3	3	2	3
Order Edentata				
Dasypus sabanicola	2	3	2	2
Dasypus novemcinctus	2	3	2	2
Cabassous unicinctus	1	1	3	2
Priodontes maximus	2	3	3	4
Tamandua tetradactyla	2	3	3	2
Myrmecophaga tridactyla	2	3	3	4
Other Lagomorpha				
Sylvilagus floridanus	2	3	1	2

Table 27A.2 *(continued)*

Species	*Forest*	*Vagility*	*Plant*	*Size*
Order Rodentia				
Sciurus granatensis	3	3	0	2
Heteromys anomalus	3	2	0	2
Rhipidomys mastacalis	3	2	0	2
Oryzomys bicolor	3	2	0	1
Oryzomys concolor	3	2	2	2
Oryzomys fulvescens	1	2	2	1
Neacomys tenuipes	3	2	2	1
Nectomys squamipes	1	2	1	2
Zygodontomys brevicauda	2	2	0	2
Holochilus sciureus	1	2	1	2
Sigmodon hispidus	2	2	1	2
Sigmodon alstoni	1	2	1	2
Coendou prehensilis	3	3	2	2
Proechimys guairae	3	3	2	2
Echimys semivillosus	3	2	5	2
Dasyprocta leporina	2	3	2	2
Agouti paca	3	3	2	3
Hydrochaeris hydrochaeris	2	3	2	4
Order Carnivora				
Conepatus semistriatus	3	3	4	2
Eira barbara	3	3	4	3
Galictis vittata	3	3	3	2
Potos flavus	3	3	4	3
Procyon cancrivorus	2	3	4	3
Cerdocyon thous	2	3	3	3
Felis yagouaroundi	3	3	3	2
Felis pardalis	3	3	3	3
Felis tigrina	3	3	3	2
Felis wiedii	3	3	3	2
Felis concolor	2	4	3	4
Other Perissodactyla				
Tapirus terrestris	2	4	2	4
Older Artiodactyla				
Tayassu tajacu	2	4	2	4
Tayassu pecari	2	4	2	4
Odocoileus virginianus	2	4	1	4
Mazama americana	3	4	1	3

Table 27A.3 Mammals from the cerrado Brazil

Species	*Forest*	*Vagility*	*Plant*	*Size*
Order Marsupiala				
Chironectes minimus	3	3	3	2
Didelphis albiventris	2	3	4	2
Marmosa agilis	3	2	4	1
Marmosa murina	2	2	4	1
Marmosa cinerea	2	2	4	1
Marmosa incana	2	2	4	1
Marmosa velutina	2	2	4	1
Marmosa noctivaga	2	2	4	1
Marmosa constantiae	2	2	4	1
Monodelphis umbistriata	1	2	4	1
Monodelphis americana	3	2	4	1
Monodelphis domestica	1	2	4	1
Monodelphis kunsi	1	2	4	1
Caluromys philander	3	3	4	2
Caluromys lanatus	3	3	4	2
Lutreolina crassicaudata	2	3	3	2
Metachirus nudicaudatus	3	3	4	2
Philander opossum	2	3	4	2
Order Chiroptera				
Rhynchonycteris naso	3	5	3	1
Peropteryx macrotis	3	5	3	1
Pteronotus parnellii	2	5	3	1
Lonchorhina aurita	3	5	3	1
Micronycteris minuta	3	5	4	1
Phyllostomus hastatus	3	5	6	2
Phyllostomus discolor	3	5	6	1
Mimon crenulatum	3	5	3	1
Chrotopterus auritus	3	5	3	2
Anoura caudifer	2	5	6	1
Anoura geoffroyi	2	5	6	1
Glossophaga soricina	2	5	6	1
Lonchophylla dekeyseri	3	5	6	1
Carollia perspicillata	2	5	5	1
Rhinophylla pumilio	3	5	5	1
Artibeus cinereus	3	5	5	1
Artibeus lituratus	2	5	5	2
Artibeus planirostris	3	5	5	1
Chiroderma doriae	3	5	5	1
Sturnira lilium	2	5	5	1
Vampyrops lineatus	2	5	5	1
Chiroderma villosum	3	5	5	1
Desmodus rotundus	2	5	3	1
Furipterus horrens	3	5	3	1
Eptesicus brasiliensis	3	5	3	1
Myotis nigricans	2	5	3	1
Lasiurus borealis	2	5	3	1

Table 27A.3 *(continued)*

Species	*Forest*	*Vagility*	*Plant*	*Size*
Molossops planirostris	3	5	3	1
Molossops temmincki	2	5	3	1
Molossus molossus	2	5	3	1
Nyctinomops laticaudatus	1	5	3	1
Eumops glaucinus	2	5	3	1
Order Primates				
Aotus trivirgatus	3	3	4	2
Cebus apella	3	3	4	2
Alouatta caraya	3	3	2	3
Callithrix jacchus	2	3	4	2
Callithrix penicillata	2	3	4	2
Callithrix argentata	2	3	4	2
Order Edentata				
Cabassous unicinctus	1	1	3	2
Cabassous tatouay	1	1	3	2
Dasypus novemcinctus	2	3	2	2
Dasypus septemcinctus	2	3	2	2
Tolypeutes matacus	1	3	2	2
Tolypeutes tricinctus	1	3	2	2
Euphractus sexcinctus	2	3	2	2
Priodontes maximus	2	3	3	4
Myrmecophaga tridactyla	2	3	3	4
Tamandua tetradactyla	2	3	3	2
Bradypus variegatus	3	3	1	2
Order Lagomorpha				
Sylvilagus brasiliensis	3	3	1	2
Other Rodentia				
Akodon cursor	3	2	1	1
Akodon reinhardti	1	2	1	1
Bolomys lasiurus	1	2	1	1
Calomys callosus	2	2	1	1
Calomys tener	2	2	1	1
Holochilus brasiliensis	1	2	1	2
Neacomys spinosus	3	2	2	1
Blarionomys breviceps	3	1	3	1
Juscelinomys candango	1	1	1	1
Kunsia fronto	2	1	1	2
Kunsia tomentosus	2	1	1	2
Wiedomys pyrorrhinos	2	2	2	1
Nectomys squamipes	1	2	1	2
Oryzomys bicolor	3	2	0	1
Oryzomys capito	3	2	2	2
Oryzomys ratticeps	3	2	2	2
Oryzomys chacoensis	3	2	2	1
Oryzomys concolor	3	2	2	2
Oryzomys fornesi	2	2	2	1
Oryzomys nigripes	3	2	2	1

Table 27A.3 *(continued)*

Species	*Forest*	*Vagility*	*Plant*	*Size*
Oryzomys subflavus	1	2	2	2
Oryzomys lamia	2	2	2	1
Oryzomys utiaritensis	2	2	2	1
Oxymycterus roberti	1	2	3	2
Oxymycterus rutilans	2	2	3	2
Plectomys paludicola	2	2	2	1
Pseudoryzomys simplex	3	2	2	1
Rhipidomys mestacalis	3	2	0	2
Coendou prehensilis	3	3	2	2
Sphiggurus villosus	3	3	2	2
Cavia aperea	2	2	1	2
Galea flavidens	1	2	1	2
Galea spixii	1	2	1	2
Dasyprocta mexicana	2	3	2	2
Kerodon rupestris	1	2	1	2
Hydrochaeris hydrochaeris	2	3	2	4
Dasyprocta azarae	2	3	2	2
Dasyprocta punctata	2	3	2	2
Dasyprocta prymnolopha	2	3	2	2
Agouti paca	3	3	2	3
Carterodon sulcidens	1	2	2	2
Talpomys lasiotis	2	2	2	2
Clyomys laticeps	3	2	2	2
Echimys braziliensis	2	2	2	2
Proechimys longicaudatus	3	3	2	2
Trichomys apereoides	1	2	2	2
Ctenomys brasiliensis	2	1	1	2
Ctenomys minutus	2	1	1	2
Ctenomys natterei	2	1	1	2
Euryzygomatomys guiara	1	1	1	2
Order Carnivora				
Cerdocyon thous	2	3	3	3
Lycalopex vetulus	1	3	3	3
Chrysocyon brachyurus	2	3	4	4
Speothos venaticus	3	3	3	3
Nasua nasua	2	3	4	3
Procyon cancrivorus	2	3	4	3
Potos flavus	3	3	4	3
Eira barbara	3	3	4	3
Galictis vittata	3	3	3	2
Galictis cuja	2	3	3	2
Conepatus chinga	2	3	4	2
Conepatus semistriatus	3	3	4	2
Felis yagouaroundi	3	3	3	2
Felis colocolo	2	3	3	3
Felis wiedii	3	3	3	2
Felis tigrina	3	3	3	2

Table 27A.3 *(continued)*

Species	*Forest*	*Vagility*	*Plant*	*Size*
Felis pardalis	3	3	3	3
Felis concolor	2	4	3	4
Felis onca	2	4	3	4
Order Perissodactyla				
Tapirus terrestris	2	4	2	4
Order Artiodactyla				
Tayassu tajacu	2	4	2	4
Tayassu pecari	2	4	2	4
Blastocerus dichotomus	1	4	1	4
Ozotoceros bezoarticus	1	4	1	4
Mazama americana	3	4	1	3
Mazama gouazoubira	2	4	1	3

PART SIX

Modelling of the Boundary

28

Modelling energy flows and surface temperatures over forest and savanna

JOHN GRACE

SUMMARY

1. A 'big-leaf' energy balance model is used in conjunction with existing data to compare the fluxes of energy over forest and savanna.
2. The model requires climatological, structural and physiological data as inputs, and it calculates surface temperature, transpiration or evaporation rate, and the rate of exchange of sensible heat (convection) between the vegetation and the atmosphere.
3. In dry weather, tropical forest gives up both heat and water to the atmosphere at a somewhat faster rate than savanna. However, its characteristics are not very different from temperate forest.
4. When wet, the forest canopy is a very strong source of water vapour, and a strong sink for sensible heat. It is in these conditions that the forest can be said to humidify and cool the atmosphere.

INTRODUCTION

There is a general view that tropical forests influence, even control, local and regional climates; and that trees are great humidifiers and coolers of the atmosphere. However, to examine this assertion as a serious scientific hypothesis is not straightforward, as it is difficult to arrange a field experiment on a scale large enough, and with proper controls, to come to an unequivocal conclusion. Nevertheless, some progress towards testing this hypothesis can, and has, been made by modelling heat and mass transfer from stands of vegetation, using field data from a limited number of sites to provide parameters for such models. The interplay between data collection and model building is critical here, some models being very complex with far more parameters than have been obtained for any particular vegetation type. The modelling approach used in the present paper is necessarily parsimonious, as more elaborate models (e.g. Sellers, 1987) will not be useful until field measurements have provided the parameters to run them.

Here, an attempt will be made to assemble data relevant to the tropical forest and savanna, and to use these data as inputs for a simple model of heat and mass transfer to explore the differences between the two vegetation types, and especially to test the hypothesis that tropical forests humidify and cool the atmosphere.

CONCEPTUAL FRAMEWORK

The vegetation is envisaged as a 'big leaf', characterized by its reflectance and surface resistances to heat and mass transfer (Fig. 28.1). These resistances are of two types: an essentially physiological resistance r_s which depends on the extent of opening of stomatal pores, and an aerodynamic resistance r_a which depends on the wind speed and struc-

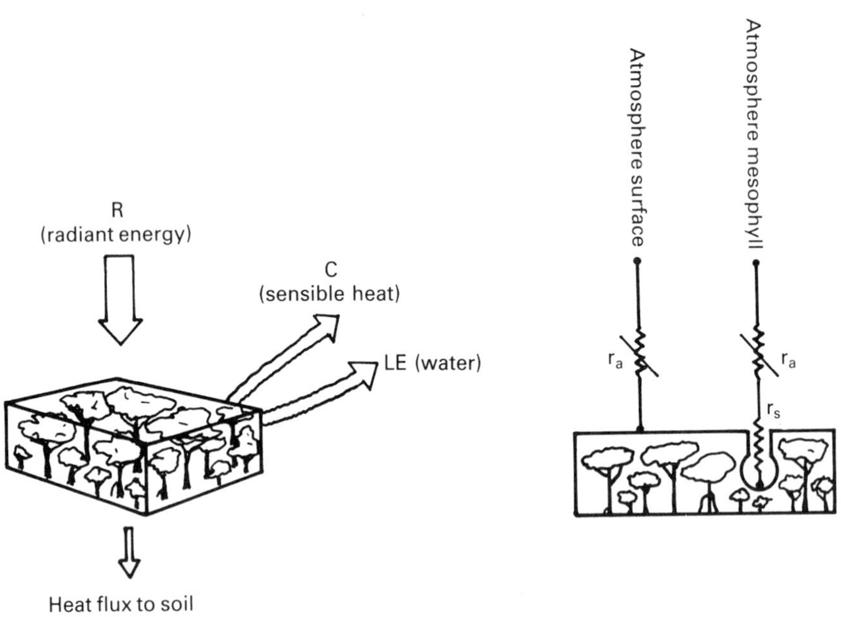

Fig. 28.1 The partitioning of radiation R between convection C and evaporation E in the 'big leaf' model, and the controlling influence of the aerodynamic resistance r_a and the physiological resistance of the canopy r_s.

ture of the vegetation. Depending on the magnitude of these resistances, net radiation R is partitioned between evaporation E and sensible heat C. The evaluation of E and C is achieved by an iterative process in which the surface temperature is adjusted until $C + E = R$. The model in its original form is described by Grace (1983) but has since been enhanced by the inclusion of a free convection term, and tested against field data collected at a forest boundary in Scotland (Grace, 1989a,b; Grace *et al.*, 1989). The inputs to the model are climatological (radiation, temperature, wind speed and humidity), structural (vegetational height, and hence roughness length) and physiological (the canopy stomatal resistance). The outputs are: surface temperature, evaporation rate, and the rate of loss (or gain) of sensible heat. Variations of this big leaf model are widely used and very flexible. Where the vegetation is patchy, as it often is at the forest–savanna boundary, two transfer resistances in parallel may be used (Shuttleworth and Wallace, 1985).

DATA SOURCES

Relevant physiological and structural properties of tropical rainforest have been determined at sites in Thailand and more recently in the Amazon (Pinker *et al.*, 1980a,b; Shuttleworth *et al.*, 1984a,b). The latter are perhaps the best available data, but are strictly relevant to one site, and all the data were collected on dry days. In this study the minimal canopy resistance was $80\,\mathrm{s\,m^{-1}}$, occurring in the late morning. A surprising feature of the data is their similarity to those of other forest vegetation (Shuttleworth, 1989). It seems, for example, that the diurnal course of canopy resistance in the Amazonian forest is not very different from that of a spruce forest in Scotland. Not all tropical forests may be like this: one study of a teak plantation suggested a minimal canopy resistance of as little as $16\,\mathrm{s\,m^{-1}}$ (Grace *et al.*, 1982).

It is much more difficult to find data relevant to savanna, especially as the term savanna represents a wide range of physiognomic type, from grassland to dense woodland; and perhaps because it is inherently more patchy and seasonally variable. Relevant studies are currently in progress in the Sahel, but for the present purpose it has been necessary to use data of prairie (Ripley and Redmann, 1976). In this case, the canopy resistance was in the range $300–500\,\mathrm{s\,m^{-1}}$ for much of the year. This may not be far wrong for grassy savanna, but may underestimate the evaporation rate for savanna woodland, particularly where the trees are deep-rooted. A second difficulty in modelling the savanna arises on formulating the relationship between aerodynamic resistance and the structural composition of savanna vegetation. Data compiled by Garrett (1977) showed that for many practical purposes r_a could be estimated from vegetational height alone, and this is what the model does. However, when the vegetation is particularly sparse, an additional parameter, incorporating the spacing of the plants and their silhouette area is required.

One of the most notable features of tropical forest is its great height. Tall vegetation generally absorbs more radiation than short vegetation because of the so-called velvet pile effect. A ray entering a gap in a tall forest with its myriad leaves all capable of reflection, is much less likely to emerge than one entering a gap in, say, a field of potatoes. This view is supported by models of radiation penetration (Unkasevic, 1989). Of all vegetation types so far examined, tropical rainforest and coniferous forest have the lowest reflectance of short-wave (solar) radiation (Ojo, 1977; Pinker, 1982; Shuttleworth *et al.*, 1984a). At high solar angles both reflect 12%, much less than the corresponding figure for savanna: Ojo (1977) gives 20–22% for Guinea savanna and 19–20% for Sudan savanna. Similar data come from results of an earlier study conducted in Nigeria and reproduced in Table 28.1 (Oguntoyinbo, 1974). For the present purposes of estimating the energy absorption of tropical vegetation, it is important to realize that this variation in reflectance significantly affects the relationship between solar and net radiation (Fig. 28.2). In full sunlight (about $800\,W\,m^{-2}$), tropical forest might be expected to absorb as much as $100\,W\,m^{-2}$ more energy than grassy savanna. The data for a

Table 28.1 Mean reflection coefficients over different vegetation covers in Nigeria (Oguntoyinbo 1974)

Description of vegetation cover	*Reflection coefficient*
Mangrove and freshwater swamps	0.12
Tropical rainforest	0.13
Derived savanna:	
forest	0.12
mixed deciduous and grass	0.14
grass (green)	0.17
grass (dry)	0.19
Savanna:	
pure, green	0.17
pure, dry	0.22
mixed, green	0.17
mixed, dry	0.20
Bahama grass:	
closely mown lawn grass (green)	0.21
Water surface (river Niger):	
clear	0.06
dirty	0.12

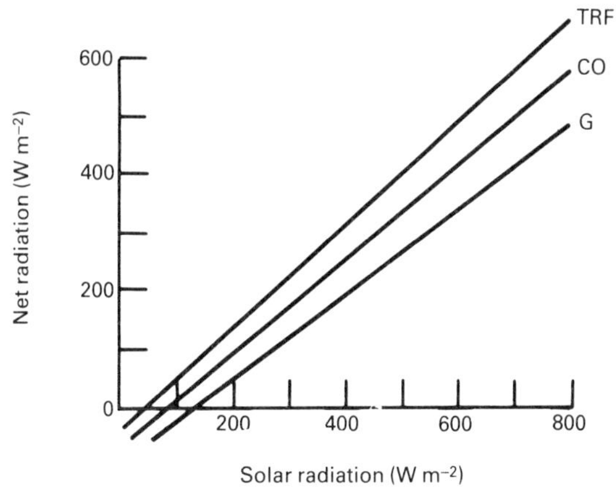

Fig. 28.2 Empirical relationships between solar radiation incident on the vegetation, and the net radiation absorbed which is available for evaporation and convection. Each line represents a fitted regression: TRF, tropical rain forest; CO, citrus orchard; G, prairie grassland. Data sources: Shuttleworth *et al.* (1984b), Kalma and Fuchs (1976), Ripley and Redmann (1976).

citrus orchard are included in Fig. 28.2 as this system is structurally similar to savanna woodland.

Climatological regimes for the forest and savanna regions of West Africa were estimated by using data for two stations in Nigeria, Lagos and Sokoto. Data for the region of West Africa are available in Davies (1966), Ojo (1977) and Muller (1982). Lagos (latitude 6°N) is in the evergreen tropical forest and mangrove zone, with a considerable maritime influence. Sokoto (latitude 13°N, elevation 345 m) is in the savanna region, classified as Sudan savanna by White (1983). The transition from forest to savanna in this part of West Africa has been extensively reported (Hopkins, 1974; Adejuwon and Adesina, 1988), and is usually related to rainfall, fire and, to a lesser degree, human influences (see Chapter 16).

Comparison of the climatological data for the two stations shows the strongly seasonal nature of the savanna region (Fig. 28.3). At Sokoto the annual course of temperature shows a characteristic double peak, the trough in August coinciding with the rainfall maximum. Six months of the year are without rain, and in these months the relative humidity is at or below 30%. In contrast, the rainforest region shows a rather small seasonality in humidity and temperature, and August is a notably dry month, a 'little dry season' in the middle of the wet season.

Relative humidity is not a good predictor of evaporation when comparing sites at different temperatures. Thus, the saturation deficit of the atmosphere, expressed as a vapour pressure in kilopascals (SVPD)

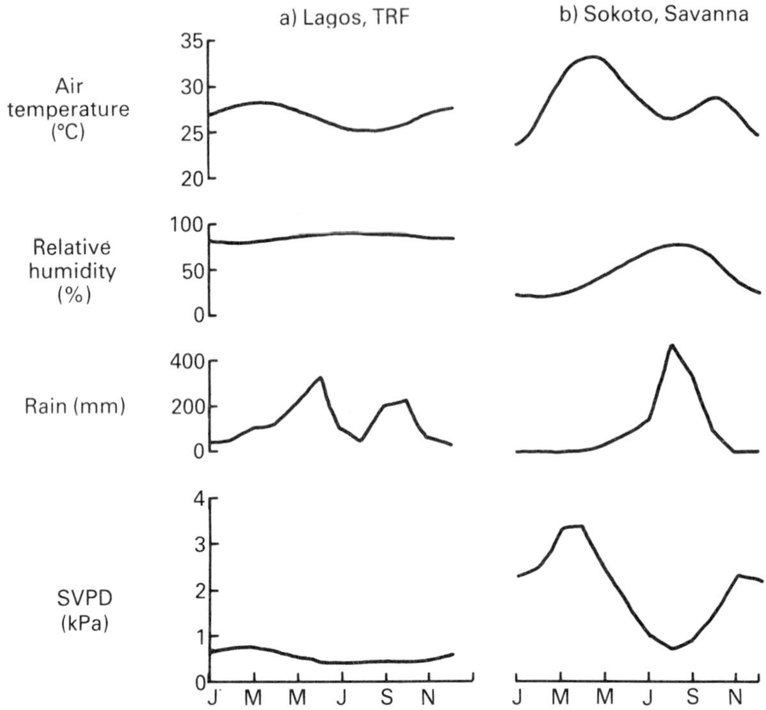

Fig. 28.3 Climatological data for Lagos (rainforest zone) and Sokoto (savanna zone), plotted from tables in Muller (1982).

has been appended to the published climatological data. This shows that in the dry season, the driving gradient for evapotranspiration is 4–5 times greater in the savanna zone than in the rainforest zone.

RESPONSE TO MODEL PARAMETERS

In the standard run, the height of the rainforest was set to 40 m, the water vapour pressure was 3.0 kPa, the wind speed was set at $3\,m\,s^{-1}$, the canopy resistance was $80\,s\,m^{-1}$ and the air temperature was 27 °C. For savanna, height was 2 m, vapour pressure was set to represent the onset of dry season at 0.7 kPa, and the stomatal resistance was $500\,s\,m^{-1}$. The response of the two systems to net absorbed radiation is practically linear (Fig. 28.4). In full sunlight, it is expected that the net radiation absorbed by forest and savanna would be 500 and $400\,W\,m^{-2}$ respectively and that these figures would be associated with transpiration rates of 155 and

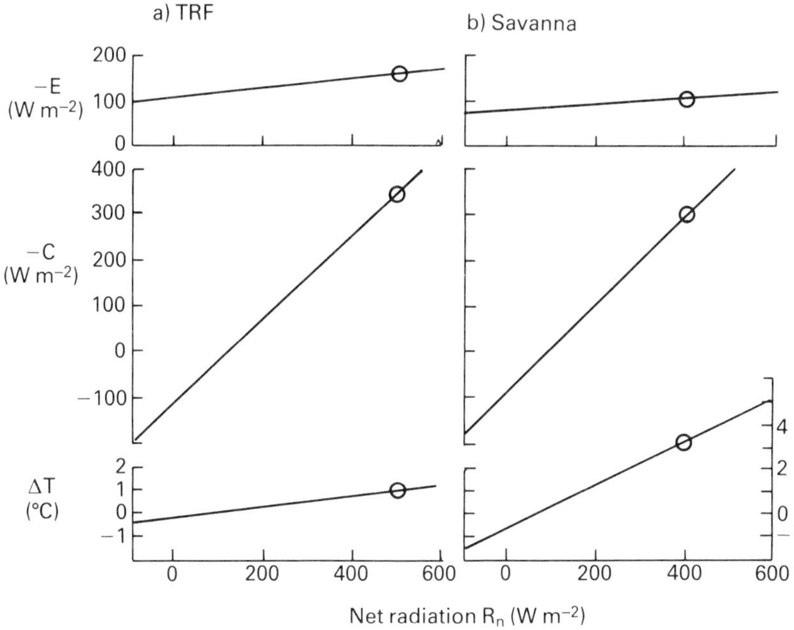

Fig. 28.4 Calculated response of rainforest and savanna to net radiation for the standard run (see text). The model outputs are the transpiration rate (*E*), the convection rate (*C*) and the surface temperature relative to the air temperature *AT*. The circles denote the regions of the graph corresponding to 'bright sunshine'. Note that an evaporation rate of $1\,mm\,h^{-1}$ is approximately equal to $244\,W\,m^{-2}$ at 25 °C.

$100\,W\,m^{-2}$ respectively. At this point the forest would be giving up sensible heat to the atmosphere at $350\,W\,m^{-2}$ with a corresponding figure of $300\,W\,m^{-2}$ by the savanna. Thus, the rainforest would be a source of both water and heat, and a stronger source than the savanna. The forest would not cool the air at all. The surface temperature is shown as the difference between the surface and the air. The model calculation shows that the savanna is capable of being considerably warmer than the air, whilst the rainforest is rarely more than one degree different from the air above it.

It is instructive to rerun the model for the rainforest after setting the stomatal conductance to zero to simulate a wet canopy. In this case the only resistance to the diffusion of water vapour is the aerodynamic one, and this is very small for all forests. Very high rates of evaporation occur, exceeding $1\,kW\,m^{-2}$ when the sun shines, and the vegetation becomes a sink for heat, i.e. heat flows from the atmosphere to the forest (Fig. 28.5).

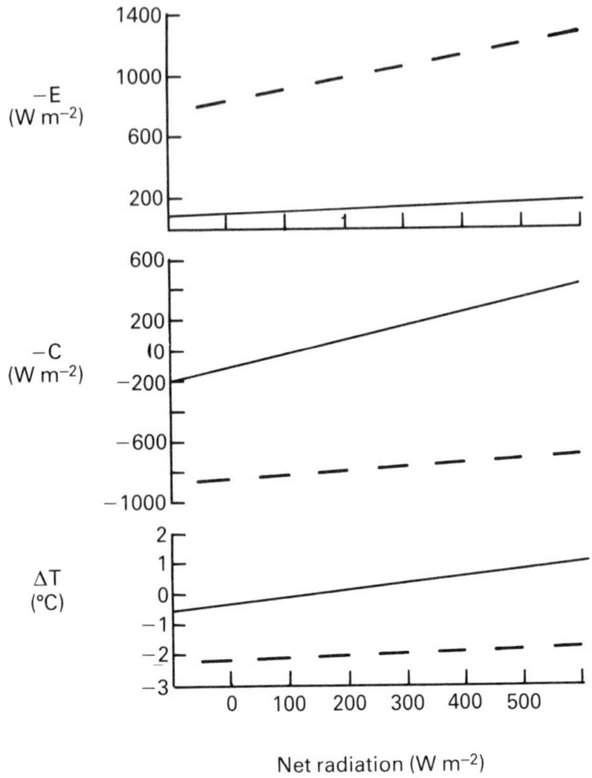

Fig. 28.5 The wet forest. The solid lines are as for the rain forest in Fig. 28.4, the broken lines are what happens when the stomatal resistance is set to zero to simulate unrestrained evaporation of water from the canopy.

What really happens is that for most of the time wet weather is cloudy, and the saturation deficit is smaller than usual, approaching zero; until of course the rain clouds disperse. Thus, the actual evaporation over the day would depend strongly on the intermittency of rain.

In nature, the stomatal resistance of the canopy varies diurnally (Shuttleworth *et al.*, 1984b). The result of varying the canopy resistance is shown in Fig. 28.6. The effective range of r_s, judged from the available literature, is shown by the solid lines. It is noteworthy that the rainforest transpiration is very sensitive to r_s, the solid line embracing all values between a teak plantation and the Amazonian forest. Note that these calculations are again appropriate to bright sunshine. There is much uncertainty about the true value of r_s, and for much of the time the rainforest is likely to have partially wet leaves, thus giving a lower resistance than the $85\,s\,m^{-1}$ that comes from the study by Shuttleworth *et al.*

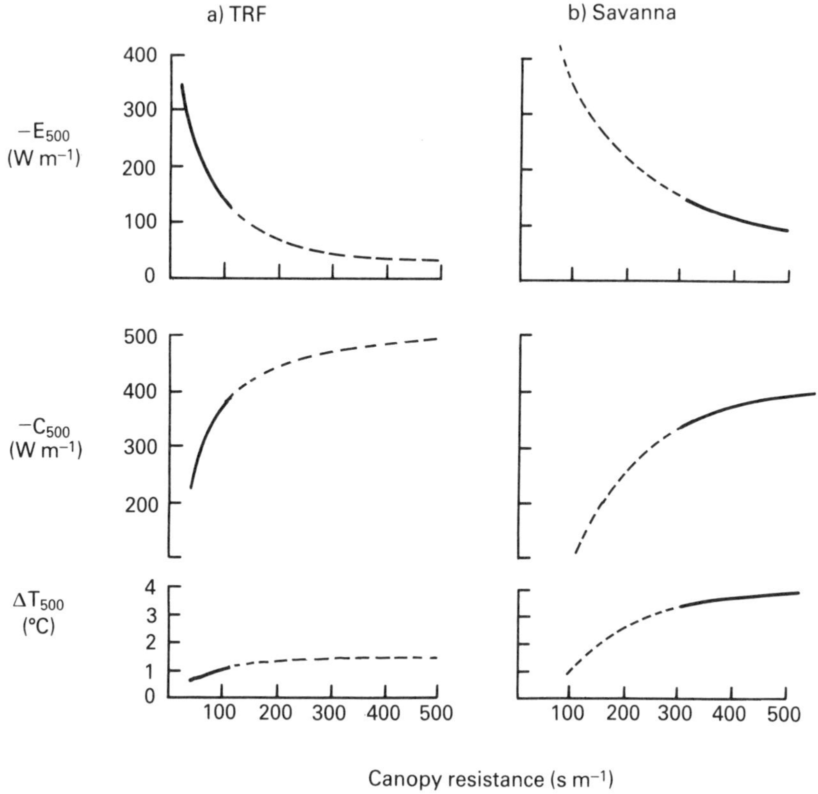

Fig. 28.6 Stomatal control. The effect of varying the stomatal resistance on the rates of transpiration and convection (*E* and *C*) using the parameters of the standard run and taking the output appropriate for a net absorbed radiation of $500\,W\,m^{-2}$.

(1984a). However, within this wide range of values, the forest is always giving up heat to the atmosphere. The surface temperatures of the rainforest are close to air temperatures over the effective range of canopy resistance.

The seasonal variation in transpiration, using the model in conjunction with climatological data for Lagos and Sokoto, shows a remarkable similarity between forest and savanna (Fig. 28.7). Although the driving gradient for evapotranspiration (SVPD) is much greater at Sokoto, the higher stomatal resistance for the savanna vegetation results in a transpiration rate at Sokoto which is about the same as the rainforest. The broken line shows the transpiration obtained by the model if the stomatal resistance is set to $50\,s\,m^{-1}$. This value might conceivably be achieved

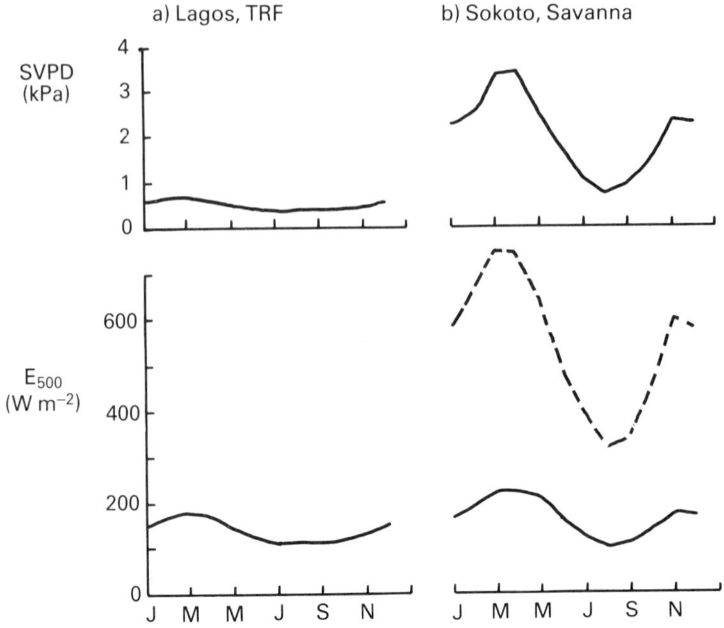

Fig. 28.7 Seasonal fluctuations in transpiration rates at $500\,W\,m^{-2}$ in the rainforest zone (Lagos) and the savanna zone (Sokoto). For the calculated transpiration, the solid lines are obtained by setting r_s to 80 and $300\,s\,m^{-1}$ respectively. The broken line is obtained by reducing the savanna's stomatal conductance to $50\,s\,m^{-1}$.

following the rains, when foliage is newly formed and green. The contrast is evident: because the SVPDs are so large, there is an enormous potential for transpiration to occur. Note also that the seasonal trends in transpiration closely follow those in SVPD.

The influence of vegetational height on transpiration and surface temperature is important only in the range of 0.4 m (Fig. 28.8). This is because the relationship between r_a and height is logarithmic. In the model the relationship used is the familiar one applying strictly to the transfer of momentum in neutral meteorological conditions (Monteith, 1973), although in this model the relationship is appreciably modified by the Grashof number at low windspeed to accommodate natural as opposed to forced convection (Grace, 1989b). Very short vegetation always shows a high degree of uncoupling from the atmosphere, and it may be that savanna grassland behaves in that way in the early stages of regrowth after fire.

The effect of wind speed is modelled using the same relationship as for height. It is at low wind speed that buoyancy becomes potentially

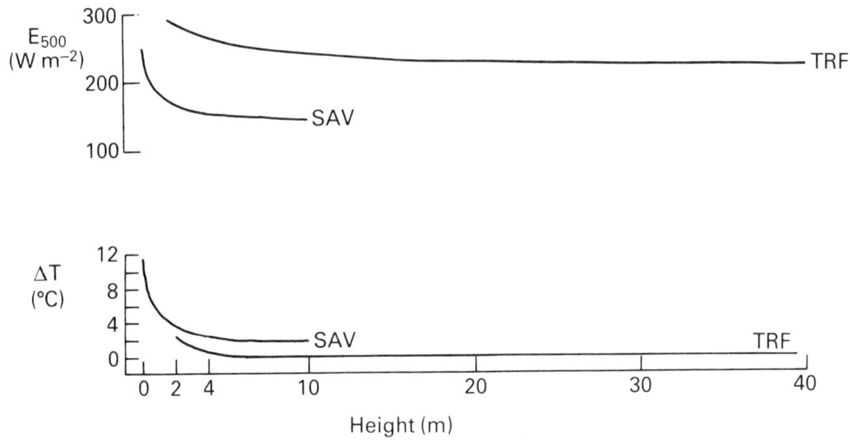

Fig. 28.8 Influence of varying vegetational height on transpiration at ΔT. Lines refer to tropical rainforest (TRF) and savanna (SAV).

Table 28.2 Parameter values used in the simulation of changes along a transect. Lagos and Sokoto correspond to stations 1 and 6 respectively

Distance (arbitrary units)	1	2	3	4	5	6	7	8
Height (m)	40	35	30	10	4	2	0.5	0.5
Water vapour pressure (kPa)	30	28	25	15	10	7	7	5
Wind speed ($m\,s^{-1}$)	1	1	1	1	1	1	1	1
Stomatal resistance ($s\,m^{-1}$)	80	100	250	300	400	500	500	5000
Air temperature (°C)	27	27	26	25	24	24	24	24

important. When the model was tested against field data on the temperatures at a tree-line ecotone in Scotland, the largest discrepancies between observed and calculated were at the lowest wind speeds. Low wind speed, like short stature, has the effect of decoupling (Fig. 28.9). For savanna, quite extreme surface temperatures and high rates of transpiration are calculated at wind speeds in the region of $0.1\,m\,s^{-1}$.

MODELLING THE FOREST–SAVANNA BOUNDARY

Along a transect from the forest to the savanna all the variables mentioned above are liable to change independently of each other. In the absence of real data, several simulations were made using artificial data sets. Here, the result of one such run of the model is reported. The imaginary transect is very long and the artificial data set (Table 28.2)

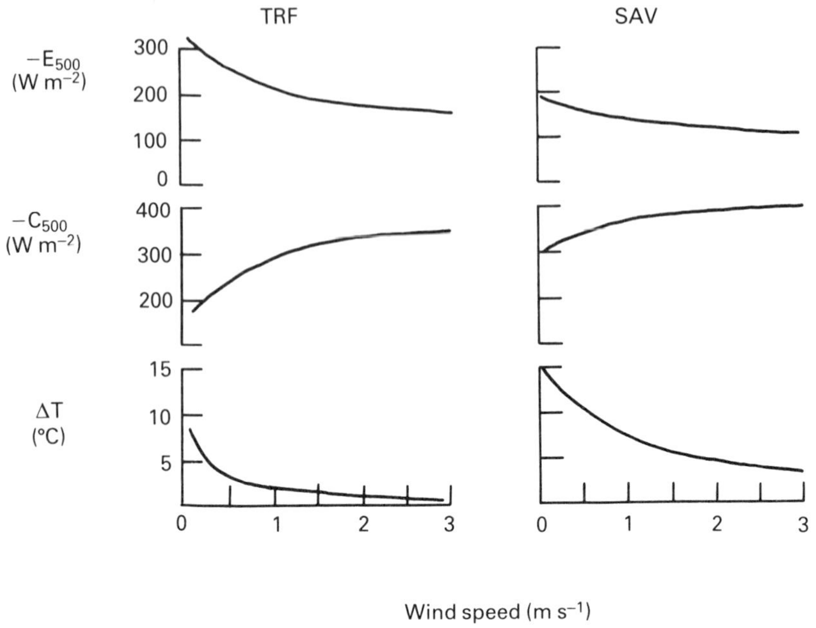

Fig. 28.9 Influence of wind speed on the calculated rates of transpiration, convection and ΔT.

includes desert. The scale of distance is not necessarily linear. There are several caveats to be made. First, when the cover of the vegetation becomes sparse, there will be a significant flux of heat to the soil, not accounted for in the current version of the model. To model this would require a storage term, enabling the soil to absorb heat by day and give it up at night. Second, the reflection of short wave by desert regions is likely to be up to 25%, and so the absorbed radiation will be correspondingly less. Third, the stomatal conductance of the desert region is assumed to be $5000\,s\,m^{-1}$, a figure which has been arrived at by guesswork.

The wind speed for this simulation has been decreased from the standard run to $1.0\,m\,s^{-1}$. The reason for this is that the purpose of the run is to examine the trends in the outputs to see if they reach any extreme values, liable to cause stress to the vegetation. Extreme values are more likely to occur at low wind speeds, when the vegetation is less closely coupled to the atmosphere.

The simulation shows a gradient of ΔT, with desert values exceeding 30 °C (Fig. 28.10). With an air temperature of 24.5 °C, the surface temperature of the desert is not unreasonably high, giving some confidence

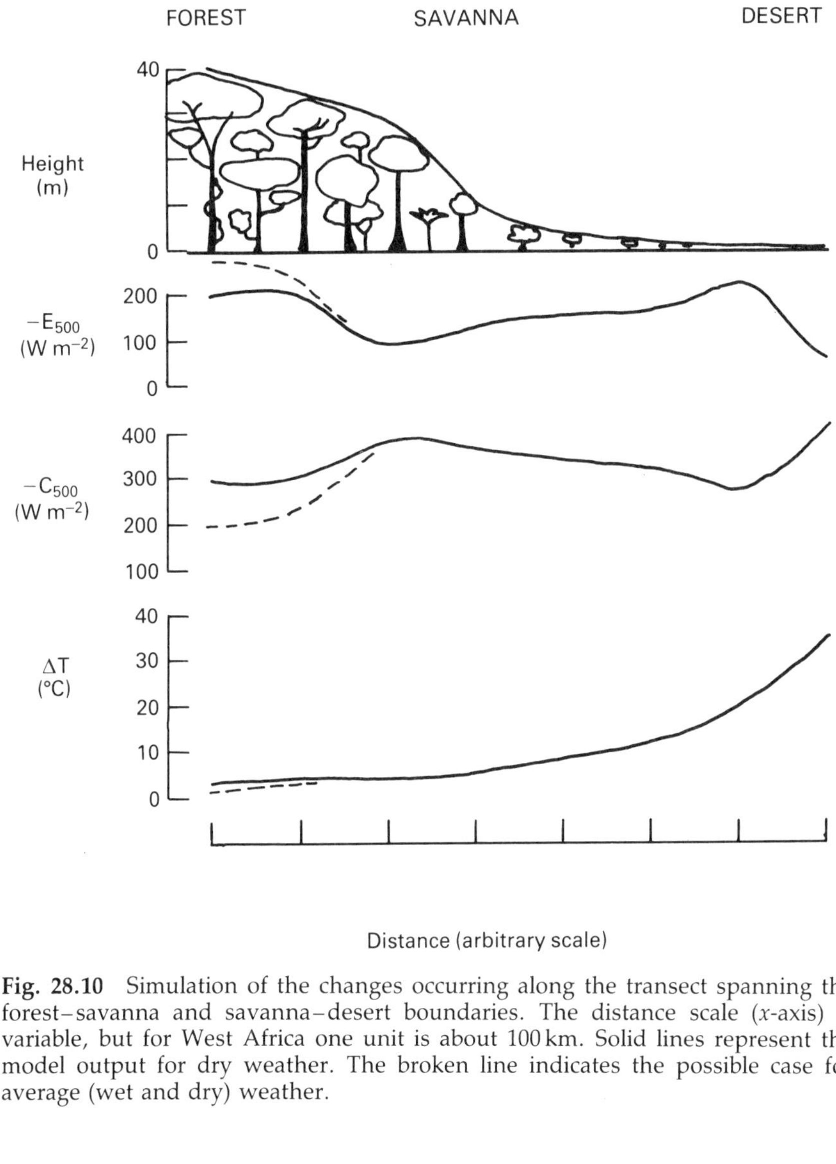

Fig. 28.10 Simulation of the changes occurring along the transect spanning the forest–savanna and savanna–desert boundaries. The distance scale (*x*-axis) is variable, but for West Africa one unit is about 100 km. Solid lines represent the model output for dry weather. The broken line indicates the possible case for average (wet and dry) weather.

in the guessed value of stomatal conductance. The trends in *E* and *C* with distance show intriguing non-linearity. At the forest–savanna boundary there is a singularity where the partitioning tends strongly towards sensible heat and transpiration rates are low. At this point, the vegetation gives up heat to the atmosphere at a rate which nearly equals that of the desert.

DISCUSSION

The model is a useful tool in understanding the first order interplay between structural, physiological and climatic conditions. Its main limitation is that it does not model the feedbacks that are likely to occur in nature, and which will affect the parameters over time. Transpiration and evaporation from the real canopy humidifies the atmosphere, thus increasing the humidity of the air mass within the planetary boundary layer (McNaughton, 1976; McNaughton and Jarvis, 1983), and consequently causing a decline in transpiration. There may be consequent physiological adjustment: as the humidity increases, the stomatal resistance is likely to decline (Schulze and Hall, 1982; Roberts *et al.*, 1990). If one or both types of feedback occur, the vegetation will tend to behave conservatively, responding less to climatological variables than would be expected from the model. Already, models of the energy balance of the vegetation are being linked to models of boundary layer development, at a regional scale (de Bruin and Jacobs, 1989). Soon, it will be possible to estimate the crucial feedbacks from atmosphere to vegetation.

The original hypothesis, that the trees of rainforests are great humidifiers and coolers of the atmosphere, can now be examined in the light of the results. It seems from the literature that the canopy in its dry condition behaves in a similar way to coniferous forest: it has similar values for aerodynamic and stomatal conductance, and the two canopies have the same short-wave reflectance. There has perhaps been a misunderstanding in the past about water relations of tropical rainforest trees. Because it is a rainforest, it has been assumed that water stress does not occur. Yet at the top of the canopy, the saturation deficits of the air and the negative water potentials of foliage can at times be appreciable (Grace *et al.*, 1982; Chiarello, 1984; Robichaux *et al.*, 1984; Shuttleworth *et al.*, 1985). Thus, the leaves sometimes display stomatal resistances just as high as those recorded for other forests (Shuttleworth, 1989), and the leaf area per land area may not be very different from other forests. Consequently, the transpiration from the forest is rather low in relation to that reported for other tropical systems such as the maize and rice crop (Uchijima, 1976), and the convection rate is high. The forest in its dry state does not cool the atmosphere, but rather it warms it. It does humidify the atmosphere, but not especially so. However, in the wet condition the situation is transformed. The great aerodynamic roughness causes very fast evaporation of water. In the wet season, when rains may occur every day, the water balance of the vegetation is difficult to predict because it will depend on the intermittency of rain, and on the storage capacity of the canopy. This storage capacity may be expected to be sensitive to the structure of the canopy, and so be affected by disturbances to the canopy which open up gaps. At least for some of the time, and on some days for all of the time, the trees will indeed behave as coolers and humidifiers; though this is the case for all forests. It is unfortunate that current technology permits field measurements to be made only during dry hours, whereas the most interesting events may occur in showers and storms.

The transect simulation suggests that there may be an unexpected zone of low transpiration and high convection at the forest–savanna boundary (Fig. 28.10). If so, could this be the cause of the sharp transition that sometimes occurs? The influence of this might be felt when the wind blows from right to left in Fig. 28.10. The air passing over the transition zone would become particularly warm and less humid than usual. Some weather conditions, for example those associated with the harmattan winds, are known to stress the trees of the rainforest (Longman and Jenik, 1987), but these are associated with north-easterly air masses which have passed over the desert regions for hundreds or thousands of kilometres. It seems unlikely that the relatively narrow zone would contribute more than a small part of this effect.

Sharp forest savanna boundaries do occur, though they are often associated with gallery forest or are a consequence of burning (Hopkins, 1974). The microclimate can be expected to adjust fairly rapidly as air flows from the savanna to the forest. There are several theoretical or observational studies in which the adjustment of the wind and temperature profile on passage from grass to forest has been examined (Raynor, 1971; McNaughton, 1976; Kapos, 1989). Most of the 'edge effect' occurs over a very short horizontal distance, $10h$, where h is the height of the trees. At the edge, one would expect considerable stress, both from the direct force of the wind during tropical storms and also a drying influence if the air were much less humid over the savanna region. Similar stresses are invoked to explain a sharp-edge phenomenon in boreal forests: the wave-regeneration forest (Robertson, 1987).

The task of modelling the impact of vegetation on the regional climate has attracted much attention recently, with the development of general circulation models (GCMs) in which very large models of atmospheric circulation are 'carpeted' with vegetation submodels similar to the one used here. The paucity of data for savanna and degraded forest is currently limiting progress: roughness length and albedo can usually be estimated satisfactorily but the stomatal resistance – its diurnal and seasonal characteristics and its response to atmospheric and soil water – is much more elusive. A sceptical review of current progress is provided by Henderson-Sellers (1987). In her model of the Amazon, an increase in short-wave reflectance and a decline in roughness, to simulate deforestation, does lead to substantial regional changes in rainfall, cloud cover and evapotranspiration, but no change in regional temperatures.

Finally, it should be realized that models must be verified, and the task of doing this can be both difficult and expensive. Remote sensing can provide information in the thermal and microwave wavebands which may ultimately be useful not only in verifying models but also in the direct determinations of canopy resistances.

REFERENCES

Adejuwon, J.O. and Adesina, F.A. (1988) Vegetation patterns along the forest–savanna boundary in Nigeria. *Singapore Journal of Tropical Geography*, **9**, 18–32.

Chiariello, N. (1984) Leaf energy balance in the wet lowland tropics, in *Physiological Ecology of Plants in the Wet Tropics*, (eds E. Medina, H.A. Mooney and C. Vazquez-Yanes) Junk, The Hague.

Davies, A.J. (1966) An assessment of evapotranspiration for Nigeria. *Geografiska Annalter*, **48**, 139–56.

de Bruin, H.A.R. and Jacobs, C.M.J. (1989) Forests and regional scale processes. *Philosophical Transactions of the Royal Society, London*, **B324**, 393–406.

Garrett, J.R. (1977) *Aerodynamic Roughness and Mean Monthly Surface Stress over Australia*, CSIRO, Division of Atmospheric Physics, Technical Paper, No. 29, pp. 1–19.

Grace, J. (1983) *Plant–Atmosphere Relationships*, Chapman & Hall, London.

Grace, J. (1989a) Tree lines. *Philosophical Transactions of the Royal Society*, **B324**, 233–45.

Grace, J. (1989b) Heat transfer and meristem temperature in vegetation near the treeline, in *Heat and Mass Transfer '89. The Fourth Australian Conference on Heat and Mass Transfer*, (ed. R.B. Keay), University of Christchurch, New Zealand, pp. 193–8.

Grace, J., Allen, S. and Wilson, C. (1989) Climate and the meristem temperatures of plant communities near the tree-line. *Oecologia*, **79**, 198–204.

Grace, J., Okali, D.U.U. and Fasehun, F.E. (1982) Stomatal conductance of two tropical trees during the wet season in Nigeria. *Journal of Applied Ecology*, **19**, 659–70.

Henderson-Sellers, A. (1987) Effects of change in land use on climate in the humid tropics, in *The Geophysiology of Amazonia*, (ed. R.E. Dickinson), John Wiley, New York, pp. 463–93.

Hopkins, B. (1974) *Forest and Savanna: An Introduction to Terrestrial Ecology*, Heinemann, London.
Kalma, J.D. and Fuchs, M. (1976) Citrus orchards. In *Vegetation and the Atmosphere*, Vol. 2, (ed. J.L., Monteith), Academic Press, London.
Kapos, V. (1989) Effects of isolation on water status of forest patches in the Brazilian Amazonia, *Journal of Tropical Ecology*, **5**, 173–85.
Longman, K.A. and Jenik, J. (1987) *Tropical Forest and its Environment*, Longman, Harlow.
McNaughton, K.G. (1976) Evaporation and advection II. Evaporation separating regions having different surface resistances and available energies. *Quarterly Journal of the Royal Meteorological Society*, **102**, 193–202.
McNaughton, K.G. and Jarvis, P.G. (1983) Predicting the effects of vegetation changes on transpiration and evaporation, in *Water Deficits and Plant Growth Volume 7*, (ed. T.T. Kozlowski), Academic Press, London, pp. 1–47.
Monteith, J.L. (1973) *Principles of Environmental Physics*, Arnold, London.
Muller, M.J. (1982) *Selected Climatic Data*, Junk, The Hague.
Oguntoyinbo, J.S. (1974) Land use and reflection coefficient (albedo) map for southern parts of Nigeria. *Agricultural Meteorology*, **13**, 227–37.
Ojo, O. (1977) *The Climates of West Africa*, Heinemann, London.
Pinker, R.T. (1980) The energy balance of a tropical evergreen forest. *Journal of Applied Meteorology*, **19**, 1341–9.
Pinker, R.T. (1982) The diurnal asymmetry in the albedo of tropical forest vegetation. *Forest Science*, **28**, 297–304.
Pinker, R.T., Thompson, O.E. and Eck, T.F. (1980a) The albedo of a tropical evergreen forest. *Quarterly Journal of the Royal Meteorological Society*, **106**, 551–8.
Pinker, R.T., Thompson, O.E. and Eck, T.F. (1980b) The energy balance of a tropical evergreen forest. *Journal of Applied Meteorology*, **19**, 1341–50.
Raynor, G.S. (1971) Wind and temperature structure in a coniferous forest and a contiguous field. *Forest Science*, **3**, 351–63.
Ripley, E.A. and Redmann, R.E. (1976) Grassland, in *Vegetation and the Atmosphere*, Vol. 2, (ed. J.L. Monteith), Academic Press, London, 349–98.
Roberts, J., Cabral, O.M.R. and de Aguiar, L.F. (1990) Stomatal and boundary-layer conductance in an Amazonian terra firme rain forest. *Journal of Applied Ecology*, **27**, 336–53.
Robertson, A. (1987) The centroid of tree crowns as an indicator of abiotic processes in a balsam fir wave forest. *Canadian Journal of Forestry Research*, **17**, 746–55.
Robichaux, R.E., Rundel, P.W., Stemmerman, L. *et al.* (1984) Tissue water

deficits and plant growth in wet tropical environments, in *Physiological Ecology of Plants in the Wet Tropics*, (eds E.Medina, H.A. Mooney and C. Vazquez-Yanes), Junk, The Hague, pp. 99–112.

Schulze, E.-D. and Hall, A.E. (1982) Stomatal responses, water loss and CO_2 assimilation rates of plants in contrasting environments, in *Encyclopaedia of Plant Physiology*, Vol. 12a, (eds O.L. Lange, P.S. Nobel, C.D. Osmond and H. Ziegler), Springer-Verlag, Berlin, pp. 181–230.

Sellers, P.J. (1987) Modelling effects of vegetation on climate, in *The Geophysiology of Amazonia*, (ed. R.E. Dickinson), John Wiley, New York.

Shuttleworth, W.J. (1989) Micrometeorology of temperate and tropical forest. *Philosophical Transactions of the Royal Society*, **B324**, 299–334.

Shuttleworth, W.J. and Wallace, J.S. (1985) Evaporation from sparse crops – an energy combination theory. *Quarterly Journal of the Royal Meteorological Society*, **111**, 839–55.

Shuttleworth, W.J., Gash, J.H.C., Lloyd, C.R. *et al.* (1984a) Observations of radiation exchange above and below Amazonian forest. *Quarterly Journal of the Royal Meteorological Society*, **110**, 1163–9.

Shuttleworth, W.J., Gash, J.H.C., Lloyd, C.R. *et al.* (1984b) Eddy correlation measurements of energy partition for Amazonian forest. *Quarterly Journal of the Royal Meteorological Society*, **110**, 1143–62.

Shuttleworth, W.J., Gash, J.H.C., Lloyd, C.R. *et al.* (1985) Daily variations of temperature and humidity within and above Amazonian forest. *Weather*, **40**, 102–8.

Uchijima, Z. (1976) Maize and rice, in *Vegetation and the Atmosphere*, Vol. 2, (ed. J.L. Monteith), Academic Press, London, pp. 33–64.

Unkasevic, M. (1989) Some improvements in calculating the plant stand surface albedo and its influence on ground surface temperature. *International Journal of Biometeorology*, **33**, 184–95.

White, F. (1983) *The Vegetation of Africa*, Unesco, Rome.

29

Modelling the location of woody–grassland boundaries

W.N. ELLERY, M.T. MENTIS and R.J. SCHOLES

SUMMARY

1. A novel modelling approach has been used to synthesize, test and improve current understanding at the subcontinental scale of the location of boundaries between the grassland biome and those with woody elements (forest, *fynbos*, *nama-karoo* and savanna) in South Africa, and also at the landscape scale of the determinants of the distribution of wooded patches of different affinity within the grassland biome itself. At the landscape scale the determinants of the distribution of forest patches in the grassland biome are emphasized.
2. The development of predictive and therefore testable models has been useful in identifying weaknesses in current wisdom, has illustrated the importance of climate in determining the distribution of vegetation at a subcontinental (biome-wide) scale, and of edaphic conditions and disturbance regimes at a landscape scale.
3. The hierarchical nature of the modelling approach used makes it useful in modelling the determinants of vegetation distribution at different scales of resolution, a prerequisite for predicting the consequences of global changes in climate and disturbance regimes for vegetation change.

INTRODUCTION

The aim of this chapter is to outline an approach being used to gain an understanding of the functional response of vegetation to the environment in South Africa. The use of modelling to summarize understanding that until now has been diffuse and non-integrated, and to present it in a predictive and testable way, is outlined. This is considered a strength of the present exercise, which was initially aimed at describing the location of boundaries between grassland (less than 5% aerial cover by woody elements) and forest, *fynbos* (sclerophyllous vegetation), *nama-karoo* (an arid grassland with shrubs) and savanna (grassland with greater than 5% cover by trees) biomes at a subcontinental scale, and at a landscape scale, of forest patches within the grassland biome. The modelling approach specifically addresses the problem of scale, which is inherent in modelling vegetation distribution at different scales of resolution. The nature of knowledge and the problems of scale addressed in the present exercise appear pertinent to the modelling of the forest–savanna boundary itself, particularly in developing a biologically meaningful understanding of factors that determine its distribution and behaviour.

The functional classification of vegetation attempts to reduce floristically diverse vegetation types to a smaller number of types, each exhibiting characteristic structure and process. The approach follows the general purpose of classification, which is to create classes that summarize understanding, communicate knowledge, indicate similarity and enable prediction. The quality of a classification should match the specific purpose for which it is intended. In the present case it is to provide a basis for the prediction of vegetation change in response to global changes in climate, and also to scenarios of man-induced disturbance. These issues are being widely debated world-wide as well as in South Africa – as illustrated by recent publications such as Sunter (1987) and Huntley *et al.* (1989). The plant taxa of conventional taxonomy for this purpose are at present unwieldy, and assume that differences between taxa are greater than the variability within taxa. Despite differences in forest, grassland and savanna species on different continents, there are many intercontinental parallels in the structure and function of these vegetation types. The study of functional aspects makes comparison and interpolation between widely separated study sites possible.

Despite the existence of several approaches to the functional classification of vegetation, others are needed. A single multipurpose classification is unlikely to be equally successful for all purposes. Different classifications are needed to suit different purposes and scales of operation. It is therefore argued that the ideal is to have several classifications specific to our South African purposes and operating at different scales. The present study forms part of a broader study aimed at developing functional classifications with increasing resolution, from the strategic differentiation of biomes at the subcontinental scale, through landscape and plant community processes at an intermediate scale, to tactical modelling of individual species responses for localized application.

Modelling vegetation distribution at different scales of resolution

A number of attempts have been made to differentiate the vegetation types of South Africa at a subcontinental scale, the most recent being Rutherford and Westfall (1986). Their study resulted in a relatively objective categorization of the biomes of South Africa based on the dominance and co-dominance of life-forms. The biomes they identified were desert, forest, fynbos, grassland, nama-karoo, savanna and succulent karoo. Rutherford and Westfall (1986) further tried to discriminate these biomes based on climatic indices, assuming that climate is the determinant of vegetation distribution at this scale (Odum, 1971; Box, 1981a). The initial objective of the present study was to test the discrimination of South Africa's biomes based on climatic indices as described by Rutherford and Westfall (1986), and to identify its shortcomings.

Although biomes are indicated on maps as homogeneous and as having abrupt boundaries, the flora of a given biome contains elements from neighbouring biomes. The inclusion of these is considered to be due to the superimposition of smaller-scale climatic influences, edaphic conditions and disturbance regimes on the congruence of predominant life-forms and climate identified at the subcontinental scale. The second objective of the present study therefore was to identify relationships between the inclusion of elements from neighbouring biomes and smaller-scale processes within the grassland biome in South Africa. For the purposes of this paper the approach that has been used will be described for forest patches within the grassland biome.

At the local scale the aggregate behaviour of individual species represents the corporate response of the community. These viewpoints underpin competition–stress–disturbance models and life-history and demographic studies. The essence of the approach is to identify and understand the characteristics of individual species and their interactions with their cohabiters. This subject has received scant attention in Southern Africa, although it is a desirable goal if the consequences to vegetation of natural and man-made disturbances are to be understood at a scale useful to land-use managers.

Selection of an appropriate modelling approach

At the outset a modelling approach of consolidating current wisdom, model testing and model improvement was adopted, in accordance with the 'bootstrapping and successive approximation' approach of Starfield and Bleloch (1986). Knowledge about the distribution of vegetation at a subcontinental scale in South Africa is generally widely scattered and poorly integrated. Many relationships are qualitative and lack any satisfactory theoretical or quantitative basis. Despite these shortcomings, the domain of knowledge was well bounded and the study objectives were well defined. Knowledge of this nature is amenable to modelling in a group of models known as 'expert systems'.

An expert system is generally encoded in a computer program which is capable of holding an apparently intelligent conversation with the user. The essential ingredients of an expert system are the knowledge base and the inference engine. The knowledge base comprises expert wisdom. The inference engine enables the knowledge base to be manipulated in an apparently intelligent way. Several inference engines are commercially available as computer packages or shells.

Depending on the expert system shell, knowledge bases can operate with explicit logic, or by implicit reasoning that can be made explicit by optimal success. Ecologically useful shells can perform mathematical operations, and interact with outside databases and executable files of almost any type. Unlike first-wave computer simulation of ecological problems, contemporary systems of artificial intelligence – with a mix of qualitative and quantitative logic, and access to outside files – appear to mimic human thought processes closely.

METHODS

A literature review of vegetation surveys in South Africa at a subcontinental and landscape scale was undertaken with an emphasis on the grassland biome. An attempt to synthesize the information in an expert system was made. The expert system itself was arranged in two tiers. The first, a preliminary attempt to predict the distribution of South Africa's biomes based on climatic indices, identified the 'ability of an area to support' vegetation of a given biome type. The second, a preliminary attempt at the landscape scale classification, predicted whether the vegetation at a site within the grassland biome was in fact grassland or vegetation represented in one of the adjacent biomes (forest, fynbos, nama-karoo and savanna). The expert system model was tested at both scales using independently collected data, and shortcomings were identified. The identification of these problems formed the basis of the second phase of research which was to improve the model at both the subcontinental and landscape scales.

Problems at the subcontinental scale related to the use of climatic indices which had limited biological meaning, and also to the fact that confusion existed in the literature regarding the extent and nature of anthropogenic modification of vegetation at a subcontinental scale. The latter issues have received increasing attention recently (Feely, 1987;

Meadows and Meadows, 1988; Hoffman and Cowling, 1989; Chapter 14), and will not be dealt with here – except to mention that there is not much direct evidence of anthropogenic change; therefore it ought not to be assumed unless it can be more convincingly demonstrated. The results presented here deal with the approach that has been used to develop biologically meaningful climatic indices in the differentiation of South Africa's biomes.

At the landscape scale problems were experienced owing to discrepancies in the scale of survey, and the non-independence and non-exclusiveness of many of the discriminating variables described by different authors. Within the grassland biome, the expert system model developed at the landscape scale, based on a literature survey, was tested on 70 sites for which compositional, structural and environmental data were collected. Shortcomings in the initial model were identified and, based on the data set, the model was improved. It is currently being tested and improved in a similar fashion to that done previously.

Development of biologically meaningful climatic indices

At the subcontinental scale the approach focussed on the development of biologically meaningful indices. The mean duration of the growth season per annum was estimated for 163 weather stations in known biome types using a water balance model which estimated plant available moisture based on daily rainfall sequences and daily potential evapotranspiration using Linacre's (1977) equation. Output from this component of the water balance model was the mean number of 'days of growth opportunity' per annum; this being the mean number of days per annum on which potential evapotranspiration is fully satisfied by available soil water. The mean temperature on each growth day was recorded, the mean of all these values for a station being reported as 'growth temperature'. The temperature on days when moisture was not available for plant growth was also recorded, the mean value being reported as 'non-growth temperature'. The temperature difference between growth and non-growth temperature (growth − non-growth temperature) indicates the season of rainfall (positive in summer and negative in winter rainfall areas) and the intensity of seasonality (large absolute values indicate intense seasonality while small absolute values indicate weak seasonality). Each weather station ascribed to a biome was located within the space defined by these three variables, and rules were devised to achieve the maximum degree of discrimination between biomes.

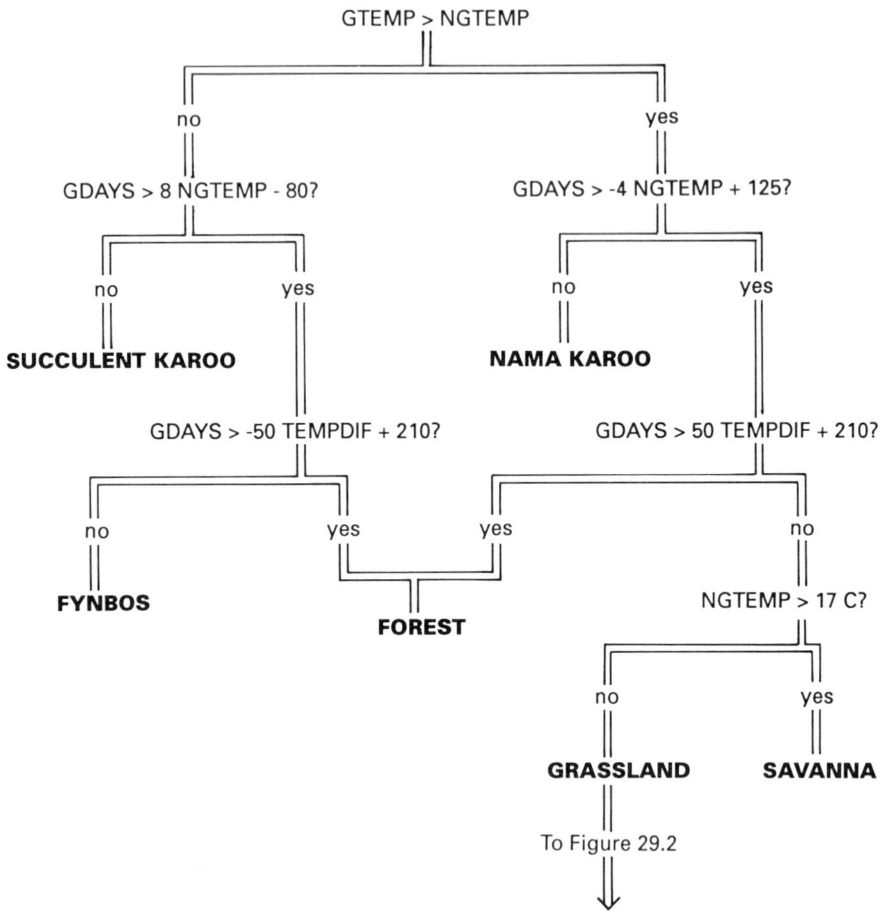

Fig. 29.1 A decision tree depicting the expert system which discriminates the biomes of Southern Africa based on climatic indices.

RESULTS

Subcontinental differentiation of South Africa's biomes

The decision tree (Fig. 29.1) which describes relationships between the distribution of biome ('potential to support' a given biome) in relation to climatic indices is derived from results presented elsewhere (Ellery *et al.*, 1991). The degrees of success of the discriminanting functions indicated in the decision tree (Fig. 29.1) are given in Table 29.1. The decision tree is considered the most parsimonious description of biome differentiation,

Table 29.1 The prediction of biome types for 163 sites based on the decision tree presented in Fig. 29.1

Classification of site in known biome type	*Classification into biome type based on the decision tree (Fig. 29.1).*					
	Forest	*Fynbos*	*Grass-land*	*Nama-karoo*	*Savanna*	*Succul. karoo*
Forest ($n = 3$)	3	0	0	0	0	0
Fynbos ($n = 24$)	1	21	1	0	0	1
Grassland ($n = 55$)	0	0	54	0	1	0
Nama-karoo ($n = 34$)	0	0	1	31	2	0
Savanna ($n = 37$)	0	0	2	0	35	0
Succ. karoo ($n = 10$)	0	0	0	0	0	10

and being stated in this way, the model is predictive and therefore explicitly testable.

The first question in the decision tree on which the expert system was based relates to the seasonality of rainfall. Where the growth and non-growth season temperatures are equal, rainfall is non-seasonal. Summer rainfall occurs in areas where the growth season temperature (GTEMP) is greater than the non-growth season temperature (NGTEMP), and winter rainfall in areas where the growth season temperature is lower than the non-growth season temperature. This question separates the winter and summer rainfall biomes. Fynbos and succulent karoo are restricted entirely to winter rainfall areas while grassland, nama-karoo and savanna occur in summer rainfall areas. Forest occurs in areas which receive rainfall all year.

The question at the second level of the decision tree describe relationships between the duration of the growing season (GDAYS) and the temperature during the non-growth season (NGTEMP). These questions separate the more arid biomes (nama-karoo and succulent karoo in the summer and winter rainfall areas respectively). The separation of these biomes from the more mesic ones suggests that moisture availability interacts with the temperatures during the non-growth season. In addition, the slope of the lines separating biomes in the winter and summer rainfall areas is positive and negative respectively. This is considered to be biologically meaningful. In the winter rainfall areas high temperatures in the non-growth season limit the distribution of the more mesic of these biomes (fynbos), and for this reason it requires increased moisture availability at higher non-growth temperatures. Conversely, in the summer rainfall areas it is low temperatures during the non-growth season that limit the distribution of the more mesic of these biomes (grassland and savanna). For this reason, increased moisture availability is required at lower non-growth season temperatures to support grassland and savanna in the summer rainfall areas.

Questions at the next level of division describe the relationship between the degree of seasonality of rainfall (TEMPDIF) and the distribution of forest in relation to fynbos (in the winter rainfall area), savanna and grassland (in the summer rainfall area). Forest requires a greater number of growth days as the degree of seasonality of rainfall increases. This reflects the requirement by forest elements for a relatively constant moisture supply during the year.

The final question in the decision-tree discriminates between grassland and savanna based on the mean temperature during the non-growth season (NGTEMP). In the summer rainfall areas it appears that woody elements require warmer non-growth season temperatures than grasses to become established.

Landscape-scale differentiation of grassland and forest

Biomes are recognized to contain vegetation elements from adjacent biomes depending on local conditions and disturbance. Having identified the potential of a region to support a given biome, it was necessary to consider local modifying factors. The decision tree describing the relationship between the distribution of forest patches in the grassland biome of South Africa is presented in Fig. 29.2. Forest patches in the grassland biome are confined to regions that experience a mean number of growth days per annum greater than 180. This indicates the importance of a minimum moisture requirement for forest growth.

The effect of soil type on the distribution of forest patches in the grassland biome is evident in the second question. Pure grasslands occur in areas with shallow soils, either because of an indurated B-horizon, or because of rock that is impervious to water movement (Table 29.2). Duplex soils (those with an abrupt textural change between the A-horizon and an argillic B-horizon), vertisols and soils which show evidence of waterlogging also support grassland regardless of management. In contrast, soils with a poorly developed illuvial B-horizon generally had woody elements (Table 29.2). These soils supported pure grasslands in areas that were frequently burned (at intervals of 1–3 years). The effect of fire is indicated in the next question in the decision tree. Of the soils with a well developed illuvial B-horizon, seven had woody elements and twelve had grassland. Of the seven with woody elements, five were dominated by *Acacia karoo*, which is invading the grassland biome extensively. It is possible that within historic time the stands dominated by this species were grassland, suggesting that the presence of a well devel-

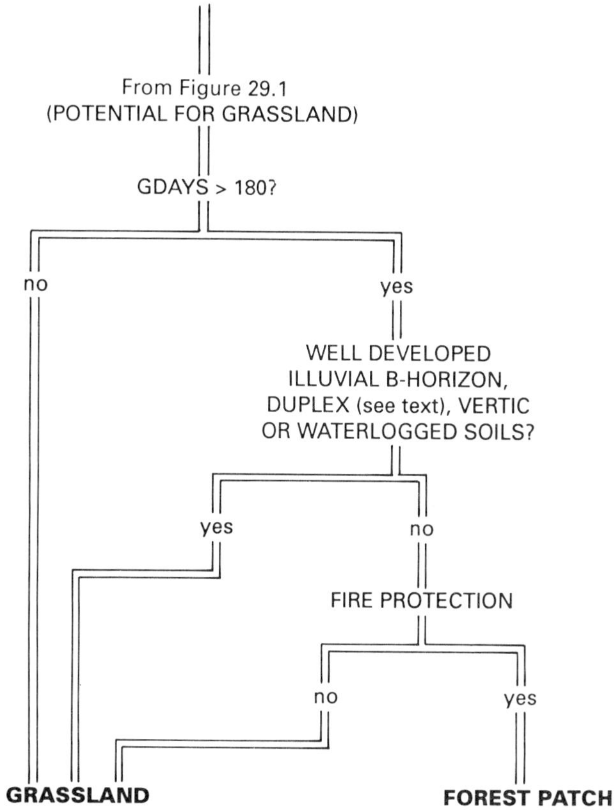

Fig. 29.2 A decision tree depicting a portion of the expert system which predicts the occurrence of woody patches of different affinity within the grassland biome of South Africa. The portion shown predicts the distribution of forest patches.

Table 29.2 The occurrence of wooded patches within the grassland biome of South Africa in relation to soil type

Shallow-impervious, duplex (see text), vertic or water-logged soils (n = 21)	*Soils with well developed illuvial B-horizon* (n = 19)	*Soils with poorly developed illuvial B-horizon* (n = 30)
2	7	26

oped illuvial B-horizon may be associated with grassland in an undisturbed situation.

The same level of detail has been achieved in respect of other biome inclusions within the grassland biome in South Africa, namely fynbos, nama-karoo and savanna.

DISCUSSION

Improvements in current wisdom

The present study has enabled the synthesis of current wisdom and the development of a more biologically meaningful model of the distribution of functionality in vegetation types in South Africa at the subcontinental scale, as well as the distribution of inclusions of some of these functionally distinct vegetation types at a landscape scale within the grassland biome itself. While our approach may be seen to overlap with other approaches, we feel that our strength is that we explicitly address the problem of scale in the modelling exercise, focus on developing biologically meaningful indices and present our results in a predictive and therefore testable way.

At the subcontinental (biome-wide) scale, climate is suggested to be the determinant of the distribution of vegetation. This is supported by numerous attempts to relate the distribution of different life-form spectra to climate on a world-wide scale (Raunkaier, 1934; Walter, 1979; Odum, 1971; Box, 1981a,b) as well as on a subcontinental scale (Fitzpatrick and Nix, 1970; Schulze and McGee, 1978; Phillips, 1983; Rutherford and Westfall, 1986). It does, however, appear to be in conflict with the widely held view that fire (Acocks, 1975; White, 1983; Tainton and Mentis, 1984) or soils (Feely, 1987) are the overriding determinants of many of the grasslands of Southern Africa, particularly those that are classically considered seral to forest.

The present study provides some insight into these conflicts. Within the grasslands mapped by a number of authors (Acocks, 1975; White, 1983) as seral to forest, rainfall is highly seasonal and occurs in summer. The presence of a marked dry season in winter may result in vegetation becoming sufficiently desiccated to burn in natural or man-initiated fires. It is furthermore evident that areas with a highly seasonal summer rainfall are subject to high lightning frequencies in South Africa (Manry and Knight, 1986). For this reason it appears that as the degree of seasonality of rainfall increases, increased moisture availability is required to support forest (Fig. 29.2), and that fire exclusion is the important anthropogenic factor that may enable the establishment of woody plants in those areas that experience summer rainfall. In contrast to the anthropogenic view, it is therefore suggested that climate is of overriding importance in determining the distribution of Southern Africa's grasslands and forests.

In contrast to these views, Feely (1987) suggested that the presence of a well developed illuvial B-horizon limited the development of woody patches within the grassland biome. Certainly, edaphic conditions are important determinants of the distribution of vegetation. It is questionable, however, whether edaphic conditions *per se* determine the distribution of vegetation at a biome-wide scale, or, as is suggested in the present study, at a smaller scale (Fig. 29.2). Many soils in biomes other than the grassland biome have well developed illuvial B-horizons, and only in exceptional circumstances do these preclude the establishment of woody elements.

At the landscape scale, the distribution of forest patches is confined to areas that have greater than a certain number of growth days per annum. Rutherford and Westfall (1986) recognized a minimum annual rainfall requirement of 800 mm for the establishment of forest patches. The units presented in this study (days of growth opportunity) are considered to be more biologically meaningful as they incorporate the effects of evapotranspiration on the moisture balance, unlike a simple measure of rainfall.

The second question in the decision tree describes the relationship that exists within the grassland biome between the distribution of forest patches and soil type. Saturation of a portion of the profile, or the presence of duplex soils, shallow hardpan horizons, or vertisols, have been recognized elsewhere to limit woody plant establishment (Tinley, 1982). It is only possible at this stage, however, to speculate on the significance of a well developed illuvial B-horizon on the distribution of woody elements in the grassland biome. It is possible that these soils are susceptible to excessive wetting in the wet season, or to excessive drying in the dry season, or possibly to both. The only satisfactory way of resolving this issue in the immediate future appears to be to model in detail the soil-water relations of each site visited in the grassland biome to date, and to correlate these with the extent to which each is colonized by woody plants. It may be that soils with a well developed B-horizon are in positions that are more susceptible to excessively hot fires or to different grazing regimes, and that the observed correlation has little to do with soil type *per se*. It may also be that the type of vegetation is the cause (and not the consequence) of the observed soil type. These problems are highlighted here as issues that need to be addressed before a satisfactory model of the location of woody–grassland boundaries at a smaller scale than the subcontinental differentiation of biomes can be presented.

The significance of the approach

The decision tree (Fig. 29.1) is a representation of an expert system which may appear so simple that it hardly needs to be computerized. Its initial development was based on a review of the literature. Owing to the diffuse nature and lack of integration of knowledge about the distribution of vegetation at the subcontinental scale, the initial model was somewhat more complex and confusing, for which reason a computer-based knowledge-base was recommended. The decision tree depicted in Fig. 29.2 is only a small portion of the entire knowledge-base that, in addition to the distribution of forest patches, models the distribution of patches of fynbos, nama-karoo and savanna trees within the grassland biome. The entire knowledge-base is too complex to represent in such a decision tree.

The explicit representation of the model is predictive and therefore testable. This makes it possible to identify its shortcomings and areas that warrant further investigation.

The hierarchical nature of expert system modelling makes it very well suited to the sort of study in which processes operating at different scales are to be incorporated. In the face of increased natural and man-made disturbance on a local and global scale, and the desire to model the consequences of these for vegetation, it is becoming increasingly necessary to develop understanding of the determinants of vegetation distribution at different scales of resolution. An expert system modelling approach may be a useful way of consolidating current understanding, and of exploring the consequences to vegetation of different scenarios of change. This has been attempted for South African vegetation at the subcontinental scale (W.N. Ellery *et al.*, 1991), although it may be argued that the scale and time-span of interest to land-use managers is smaller and more rapid respectively than was addressed in that study. The approach in the present study is from the broad scale to more detailed scales of resolution, and it is envisaged that ultimately such understanding will be achieved.

ACKNOWLEDGEMENTS

The authors gratefully acknowledge financial assistance provided by the Council for Scientific and Industrial Research.

REFERENCES

Acocks, J.P.H. (1975) Veld types of South Africa. *Memoirs of the Botanical Survey of South Africa*, No. 40.

Box, E.O. (1981a) *Macroclimate and Plant Forms: An Introduction to Predictive*

Modelling in Phytogeography, Junk, The Hague.
Box, E.O. (1981b) Predicting physiognomic vegetation types with climatic variables. *Vegetatio*, **45**, 127–39.
Ellery, W.N. and Mentis, M.T. (1989) How old are South Africa's grasslands? Submitted to be published in the present conference proceedings.
Ellery, W.N., Scholes, R.J. and Mentis, M.T. (1991) An initial approach to predicting the sensitivity of the South African grassland biome to climate change. *South African Journal of Science*, **87**, 499–503.
Feely, J.M. (1987) The early farmers of Transkei, Southern Africa. *Cambridge Monographs in Archaeology*, No. 24.
Fitzpatrick, E.A. and Nix, H.A. (1970) The climatic factor in Australian grasslands, in *Australian Grasslands*, (ed. R.M. Moore), Australian National University Press, Canberra, pp. 3–26.
Hoffman, T.M. and Cowling, R. (1989) An expanding karoo: Fact or fiction? Poster presented at IGBP Conference on Global Change, Cape Town, December 1989.
Huntley, B.J., Siegfried, W.R. and Sunter, C. (1989) *South African Environments into the 21st Century*, Human & Rouseau, Cape Town.
Linacre, E.T. (1977) A simple formula for estimating evaporation rates in various climates, using temperature data alone. *Agricultural Meteorology*, **18**, 409–24.
Manry, D.E. and Knight, R.S. (1986) Lightning density and burning frequency in South African vegetation. *Vegetatio*, **66**, 67–76.
Meadows, M.E. and Meadows, K.F. (1988) Late quaternary vegetation history of the Winterberg mountains, eastern Cape, South Africa. *South African Journal of Science*, **84**, 253–59.
Odum, E.P. (1971) *Fundamentals of Ecology*, Saunders, Philadelphia.
Phillips, J.F.V. (1983) A contribution to the concept and the classification of the bioclimatic units in sub-Saharan Africa. *Bothalia*, **14**, 713–19.
Raunkaier, C. (1934) *The Life Forms of Plants and Statistical Plant Geography*, Oxford University Press, Oxford.
Rutherford, M.C. and Westfall, R.H. (1986) Biomes of Southern Africa – an objective categorization. *Memoirs of the Botanical Survey of South Africa*, No. 40.
Schulze, R.E. and McGee, O.S. (1978) Climatic indices and classifications in relation to the biogeography of southern Africa, in *Biogeography and Ecology of Southern Africa*, (ed. M.J.A. Werger), Junk, The Hague, pp. 19–52.
Starfield, A.M. and Bleloch, A.L. (1986) *Building Models for Conservation and Wildlife Management*, Macmillan, New York.
Sunter, C. (1987) *The World and South Africa in the 1990s*, Human & Rouseau, Cape Town.

Tainton, N.M. and Mentis, M.T. (1984) Fire in grassland, in *Ecological effects of fire in South African Ecosystems*, (eds P. de V. Booysen and N.M. Tainton), Springer-Verlag, Berlin, pp. 115–47.

Tinley, K.L. (1982) The influence of soil moisture balance on ecosystem patterns in Southern Africa, in *Ecology of Tropical Savannas*, (eds B.J. Huntley and B.H. Walker), Springer-Verlag, Berlin, pp. 175–92.

Walter, H. (1979) *Vegetation of the Earth and Ecological Systems of the Geobiosphere*, Springer, New York.

White, F. (1983) *The Vegetation of Africa. A Descriptive Memoir to Accompany the Unesco/AETFAT/UNSO Vegetation Map of Africa*, Unesco, Paris.

30

A conceptual model relating environmental factors and vegetation formations in the lowlands of tropical South America

GUILLERMO SARMIENTO

SUMMARY

1. A resumé is given of previous models relevant to forest and savanna distribution, particularly in South America.
2. Previous works on the classification of South American lowland forests and savannas are considered.
3. The environmental determination of tropical lowland ecosystems is discussed, paying particular attention to climate and soil, as well as to human action and long-term processes. The environmental factors determining the occurrence of forests and savannas in South America operate through the combined effects of drought, water excess and deficiency of nutrients. A preliminary model is presented where tropical lowland formations are placed in environmental space between these three axes. While seasonal water availability has long been considered one of the key factors in the determination of tropical ecosystems, the importance of nutrient shortage has only been appreciated more recently. The role of nutrient stocks in soils and vegetation are considered, emphasizing the sharp differences between forests and savannas in biomass, mineral element allocation and stocks.
4. Replacement of forests by savannas, or vice versa, is also examined as a natural and/or human-induced process (the two operating on different time-scales), that leads to crucial changes in nutrient availability and cycling.

INTRODUCTION

To establish qualitative or quantitative conceptual models that explain and predict how the physico-chemical environment interrelates with the distribution of ecosystems has always been a key objective of biogeography.

Walter (1964, 1973) produced a simple model representing the distribution of vegetation zones on 'an average continent' that gives emphasis to climate. This model shows how the equatorial rainforest is bordered in both hemispheres by a belt of tropical deciduous forest (and moist savannas), whereas a longitudinal band of orographically determined tropical rainforest occupies the equatorial eastern coasts of this ideal continent. The tropical thorn bush (and dry savannas) limit deciduous forest towards the western subtropical zones as a transition to steppes or semi-deserts and hot deserts. According to Walter, within the tropical deciduous forest belt, moist savannas are largely anthropogenic or of edaphic origin. It may be seen then how lowland tropical vegetation could be ordinated from rainforest to desert along a climatic axis of increasing drought, with the possible exception of those savannas not determined by soil or man.

Sarmiento (1968) arranged all plant formations in the Americas within a climatic plane with mean annual temperature and rainfall as the two environmental axes. The aim of that model was to show how each vegetation type, with some exceptions, has a definite trend towards the occupancy of a well delimited area within a bidimensional climatic space. In this space, tropical American savannas occupy a restricted but well defined climatic zone between the tropical rainforest and the deciduous or tropophyllous tropical forest, where annual rainfall ranges from 1200 to 1500 mm and mean annual temperature from 21 to 28 °C. We now know that in this climatic belt, besides tropical savannas, both rainforests and deciduous forests may occur. However, moist savannas appear in this area more frequently than the two tropical forest types.

Walter (1973), on a smaller geographical scale, related the tropical plant formations in India to a gradient of increasing dryness, taking annual rainfall and length of drought as his environmental axes. Each vegetation formation occupies a well defined zone in this space, though dry savannas and thorn bushes overlap within the same area. Box (1981) produced a simple predictive model on the determination of 77 plant-forms by eight climatic parameters, validating the model with 74 sites world-wide. As would be expected, tropical plant-forms, such as rain-forest trees, monsoon deciduous trees, tall tussock grasses, etc., showed wide and overlapping temperature and rainfall ranges, seriously restricting the usefulness of simple climatic parameters to set apart the ecological spaces of these plant-forms.

Sarmiento and Monasterio (1975), in their discussion on factors associated with the occurrence of tropical American savannas, presented a model on the ideal distribution of ecosystems under tropical wet and dry climates, relating plant formations to two environmental axes: soil humidity at the end of the dry season and soil water at the end of the rainy season. The discriminatory power of these two environmental factors was high for most formations but it failed to separate some types of savannas from deciduous forests.

Outside these conceptual models aimed at clarifying the environmental determination of tropical vegetation, little further progress has been attained. The savanna problem in particular has not been satisfactorily solved. In this chapter we want to discuss the determination of South American tropical forests and savannas by climate and soil factors. Besides focusing our attention on the variety of savanna ecosystems in this continent, we will also try to relate neotropical lowland vegetation zones to some broad features of their environments.

ENVIRONMENTAL FACTORS DELIMITING TROPICAL FORESTS AND SAVANNAS

Classification of lowland forests and savannas

Beard (1944, 1955) placed plant formations in five formation series: seasonal, montane, dry evergreen, seasonal swamp and swamp. Apart from the montane series determined by altitude and the woodlands and scrubs characterizing the driest areas, most of Beard's units are moist forests whose variety is related to major environmental gradients. In the case of his seasonal series, ranging from the rainforest *sensu stricto* to the seasonal evergreen, semi-deciduous and deciduous forest, the major environmental gradient is one of increasing seasonal drought, paralleling the change from continually wet to tropical wet and dry climates. On the other hand, dry evergreen formations appear as climax types under constantly subhumid conditions. Seasonal waterlogging determines the seasonal swamp series, while the swamp series corresponds to habitats under permanent water excess. This simple scheme was followed in later studies of tropical American vegetation (Hueck, 1966; Walter, 1973; Schnell, 1987).

Beard's system recognizes only a few non-forest formations: palm and herbaceous swamps in the swamp formation series, and palm marshes and savannas in the seasonal swamp series. Occurrence of savannas is explained by the occurrence of seasonal waterlogging. In our opinion, Beard's model still holds, except perhaps with regard to the ecological explanation of savannas. Later classifications of tropical plant formations, at regional or at continental level, do not provide more detailed interpretations of relationships between environment and vegetation.

The physiognomic-ecological classification of Mueller-Dombois and Ellenberg (1974) divides tropical forests into open and dense formations, further subdividing each type according to seasonality following Beard's terminology. On the other side, their predominantly herbaceous formations include a wide range of vegetation types, such as savannas, steppes, meadows, sedge swamps and salt swamps. Savannas and related tropical and subtropical grasslands are subdivided according to height of grasses and to the distribution patterns and height of woody species. Their comprehensive system may be applied to an ecological interpretation of tropical vegetation units, but this task has not yet been undertaken.

In Brazil, Veloso *et al.* (1975) established an *ecological system* of vegetation classification to be used in the Amazonian region. Three main types of stable formations were recognized: forests, savannas (*cerrados*) and steppes (*caatingas*). Two major forest types were distinguished: open forest and dense forest, the former on hydromorphic soils, the latter on well drained soils; each was subdivided according to seasonality into ombrophyllous, seasonal and deciduous forest. Both savannas and steppes were classified by physiognomic features following the regional names. This system has the obvious advantage over previous classifications that savannas are unequivocally considered as stable plant formations of regional siginificance, as might be expected in a classification of the vegetation of the South American country with the greatest area of savannas. However, this so-called ecological system is only partly based on ecological features, since it considers physiognomy at least as important for recognition of plant formations as habitats and vegetation seasonality.

Rizzini (1979), in his comprehensive treatise on the phytogeography of Brazil, discusses the ecological characteristics of plant species in all major formations in that country: rainforests, dry forests, evergreen sclerophyllous forest (*cerradão*) and savannas. He considers that savannas are determined by soil conditions (drought and scarcity of nutrients) and by recurrent fires.

In order to clarify the environmental conditions associated with savanna ecosystems, it is first necessary to describe the differentiation of some major savanna types. The first basic distinction is between dry and moist savannas which are quite unrelated ecosystems. Moist savannas are the only type found in tropical America, and, as Walker (1985) points out, are more akin to rainforests than to dry savannas. Dry savannas, on the other hand, are floristically and ecologically related to thorn woodlands and to various semi-arid scrubs. Dry savannas are widespread in Africa and Australia but they do not occur in the Americas, since in the neotropics, when rainfall is less than 700 mm per year, low woody formations lacking a layer of perennial grasses (such as the Brazilian caatinga) appear as the regional climax. In subtropical areas of South America, such as the Chaco, mixed or grassy formations occur under semi-arid climates, but these subtropical ecosystems bear little resemblance to the moist neotropical savannas (Sarmiento, 1990).

Within moist savannas, at least three major ecological types have to be distinguished according to soil-water regime, vegetation seasonality and functioning (Sarmiento and Monasterio, 1975; Sarmiento, 1984). In *seasonal savannas* the dominant perennial grasses overcome an extended period of water shortage induced by rainfall seasonality, although water excess is absent or unimportant. In *hyperseasonal savannas* water excess and water shortage occur during each annual cycle. The topsoil becomes water-saturated during the rains, and dries out completely during the dry season. *Semiseasonal savannas* just suffer from an extended period of water excess without any long period of water deficiency, thus approaching the soil-water regime of seasonal marshes (although the waterlogging may not be as pronounced nor as extended as it is in marshes).

These three types of savanna frequently coexist on the topographic and soil catenas of savanna landscapes, with seasonal savannas on the higher positions with coarse, well drained soils (alluvial levées, sand dunes, high plains, etc.), hyperseasonal savannas on silty soils of intermediate topographic positions, and semiseasonal savannas, together with seasonal marshes and swamps, in the bottomlands, with heavy clay soils.

Having thus briefly considered the main ecological types of forests and savannas, we now discuss their environmental boundaries.

DO FORESTS AND SAVANNAS OCCUR IN SIMILAR OR IN DIFFERENT ENVIRONMENTS?

The lowlands of tropical South America, below 1300–1500 m, show a great diversity of climates and soils, as we might expect for an area of about 11.1 million km^2. Given that mean monthly temperatures are relatively constant throughout the year, with annual means, ranging according to latitude and elevation, between 18 and 28 °C, the two most relevant factors inducing climatic variability are total rainfall and its seasonality. Climatic factors, together with parent material, geological history, topography and vegetation are responsible for the characteristics and evolution of the soil profile. Plant formations in turn are related to climate and soil.

Forests and savannas represent alternative possibilities for occupying a given area in the humid, tropical lowlands. Which of these two plant formation actually occurs in any area will depend on the overall habitat conditions.

The conceptual model implicit in the foregoing statements is that forests and savannas occupy different multidimensional ecological spaces, or better ecological hyperspaces, whose *n* dimensions or axes are the environmental factors acting upon each ecosystem. Under this assumption, both types of ecosystem will not coexist in the same area, unless there are different habitat conditions (e.g. variations in soil type).

Several possibilities may be envisaged concerning these environmental hyperspaces. If their respective hyperspaces do not have common points along some axes, forests and savannas will appear as two systems without any frontier or common border between their environments, even if they overlap along other axes (Fig. 30.1a). This is, of course, a rather extreme assumption concerning the environmental delimitation of these formations. In Fig. 30.1b, the first axis separates forests from savannas, whereas the second axis does not. Another possibility is that if forests and savannas do overlap along all axes, then both hyperspaces will still be partly different but will have a common frontier, or tension area, of environmental indetermination (Fig. 30.1c).

Other conceptually different models depict the cases where the environmental conditions of one type of ecosystem are totally included within the space of the other (Fig. 30.1d,e), or the most extreme case of environmental indetermination occurs when both types of ecosystems share exactly the same ecological hyperspace (Fig. 30.1f). If these last three models approach real situations, then we have to appeal to non-environmental factors as determinants, or to assume that disturbance has been responsible for the existence of forests or savannas. Most probably the truth lies in between, with each system occurring in an exclusive part of the environmental space but with overlapping fields with the other system.

THE DISCRIMINATING POWER OF SOME ENVIRONMENTAL AXES

We shall consider boundaries along certain environmental axes as an initial step in the analysis of forest and savanna environments, taking into account a few important environmental factors, such as annual rainfall, mean temperature, number of dry and wet months, soil fertility, etc.,

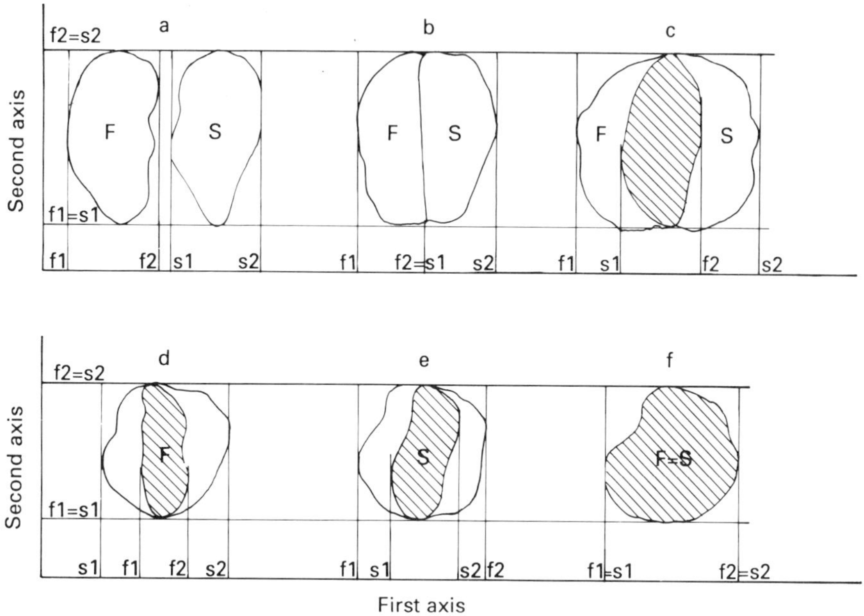

Fig. 30.1 Several possibilities for distribution of forests (F) and savannas (S) in a two-axis environmental space. According to this case, the first axis may or may not separate both types of ecosystems, while the second axis does not show any discriminating power.

in order to assess their power in discriminating the two ecosystems. Climatic data from 84 localities, embracing all South American countries with tropical lowlands and representative of the full range of climatic variation found in this continent, served as a basis for the diagrams, while the boundaries between plant formations reflect both published data and the author's own field experience. The conclusions should be regarded as a preliminary hypothesis.

Figure 30.2 shows the distribution of forests and savannas on the two-dimensional space or plane of annual rainfall and mean temperature. A total overlapping is evident except in the area of highest rainfall (above 2.5 m) where savannas do not occur, leaving only evergreen tropical forests. For the rest, the overlapping represents the coexistence of seasonal savannas and semi-deciduous forests under conditions of intermediate rainfall (1.5–2.5 m), and of seasonal savannas and deciduous forests in the somewhat less rainy areas. With less than 0.6 or 0.7 m of rainfall, already under tropical dry climates xerophytic formations replace deciduous forests and savannas, while below an annual mean tempera-

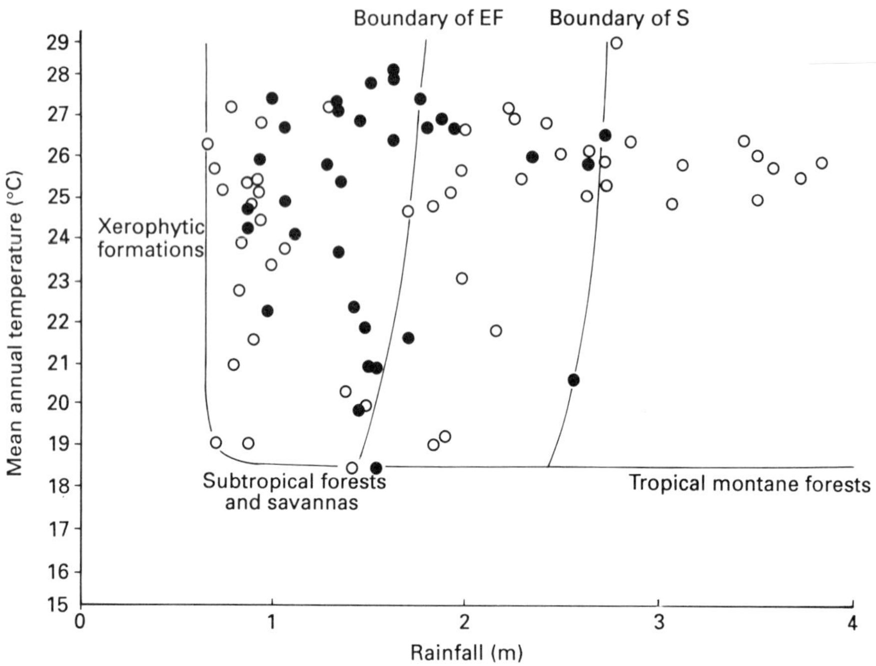

Fig. 30.2 Distribution of forests (○) and savannas (●) in the environmental space whose axes are annual rainfall and mean annual temperature, showing the boundary of savannas towards the wettest climates and the boundary of evergreen tropical forests towards the less rainy climates. EF, evergreen forest; S, savanna.

ture of 18 or 19 °C tropical formations give place either to subtropical ecosystems or to tropical montane forests. To summarize, then, the discriminating or predictive power of the temperature axis to set apart savannas and forests seems to be null, while that of the rainfall axis appears to be quite low.

Considering now the distribution of forests and savannas in the plane determined by annual rainfall and by the number of dry months (taking as 'dry' a month with less than 50 mm rainfall), the almost total superposition in areas of intermediate rainfall can be seen in Fig. 30.3. This overlapping is mainly due to the fact that the environments of tropical semi-deciduous and tropical deciduous forest are indistinguishable from that of seasonal savanna (SS). Figure 30.3 also shows that forests may occur in wetter as well as in drier areas than savannas, and also that neither of these two contrasting formations occurs where the dry season extends for more than eight consecutive months.

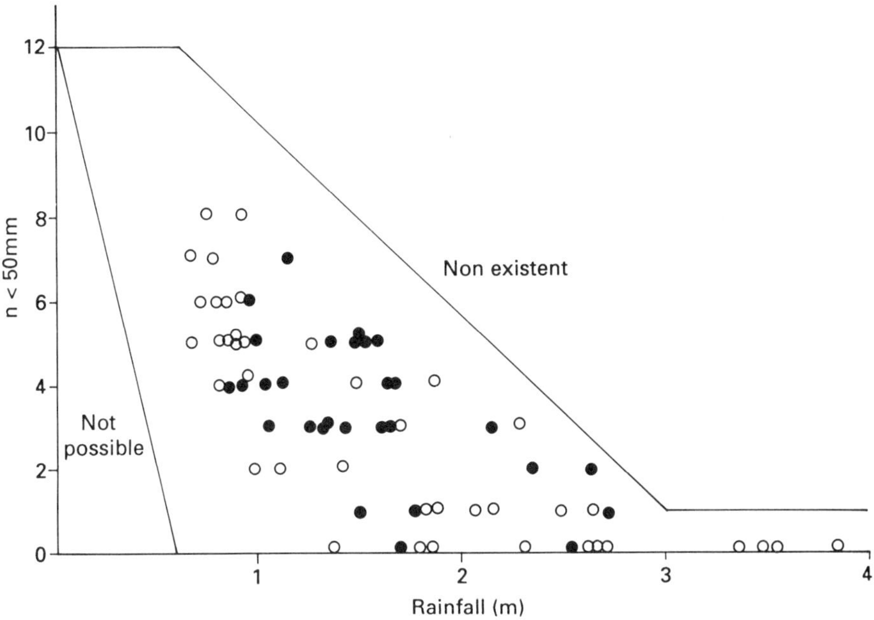

Fig. 30.3 Distribution of forests (○) and savannas (•) in the environmental space determined by annual rainfall and the number of months with less than 50 mm rainfall. Evergreen forests predominate towards the lower-right part, while deciduous forests occupy the upper-left part. SS, seasonal savanna.

If both the number of dry months and of months with water-saturation or waterlogging are taken into account (Fig. 30.4), the savannas become clearly divided into two main ecological types: seasonal savannas where the period of water excess is short or non-existent, and hyperseasonal savannas with both contrasting stresses alternating during each annual cycle. The field of seasonal savannas greatly overlaps with that of deciduous forests and somewhat less with evergreen forests, while swamp forests separate from other formations on this environmental plane.

As a last example of a two-dimensional analysis, the distribution of forests and savannas in the plane given by a climatic axis (number of dry months) and an axis of nutrient availability (sum of exchangeable bases in topsoil) is given in Fig. 30.5. On this plane the habitats of deciduous forests and seasonal savannas separate into a strongly nutrient-deficient area occupied by seasonal savannas and a field with richer soils where deciduous forests occur, while evergreen forests extend throughout the whole range of soil fertility, but only in the less extreme part of the

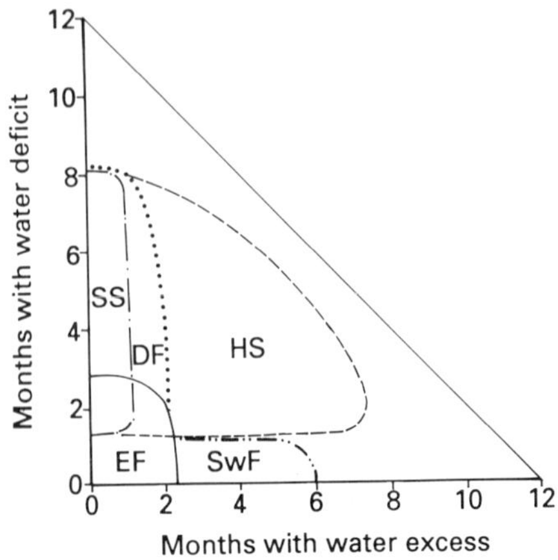

Fig. 30.4 Distribution of tropical lowland ecosystems in the environmental space determined by number of months with water excess and with water deficit. DF, deciduous forest; EF, evergreen forest; HS, hyperseasonal savanna; SS, seasonal savanna; SWF, swamp forest. Note the overlapping of SS and DF.

drought gradient. These few examples clearly illustrate that it is necessary to employ more than two axes to discriminate forest and savanna environments in the lowland tropics.

MODIFICATIONS OF THE ORIGINAL ENVIRONMENTS

The relationships between the physico-chemical environment and tropical biotas are still more complex, since once an ecosystem becomes established, be it forest or savanna, the modifications to the habitat initiated during its development will continue, tending to widen still more the original environmental differences. Thus, for instance, each ecosystem has characteristic patterns of carbon and nutrient balance in the vegetation biomass, of litter production and decomposition, and of soil organic matter accumulation, as well as divergent water budgets produced by differences in such processes as throughfall, stem flow, transpiration, etc. When recurrent fires become a normal ecological event, they also alter plant growth and nutrient cycling (Table 30.1). Consequently, it becomes difficult to know whether environmental conditions in forest and savannas were different before maturity, or whether their environments underwent a secondary, *a posteriori*, differentiation through the action of vegetation and ecosystem processes.

These biologically induced changes in habitats are gradual and accompany the development of each ecosystem, but after a certain time,

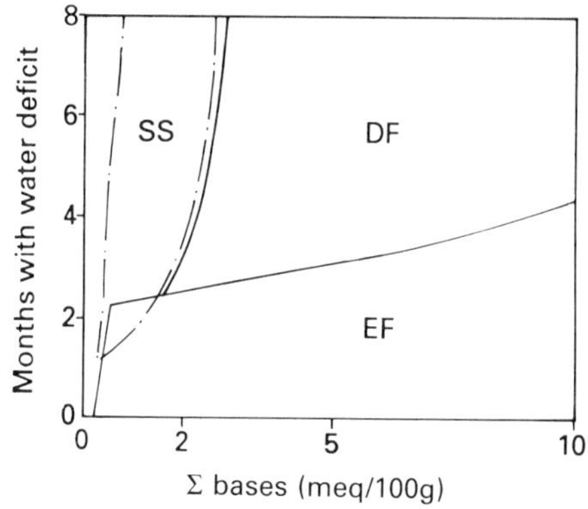

Fig. 30.5 Distribution of tropical lowland ecosystems in the environmental space determined by the sum of exchangeable bases and the number of dry months. On this plane seasonal savannas (SS) become almost entirely separated from deciduous (DF) and evergreen forests (EF).

Table 30.1 Some contrasting characteristics of tropical forests and savannas

	Forests	*Savannas*
Vegetation biomass	Higher stocks of carbon, nitrogen and mineral elements in the biomass. Higher stocks in the aerial parts	Lower stocks of carbon, nitrogen and mineral elements in the biomass. Higher stocks in the subterranean ground parts
Litter accumulation	Litter accumulation and decomposition on the soil surface (leaf litter)	Subterranean decomposition of plant materials (root litter)
Soil organic matter	Concentrated in the topsoil. Occurrence of organic surface horizons. High CEC in topsoil due to SOM	More gradual decrease with depth. Deeper organo-mineral horizons. Lower cation exchange capacity (CEC) in topsoil due to less soil organic matter (SOM)
Fire	Exceptional. It starts vegetation succession	Almost yearly. Little damage to soil and vegetation. It starts regrowth. Losses of volatiles

perhaps thousands of years, the original conditions become strongly altered and many positive feedback mechanisms begin to operate, securing the maintenance of a new equilibrium between the biotic and the abiotic components of the steady-state ecosystems. We will refer again to these changes when discussing the dynamics of the whole landscape in the next section.

ECOSYSTEM REPLACEMENTS: NATURAL AND HUMAN-INDUCED

The occurrence of natural savannas on previously forested areas, without changes in the original environment, does not seem to have been documented anywhere in the neotropics. The palynological record in tropical American lowlands for the Holocene and the Upper Pleistocene (Wijmstra and van der Hammen, 1966; van der Hammen, 1974, 1983) certainly shows almost continuous replacements of forests and savannas in both directions, but at time-scales of hundreds or thousands of years. These replacements have been explained as induced by climatic changes associated with glacial and interglacial periods, or by geomorphological processes operating in the plains that suddenly or gradually modify soil and water conditions. I have no information about forests converted to savannas, or vice versa, during historical times. Certainly, many previously forested areas have been, and continue to be, deforested, but in these cases the secondary plant formations replacing the original forests differ widely from the primary savannas that still persist in neighbouring areas, and they can be easily recognized as man-induced, unstable systems by their composition or structure.

It has already been shown that neither seasonal drought nor water excess or poor soil may be itself explain the alternative occurrence of forests and savannas. Furthermore, tropical forests have been widely converted into secondary grasslands or bushlands, and these successional formations are frequently subjected to the same types of influences in recurrent fires and grazing pressure as the primary savannas. But it is still not known if, given enough time (10–100 years?), these secondary communities under such pressures will evolve into ecosystems similar to primary savannas, or whether they will remain as disclimaxes.

Independently of human action, the natural evolution of forest ecosystems in the long term (10^4–10^5 years) leads to irreversible environmental changes involving water and nutrient cycles that may favour the replacement of long-established forests by savannas. Some of the processes involved in this senescence of the landscape are related to latosolization, podzolization, cuirassment, extreme leaching and depletion of soil cations, etc. Under these conditions of decreasing soil fertility, woody vegetation accumulates in its long-lived biomass a significant part of the total stock of nutrients in the ecosystem, at the same time that the overall stock in the ecosystem decreases. Moreover, one of the consequences of nutrient stress on vegetation seems to be the development of scleromorphic traits in the dominant trees together with an opening of the forest canopy. Both features favour occurrence of natural fires.

These features render the ecosystems quite fragile and less able to recover from natural disturbances, such as occasional fires, and in the long term may lead to the gradual replacement of these extremely oligotrophic forests by savannas that can cope more successfully with poor soils and fires. Probably such processes, operating on white sands and other very poor parent materials in various regions of tropical South America and the Caribbean, produced the present landscapes where highly scleromorphic forests intermix with open woodlands and savanna bushes (van Donselaar, 1965; Hoock, 1971; Rizzini, 1979).

These trends towards *savannization* may be reversed by geological and geomorphological processes that produce a rejuvenation of the landscape. An uplift, for instance, induces a renewal in erosion by running waters that increase availability of nutrients, since new, deep soil layers can be attained by tree roots, favouring in this way the reconversion of savanna areas to forest. Similar changes may be brought about by a climatic change towards increasing rainfall. Such landscapes of contracting savannas and expanding forest appear in various neotropical regions, such as the Venezuelan llanos and the Brazilian cerrados (Sarmiento and Monasterio, 1975; Cole, 1986). In Fig. 30.6, the combined action of all these factors and processes on the equilibrium between forests and savannas is represented in a graphical form.

ENVIRONMENTAL CONSTRAINTS, PLANT RESPONSES AND THE DYNAMICS OF FORESTS AND SAVANNAS

Tropical lowland forests are dominated by tall trees, either evergreen or deciduous. Considering the whole range of plant formations in the broad definition of 'tropical lowland forest' these tall trees may vary widely in height, but even under less favourable conditions the forest canopy reaches at least 12 or 15 m, while on more favourable habitats the canopy may reach 35 or 40 m. As a direct consequence of the enormous size of forest trees, the total above-ground vegetation biomass normally attains 300–500 t dryweight ha^{-1}.

Savannas have a different structural pattern and are always dominated by perennial grasses and sedges, with low trees as the growth-form second in importance. Compared with forest trees, savanna low woody species together with grasses have a much smaller biomass, which usually ranges from a maximum of about 100 t ha^{-1} in woody savannas to less than 20 t ha^{-1} in savanna grasslands (Sarmiento, 1984).

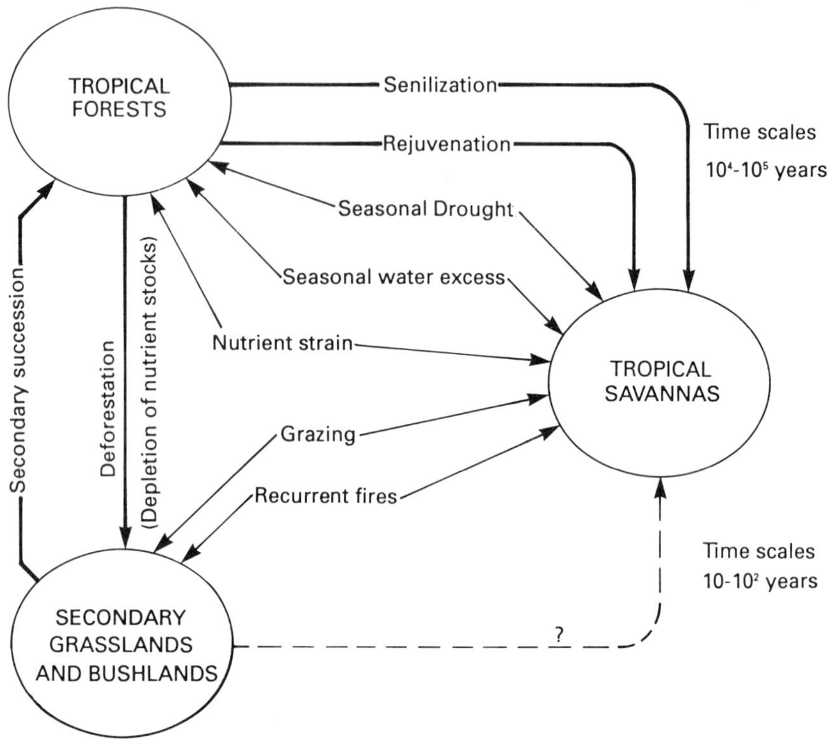

Fig. 30.6 Major factors acting upon tropical forest and savannas, together with processes involved in the conversion of one to another and to secondary formations.

As the total vegetation biomass is about ten times higher in forest than in savannas, the stocks of mineral elements follow this same pattern. Thus for instance, savanna formations in the Venezuelan llanos have in the grass layer (above- and below-ground) about 20 kg ha^{-1} of phosphorus and 60–140 kg ha^{-1} of potassium. The woody biomass in an open savanna woodland has roughly the same quantities. In contrast a semi-evergreen rainforest in the same area contains 290 kg ha^{-1} of phosphorus and 1820 kg ha^{-1} of potassium just in the above-ground biomass (Hase and Folster, 1982).

Finally, in humid tropical forests a direct transference of nutrients from decaying plant parts to roots, via mycorrhizae and decomposing micro-organisms, seems to occur, short-circuiting the mineral soil and securing a most efficient recycling of critical nutrients (Herrera *et al.*, 1978). That is, even if forests and savannas are both subjected to severe nutrient limitations as a consequence of their highly dystrophic soils, forest trees have developed various mechanisms that increase the efficiency of nutrient storage and recycling that tend to create an important stock of mineral elements within the living vegetation. The disruption of these mechanisms may induce the natural replacement of forest by more tolerant savanna ecosystems.

CONCEPTUAL MODELS RELATING VEGETATION TO HABITAT IN TROPICAL AMERICA

A major difficulty in understanding the environmental conditions determining tropical American savannas derives from not having, in previous models, considered nutrient shortage as a constraint acting on species occurring on poor tropical soils. In the first experimental studies on the Brazilian cerrado, occurrence of nutrient-poor soils was recognized to be among the principal determinants of these savannas (Alvim and Araujo, 1952; Arens, 1958, 1963). This interpretation was later extended to the seasonal savannas in tropical America (Sarmiento, 1984).

The seasonal savanna associated with well drained soils stands as a characteristic ecosystem on some of the poorest soils of tropical America. But obviously, if only soil nutrient stocks were considered, many or even most of the neotropical humid forests would overlap with seasonal savannas as ecosystems occurring on dystrophic soils. The points emphasized here are, first, that the whole nutrient capital of the ecosystem must be taken into account, and that this capital is much more important in forest than in savannas because of the storage of mineral elements in the forests biomass. Second, that nutrient cycles are very tight and therefore an impact on nutrient stocks may switch the system towards a new equilibrium, that is from forests, with higher stocks, to savannas, with lower stocks and requirements.

On the basis of the foregoing considerations, tropical American lowland ecosystems were arranged in a three-dimensional space where drought, water excess and low nutrient levels are the three axes of environmental variability (Figs 30.7 and 30.8). Each axis in turn represents a sort of complex gradient resulting from the joint action of simpler factors. This arrangement, although purely qualitative, may serve as a conceptual model, suggesting the principal environmental constraints impinging upon these ecosystems and their species.

Seasonal drought is the main environmental factor determining the deciduous forest and, to a lower degree also, the semi-deciduous forest, whereas the thorn forest and related arid formations occupy the extreme of the drought axis (Fig. 30.7). Nutrient deficiency is the most severe

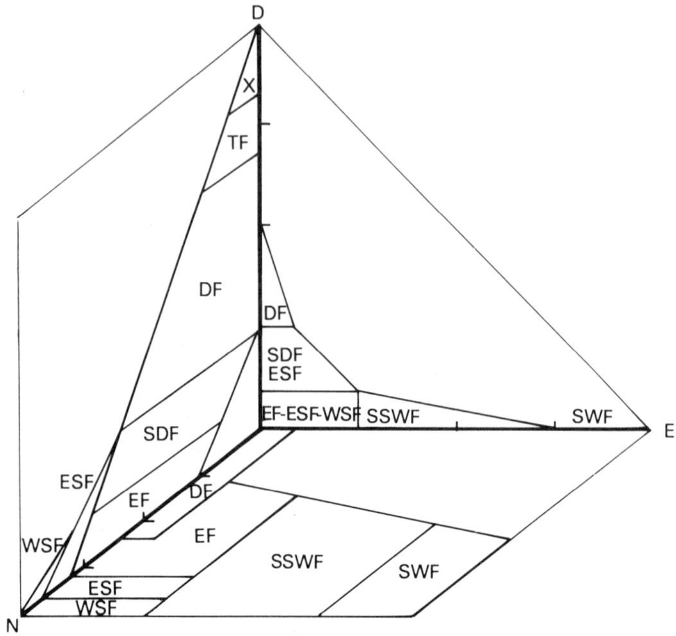

Fig. 30.7 Distribution of lowland tropical forests in the environmental space determined by water excess, drought and nutrient levels. The projections on the three environmental planes are shown. The *E* and *D* axes vary from 0 to 12 months of water excess or drought respectively, while the *N* axis has been divided into four parts corresponding to eutrophic, mesotrophic, dystrophic and hyper-dystrophic soils. DF, deciduous forest; EF, tropical evergreen forest; ESF, evergreen sclerophyllous forest; SDF, semi-deciduous forest; SSWF, seasonal swamp forest; SWF, swamp forest; TF, thorn forest; X, xerophytic vegetation.

environmental constraint in evergreen sclerophyllous forest, as well as in forests and scrubs on white sands. To a lesser degree the tropical evergreen forest, or rainforest *sensu stricto*, as well as the semi-deciduous forest, are also subjected to nutrient limitations. Water excess is the environmental pulse modelling the two types of alluvial Amazonian forests: *várzea* and *igapó*; the former corresponds roughly to Beard's seasonal swamp forest, the latter to his swamp forest.

Savannas occupy different positions in Fig. 30.8, showing how they are simultaneously influenced by more than one major environmental constraint. The locations of seasonal savanna, semiseasonal savanna and hyperseasonal savanna illustrate the role of drought, water excess and nutrients as environmental factors modelling the structural and functional features of these ecosystems. Fire seems to be an unavoidable con-

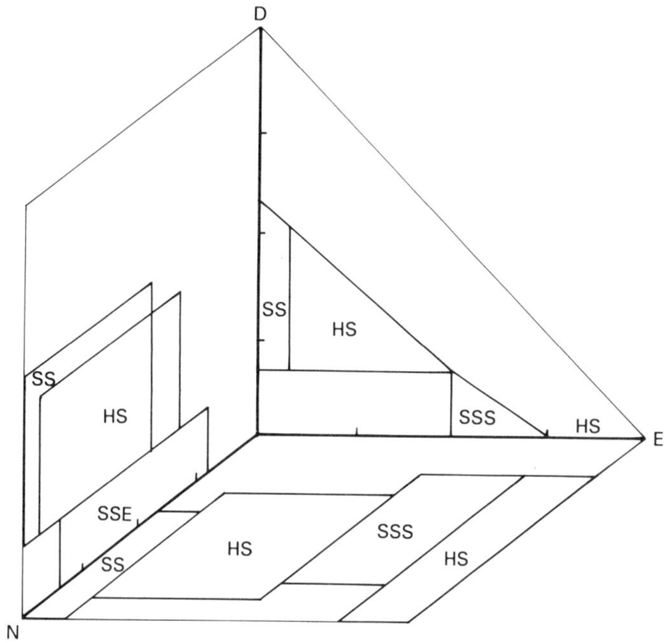

Fig. 30.8 Distribution of tropical savannas in the environmental space determined by water excess, drought and nutrient levels. The projections on the three environmental planes are shown. Notice how the three axes have a joint determination on savanna ecosystems. HS, hyperseasonal savanna; SS, seasonal savanna; SSS, semiseasonal savanna.

sequence of savanna structure and seasonality, reinforcing the effects of drought and nutrients.

When enough quantitative data concerning the environmental features of tropical ecosystems are available, more precise statistical approaches and multivariate analysis will be possible, and then sounder models relating biotic and abiotic characteristics may be elaborated. For the time being, our preliminary approach may operate as a guide to further research.

REFERENCES

Alvim, P. de T. and Araujo, W. (1952) El suelo como factor ecológico en el desarrollo de la vegetación en el centrooeste del Brasil. *Turrialba*, **2**, 153–60.

Arens, K. (1958) O cerrado como vegetaçao oligotrófica. *Bol. Fac. Fil. Ciênc. Letr. USP Botánica*, **15**, 58–77.

Arens, K. (1963) As plantas lenhosas dos campos cerrados como flora adaptada as deficiências minerais do solo, in *Simpósio sôbre o Cerrado*, (ed. M.G. Ferri), Editora da Universidade de São Paulo, São Paulo, pp. 287–99.

Beard, J.S. (1944) Climax vegetation in tropical America. *Ecology*, **25**, 127–58.

Beard, J.S. (1955) The classification of tropical American vegetation types. *Ecology*, **36**, 89–100.

Box, E.O. (1981) *Macroclimate and Plant Forms: An Introduction to Predictive Modelling in Phytogeography*, Junk, The Hague.

Cole, M.M. (1986) *The Savannas. Biogeography and Geobotany*, Academic Press, London.

Hase, H. and Folster, H. (1982) Bioelement inventory of a tropical (semi-) evergreen seasonal forest on eutrophic alluvial soils, western llanos, Venezuela. *Acta Oecologica, Oecologia Plantarum*, **3**, 331–46.

Herrera, R., Mérida, T., Stark, N. and Jordan, C. (1978) Direct phosphorus transfer from leaf litter to roots. *Naturwissenschaften*, **65**, 208–9.

Hoock, J. (1971) *Les Savanes Guyanaises: Kourou*, ORSTOM, Paris.

Hueck, K. (1966) *Die Wälder Sudamerikas*, Gustav Fischer Verlag, Stuttgart.

Mueller-Dombois, D. and Ellenberg, H. (1974) *Aims and Methods of Vegetation Ecology*, John Wiley, New York.

Rizzini, C.T. (1979) *Tratado de Fitogeografia do Brasil*, Hucitec, São Paulo.

Sarmiento, G. (1986) Correlación entre los tipos de vegetación de América y dos variables climáticas simples. *Boletín de la Sociedad Venez. de Ciencias Naturales*, **113–114**, 454–470.

Sarmiento, G. (1984) *The Ecology of Neotropical Savannas*, Harvard University Press, Cambridge, Massachusetts.

Sarmiento, G. (1990) Ecologia comparada de ecosistemas de sabana en América del Sur, in *Las Sabanas Americanas. Aspectos de su Biogeografia, Ecologia y Utilización*, (ed. G. Sarmiento), Fondo Editorial Acta Cientifica Venezolana, Caracas, pp. 15–56.

Sarmiento, G. and Monasterio, M. (1975) A critical consideration of the environmental conditions associated with the occurrence of savanna ecosystems in tropical America, in *Tropical Ecological Systems*, (eds E. Medina and F.B. Golley), Springer-Verlag, New York, pp. 223–50.

Schnell, R. (1987) *La Flore et la Végétation de l'Amérique Tropicale*, vol. 1; Masson, Paris.

van der Hammen, Th. (1974) The Pleistocene changes of vegetation and climate in tropical South America. *Journal of Biogeography*, **1**, 3–26.

van der Hammen, Th. (1983) The paleoecology and paleogeography of savannas, in *Ecosystems of the World, 13: Tropical Savannas*, (ed. F. Bourlière), Elsevier, Amsterdam, pp. 79–108.

van Donselaar, J. (1965) An ecological and phytogeographic study of northern Surinam savannas. *Wentia*, **14**, 1–163.

Veloso, H.P., Goés Filho, L., Barros-Silva, S., Ferreira, H. de C. *et al.* (1975) Vegetaçao, in *Projeto RADAMBRASIL. Levantamento de Recursos Naturais*, vol. 8, Departamento Nacional de Produçao Mineral, Rio de Janeiro, pp. 307–403.

Walker, B.H. (1985) Structure and function of savannas: an overview, in *Ecology and Management of the World's Savannas*, (eds J.C. Tothill and J.J. Mott), The Australian Academy of Science, Canberra, pp. 83–91.

Walter, H. (1964) *Die Vegetation der Erde in Oko-physiologischer Betrachtung. I. Die Tropischen und Subtropischen Zonen*, 2nd edn, E. Ulmer, Stuttgart.

Walter, H. (1973) *Vegetation of Earth, in Relation to Climate and the Eco-physiological Conditions*, English Universities Press, London.

Wijmstra, T.A. and van der Hammen, Th. (1966) Palynological data on the history of tropical savannas in northern South America. *Leidsche Geologische Mededelingen*, **38**, 71–90.

Index